CERAMIC PROCESSING and SINTERING

MATERIALS ENGINEERING

1. Modern Ceramic Engineering: Properties, Processing, and Use in Design. Second Edition, Revised and Expanded, *David W. Richerson*
2. Introduction to Engineering Materials: Behavior, Properties, and Selection, *G. T. Murray*
3. Rapidly Solidified Alloys: Processes · Structures · Applications, *edited by Howard H. Liebermann*
4. Fiber and Whisker Reinforced Ceramics for Structural Applications, *David Belitskus*
5. Thermal Analysis of Materials, *Robert F. Speyer*
6. Friction and Wear of Ceramics, *edited by Said Jahanmir*
7. Mechanical Properties of Metallic Composites, *edited by Shojiro Ochiai*
8. Chemical Processing of Ceramics, *edited by Burtrand I. Lee and Edward J. A. Pope*
9. Handbook of Advanced Materials Testing, *edited by Nicholas P. Cheremisinoff and Paul N. Cheremisinoff*
10. Ceramic Processing and Sintering, *M. N. Rahaman*

Additional Volumes in Preparation

CERAMIC PROCESSING and SINTERING

M. N. RAHAMAN

Department of Ceramic Engineering
University of Missouri—Rolla
Rolla, Missouri

Marcel Dekker, Inc. New York•Basel•Hong Kong

Library of Congress Cataloging-in-Publication Data

Rahaman, M. N.
  Ceramic processing and sintering / M.N. Rahaman.
    p.   cm. — (Materials engineering ; 10)
  Includes index.
  ISBN 0-8247-9573-3 (alk. paper)
  1. Ceramics. 2. Sintering.   I. Title.   II. Series: Materials
engineering (Marcel Dekker, Inc.) ; 10.
TP807.R28   1995
666—dc20   95-32195
                                                                            CIP

The publisher offers discounts on this book when ordered in bulk quantities. For more information, write to Special Sales/Professional Marketing at the address below.

This book is printed on acid-free paper.

Copyright © 1995 by Marcel Dekker, Inc. All Rights Reserved.

Neither this book nor any part may be reproduced or transmitted in any form or by any means, electronic or mechanical, including photocopying, microfilming, and recording, or by any information storage and retrieval system, without permission in writing from the publisher.

Marcel Dekker, Inc.
270 Madison Avenue, New York, New York 10016

Current printing (last digit):
10 9 8 7 6 5 4 3 2 1

PRINTED IN THE UNITED STATES OF AMERICA

To
Vashanti, Lennard, and Ronald

Preface

Ceramics have been used since the earliest civilizations. The field of ceramic materials has its roots in the more traditional aspects of the subject such as clay-based ceramics and glasses. However, during the past few decades, new developments in the use of ceramics in more advanced technological applications have attracted considerable attention. In addition to the discovery of ceramic superconductors, the use of ceramics for heat-resistant tiles in the space shuttle, for optical fibers, and for components in high-temperature engines has generated considerable interest in the field.

The increasing use of ceramics in more advanced technological applications has resulted in a heightened demand for improvements in properties and reliability. In recent years there has been the realization that such improvements can be achieved only through careful attention to the fabrication process. The engineering properties of a polycrystalline ceramic are controlled by the microstructure, which in turn depends on the processing method used to fabricate the body. Therefore, the fabrication processes govern the production of microstructures with the desired properties. It is often stated that materials science is a field at the interface between the physical sciences (physics, chemistry, and mathematics) and engineering (such as electrical, mechanical, and civil engineering). In this view, the approach to the processing of ceramics is concerned with the understanding of fundamental issues and the application of that knowledge to the production of microstructures that have useful properties.

This book is concerned primarily with the processing of polycrystalline ceramics. Because of its importance and widespread use, the fabrication of ceramics by the firing of consolidated powders forms the focus of the book. The production of ceramics (and glasses) by the less conventional sol-gel route has been attracting considerable attention. A brief treatment of sol-gel processing is also included. The approach is to outline the fundamental issues of each process and show how they are applied to the practical fabrication of ceramics. Each fabrication route involves a number of processing steps, and each step has the potential for producing microstructural flaws that degrade the properties of the fabricated material. An important feature of the treatment is the attempt to show the importance of each step as well as the interconnection between the various steps in the overall fabrication route. Chapter 1 provides an introductory overview of the various methods that can be used for the production of ceramic materials. In the production of ceramics from powders and, to a lesser extent, the sol-gel route, the overview also forms the basis for the more detailed considerations later in the book. Chapters 2 to 12 form a logical development from the start of the fabrication route to the final fabricated microstructure.

My intention has been to prepare a text that is suitable for a one-semester (or two-quarter) course in the processing of ceramics at the senior undergraduate level or the introductory graduate level. A background in the concepts and processing of traditional ceramics, typically obtained in lower-level undergraduate classes, is assumed. For a processing course, it may be advisable to omit Chapter 11, which covers some difficult issues of sintering in depth. The second half of the book (Chapters 7 to 12) may also be suitable for a one-semester course in sintering and microstructural control of ceramics at the introductory graduate level. It is hoped that the book will also be useful to researchers in industry who are involved in the production of ceramics or who wish to develop a background in the processing and sintering of ceramics.

I am greatly indebted to Dr. G. W. Scherer, who reviewed most of the chapters. His comments and constructive criticism saved me from perpetuating many mistaken ideas. Any remaining mistakes, however, are my own and I accept responsibility for them. I also wish to thank the many authors and publishers who have allowed me pemission to reproduce their figures in this book. Last but not least, I wish to thank my wife, Vashanti, for her unfailing support and her forbearance during my preoccupation with the completion of this book.

M. N. Rahaman

Contents

Preface		v
1	**Ceramic Fabrication Processes: An Introductory Overview**	**1**
	1.1 Introduction	1
	1.2 Ceramic Fabrication Processes	5
	1.3 Production of Polycrystalline Ceramics from Powders: An Overview	19
	1.4 A Case Study in Processing: The Fabrication of TiO_2 from Powders	33
	1.5 Concluding Remarks	36
	References	36
	Further Reading	37
2	**Synthesis of Powders**	**38**
	2.1 Introduction	38
	2.2 Desirable Powder Characteristics	38
	2.3 Powder Preparation Methods	40
	2.4 Powder Preparation by Mechanical Methods	42
	2.5 Powder Preparation by Chemical Methods	54
	2.6 Concluding Remarks	89
	Problems	90
	References	91
	Further Reading	93
3	**Powder Characterization**	**94**
	3.1 Introduction	94
	3.2 Physical Characterization	95
	3.3 Chemical Composition	122

	3.4	Phase Composition	127
	3.5	Surface Characterization	130
	3.6	Concluding Remarks	142
		Problems	143
		References	144
4	**Science of Colloidal Processing**		**146**
	4.1	Introduction	146
	4.2	Types of Colloids	147
	4.3	Attractive Surface Forces	148
	4.4	Lyophobic Colloids	152
	4.5	Electrostatic Stabilization	153
	4.6	Polymeric Stabilization	172
	4.7	Structure of Consolidated Colloids	180
	4.8	Rheology of Colloidal Suspensions	183
	4.9	Stabilization of Suspensions with Polyelectrolytes: Electrosteric Stabilization	191
	4.10	Concluding Remarks	198
		Problems	198
		References	199
		Further Reading	200
5	**Sol-Gel Processing**		**201**
	5.1	Introduction	201
	5.2	Types of Gels	203
	5.3	Metal Alkoxides (Metal-Organic Compounds)	207
	5.4	The Sol-Gel Process for Metal Alkoxides	214
	5.5	Sol-Gel Preparation Techniques	246
	5.6	Applications of Sol-Gel Processing	256
	5.7	Concluding Remarks	261
		Problems	261
		References	262
		Further Reading	263
6	**Powder Consolidation and Forming of Ceramics**		**264**
	6.1	Introduction	264
	6.2	Packing of Particles	265
	6.3	Powder Consolidation Methods: A Preview	279
	6.4	Importance of Additives in the Forming Process	279
	6.5	Selection of Additives	280
	6.6	Dry and Semidry Pressing Methods	290
	6.7	Casting Methods	298

	6.8 Plastic-Forming Methods	309
	6.9 Drying of Cast or Extruded Articles	316
	6.10 Binder Removal	320
	6.11 Microstructural Characterization of the Green Body	328
	6.12 Concluding Remarks	328
	Problems	328
	References	329
	Further Reading	330
7	**Sintering of Ceramics: Fundamentals**	**331**
	7.1 Introduction	331
	7.2 Sintering Studies: Some General Considerations	332
	7.3 Driving Forces for Sintering	334
	7.4 Diffusion in Solids	336
	7.5 Defects and Defect Chemistry	347
	7.6 The Chemical Potential	357
	7.7 Diffusional Flux Equations	365
	7.8 Vapor Pressure over a Curved Surface	366
	7.9 Diffusion in Ionic Crystals: Ambipolar Diffusion	367
	7.10 Concluding Remarks	371
	Problems	372
	References	373
8	**Theory of Solid-State and Viscous Sintering**	**374**
	8.1 Introduction	374
	8.2 Sintering of Polycrystalline and Amorphous Materials: A Preview	376
	8.3 Theoretical Analysis of Sintering	382
	8.4 Scaling Laws	383
	8.5 Analytical Models	389
	8.6 Numerical Simulations of Sintering	414
	8.7 Phenomenological Sintering Equations	417
	8.8 Sintering Diagrams	419
	8.9 Sintering with an Externally Applied Pressure: Hot Pressing	422
	8.10 The Stress Intensification Factor	430
	8.11 The Sintering Stress	433
	8.12 An Alternative Derivation of the Sintering Equations	437
	8.13 Concluding Remarks	442
	Problems	442
	References	443
	Further Reading	444

Contents

9 Grain Growth and Microstructural Control — **445**
- 9.1 Introduction — 445
- 9.2 Grain Growth: Preliminary Considerations — 447
- 9.3 Ostwald Ripening: The LSW Theory — 454
- 9.4 Topological and Interfacial Tension Requirements — 458
- 9.5 Normal Grain Growth in Dense Solids — 460
- 9.6 Abnormal Grain Growth in Dense Solids — 471
- 9.7 Effect of Inclusions and Dopants on Boundary Mobility — 473
- 9.8 Grain Growth in Porous Solids — 481
- 9.9 Grain Growth in Very Porous Solids — 493
- 9.10 Pore Evolution During Sintering — 496
- 9.11 Thermodynamic Aspects of Pore–Boundary Interactions — 499
- 9.12 Interaction Between Densification and Coarsening — 504
- 9.13 Fabrication Routes for the Production of Ceramics with High Density and Controlled Grain Size — 508
- 9.14 Concluding Remarks — 511
- Problems — 511
- References — 512
- Further Reading — 514

10 Liquid-Phase Sintering — **515**
- 10.1 Introduction — 515
- 10.2 Elementary Features of Liquid-Phase Sintering — 517
- 10.3 Microstructure Produced by Liquid-Phase Sintering — 520
- 10.4 The Stages in Liquid-Phase Sintering — 521
- 10.5 The Controlling Kinetic and Thermodynamic Factors — 525
- 10.6 Is the Liquid Squeezed Out From Between the Grains? — 537
- 10.7 The Basic Mechanisms of Liquid-Phase Sintering — 540
- 10.8 Hot Pressing with a Liquid Phase — 569
- 10.9 Phase Diagrams and Their Use in Liquid-Phase Sintering — 570
- 10.10 Vitrification — 574
- 10.11 Concluding Remarks — 579
- Problems — 580
- References — 581
- Further Reading — 582

Contents

11 Problems of Sintering — 583
- 11.1 Introduction — 583
- 11.2 Inhomogeneities and Their Effects on Sintering — 585
- 11.3 Constrained Sintering: I. Rigid Inclusions — 595
- 11.4 Constrained Sintering: II. Thin Films — 624
- 11.5 Solid Solution Additives and the Sintering of Ceramics — 637
- 11.6 Sintering with Chemical Reaction: Reaction Sintering — 653
- 11.7 Viscous Sintering with Crystallization — 665
- 11.8 Concluding Remarks — 674
- Problems — 675
- References — 676

12 Densification Process Variables and Densification Practice — 679
- 12.1 Introduction — 679
- 12.2 Conventional Sintering — 680
- 12.3 Microwave Sintering — 725
- 12.4 Pressure Sintering — 735
- 12.5 Concluding Remarks — 750
- Problems — 750
- References — 751
- Further Reading — 752

Appendix *Values of Constants, SI Units, and Conversion of Units* — *753*
Index — *757*

1
Ceramic Fabrication Processes: An Introductory Overview

1.1 INTRODUCTION

The subject of ceramics covers a wide range of materials. Recent attempts have been made to divide it into two parts: traditional ceramics and advanced ceramics. The use of the term "advanced" has, however, not received general acceptance, and other terms, including "technical," "special," "fine," and "engineering" will also be encountered. *Traditional ceramics* bear a close relationship to those materials that have been developed since the earliest civilizations. They are pottery, structural clay products, and clay-based refractories, with which we may also group cements and concretes and glasses. Whereas traditional ceramics still represent a major part of the ceramics industry, interest in recent years has focused on *advanced ceramics*, ceramics that, with minor exceptions, have been developed within the last 50 years or so. Advanced ceramics include ceramics for electrical, magnetic, electronic, and optical applications (sometimes referred to as *functional ceramics*) and ceramics for structural applications at ambient as well as elevated temperatures (*structural ceramics*). Although the distinction between traditional and advanced ceramics may be referred to in this book occasionally for convenience, we do not wish to overemphasize it. There is much to be gained through continued interaction between the traditional and advanced sectors.

Chemically, with the exception of carbon, ceramics are nonmetallic, inorganic compounds. Examples are the silicates such as kaolinite ($Al_2Si_2O_5(OH)_4$) and mullite ($Al_6Si_2O_{13}$), simple oxides such as alumina (Al_2O_3) and zirconia (ZrO_2), complex oxides other than the silicates such as barium titanate ($BaTiO_3$), and the superconducting material $YBa_2Cu_3O_{6+\delta}(0 \leq \delta \leq 1)$. In addition, there are nonoxides, including carbides such as silicon carbide (SiC) and boron carbide (B_4C), nitrides such as silicon nitride (Si_3N_4) and boron nitride (BN), borides such as titanium diboride (TiB_2), silicides such as molybdenum disilicide ($MoSi_2$), and halides such as lithium fluoride (LiF). There are also compounds based on nitride-oxide or oxynitride systems (e.g., β′-sialons with the general formula $Si_{6-z}Al_zN_{8-z}O_z$, where $0 < z < \approx 4$).

Structurally, all materials are either *crystalline* or *amorphous* (also referred to as *glassy*). The difficulty and expense of growing single crystals means that, normally, crystalline ceramics (and metals) are actually *polycrystalline*—they are made up of a large number of small crystals, or grains, separated from one another by grain boundaries. In ceramics as well as in metals, we are concerned with two types of structures, both of which have a profound effect on properties. The first type of structure is at the atomic scale: the type of *bonding* and the *crystal structure* (for a crystalline ceramic) or *amorphous structure* (if it is glassy). The second type of structure is at a larger scale: the *microstructure*, which refers to the nature, quantity, and distribution of the structural elements or phases in the ceramic (e.g., crystals, glass, and porosity).

It is sometimes useful to distinguish between the intrinsic properties of a material and the properties that depend on the microstructure. The *intrinsic properties* are determined by the structure at the atomic scale and are properties that are not susceptible to significant change by modification of the microstructure, properties such as the melting point, elastic modulus, coefficient of thermal expansion, and whether the material is brittle, magnetic, ferroelectric, or semiconducting. In contrast, many of the properties critical to the engineering applications of materials are strongly dependent on the microstructure (e.g., mechanical strength, dielectric constant, and electrical conductivity).

Intrinsically, ceramics usually have high melting points and are therefore generally described as highly refractory. They are also usually hard, brittle, and chemically inert. This chemical inertness is usually taken for granted, for example, in ceramic and glass tableware and in the bricks, mortar, and glass of our houses. However, when used at high temperatures, as in the chemical and metallurgical industries, this chemical inertness is severely tried. The electrical, magnetic, and dielectric behavior covers a wide range—for example, in the case of electrical behavior, from insulators to conductors. The applications of ceramics are many. Usually, for a given application one property may be of particular importance, but, in fact, all relevant properties need to be considered. We are therefore usually interested in combinations of properties. For traditional ceramics and glasses, familiar applications include structural building materials (e.g., bricks and roofing tile), refractories for furnace linings, tableware

Table 1.1 Applications of Advanced Ceramics Classified by Function

Function	Ceramic	Application
Electric	Insulation materials (Al_2O_3, BeO, MgO)	Integrated circuit substrate, package, wiring substrate, resistor substrate, electronics interconnection substrate
	Ferroelectric materials ($BaTiO_3$, $SrTiO_3$)	Ceramic capacitor
	Piezoelectric materials (PZT)	Vibrator, oscillator, filter, etc. Transducer, ultrasonic humidifier, piezoelectric spark generator, etc.
	Semiconductor materials ($BaTiO_3$, SiC, ZnO-Bi_2O_3, V_2O_5 and other transition metal oxides)	NTC thermistor: temperature sensor, temperature compensation, etc. PTC thermistor: heater element, switch, temperature compensation, etc. CTR thermistor: heat sensor element Thick-film thermistor: infrared sensor Varistor: noise elimination, surge current absorber, lighting arrestor, etc. Sintered CdS material: solar cell SiC heater: electric furnace heater, miniature heater, etc.
	Ion-conducting materials (β-Al_2O_3, ZrO_2)	Solid electrolyte for sodium battery ZrO_2 ceramics: oxygen sensor, pH meter fuel cells
Magnetic	Soft ferrite	Magnetic recording head, temperature sensor, etc.
	Hard ferrite	Ferrite magnet, fractional horse power motors, etc.
Optical	Translucent alumina	High-pressure sodium vapor lamp
	Translucent Mg-Al spinel, mullite, etc.	For a lighting tube, special-purpose lamp, infrared transmission window materials
	Translucent Y_2O_3-ThO_2 ceramics	Laser materials
	PLZT ceramics	Light memory element, video display and storage system, light modulation element, light shutter, light valve

Table 1.1 *Continued*

Function	Ceramic	Application
Chemical	Gas sensor (ZnO, Fe_2O_3, SnO_2)	Gas leakage alarm, automatic ventilation fan, hydrocarbon, fluorocarbon detectors, etc.
	Humidity sensor ($MgCr_2O_4$-TiO_2)	Cooking control element in microwave oven, etc.
	Catalyst carrier (cordierite)	Catalyst carrier for emission control
	Organic catalyst	Enzyme carrier, zeolites
	Electrodes (titanates, sulfides, borides)	Electrowinning aluminum, photochemical processes, chlorine production
Thermal	ZrO_2, TiO_2	Infrared radiator
Mechanical	Cutting tools (Al_2O_3, TiC, TiN, others)	Ceramic tool, sintered CBN; cermet tool, artificial diamond; nitride tool
	Wear-resistant materials (Al_2O_3, ZrO_2)	Mechanical seal, ceramic liner, bearings, thread guide, pressure sensors
	Heat-resistant materials (SiC, Al_2O_3, Si_3N_4, others)	Ceramic engine, tubine blade, heat exchangers, welding burner nozzle, high-frequency combustion crucibles
Biological	Alumina ceramics implantation, hydroxyapatite bioglass	Artificial tooth root, bone, and joint
Nuclear	Nuclear fuels (UO_2, UO_2-PuO_2)	
	Cladding materials (C, SiC, B_4C)	
	Shielding materials (SiC, Al_2O_3, C, B_4C)	

Source: Ref. 1, with permission.

and sanitaryware, electrical insulation (e.g., electrical porcelain and steatite), glass containers, and glasses for building and transportation vehicles. The applications for which advanced ceramics have been developed or proposed are already very diverse, and this area is expected to continue to grow at a reasonable rate. Table 1.1 illustrates some of the applications for advanced ceramics [1].

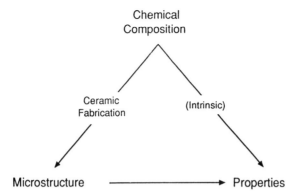

Figure 1.1 The important relationships in ceramic fabrication.

The important relationships between chemical composition, atomic structure, fabrication, microstructure, and properties of polycrystalline ceramics are illustrated in Fig. 1.1. The intrinsic properties must be considered at the time of materials selection. For example, the phenomenon of ferroelectricity originates in the perovskite crystal structure, of which $BaTiO_3$ is a good example. For the production of a ferroelectric material, we may therefore wish to select $BaTiO_3$. The role of the fabrication process, then, is to produce microstructures with the desired engineering properties. For example, the measured dielectric constant of the fabricated $BaTiO_3$ will depend significantly on the microstructure (grain size, porosity, and the presence of any secondary phases). Normally, the overall fabrication method can be divided into a few or several discrete steps, depending on the complexity of the method. Although there is no generally accepted terminology, we will refer to these discrete steps as *processing steps*. The fabrication of a ceramic body therefore involves a number of processing steps. In the next section, we examine, in general terms, some of the commonly used methods for the fabrication of ceramics.

1.2 CERAMIC FABRICATION PROCESSES

Ceramics can be fabricated by a variety of methods, some of which have their origins in early civilization. Our normal objective is the production, from suitable starting materials, of a solid product with the desired shape such as a film, fiber or monolith and with the desired microstructure. As a first attempt, we divide the main fabrication methods into three groups, depending on whether the starting materials involve a gaseous phase, a liquid phase, or a solid phase (Table 1.2). In the following sections, we examine the main features of the processing steps in these methods and, from the point of view of ease of processing, their main advantages and disadvantages.

A. Gas-Phase Reactions

The category of methods based on the use of gaseous starting materials can involve reactions between gases, reactions between a gas and a liquid, or reactions between a gas and a solid. By far the most important are chemical vapor deposition methods, where the desired material is formed by chemical reaction between gaseous species. The fabrication route involving gas–liquid reactions has been shown to offer some promise only recently and is referred to as directed metal oxidation. Reaction between a gas and a solid, referred to as reaction bonding, has been used mainly for the production of Si_3N_4 and SiC.

Chemical Vapor Deposition

Chemical vapor deposition (CVD) is a well-established technique that has been used for the production of thin films, thick films, and even monolithic bodies. Films are produced preferably by endothermic reactions, so the surfaces on which they are deposited must be heated. The reactions occur predominantly by heterogeneous precipitation onto the surfaces, and homogeneous precipitation in the gas phase, leading to the formation of particles, must be suppressed. (We will see in the next chapter that homogeneous precipitation in the gas phase is an important method for the preparation of some ceramic powders.) The fabrication of monolithic bodies basically involves prolonging the deposition process so that the desired thickness is achieved. A wide variety of chemical compositions, consisting of nonoxide as well as oxide compositions, can be fabricated by CVD. Table 1.3 shows some of the important reactions used for the fabrication

Table 1.2 Common Ceramic Fabrication Methods

Starting materials	Method	Product
Gases	Chemical vapor deposition	Films, monoliths
Gas–liquid	Directed metal oxidation	Monoliths
Gas–solid	Reaction bonding	Monoliths
Liquids	Sol-gel process	Films, fibers
	Polymerization	Films, fibers
Solids (powders)	Melt casting	Monoliths
	Sintering of powders	Films, monoliths

Ceramic Fabrication Processes

Table 1.3 Some Important CVD Reactions for the Fabrication of Ceramics

Reaction	Temperature (K)	Application
$2C_xH_y \rightarrow 2xC + yH_2$	1100–2700	Pyrolytic carbon and graphite
$SiCl_4 + 2H_2O \rightarrow SiO_2 + 4HCl$	700–1300	Films for semiconductor devices, optical fibers
$TiCl_4 + O_2 \rightarrow TiO_2 + 2Cl_2$	1200–1500	Films for electronic devices
$3SiCl_4 + 4NH_3 \rightarrow Si_3N_4 + 12HCl$	1200–1800	Films for semiconductor devices
$3SiH_4 + 4NH_3 \rightarrow Si_3N_4 + 6H_2$	1000–1800	Films for semiconductor devices
$CH_3Cl_3Si \rightarrow SiC + 3HCl$	1200–1500	Composites
$3HSiCl_3 + 4NH_3 \rightarrow Si_3N_4 + 9HCl + 3H_2$	1100–1400	Composites
$NH_3 + BCl_3 \rightarrow BN + 3HCl$	1000–1300	Monoliths
$TiCl_4 + 2BH_3 \rightarrow TiB_2 + 4HCl + H_2$	1200–1500	Monoliths, composites
$W(CO)_6 \rightarrow WC + CO_2 + 4CO$	600–1100	Coatings

of ceramics together with the temperature range for each reaction and the applications of the fabricated articles [2].

The apparatus used for CVD depends on the reaction being used, the reaction temperature, and the configuration of the substrate. Figure 1.2 shows examples of reactors for the deposition of films on substrates such as silicon wafers [3]. The general objective for any design is to provide uniform exposure of the substrate to the reactant gases. Chemical vapor deposition has a number of process variables that must be manipulated to produce a deposit with the desired properties. These variables include the flow rate of the reactant gases, the nature and flow rate of any carrier gases, the pressure in the reaction vessel, and the temperature of the substrate. The pressure in the reaction vessel influences the concentration of the reactant gases, the diffusion of reactants toward the substrate, and the diffusion of the products away from the surface. The higher diffusivity at lower pressure leads to the formation of films with better uniformity, so that most CVD reactors are operated in the pressure range of 1–15 kPa. The temperature of the substrate influences the deposition rate and is the main factor controlling the structure of the deposit.

The technology for chemical vapor deposition has recently been attracting much interest as a fabrication route for ceramic composites. In the

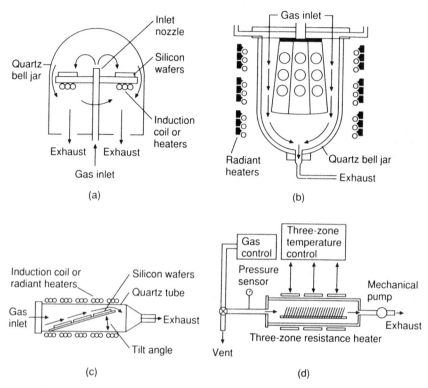

Figure 1.2 Typical reactors used in chemical vapor deposition: (a) pancake reactor; (b) barrel reactor; (c) horizontal reactor; (d) low-pressure (LPCVD) reactor. (From Ref. 3, with permission.)

reaction between $SiCl_4$ and NH_3 outlined in Table 1.3 for the production of Si_3N_4, the addition of gaseous B_2H_6 leads to the deposition of a composite of Si_3N_4 and BN. For ceramics reinforced with continuous fibers, one approach that has shown considerable promise is chemical vapor infiltration (CVI). The fibers, preformed into the shape and dimensions of the finished body, are placed into the reactant gases and held at the desired temperature so that the deposited material is formed in the interstices between the fibers. Although successful fabrication leading to a composite with uniform structure and reasonably high density requires fairly precise control of the process variables, the CVI route has led to the development of a number of ceramic composites with useful properties for structural applications at high temperatures [4]. These include carbon fiber reinforced SiC and SiC fiber reinforced SiC. Composites containing as high

Ceramic Fabrication Processes

as 45 volume percent (vol %) of fibers have been fabricated with an open porosity of ≈10%. The measured fracture toughness remained unchanged at ≈30 MPa·m$^{1/2}$ up to 1400°C, which is considerably better than that of unreinforced SiC, which has a fracture toughness of ≈5 MPa·m$^{1/2}$.

Table 1.3 indicates that the reaction temperatures for the CVD fabrication of most of the highly refractory ceramics listed are rather low. CVD methods therefore provide a distinct advantage of fairly low fabrication temperatures for ceramics and composites with high melting points that are difficult to fabricate by other methods or require very high fabrication temperatures (e.g., SiC, discussed above). The low reaction temperatures also increase the range of materials that can be coated by CVD, especially for the highly refractory coatings. However, a major disadvantage is that the material deposition rate by CVD is very slow, typically in the range of ≈1–100 μm/h. The production of monolithic bodies can therefore be very time-consuming and expensive. Another problem that is normally encountered in the fabrication of monolithic bodies by CVD is the development of a microstructure consisting of fairly large, columnar grains, which leads to fairly low intergranular strength. These difficulties encountered in the fabrication of monolithic bodies limit CVD methods predominantly to the formation of thin films and coatings. A modification of the CVD process has also been used successfully for the production of optical waveguide fibers [2].

Directed Metal Oxidation

Fabrication routes involving reactions between a gas and a liquid are generally impractical for the production of ceramic bodies because the reaction product usually forms a solid protective coating, thereby separating the reactants and effectively stopping the reaction. However, a novel method employing directed oxidation of a molten metal by a gas has been used by the Lanxide Corporation for the production of porous and dense materials as well as composites [5]. Figure 1.3 illustrates how the directed oxidation is carried out. In Fig. 1.3a a molten metal (e.g., an aluminum alloy) is being oxidized by a gas (e.g., air). If the temperature is in the range of 900–1350°C and the aluminum alloy contains a few percent of magnesium and a group IVA element (e.g., Si, Ge, Sn, or Pb), the oxide coating is no longer protective. Instead, it contains small pores through which molten metal is drawn up to the top surface of the film, thereby continuing the oxidation process. The reaction product continues to grow at a rate of a few centimeters per day until the desired thickness is obtained.

For the production of composites, a filler material (e.g., particulate reinforcement or fibers) is shaped into a preform of the size and shape

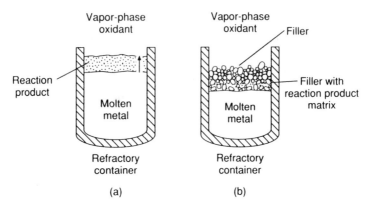

Figure 1.3 Schematic diagram of (a) the formation of a matrix of oxide and unreacted metal by directed oxidation of molten metal and (b) oxidation in the presence of a filler. (From Ref. 5, with permission.)

desired of the product. The filler and the metal alloy are then heated to the growth temperature at which the oxidation process occurs outward from the metal surface and into the preform (Fig. 1.3b), such that the oxidation product becomes the matrix of the composite. A micrograph of an SiC fiber preform that has been filled with an Al_2O_3/Al matrix by directed oxidation of molten aluminum is shown in Fig. 1.4.

The term *directed metal oxidation* is taken to include all reactions in which the metal gives up or shares its electrons. This method has been used to produce composites with matrices not only of oxides but also of nitrides, borides, carbides, and titanates. Composite systems produced by directed metal oxidation include matrices of Al_2O_3/Al, AlN/Al, ZrN/Zr, TiN/Ti, and Zr and fillers of Al_2O_3, SiC, $BaTiO_3$, AlN, B_4C, TiB_2, ZrN, ZrB_2, and TiN. A distinct advantage of the method is that growth of the matrix into the preform involves little or no change in dimensions. The problems associated with shrinkage during densification in other fabrication routes are therefore avoided. Furthermore, large components can be produced readily with good control of the component dimensions.

Reaction Bonding
Chemical reactions between a gas and a solid are associated mainly with the fabrication of Si_3N_4 and SiC bodies. The process is referred to as *reaction bonding* [6]. In the production of Si_3N_4, silicon powder is consolidated by one of a variety of methods (e.g., pressing in a die, slip casting, and injection molding) to form a billet or a shaped body; this is then

Ceramic Fabrication Processes

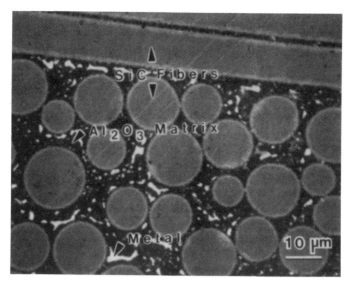

Figure 1.4 Optical micrograph of an Al_2O_3/Al matrix reinforced with SiC fibers produced by directed metal oxidation. (From Ref. 5, with permission.)

preheated in argon at ≈1200°C, after which it can be machined to the required component shape and dimensions. Finally, the component is heated, usually in N_2 gas, at temperatures in the region of 1200–1400°C, and reaction bonding occurs to produce reaction-bonded silicon nitride (RBSN). Typically, the RBSN has a porosity of ≈20%, and because of this the strength of RBSN is inferior to that of dense Si_3N_4 produced by other methods (e.g., hot pressing). An attractive feature of the fabrication route is that very little shrinkage occurs during the reaction bonding process. RBSN bodies with a high degree of dimensional accuracy and with complex shapes can be prepared fairly readily without the need for expensive machining after firing.

B. Liquid Precursor Methods

Methods in which a solution of metal compounds is converted into a solid body are sometimes referred to as liquid precursor methods. A liquid precursor route that has attracted intense interest since the mid-1970s is the sol-gel process [7]. Chemical compositions consisting of simple or complex oxides are produced by this route. Another route that has attracted a fair degree of interest in the past 20 years is polymer pyrolysis

[8], in which nonoxides (mainly Si_3N_4 and SiC) are produced by pyrolysis of suitable polymers.

Sol-Gel Processing

In the *sol-gel process*, a solution of metal compounds or a suspension of very fine particles in a liquid (referred to as a *sol*) is converted into a semirigid mass (the *gel*). Two different sol-gel processes can be distinguished, depending on whether a sol or a solution is used. A basic flow chart outlining the processing steps for the two methods is shown in Fig. 1.5. Starting with a sol, the gelled material consists of identifiable colloidal particles that have been joined together by surface forces to form a network (Fig. 1.6b). When a solution is used, typically a solution of metalor-

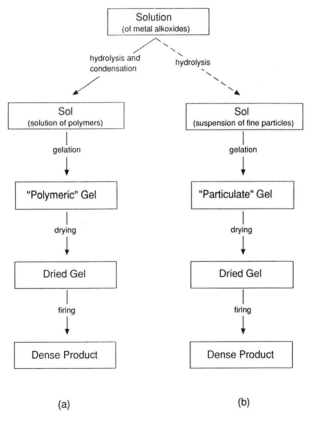

Figure 1.5 Basic flow charts for sol-gel processing using (a) a solution and (b) a suspension of fine particles.

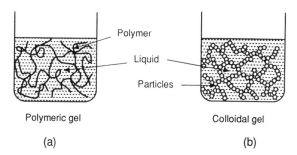

Figure 1.6 Schematic diagram of the structure of (a) a polymeric gel from a solution and (b) a particulate gel formed from a suspension of fine particles.

ganic compounds (such as metal alkoxides), the gelled material in many cases may consist of a network of polymer chains (Fig. 1.6a). This "solution sol-gel process" is attracting considerable attention for the fabrication of ceramics. We shall consider the sol-gel process in more detail later (Chapter 5), but for the remainder of this section we outline the main sequence of steps in the fabrication of ceramics by the solution sol-gel route.

As indicated above, the starting material normally consists of a solution of metal alkoxides in an appropriate alcohol. Metal alkoxides have the general formula $M(OR)_x$ and can be considered as either a derivative of an alcohol, ROH, where R is an alkyl group, in which the hydroxyl proton is replaced by a metal, M, or a derivative of a metal hydroxide, $M(OH)_x$. To this solution, water is added, either in the pure state or diluted with more alcohol. Under constant stirring at temperatures slightly above room temperature (normally ≈50–90°C) and with suitable concentration of reactants and pH of the solution, hydrolysis and condensation reactions may occur, leading to the formation of polymer chains. Taking the example of a tetravalent metal, the reactions may be expressed as follows.

Hydrolysis:

$$M(OR)_4 + H_2O \rightarrow M(OR)_3OH + ROH \tag{1.1}$$

Condensation:

$$M(OR)_3OH + M(OR)_4 \rightarrow \underset{\text{(polymer)}}{(RO)_3\text{—M—O—M—}(OR)_2} + ROH \tag{1.2}$$

Polymerization of the species formed by the hydrolysis and condensation reactions together with interlinking and cross-linking of the polymer chains eventually leads to a marked increase in the viscosity of the reaction mixture and the production of a gel. The gel has a continuous solid network and a finite elastic shear modulus. Normally, excess water and alcohol are used in the reactions so that the amount of solid matter in the gel (i.e., the solids content of the gel) can be quite low, <5–10% in many cases. The remainder of the gel volume consists of liquid that must be removed prior to firing.

Drying of the gels can be the most time-consuming and difficult step in the overall fabrication route, especially when a monolithic material is required directly from the gel. Normally, the liquid is present in fine channels, typically ≈10–50 nm in diameter. Removal of the liquid by evaporation has two main consequences: large capillary stresses are generated, and the gel undergoes considerable shrinkage. The gelled material is fairly weak, and unless special precautions are taken, cracking as well as warping of the dried gel can be a severe problem. If the evaporation is carried out slowly to control the pressure gradient in the liquid, drying can take weeks for a gel with a thickness of a few centimeters. It has been claimed that the addition of certain chemical agents such as formamide, glycerol, or oxalic acid to the solution prior to gelation can speed up the drying process considerably while avoiding cracking. These chemical agents are referred to as *drying control chemical additives* (DCCAs). The role of these additives is not clear; in addition they cause problems during the firing of gels. Alternatively, removal of the liquid under supercritical conditions in an autoclave, although expensive, eliminates the capillary stresses and produces no net shrinkage of the gel during drying.

Despite drying, the gel contains a small amount of adsorbed water and organic groups such as residual alkyl groups chemically attached to the polymer chains. These are removed below ≈500°C prior to densification of the gel at a higher temperature. In general, this densification takes place at a much lower temperature than would be required to make an equivalent material by a more conventional fabrication route (e.g., sintering of powders, outlined in the next section). This densification at a lower temperature results, in general, from the amorphous nature of the gel and the very fine porosity.

As a fabrication route for ceramics, the sol-gel process has a number of advantages. Because of the ease of purification of liquids (as the starting materials for the process), with suitable precautions materials with high purity can be produced. Materials with exceptionally good chemical homogeneity, which is very desirable, especially in the case of complex oxides, can also be produced because the mixing of the constituents occurs at a molecular level during the chemical reactions. Another advantage is the lower densification temperature. However, the disadvantages are also real. The starting materials (e.g., the metal alkoxides) can be fairly expen-

Ceramic Fabrication Processes

sive. We have already mentioned the difficulties of conventional drying, during which cracking, warping, and considerable shrinkage are common problems. Figure 1.7 illustrates the enormous amounts of shrinkage that may occur during the drying and sintering of a gelled material containing 5 wt % of solids. Mainly because of these problems in drying, the sol-gel route has seen little use for the fabrication of monolithic ceramics. Instead it has seen considerable use for the fabrication of small or thin articles such as films, fibers, and powders, and its use in this area is expected to grow substantially in the future.

Polymer Pyrolysis

We have seen that under suitable conditions the hydrolysis and condensation of metal alkoxides lead to the formation of a polymeric gel in which the polymers contain metal–oxygen bonds [Eqs. (1.1) and (1.2)]. Polymer pyrolysis is another route that involves polymerization reactions in liquid solutions. The ceramic is produced after pyrolysis of the polymerized

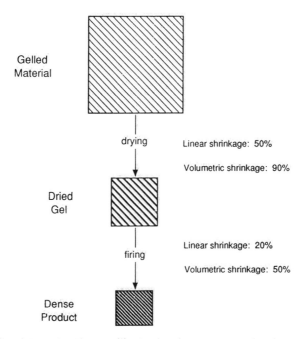

Figure 1.7 Schematic diagram illustrating the enormous shrinkages accompanying the drying by liquid evaporation and the firing of a polymeric gel. The solids content of the gelled material, the dried gel, and the final product are asssumed to be 5, 50, and 100%, respectively.

material [9]. The method has been applied mainly to the production of SiC and Si_3N_4.

In the polymer pyrolysis route, a basic requirement is that the main chain of the polymers contain Si. Two systems that have attracted considerable interest are (i) polysilazanes, $(-\overset{|}{\underset{|}{Si}}-\overset{|}{N}-)$, with Si—N bonds in the main chain, and (ii) polycarbosilanes, $(-\overset{|}{\underset{|}{Si}}-CH_2-)_n$, with CH_2 groups separating the Si atoms in the main chain.

Silicon nitride (Si_3N_4) may be produced by the pyrolysis of polysilazanes, which contain the nitrogen already bonded to the Si in the main chain, or of polycarbosilanes after suitable nitridation of the polymer between 500 and 800°C. SiC is usually prepared from polycarbosilanes. The steps leading to the production of SiC can be summarized as follows:

$$(CH_3)_2SiCl_2 \xrightarrow{Na} [-Si(CH_3)_2-]_n \xrightarrow{470°C}$$
$$(-\underset{\underset{H}{|}}{Si}CH_3CH_2-)_n \xrightarrow{1200°C} \text{silicon carbide} \quad (1.3)$$

Dimethyldichlorosilane reacts with Na in the presence of benzene to produce polysilanes, which are converted to silicon carbide by a two-step heating process.

The amount of ceramic product obtained after pyrolysis of the polymer can be anywhere from 20 to 80%, depending on the chemistry of the polymer. The pyrolysis is therefore usually accompanied by an appreciable loss in mass and considerable shrinkage. As we outlined for the sol-gel process, these factors make the fabrication of monolithic ceramics very difficult. The route is therefore more suitable for the production of films and fibers. In fact, one of the most impressive applications of the pyrolysis method has been the commercial production of SiC fibers [9]. The polymers are also being considered for use as binders in the injection molding of ceramics (Chapter 6) and as matrices for ceramic composites.

C. Fabrication from Powders

This route involves the production of the desired body from an assemblage of finely divided solids (i.e., powders) by the action of heat. It gives rise to the two most widely used methods for the fabrication of ceramics, (i) melt casting and (ii) firing of compacted powders. These two fabrication routes have their origins in the earliest civilizations.

Ceramic Fabrication Processes

Melt Casting

In its simplest form, the melt casting method involves melting a batch of powdered raw materials, followed by cooling and forming to produce a solid finished body (Fig. 1.8). This conventional melt casting technology has undergone no significant change in recent years, and extensive description of the production methods can be found elsewhere [10]. For ceramics that crystallize relatively easily, the solidification of the melt is accompanied by rapid nucleation and growth of crystals (i.e., grains). Uncontrolled grain growth is generally a severe problem that leads to the production of ceramics with undesirable properties (e.g., low strength). Another problem is that many ceramics either have a high melting point (e.g., ZrO_2, mp $\approx 2600°C$) or decompose before they reach their melting point (e.g., Si_3N_4), so that obtaining a melt is rather difficult. The melt casting method is therefore limited to the fabrication of glasses.

An important variation of the melt casting method is the *glass ceramic process* [11]. Here the raw materials are formed into a glass (e.g., by melt casting as outlined above) and subsequently heated in a two-step process, first to nucleate and then to grow the crystals throughout the glass (Fig. 1.8). Common problems are the formation of nonequilibrium phases and retention of a residual glassy phase due to incomplete crystallization. Glass ceramics are by definition $\geq 50\%$ crystalline by volume, and most are $>90\%$ crystalline. An example of a glass ceramic is cordierite, a magnesium aluminum silicate with the composition $2MgO \cdot 2Al_2O_3 \cdot 5SiO_2$. The

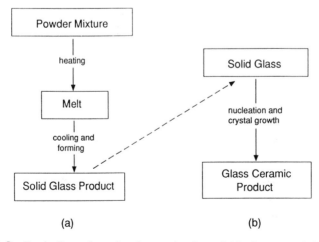

Figure 1.8 Basic flow chart for the production of (a) glasses and (b) glass ceramics by melt casting.

microstructure of a cordierite glass ceramic (Fig. 1.9) shows crystals in a glassy matrix [12]. Because the process depends initially on the formation of a glass, it is limited to chemical compositions that can form glasses and whose crystallization can be controlled.

Firing of Compacted Powders

Although in principle the firing of compacted powders can be used for the production of both glasses and polycrystalline ceramics, in practice it is hardly ever used for glasses because of the availability of more economical fabrication methods (e.g., melt casting). It is, however, by far the most widely used method for the production of polycrystalline ceramics. The processing steps are shown in Fig. 1.10. In its simplest form, this method involves the consolidation of a mass of fine particles (i.e., a powder) to form a porous, shaped powder form (sometimes called the *green body*), which is then fired (i.e., heated) to produce a dense product.

Because of its importance and widespread use, the fabrication of polycrystalline ceramics from powders forms the main focus of this book. In addition, we also consider the fabrication of ceramics from liquid starting materials by the sol-gel process. In the next section, we provide an overview of the fabrication of polycrystalline ceramics from powders that will

Figure 1.9 The microstructure of a cordierite glass ceramic showing crystals in a glass matrix. (From Ref. 12, with permission.)

Ceramic Fabrication Processes

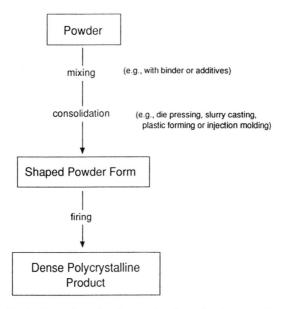

Figure 1.10 Basic flow chart for the production of polycrystalline ceramics by firing of consolidated powders.

form the basis for the more detailed considerations in subsequent chapters.

1.3 PRODUCTION OF POLYCRYSTALLINE CERAMICS FROM POWDERS: AN OVERVIEW

In the flow chart shown in Fig. 1.10 for the fabrication of ceramics from powders, we can divide the processing steps into two parts: processes prior to the firing of the green body and those that occur during firing. Until recently, most emphasis was placed on the processes that occur during firing, and the understanding gained from these studies is considerable. However, the increased attention given recently to powder preparation and forming methods has yielded clear benefits. The most useful approach (and the one adopted in this book) requires that close attention be paid to each processing step in the fabrication route if the very specific properties required of many ceramics are to be achieved. Each step has the potential for producing undesirable microstructural flaws in the body that can limit its properties and reliability. The important issues in each step and how these influence or are influenced by the other steps are outlined in the following sections.

A. Powder Preparation and Powder Characterization

In most cases the fabrication process starts from a mass of powder obtained from commercial sources. Nevertheless, knowledge of powder preparation methods is very important. Equally important are methods that can be used to determine the physical, chemical, and surface characteristics of the powder. Essentially, the characteristics of the powder depend strongly on the method used to prepare it, and these in turn influence the subsequent processing of the ceramic. The powder characteristics of greatest interest are size, size distribution, shape, degree of agglomeration, chemical composition, and purity.

Many methods are available for the preparation of ceramic powders. These range from mechanical methods involving predominantly grinding or milling for the reduction in size of a coarse, granular material (referred to as *comminution*) to chemical methods involving chemical reactions under carefully controlled conditions. Some of the chemical methods, although often more expensive than the mechanical methods, offer unprecedented control of the powder characteristics. Figure 1.11 shows an

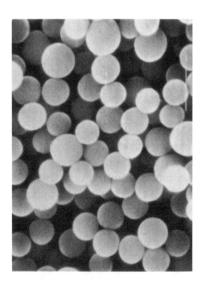

Figure 1.11 An example of a submicrometer TiO_2 powder prepared by controlled chemical precipitation from a solution. (From Ref. 13, with permission.)

example of a chemically prepared powder consisting of spherical particles of approximately the same size [13].

One of the more troublesome issues in the production of ceramics is the effect of minor variations in the chemical composition and purity of the powder on processing and properties. These variations, caused by insufficient control of the preparation procedure or introduced during subsequent handling, often go unrecorded or undetected. However, their effects on the microstructure and properties of the fabricated material can often be quite profound. For ceramics that must satisfy very demanding property requirements, one of the major advances made in the last 10–15 years has been the attention paid to powder quality. This has resulted in greater use of chemical methods for powder preparation. The use of cleanroom facilities, common in the semiconductor industry, has been suggested as an improvement in current ceramic processing practice.

A continuing trend is toward the preparation of fine powders. In principle, the enhanced activity of fine powders is beneficial for the attainment of high-density bodies at lower firing temperatures. The benefits of finer powders has been clearly demonstrated. For example, consolidated bodies of CeO_2 powders with a particle size of 10–20 nanometer (nm) have been fired to full density at 1100°C, compared with a firing temperature of 1600°C and a somewhat lower density for bodies produced from a powder with a particle size of ≈ 1 micrometer (μm). A major problem, however, is that the benefits of fine powders are normally realized only when extreme care is taken in their handling and subsequent consolidation. Generally, as the size decreases below ≈ 1 μm, the particles exhibit a greater tendency to interact, giving rise to the formation of agglomerates. One consequence of the presence of agglomerates is that the packing of the consolidated powder can be quite nonuniform. The overall effect is that during the firing stage little benefit is achieved over that of a coarse powder with a particle size corresponding to the agglomerate size of the fine powder. The use of fine powders therefore requires proper control of the handling and consolidation procedures in order to minimize the deleterious effects due to the presence of agglomerates. Such procedures may be quite demanding and expensive.

Along with the trend toward cleaner powders has been the growing use of analytical techniques for the characterization of the surface chemistry of the powders. These techniques include transmission electron microscopy (TEM), secondary ion mass spectroscopy (SIMS), Auger electron spectroscopy (AES), and X-ray photoelectron spectroscopy (XPS), also referred to as electron spectroscopy for chemical analysis (ESCA). The spectroscopic techniques have the capability of detecting constituents (atoms, ions, or molecules) down to the parts per million range.

B. Powder Consolidation

The consolidation of ceramic powders to produce a shaped powder form (generally referred to as the green body) is sometimes referred to as *forming*. The main consolidation methods include (i) dry or semidry pressing of the powder (e.g., in a die), (ii) mixing of the powder with water or organic polymers to produce a plastic mass that is shaped by pressing or deformation (referred to as *plastic forming*), and (iii) casting from slurry (e.g., *slip casting* and *tape casting*). These methods have been in use for a long time and most of them have originated in the traditional ceramics industry for the manufacture of clay-based materials.

Perhaps the greatest advance made in recent years has been the realization of the important effect of the green body microstructure on the subsequent firing stage. If severe variations in packing density occur in the green body, then, under conventional firing conditions, the fabricated body will usually contain heterogeneities, which have strong consequences for properties. This realization led to the increasing use of an approach based on colloidal techniques for the consolidation of powders from a suspension, an approach that can be regarded as essentially a refinement of the slip casting technique. The colloidal approach starts from powders with carefully controlled characteristics that have been prepared under controlled conditions by chemical methods or by fractionation of a commercial powder. The powders are dispersed in a liquid (normally water) and stabilized to prevent agglomeration through the use of electrolytes or polymers that are dissolved in the liquid. The suspension is then made to settle by itself, by filtration, or by centrifuging. The deposit forms the green body for subsequent firing. An example of the uniform arrangement formed from almost spherical, nearly monosized particles is shown in Fig. 1.12. Whereas their benefits have been demonstrated, colloidal methods have not made inroads into many industrial applications where mass production is desired and fabrication cost is a serious consideration.

C. The Firing Process

In the firing stage of the fabrication route, the shaped powder form is heated to produce the desired microstructure. The changes occurring during this stage may be fairly complex, depending on the complexity of the starting materials. In the ceramics literature, two terms have been used to refer to the heating stage: firing and sintering. Generally, the term *firing* is used when the processes occurring during the heating stage are fairly complex, as in the production of many traditional ceramics from clay-based materials. In less complex cases, the term *sintering* is used. We will distinguish between firing and sintering when this is convenient; however, we do not wish to attach much importance to this distinction.

Ceramic Fabrication Processes

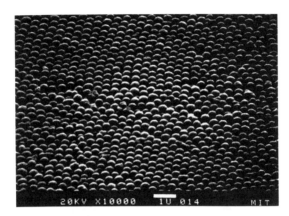

Figure 1.12 Uniformly packed submicrometer TiO_2 powder produced by consolidation from a stable suspension. (From Ref. 20, with permission.)

The less complex nature of sintering allows it to be analyzed theoretically in terms of idealized models. The theoretical analyses combined with experimental investigations during the last 50 years or so have provided a considerable understanding of sintering. The simplest case is that of a pure, single-phase material (e.g., Al_2O_3). The system is heated to a temperature that is typically less than about three-fourths (0.75) of the melting temperature, e.g., in the range of 1400–1650°C for Al_2O_3 (mp 2073°C). The powder does not melt; instead, the joining together of the particles and the reduction in the porosity (i.e., densification) of the body, as required in the fabrication process, occurs by atomic diffusion in the solid state. This type of sintering is usually referred to as *solid-state sintering*. Although solid-state sintering is the simplest case of sintering, the processes that occur and their interaction can be fairly complex.

The driving force for sintering is the reduction in surface free energy of the system consisting of the mass of consolidated particles. It turns out that the reduction in energy can be accomplished by atomic diffusion processes that lead to either *densification* of the body (by transport of matter from the grain boundary into the pores) or *coarsening* of the microstructure (by rearrangement of matter between different parts of the pore surfaces without actually leading to a decrease in pore volume). The diffusion paths for densification and coarsening are shown in Fig. 1.13 for an idealized situation of two spherical particles in contact [14]. From the point of view of achieving high densities during sintering, a major problem is that the coarsening process reduces the driving force for densification. This interaction is sometimes expressed by the statement that sintering

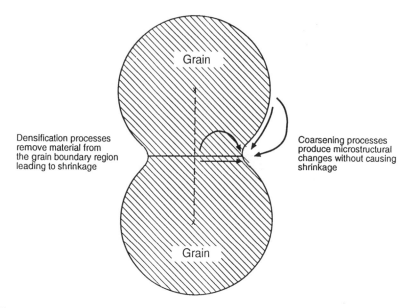

Figure 1.13 Schematic indication of the distinction between densifying and nondensifying microstructural changes resulting from atomic transport during the firing of ceramic powders.

involves a competition between densification and coarsening. The domination of densifying diffusion processes will favor the production of a dense body (Fig. 1.14a). When coarsening processes dominate, the production of a highly porous body will be favored (Fig. 1.14b).

The effects of key material and processing parameters such as temperature, particle (or grain) size, applied pressure, and gaseous atmosphere on the densification and coarsening processes are well understood. The rates of these processes are enhanced by a higher sintering temperature and by fine particle size. Densification is further enhanced by the application of an external pressure during sintering. A key issue that has received increasing attention in recent years is the effect of inhomogeneities within the sintering microstructure (e.g., density variations, grain size variations, and compositional variations). It is now fully realized that inhomogeneities can seriously hinder densification and the control of the fabricated microstructure [13]. As outlined earlier, a consequence of this realization has been the increased attention paid to powder quality and to powder consolidation by colloidal methods.

Ceramic Fabrication Processes 25

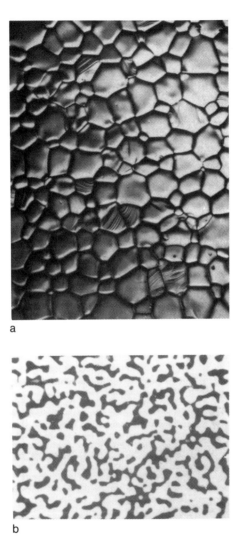

Figure 1.14 (a) The surface of an alumina ceramic from which all porosity has been removed during the firing of the powder; the microstructure consists of the cystalline grains and the boundaries (interfaces) between them. (b) The sintering of silicon results in the formation of a continuous network of solid material (white) and porosity (black). This microstructural change is not accompanied by any shrinkage. (From Ref. 14, with permission.)

A common difficulty in solid-state sintering is that coarsening may dominate the densification process, with the result that high densities are difficult to achieve. This difficulty is especially prominent in highly covalent ceramics (e.g., Si_3N_4 and SiC). One solution is the use of an additive that forms a small amount of liquid phase between the grains at the sintering temperature. This method is referred to as *liquid-phase sintering*. An example is the addition of 5–10 weight percent (wt %) of MgO to Si_3N_4 (Fig. 1.15). The liquid phase provides a high-diffusivity path for transport of matter into the pores to produce densification but the quantity of liquid is insufficient, by itself, to fill up the porosity. The presence of the liquid phase adds a further complexity to the sintering process. The analysis of the process therefore becomes somewhat more difficult than that of solid-state sintering. Another solution to the difficulty of inadequate densification is the application of an external pressure to the body during heating in the case of either solid-state or liquid-phase sintering. This modification is referred to as *pressure sintering*, and hot pressing and hot isostatic pressing are well-known examples. The applied pressure has the effect of increasing the driving force for densification without significantly affecting the rate of coarsening.

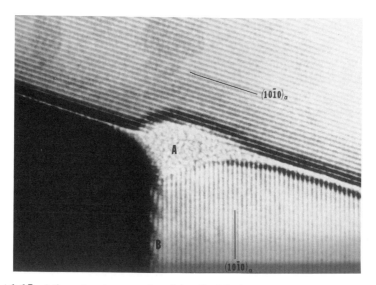

Figure 1.15 Microstructure produced by liquid-phase sintering, under an applied pressure, of Si_3N_4 with MgO additive. A continuous glassy phase, ≈ 0.8 nm thick, separates the crystalline grains. (Courtesy of D. R. Clarke.)

Ceramic Fabrication Processes

A further type of sintering is *viscous sintering*. In this case, a viscous liquid present at the sintering temperature flows under the action of capillary forces to fill up the pores of the body. A relatively simple example of viscous sintering is that of a porous glass body (e.g., consolidated glass particles). A more complex example is the fabrication of clay-based ceramics (e.g., porcelain) from a mixture of naturally occurring raw materials. Chemical reaction, liquid formation, and viscous flow of the liquid into the pores lead to a dense body that on cooling consists of a microstructure of crystalline grains and glassy phases. This rather complex case of viscous sintering in clay-based materials is referred to as *vitrification*.

D. Ceramic Microstructures

As outlined earlier, the microstructure of the fabricated article is significantly dependent on the processing method. The examination of the microstructure may therefore serve as a test of successful processing. Equally important, microstructural observations may lead to inferences about the way in which the processing methods must be modified or changed to give the desired characteristics. As we have seen from the previous section, ceramic microstructure covers a wide range. For solid-state sintering in which all the porosity is successfully removed, a microstructure consisting of crystalline grains separated from one another by grain boundaries is obtained (Fig. 1.14a). However, most ceramics produced by solid-state sintering retain some residual porosity (Fig. 1.16). The use of liquid-phase sintering leads to the formation of an additional phase at the grain boundaries. Depending on the nature and amount of the liquid, the grain boundary phase may be continuous, thereby separating each grain from the neighboring grains (Fig. 1.15), or discontinuous, e.g., at the corners of the grains.

Advanced ceramics, which must meet exacting property requirements, tend to have relatively simple microstructures. A good reason for this is that the microstructure is more amenable to control when the system is less complex. Even so, as outlined earlier, the attainment of these relatively simple microstructures in advanced ceramics can be a difficult task. For the traditional clay-based ceramics, for which the properties achieved are often less critical than the cost or shape of the fabricated article, the microstructures can be fairly complex, as shown in Fig. 1.17 for a ceramic used for sanitaryware [16].

Ceramic composites are a class of materials that have undergone rapid development in recent years because of their promising properties for structural applications at high temperatures (e.g., heat engines). A reinforcing phase (e.g., SiC fibers) is deliberately added to the ceramic (e.g.,

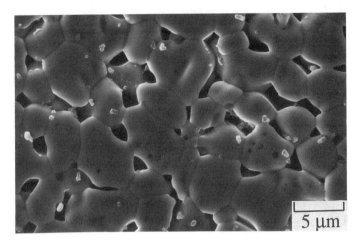

Figure 1.16 Incomplete removal of the porosity during solid-state sintering of CeO_2 results in a microstructure consisting of grains, grain boundaries, and pores.

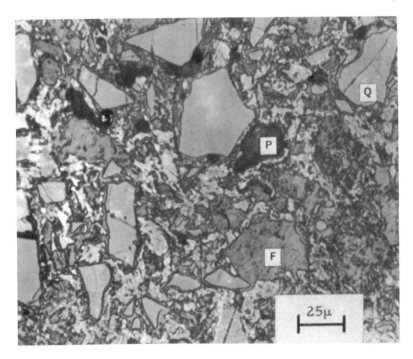

Figure 1.17 A commercial sanitaryware body produced by firing a mixture of clay, feldspar, and quartz, showing a fairly complex microstructure containing some residual feldspar (F), porosity (P), and quartz (Q). (From Ref. 16, with permission.)

Ceramic Fabrication Processes

Al$_2$O$_3$) to make it mechanically tougher (i.e., reduce its brittleness). For these materials, the complexity of the microstructure is controlled by having an ordered distribution of the reinforcing phase in the ceramic matrix. Figure 1.18 shows a polished cross section of a composite consisting of a mullite matrix reinforced with unidirectionally aligned SiC fibers [17].

We discussed earlier the strong consequences of insufficient attention to the quality of the powder and to the consolidation of the powder in the production of the green body. This is especially true for composites, where the reinforcing particles or fibers can interfere with the uniform packing of the matrix particles. Microstructural flaws that limit the properties of the fabricated body normally originate in these stages of processing. Large voids and foreign objects such as dust or milling debris are fairly common. The control of the powder quality such as the particle size distribution, shape, and composition, including minor constituents, has a major influence on the development of the microstructure. Impurities can lead to the presence of a small amount of liquid phase at the sintering temperature, which causes selected growth of large individual grains (Fig. 1.19). In such a case, the achievement of a fine uniform grain size would be impossible. As outlined earlier, much can be gained by the use of clean-room conditions. The uniformity of the packing in the consolidated material is also very important. In general, any nonuniformity in the green body is exag-

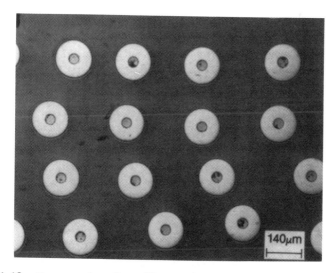

Figure 1.18 Cross section of a mullite matrix composite reinforced with a fairly uniform array of uniaxially aligned SiC fibers. (From Ref. 17, with permission.)

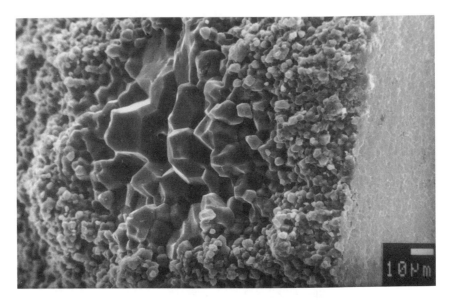

Figure 1.19 Large-grained region of microstructural heterogeneity resulting from an impurity in hot-pressed Al_2O_3. (Courtesy of B. J. Dalgleish)

gerated in the sintering process, leading to the development of cracklike voids or large pores between relatively dense regions (Fig. 1.20).

Assuming that proper precautions were taken in the processing steps prior to firing, further microstructural manipulation must be performed during sintering. Unless high porosity is a deliberate requirement, we normally wish to achieve as high a density, as small a grain size, and as uniform a microstructure as possible. Insufficient control of the sintering conditions (e.g., sintering temperature, time at sintering temperature, rate of heating, and sintering atmosphere) can lead to defects and the augmentation of coarsening, which make attainment of the desired microstructure impossible. The sintering of many materials (e.g., Si_3N_4, lead-based ferroelectric ceramics and β-alumina) require control of the atmosphere to prevent decomposition or volatilization.

Most ceramics are not single-phase solids. For a material consisting of two solid phases, our general requirement is for a uniform distribution of one phase in a uniformly packed matrix of the other phase. Close attention must be paid to the processing steps prior to firing (e.g., during the mixing and consolidation stages) if this requirement is to be achieved in the fabricated body. The consequences of nonuniform mixing for the

Ceramic Fabrication Processes

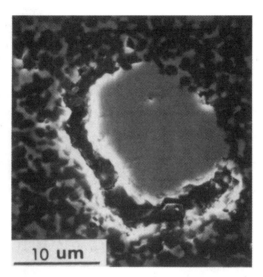

Figure 1.20 Cracklike void produced by a ZrO$_2$ agglomerate shrinking away from the surrounding Al$_2$O$_3$/ZrO$_2$ matrix during firing. (Courtesy of F. F. Lange.)

microstructural control of an Al$_2$O$_3$ powder containing fine ZrO$_2$ particles are illustrated in Fig. 1.21. As in the case of single-phase ceramics, we must also pay close attention to the firing stage. An additional effect now is the possibility of chemical reactions between the phases. As outlined earlier, Al$_2$O$_3$ can be made mechanically tougher (i.e., less brittle) by the incorporation of SiC fibers. However, oxidation of the SiC fibers leads to the formation of an SiO$_2$ layer on the fiber surfaces. The SiO$_2$ can then react with the Al$_2$O$_3$ to form aluminosilicates with a consequent deterioration of the fibers. Prolonged exposure of the system to oxygen will eventually lead to the disappearance of the fibers. A primary aim, therefore, would be the control of the atmosphere during densification in order to prevent oxidation.

In summary, although sufficient attention must be paid to the firing stage, defects produced in the processing steps prior to firing cannot normally be reduced or eliminated. In most cases, these defects are enhanced during the firing stage. In general, properties are controlled by the microstructure (e.g., density and grain size), but the defects in the fabricated body have a profound effect on those properties that depend on failure of the material (e.g., mechanical strength, dielectric strength, and thermal shock resistance). Failure events are almost always initiated at regions of

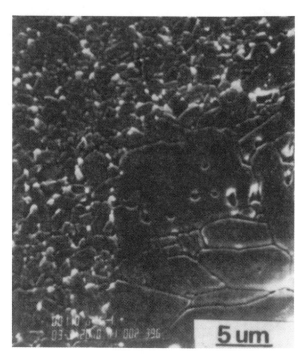

Figure 1.21 The nonuniform distribution of fine ZrO_2 particles (light phase) in Al_2O_3 (dark phase) is seen to result in a region of uncontrolled grain growth during firing. (From Ref. 15, with permission.)

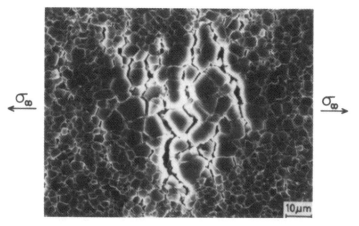

Figure 1.22 Crack nucleation at a large-grained heterogeneity during deformation of a hot-pressed Al_2O_3 body. (From Ref. 18, with permission.)

Ceramic Fabrication Processes

physical or chemical heterogeneity. An example in which mechanical failure originates at a large-grained heterogeneity is shown in Fig. 1.22 [18].

1.4 A CASE STUDY IN PROCESSING: THE FABRICATION OF TiO$_2$ FROM POWDERS

Earlier we outlined the potential benefits that may accrue from careful control of the powder quality and the powder consolidation method. To further illustrate these benefits, we consider the case of TiO$_2$. One fabrication route for the production of TiO$_2$ bodies, referred to as the *conventional route* [19], is summarized in the flow diagram of Fig. 1.23. Typically, TiO$_2$ powder available from commercial sources is mixed with small

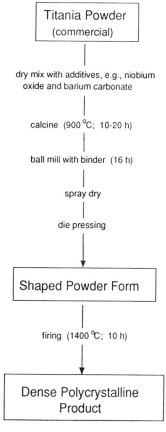

Figure 1.23 Flow chart for the production of TiO$_2$ by a conventional powder route.

amounts of additives (e.g., Nb_2O_5 and $BaCO_3$) that aid the sintering process and the mixture is calcined for 10–20 h at ≈900°C to incorporate the additives into solid solution with the TiO_2. The calcined material is mixed with a small amount of binder (to aid the powder compaction process) and milled in a ball mill to break down agglomerates present in the material. The milled powder, in the form of a slurry, is sprayed into a drying apparatus (available commercially and referred to as a spray dryer). The spray-drying process serves to produce a dried powder in the form of spherical agglomerates. After compaction, the powder is sintered for 10 h at ≈1400°C to produce a body with a density of ≈93% of the theoretical density of TiO_2.

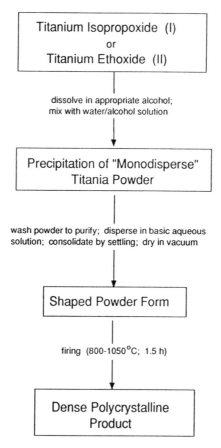

Figure 1.24 Flow chart for the production of TiO_2 by the powder route used by Barringer and Bowen [20]. (Reprinted with permission.)

Ceramic Fabrication Processes

In the fabrication route used by Barringer and Bowen [20], powders with controlled characteristics are prepared and consolidated, by colloidal methods, into uniformly packed green bodies (e.g., similar to that shown in Fig. 1.12). An abbreviated flow diagram of the fabrication route is shown in Fig. 1.24. Powders are prepared by controlled hydrolysis of titanium tetraisopropoxide (denoted powder I) or titanium tetraethoxide (powder II). The powder particles are nearly spherical and almost monosized. After washing, each powder is dispersed in a basic aqueous solution to produce a stable dispersion. Consolidation of the dispersion is accomplished by gravitational or centrifugal settling. The consolidated body is dried in a vacuum and sintered for about 90 min at 800°C (for powder I) or 1050°C (for powder II) to produce bodies with a density greater than 99% of the theoretical.

Table 1.4 gives a summary of the main results of the two fabrication methods described above. It is easily recognized that the benefits achieved by the fabrication route of Barringer and Bowen are very extraordinary. They include a substantial reduction in sintering temperature (by 350–600°C) and sintering time, a higher final density, and a much smaller grain size. Furthermore, since no milling, binder, or sintering aids are required, the purity of the fabricated bodies can be expected to be higher than for the conventional route.

In summary, the experiments of Barringer and Bowen demonstrate clearly that substantial benefits can be achieved in sintering and microstructure control when the powder quality is carefully controlled and the powder packing is homogeneous, that is, when careful attention is paid to the processing steps that precede firing. However, this fabrication route is not currently used in many industrial applications where mass production and low cost are important considerations.

Table 1.4 Processing Data for the Production of Titanium Oxide by a Conventional Route [19] and by the Method of Barringer and Bowen [20]

Process parameter	Conventional	Barringer and Bowen
Powder particle size (μm)	1	0.1 (I) 0.4 (II)
Sintering temperature (°C)	1400	800 (I) 1050 (II)
Sintering time (h)	10	1.5
Final relative density (%)	93	>99
Final grain size (μm)	50–100	0.15 (I) 1.2 (II)

1.5 CONCLUDING REMARKS

In this chapter we have examined, in general terms, the common methods for the fabrication of ceramics. By far the most widely used method for the production of polycrystalline ceramics is the sintering of consolidated powders. Although this fabrication route has its origins in antiquity, its use in the production of ceramics with the desired microstructure for many technological applications can involve severe difficulties. Each processing step has the potential for producing microstructural flaws in the fabricated body, thereby causing a deterioration in properties. Careful attention must therefore be paid to each step in order to minimize the microstructural flaws. Finally, we considered the fabrication of titania and showed the enormous benefits that can be achieved by careful attention to the processing.

REFERENCES

1. G. B. Kenney and H. K. Bowen, *Am. Ceram. Soc. Bull.*, 62(5):590 (1983).
2. J. W. Evans and L. C. De Jonghe, *The Production of Inorganic Materials*, Macmillan, New York, 1991.
3. D. W. Hess, K. F. Jensen, and T. J. Anderson, *Rev. Chem. Eng.*, 3(2):130 (1985).
4. P. J. Lamicq, G. A. Bernhart, M. M. Dauchier, and J. G. Mace, *Am. Ceram. Soc. Bull.*, 65(2):336 (1986).
5. M. S. Newkirk, H. D. Lesher, D. R. White, C. R. Kennedy, A. W. Urquhart, and T. D. Claar, *Ceram. Eng. Sci., Proc.*, 8(7–8):879 (1987).
6. A. J. Moulson, *J. Mater. Sci.*, 14:1017 (1979).
7. C. J. Brinker and G. W. Scherer, *Sol-Gel Science*, Academic, New York, 1990.
8. D. Segal, *Chemical Synthesis of Advanced Materials*, Cambridge Univ. Press, Cambridge, United Kingdom, 1989.
9. S. Yajima, K. Okamura, J. Hayashi, and M. Omori, *J. Am. Ceram. Soc.*, 59(7–8):324 (1976).
10. S. R. Scholes and C. H. Greene, *Modern Glass Practice*, R.A.N. Publishers, Marietta, OH, 1975.
11. P. W. McMillan, *Glass Ceramics*, 2nd ed., Academic, London, 1976.
12. D. G. Grossman, in *Concise Encyclopedia of Advanced Ceramics* (R. J. Brook, Ed.), MIT Press, Cambridge, MA, 1990, p. 170.
13. E. Matijevic, in *Ultrastructure Processing of Ceramics, Glasses, and Composites* (L. L. Hench and D. R. Ulrich, Eds.), Wiley, New York, 1984, pp. 334–352.
14. R. J. Brook, *Concise Encyclopedia of Advanced Ceramic Materials*, MIT Press, Cambridge, MA, 1990, pp. 1–7.
15. F. F. Lange, *J. Am. Ceram. Soc.*, 72(1):3 (1989).

16. G. P. K. Chu, in *Ceramic Microstructures* (R. M. Fulrath and J. A. Pask, Eds.), Wiley, New York, 1968, pp. 828–862.
17. R. N. Singh and A. R. Gaddipati, *J. Am. Ceram. Soc.*, *71*(2):C-100 (1988).
18. B. J. Dalgleish and A. G. Evans, *J. Am. Ceram. Soc.*, *68*(1):44 (1985).
19. M. F. Yan and W. W. Rhodes, *Am. Ceram. Soc. Bull.*, *63*(12):1484 (1984).
20. E. A. Barringer and H. K. Bowen, *J. Am. Ceram. Soc.*, *85*(12):C-199 (1982); *Am. Ceram. Soc. Bull.*, *61*:336 (1982).

FURTHER READING

The following conference proceedings, published at regular intervals, contain useful reports on processing, microstructure, properties and applications of ceramics.

Advances in Ceramics, American Ceramic Society, Westerville, OH.
Materials Research Society Symposium Proceedings, Materials Research Society, Pittsburgh, PA.
Materials Science Research, Plenum Press, New York.
Proceedings of the British Ceramic Society, Institute of Ceramics, Stoke-on-Trent, United Kingdom.

2
Synthesis of Powders

2.1 INTRODUCTION

As outlined in Chapter 1, the characteristics of the powder have a remarkable effect on subsequent processing, such as consolidation of the powder into a green body and firing to produce the desired microstructure. As a result, powder preparation is very important to the overall fabrication of ceramics. In this chapter we shall first define, in general terms, the desirable characteristics that a powder should possess for the production of successful ceramics and then consider some of the main methods used for the preparation of ceramic powders. In practice, the choice of a powder preparation method will depend on the production cost and the capability of the method for achieving a certain set of desired characteristics. For convenience, we divide the powder preparation methods into two categories: mechanical methods and chemical methods. Powder preparation by chemical methods is an area of ceramic processing that has received a high degree of interest and has undergone considerable changes in the last 20 years or so. Further new developments in this area are expected in the future.

2.2 DESIRABLE POWDER CHARACTERISTICS

Traditional ceramics generally must meet less specific property requirements than advanced ceramics. They can be chemically inhomogeneous and can have complex microstructures. Unlike the case of advanced ceramics, chemical reaction during firing is often a requirement. The starting materials for traditional ceramics therefore consist of mixtures of powders with a chosen reactivity. For example, the starting powders for an insulating porcelain can, typically, be a mixture of clay ($\approx$50 wt %), feldspar

Synthesis of Powders

($\approx$25 wt %), and silica ($\approx$25 wt %). Fine particle size is desirable for good chemical reactivity. The powders must also be chosen to give a reasonably high packing density, which serves to limit the shrinkage and distortion of the body during firing. Clays form the major constituent and therefore provide the fine particles in the starting mixture for most traditional ceramics. Generally, low-cost powder preparation methods are used for traditional ceramics.

Advanced ceramics must meet very specific property requirements, and therefore their chemical composition and microstructure must be well controlled. Careful attention must be paid to the quality of the starting powders. For advanced ceramics, the important powder characteristics are the size, size distribution, shape, state of agglomeration, chemical composition, and phase composition. The structure and chemistry of the surface are also important.

The size, size distribution, shape, and state of agglomeration have important consequences for both the consolidation step and the microstructure of the fired body. A particle size greater than $\approx$1 μm generally precludes the use of colloidal consolidation methods because the settling time of the particles is fairly short. The most profound effect of the particle size, however, is on the sintering. As we shall show later in this book, the rate at which the body densifies increases strongly with a decrease in particle size. Normally, if other factors do not cause severe difficulties during firing, a particle size of less than $\approx$1 μm allows the achievement of high density within a reasonable time (e.g., a few hours). Whereas a powder with a wide distribution of particle sizes (sometimes referred to as a *polydisperse* powder) may lead to higher packing density in the green body, this benefit is usually vastly outweighed by difficulties in the control of the microstructure during sintering. A common problem is that the large grains coarsen rapidly at the expense of the smaller grains, making the attainment of high density with controlled grain size impossible. Homogeneous packing of a powder with a narrow size distribution (i.e., a nearly *monodisperse* powder) generally allows greater control of the microstructure. A spherical or equiaxial particle shape is beneficial for controlling the uniformity of the packing.

Agglomerates lead to heterogeneous packing in the green body, which in turn leads to differential sintering (different regions of the body sintering at different rates) during the firing stage. This can cause serious problems such as the development of large pores and cracklike voids in the fired body (see Fig. 1.20). Furthermore, the rate at which the body densifies is roughly similar to that for a coarse-grained body with a particle size equivalent to that of the agglomerates. An agglomerated powder therefore has serious consequences for the fabrication of ceramics when high den-

Table 2.1 Desirable Powder Characteristics for Advanced Ceramics

Powder characteristic	Desired property
Particle size	Fine ($< \approx 1$ μm)
Particle size distribution	Narrow
Particle shape	Spherical or equiaxial
State of agglomeration	No agglomeration or soft agglomerates
Chemical composition	High purity
Phase composition	Single phase

sity coupled with a fine-grained microstructure is desired. Agglomerates are classified into two types: *soft agglomerates* in which the particles are held together by weak van der Waals forces and *hard agglomerates* in which the particles are chemically bonded together by strong bridges. The ideal situation is the avoidance of agglomeration in the powder. However, in most cases this is not possible. In such cases, we would prefer to have soft agglomerates rather than hard agglomerates. Soft agglomerates can be broken down relatively easily by mechanical methods (e.g., pressing or milling) or by dispersion in a liquid. Hard agglomerates cannot be easily broken down and therefore must be avoided or removed from the powder.

Surface impurities may have a significant influence on the dispersion of the powder in a liquid, but the most serious effects of variations in chemical composition and microstructure are encountered in the firing stage. Impurities may lead to the presence of a small amount of liquid phase at the sintering temperature, which causes selected growth of large individual grains (as shown in Fig. 1.19). In such a case, the achievement of a fine uniform grain size would be impossible. Chemical reactions between incompletely reacted phases can also be a source of problems. We would therefore like to have no chemical change in the powder during firing. For some materials, polymorphic transformation between different crystalline structures can also be a source of severe difficulties for microstructure control. Common examples are ZrO_2, for which cracking is a severe problem on cooling, and γ-Al_2O_3, where the transformation to the α phase results in rapid grain growth and a severe retardation in the densification rate. To summarize, the desirable powder characteristics for the fabrication of advanced ceramics are listed in Table 2.1.

2.3 POWDER PREPARATION METHODS

A variety of methods exist for the preparation of ceramic powders [1,2]. In this book, we divide them into two categories: mechanical methods

Synthesis of Powders

and chemical methods. Mechanical methods are generally used to prepare powders of traditional ceramics from naturally occurring raw materials. Powder preparation by mechanical methods is a fairly mature area of ceramic processing in which the scope for new developments is rather small. However, in recent years, the preparation of fine powders of some advanced ceramics by mechanical methods that use milling at high speeds has received a fair amount of interest.

Chemical methods are generally used to prepare powders of advanced ceramics from synthetic materials or from naturally occurring raw materials that have undergone a considerable degree of chemical refinement. However, some of the methods categorized as chemical include a mechanical grinding step as part of the process. The grinding step is usually necessary for the breakdown of agglomerates and for the production of the desired physical characteristics of the powder such as particle size and

Table 2.2 Common Powder Preparation Methods for Ceramics

Powder preparation method	Advantages	Disadvantages
Mechanical		
Comminution	Inexpensive, wide applicability	Limited purity, limited homogeneity, large particles
Mechanochemical synthesis	Fine particle size, good for nonoxides, low-temperature route	Limited purity, limited homogeneity
Chemical		
Solid-state reaction	Simple apparatus, inexpensive	Agglomerated powder, limited homogeneity for multicomponent powders
Liquid solutions		
Precipitation or coprecipitation: solvent vaporization (spray drying, spray pyrolysis, freeze drying, sol-gel processing); combustion (Pechini method, glycine nitrate process)	High purity, small particles, composition control, chemical homogeneity	Expensive, poor for nonoxides, powder agglomeration usually a problem
Nonaqueous liquid reaction	High purity, small particles	Limited to nonoxides
Vapor-phase reaction		
Gas–solid reaction	Inexpensive	Low purity, limited to nonoxides
Gas–liquid reaction; reaction between gases	High purity, small particles	Expensive, limited to nonoxides

Source: Adapted from Ref. 2.

particle size distribution. Powder preparation by chemical methods is an area of ceramic processing that has seen a considerable number of new developments in the past 20 years, and further new developments are expected in the future.

Table 2.2 provides a summary of the common powder preparation methods for ceramics.

2.4 POWDER PREPARATION BY MECHANICAL METHODS

The process in which small particles are produced by reducing the size of larger ones by mechanical forces is usually referred to as *comminution*. We shall first review the basic features of comminution. Following that, the preparation of fine ceramic powders by highly energetic milling at high speeds will be considered. Powder preparation by high-speed milling is referred to by various terms, including mechanochemical synthesis, mechanosynthesis, mechanically driven synthesis, mechanical alloying, and high-energy milling. Although no one term has received widespread acceptance, we use the term *mechanochemical synthesis* in this book.

A. Comminution

Comminution involves operations such as crushing, grinding, and milling. For traditional clay-based ceramics, machines such as jaw, gyratory, and cone crushers are used for coarse size reduction of the mined raw material to produce particles in the size range of 0.1–1 mm or so. We will not describe the processes used in the production of these coarse particles. Instead we will assume that a stock of coarse particles (with sizes <0.1 mm or so) is available and consider the processes applicable to the subsequent size reduction to produce a fine powder. The most common way to achieve this size reduction is by milling. One or more of a variety of mills may be used, including high-compression roller mills, jet mills (fluid energy mills), and ball mills [3]. Ball mills are categorized into various types, depending on the method used to impart motion to the balls (e.g., tumbling, vibration, and agitation).

For the following discussion, we need to define two terms that will be used from time to time. The first is the *energy utilization* of the comminution method; this is defined as the ratio of the new surface area created to the mechanical energy supplied. Second, we define the *rate of grinding* as the amount of new surface area created per unit mass of particles per unit time. Obviously, there is a connection between the two quantities. A comminution method that has a high energy utilization will also have a high rate of grinding, so the achievement of a given particle size will take a shorter time. However, for a given method we will also want to

Synthesis of Powders

understand how the rate of grinding depends on the various experimental factors.

In the milling process, the particles experience mechanical stresses at their contact points due to compression, impact, or shear with the mill medium or with other particles. The mechanical stresses lead to elastic and inelastic deformation and, if the stress exceeds the ultimate strength of the particle, to fracture of the particles. The mechanical energy supplied to the particle is used not only to create new surfaces but also to produce other physical changes in the particle (e.g., inelastic deformation, increase in temperature, and lattice rearrangements within the particle). Changes in the chemical properties (especially the surface properties) can also occur, especially after prolonged milling or under very vigorous milling conditions. Consequently, the energy utilization of the process can be fairly low, ranging from <20% for milling produced by compression forces to <5% for milling by impact. Figure 2.1 summarizes the stress mechanisms and the range of particle sizes achieved with various types of mills for the production of fine powders.

High-Compression Roller Mills

In the high-compression roller mill (Fig. 2.2), the material is stressed between two rollers. In principle, the process is similar to that of a conven-

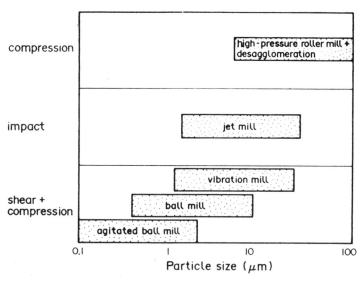

Figure 2.1 Range of particle sizes reached with different types of mills. (From Ref. 3, with permission.)

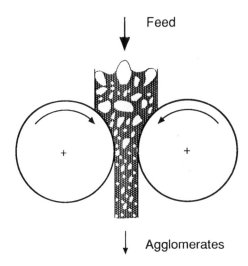

Figure 2.2 High-compression roller mill.

tional roller mill, but the contact pressure is considerably higher (in the range of 100–300 MPa). The stock of coarse particles is comminuted and compacted. This process must therefore be used in conjunction with another milling process (e.g., ball milling) to produce a powder. Although the process is unsuitable for the production of particle sizes below ≈10 μm, it has two significant advantages. First, the energy utilization is fairly good. For the production of the same size of particles from a stock of coarse particles, the use of a high-energy roller mill in conjunction with a ball mill is more efficient than the use of a ball mill alone. A second advantage is that since only a small amount of material makes contact with the rollers, the wear can be fairly low (e.g., much lower than in ball milling).

Jet Mills

Jet mills are manufactured in a variety of designs. Generally, the operation consists of the interaction of one or more streams of high-speed gas bearing the stock of coarse particles with another high-speed stream. Comminution occurs by particle–particle collisions. In some designs, comminution is achieved by collisions between the particles in the high speed stream and a wall (fixed or movable) within the mill. The milled particles leave the mill in the emergent fluid stream and are usually collected in a cyclone chamber outside the mill. The gas for the high-speed stream is usually compressed air, but inert gases such as nitrogen or argon may be used to reduce oxidation of certain nonoxide materials (e.g., Si).

Synthesis of Powders

The average particle size and the particle size distribution of the milled powder depend on a number of factors, including the size, size distribution, hardness, and elasticity of the feed particles, the pressure at which the gas is injected, the dimensions of the milling chamber, and the use of particle classification in conjunction with the milling.

Multiple gas inlet nozzles are incorporated into some jet mill designs to provide multiple collisions among the particles, thereby enhancing the comminution process. In some cases the flow of the particles in the high-speed gas stream can be used for their classification in the milling chamber. The feed particles remain in the grinding zone until they are reduced to a sufficiently fine size and are then removed from the milling chamber. An example of a design that incorporates particle size classification in the milling chamber is the spiral jet mill (Fig. 2.3). An advantage of jet mills

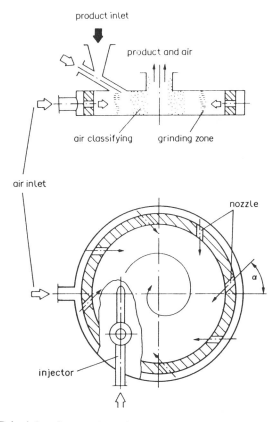

Figure 2.3 Principle of operation of a spiral jet mill. (From Ref. 3, with permission.)

is that when combined with a particle classification device they provide a rapid method for the production of a powder with a narrow size distribution for particle sizes down to ≈1 μm. A further advantage is that for some designs the particles do not come into contact with the surfaces of the milling chamber, so contamination is not a problem. Coating of the surfaces of the milling chamber and the collection vessel with, for example, Teflon® may provide further insurance against contamination.

Ball Mills

The high-compression roller mills and jet mills just described achieve comminution without the use of grinding media. In mills that incorporate grinding media (balls or rods), comminution occurs by compression, impact, and shear (friction) between the moving grinding media and the particles. Rod mills are not suitable for the production of fine powders, whereas ball milling, as shown in Fig. 2.1, can be used to produce particle sizes from ≈10 μm to as small as a fraction of a micrometer. Ball milling is suitable for wet or dry milling.

Ball milling is a fairly complex process that does not lend itself easily to rigorous theoretical analysis. The rate of grinding depends on a number of factors, including the mill parameters (diameter, speed, amount of media), the properties of the grinding media (size, hardness, shape), and the properties of the particles to be ground [4,5]. Generally, ball mills that run at low speeds contain large balls because most of the mechanical energy supplied to the particle is in the form of potential energy. Those mills that run at high speeds contain small balls because, in this case, most of the energy supplied to the particle is in the form of kinetic energy. For a given size of grinding medium, since the mass is proportional to the density, the grinding medium should consist of materials with as high a density as possible. However, in practice, the choice of the grinding medium is usually limited by cost.

The size of the grinding medium is an important consideration. Small-particle grinding media are generally better than large-particle ones. For a given volume, the number of balls increases inversely as the cube of the radius. Assuming that the rate of grinding depends on the number of contact points between the balls and the powder and that the number of contact points in turn depend on the surface area of the balls, the rate of grinding will increase inversely as the radius of the balls. However, the balls must not be too small because they must impart sufficient mechanical energy to the particles to cause fracture.

The rate of grinding also depends on the particle size. The rate decreases with decreasing particle size, and as grinding progresses and the particles become fairly fine (e.g., about 1 μm to a few micrometers) it

Synthesis of Powders

becomes more and more difficult to achieve further reduction in size. A practical grinding limit is approached (Fig. 2.4). This limit depends on a number of factors. An important factor is the increased tendency for the particles to agglomerate with decreasing particle size. A physical equilibrium is therefore set up between the agglomeration and comminution processes. Another factor is the decreased probability for the occurrence of a comminution event with decreasing particle size. Finally, the probability of a flaw of a given size existing in the particle decreases with decreasing particle size; that is, the particle becomes stronger. The limiting particle size may be reduced by wet milling as opposed to dry milling (Fig. 2.4), by the use of a dispersing agent during wet milling, and by performing the milling in stages. For staged milling, as the particles get finer they are transferred to another compartment of the mill or to another mill operating with smaller balls.

A disadvantage of ball milling is that wear of the grinding medium can be fairly high. For advanced ceramics, as discussed before, the presence of impurities in the powder is a serious concern. The best solution is to use balls with the same composition as the powder itself. However, this is possible in only a very few cases and even for these, at fairly great expense. Another solution is to use a grinding medium that is chemically inert at the firing temperature of the body (e.g., ZrO_2 balls) or can be removed from the powder by washing (e.g., steel balls). A common problem is the use of porcelain balls or low-purity Al_2O_3 balls that wear easily and introduce a fair amount of SiO_2 into the powder. Silicate liquids nor-

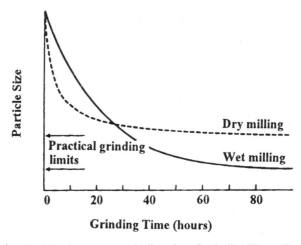

Figure 2.4 Particle size versus grinding time for ball milling. (From Ref. 4.)

mally form at the firing temperature and make microstructure control very difficult. A list of grinding balls available commercially and the approximate density of each is given in Table 2.3.

Tumbling ball mills, usually referred to simply as ball mills, consist of a slowly rotating horizontal cylinder that is partly filled with grinding balls and the particles to be ground. In addition to the factors discussed above, the speed of rotation of the mill is an important variable, as it influences the trajectory of the balls and the mechanical energy supplied to the powder. Defining the *critical speed of rotation* as the speed required to take the balls just to the apex of revolution (i.e., to the top of the mill, where centrifugal force just balances the force of gravity), we find that the critical speed (in revolutions per unit time) is equal to $(g/R)^{1/2}/(2\pi)$, where R is the radius of the mill and g is the acceleration due to gravity. In practice, ball mills are operated at $\approx 75\%$ of the critical speed so that the balls do not reach the top of the mill (Fig. 2.5).

As we outlined earlier, the ball milling process does not lend itself easily to rigorous theoretical analysis. We therefore have to be satisfied with empirical relationships. One such empirical relationship is

$$\text{Rate of milling} \approx AR^{1/2}\rho d/r \tag{2.1}$$

where A is a numerical constant that is specific to the mill being used and the powder being milled, R is the radius of the mill, ρ is the density of the balls, d is the particle size of the powder, and r is the radius of the balls. According to Eq. (2.1), the rate decreases with decreasing particle size; however, this holds up to a certain point since, as discussed earlier, a practical grinding limit is reached after a certain time. The variation of the rate of grinding with the radius of the balls must also be noted with caution; the balls will not possess sufficient energy to cause fracture of the particles if they are too small.

Table 2.3 Commercially Available Grinding Media for Ball Milling

Grinding media	Density (10^3 kg/m^3)
Porcelain	2.3
Silicon nitride	3.1
Alumina	3.6
Zirconia	5.5
Steel	7.7
Tungsten carbide	14.8

Synthesis of Powders

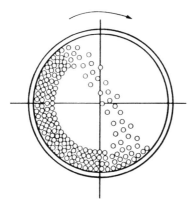

Figure 2.5 Schematic of a ball mill in cataracting motion. (From Ref. 3, with permission.)

In the milling process, the objective is to have the balls fall onto the particles at the bottom of the mill rather than onto the mill liner itself. For a mill operating at ≈75% of its critical speed, this occurs, for dry milling, for a quantity of balls filling about 50% of the mill volume and for a charge of particles filling about 25% of the mill volume. For wet milling, a useful guide is for the balls to occupy ≈50% of the mill volume and the slurry ≈40%, with the solids content of the slurry equal to ≈25–40%.

Wet ball milling has an advantage over dry milling in that its energy utilization is somewhat higher (by ≈10–20%). A further advantage, mentioned earlier, is the ability to produce a higher fraction of finer particles. Disadvantages of wet milling are the increased wear of the grinding medium, the need for drying the powder after milling, and contamination of the powder by the adsorbed vehicle.

Vibratory ball mills, or vibro-mills, consist of a drum, almost filled with a well-packed arrangement of grinding medium and the charge of particles, that is vibrated fairly rapidly (10–20 Hz) in three dimensions. The grinding medium, usually cylindrical in shape, occupies more than 90% of the mill volume. The amplitude of the vibrations is controlled so as not to disrupt the well-packed arrangement of the grinding medium. The three-dimensional motion helps to distribute the charge of particles and, in the case of wet milling, to minimize segregation of the particles in the slurry. The fairly rapid vibratory motion produces an impact energy that is much greater than the energy supplied to the particles in a tumbling ball mill. Vibratory ball mills therefore provide a much more rapid commi-

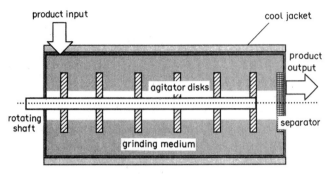

Figure 2.6 Schematic of an agitated ball mill. (From Ref. 3, with permission.)

nution process than tumbling ball mills. They are also more energy-efficient than tumbling ball mills.

Agitated ball mills, also referred to as attrition mills or stirred media mills, differ from tumbling ball mills in that the milling chamber does not rotate. Instead, the stock of particles and the grinding medium are stirred rather vigorously with a stirrer rotating continuously at frequencies of 1–10 Hz. The grinding chamber is aligned either vertically or horizontally (Fig. 2.6), with the stirrer located at the center of the chamber. The grinding medium consists of small spheres (≈ 0.2–5 mm) that make up ≈ 60–90% of the available volume of the mill. Although it can be used for dry milling, most agitated ball milling is carried out with slurries. Most agitated ball milling is also carried out continuously, with the slurry of particles to be milled fed in at one end and the milled product removed at the other end. When the agitation is fairly intense, considerable heat is produced and a means of cooling the milling chamber is required.

Agitated ball mills have a distinct advantage over tumbling ball mills and vibratory ball mills in that the energy utilization is significantly higher. They also have the ability to handle a higher solids content in the slurry to be milled. Furthermore, as we have discussed earlier, the use of a fine grinding medium improves the rate of milling. The high efficiency of the process coupled with the short time required for milling means that contamination of the milled powder is less serious than in the case of tumbling ball mills or vibratory ball mills. Contamination in agitated ball milling can be further reduced by lining the mill chamber with a ceramic material or a plastic and using ceramic stirrers and grinding media.

B. Mechanochemical Synthesis

We remarked earlier that both the physical and chemical characteristics of the particles can undergo significant changes, especially with prolonged milling or milling under vigorous conditions. Usually, our interest lies mainly in the achievement of certain physical characteristics such as particle size and particle size distribution. However, the exploitation of the chemical changes during milling for the preparation of powders has received considerable interest in recent years. Grinding enhances the chemical reactivity of powders. Rupture of the bonds during particle fracture results in surfaces with unsatisfied valences. This, combined with the high surface energy, favors reaction between mixed particles or between the particles and their surroundings.

Mechanochemical synthesis is a high-energy ball milling process in which the repeated fracture and welding events arising from impact and compression between the milling medium and the particles result in the production of fine powders of a chemical compound from a mixture of particles. First used successfully for the production of Ni and Al alloys strengthened by oxide dispersions, the method has since been used to prepare a variety of powders, including oxides, carbides, nitrides, borides, and silicides [6].

The mill consists of a cylindrical vial, containing a mixture of milling balls and the charge of particles, which undergoes large-amplitude vibrations in three dimensions at a frequency of ≈ 20 Hz. The charge occupies $\approx 20\%$ of the volume of the vial, and the amount of milling medium (in the form of balls 5–10 mm in diameter) makes up 2–10 times the mass of the charge. Although the reaction depends on a number of factors, the milling is normally carried out for a few tens of hours for the set of conditions indicated here. The method therefore involves high-intensity vibratory milling for very extended periods.

A distinct advantage of the mechanochemical synthesis method is the ease of preparation of powders that can otherwise be difficult to produce. For example, most metal carbides are formed by the reaction between metals or metal hydrides and carbon at high temperatures (in some cases as high as 2000°C). Furthermore, some carbides and silicides have a narrow compositional range that is difficult to produce by other methods. A disadvantage is the incorporation of impurities from the mill and milling medium into the powder.

The mechanism of mechanochemical synthesis is not clearly understood at the present time. One possibility is the occurrence of the reaction by a true solid-state diffusion mechanism. Since diffusional processes are thermally activated, this would require a lowering of the reaction temperature (more correctly a lowering of the activation energy) by the milling process, a considerable increase in the temperature existing in the mill, or some combination of the two. A further consequence of thermally activated processes is that the rate of the reaction varies fairly smoothly with temperature. In the case of the mechanochemical synthesis of $MoSi_2$ from a mixture of Mo and Si powders, recent results appear to rule out a true solid state reaction as the predominant mechanism [7]. Whereas considerable heating of the mill occurs, the temperature was found to be significantly lower than that required for reaction by a true solid state mechanism. Furthermore, extensive formation of MoSi$_2$ occurred quite abruptly after a specific time (Fig. 2.7).

Prior to the extensive formation of MoSi$_2$, the average particle sizes of the Mo and Si were very fine ($\approx$21.5 and 8.5 nm, respectively), with the largest powder agglomerates being <5 µm. However, following the reaction, the MoSi$_2$ powder consisted of agglomerates with sizes of up to $\approx$100 µm made up of MoSi$_2$ particles of $\approx$0.3 µm in diameter (Fig. 2.8). The highly agglomerated nature of the reacted powder suggests that considerable heating accompanies the formation of MoSi$_2$. A second possibility is that the formation of MoSi$_2$ occurs by local melting. The evidence for local melting is not clear. A third possibility is the occurrence of the reaction by a form of self-propagating process at high temperatures [7]. For a self-propagating process to occur, a source of energy must be avail-

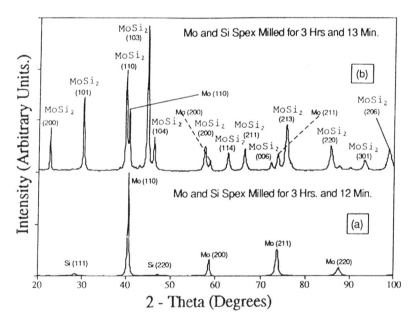

Figure 2.7 X-ray diffraction pattern of a stoichiometric mixture of Mo and Si powders after high milling for (a) 3 h 12 min and (b) 3 h 13 min, showing a fairly abrupt formation of MoSi$_2$. (From Ref. 7.)

Synthesis of Powders

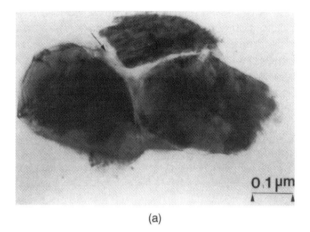

(a)

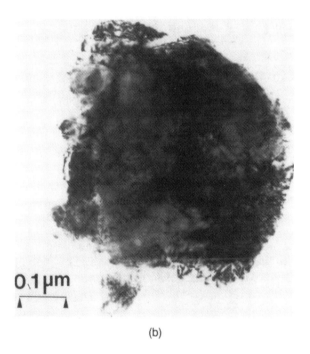

(b)

Figure 2.8 Transmission electron microscopic (TEM) image of MoSi$_2$ in a sample milled for 3 h 13 min showing (a) three particles separated by Mo (arrow) and (b) a heavily dislocated particle. (From Ref. 7.)

able to raise the adiabatic temperature of the system to that required for it to become self-sustaining. In the present case, the surface energy of the very fine Mo and Si powders (prior to extensive reaction) is quite enormous. This, coupled with the stored strain energy (predominantly in the Mo), may provide such a source of energy for sustaining the reaction. The surface energy of the particles alone (≈ 500 J/mol) is 5–10% of the heat of formation of $MoSi_2$ (7.5 kJ/mol at 298 K). A critical step for the formation reaction in mechanochemical synthesis appears to be the generation of a fine enough particle size that the available surface and strain energy is sufficient to make the reaction self-sustaining.

2.5 POWDER PREPARATION BY CHEMICAL METHODS

A wide range of chemical methods exist for the preparation of ceramic powders. For convenience, we shall start by dividing the methods into three fairly broad categories: (i) preparation by solid-state reactions, (ii) preparation from liquid solutions, and (iii) preparation by vapor-phase reactions.

A. Solid-State Reactions

Chemical decomposition reactions, in which a solid reactant is heated to produce a new solid plus a gas, are commonly used for the production of powders of simple oxides from carbonates, hydroxides, nitrates, sulfates, acetates, oxalates, alkoxides, and other metal salts. An example is the decomposition of magnesium carbonate to produce magnesium oxide and carbon dioxide gas:

$$MgCO_3(s) \rightarrow MgO(s) + CO_2(g) \tag{2.2}$$

Chemical reactions between solid starting materials, usually in the form of mixed powders, are common for the production of powders of complex oxides such as titanates, ferrites, and silicates. The reactants normally consist of simple oxides, carbonates, nitrates, sulfates, oxalates, or acetates. An example is the reaction between barium carbonate and titanium oxide to produce barium titanate:

$$BaCO_3(s) + TiO_2(s) \rightarrow BaTiO_3(s) + CO_2(g) \tag{2.3}$$

These reactions, involving decomposition of solids or chemical reaction between solids, are referred to in the ceramic literature as *calcination*.

In the case of *decomposition*, a large body of literature on different reactions exists due to the industrial and scientific interest in the subject [8]. We will not repeat this discussion but will instead focus on the issues

Synthesis of Powders

pertinent to the production of powders. Considering the decomposition of $MgCO_3$ defined by Eq. (2.2), the standard heat of reaction at 298 K is 105 kJ/mol. The reaction is strongly endothermic, which is typical for decomposition reactions. This means that heat must be supplied to the reactant to sustain the decomposition. There is also an equilibrium partial pressure of CO_2, i.e., P_{CO_2}, over the $MgCO_3$ at every temperature. From data for the standard free energy of the reaction [9], we can find the temperature at which P_{CO_2} equals the partial pressure of CO_2 gas in the atmosphere (30 Pa). This determines the *decomposition temperature* of $MgCO_3$ in air, which is 480 K. The fact that $MgCO_3$ does not decompose in air at 480°C indicates that the decomposition is governed by kinetic factors and not by thermodynamics.

Depending on the chemical nature of the reactant and the temperature, the kinetics of the decomposition may be controlled by any one of three processes: (i) the reaction at the surface, (ii) heat transfer to the reaction surface, or (iii) gas diffusion or permeation from the reaction surface through the porous product layer (Fig. 2.9). Various expressions have been developed for predicting the rate of the reaction. The exact form of the expression depends on the geometry of the particle and the preferred direction of the reaction. Assuming constant temperature and spherical reactant particles for which the reaction initiates on the surface and moves inward uniformly at a fixed rate, the reaction kinetics are given by

$$1 - (1 - \alpha)^{1/3} = Kt \tag{2.4}$$

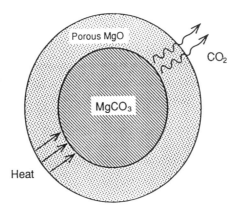

Figure 2.9 Schematic of the decomposition of magnesium carbonate.

where α is the fraction of the reactant decomposed after a time t and K is the thermally activated rate constant obeying an Arrhenius relationship.

The characteristics of the powder produced by the decomposition depend on a number of material and processing factors, including the chemical nature of the reactants, the initial size of the reactant particles, the atmospheric conditions, the decomposition temperature, and the decomposition time. A feature of decomposition reactions is the production of an extremely fine particle size from a normally coarse reactant. In the case of the production of MgO from $MgCO_3$ or from $Mg(OH)_2$, the particle size of the MgO can be as low as 2–3 nm when the decomposition is carried out in vacuum. The decomposition is also pseudomorphic; i.e., the initial size and shape of the reactant particles are retained. The product therefore consists of agglomerates of very fine particles separated by porosity. In the case of MgO, the surface area of the powder produced from $Mg(OH)_2$ or from $MgCO_3$ decreases strongly with a decrease in the initial particle size of the reactant; normally reactant particle sizes of less than a few tenths of a micrometer are required for the production of very high surface area MgO.

The atmospheric conditions are normally such that the decomposition proceeds to completion. However, in some cases the atmosphere can have a strong influence on the characteristics of the powder produced. For the case of MgO produced from $Mg(OH)_2$ or from $MgCO_3$, if the reaction is carried out in the ambient atmosphere rather than in a vacuum, then high surface area powders are not produced. This is because the water vapor in the atmosphere catalyzes the sintering of the very fine particles to produce larger particles. High decomposition temperatures and long decomposition times can lead to the sintering of the particles and agglomerates, giving a highly agglomerated, low surface area powder. Although attempts are usually made to optimize the decomposition time and temperature, agglomerates are invariably present.

In the case of *chemical reactions between solids*, powder reactions involve a far higher number of variables than reactions with single-crystal reactants. For solid-state reactions between two single crystals (Fig. 2.10a), the reaction is diffusion-controlled as long as the parabolic rate law applies:

$$x = K't^{1/2} \tag{2.5}$$

where x is the thickness of the reaction product formed after a time t and K' is a constant at a fixed temperature. (K' increases with temperature according to the Arrhenius relation.) For powder reactions, making the assumptions that the reaction product forms coherently and uniformly on the reactant particles (Fig. 2.10b) and that the particles are spherical and

Synthesis of Powders

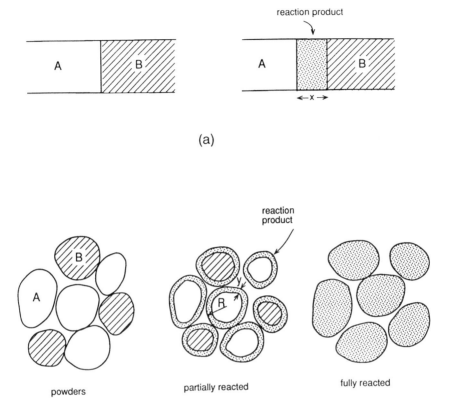

Figure 2.10 Schematic of solid-state reaction in (a) single crystals and (b) mixed powders.

have the same size, the volume of unreacted material at time t is

$$V = (4/3)\pi(R - y)^3 \tag{2.6}$$

where R is the initial radius of the particle and y is the thickness of the reaction layer. The volume of unreacted material is also given by

$$V = (4/3)\pi R^3(1 - \alpha) \tag{2.7}$$

where α is the fraction of the volume that has already reacted. Combining Eqs. (2.6) and (2.7),

$$y = R[1 - (1 - \alpha)^{1/3}] \tag{2.8}$$

Assuming that y grows according to the parabolic relationship given by Eq. (2.5), then the reaction rate is

$$[1 - (1 - \alpha)^{1/3}]^2 = Kt/R^2 \tag{2.9}$$

Many investigators have reported a parabolic reaction rate, at least for the initial stage of the powder reaction, which is usually taken to mean that the reaction is diffusion-controlled. However, for powder reactions, a more complicated relationship would be expected in view of the high number of variables involved. One powder system that has been investigated extensively is the reaction between ZnO and Al_2O_3 to produce $ZnAl_2O_4$ [10]. In this case it has been demonstrated that a mechanism involving vapor transport of ZnO can lead to a parabolic growth law in the early stages of the reaction. A high surface diffusivity may produce a similar effect.

Equation (2.9) involves two oversimplifications that limit its applicability and the range over which it adequately predicts reaction rates. First, it is valid for only small thicknesses of the reaction product, and second, it does not take into account any change in molar volume between the reactant and the product. When these two effects are corrected for [11], then the reaction kinetics are given by

$$[1 + (z - 1)\alpha]^{2/3} + (z - 1)(1 - \alpha)^{2/3} = z + (1 - z)\frac{Kt}{R^2} \tag{2.10}$$

For powder reactions, the variables that can influence the reaction rate are fairly numerous; they include the chemical nature of the reactants and the product, the particle size, the particle size distribution, the relative sizes of the powders in the mixture, the uniformity of the mixing, the reaction atmosphere, the temperature, and the time. For example, the reaction rate will decrease with an increase in particle size of the reactants because, on average, the diffusion distances will increase. For coherent reaction layers, the dependence of the reaction kinetics on particle size is given by Eq. (2.9) or (2.10). As outlined above for decomposition reactions, the reaction rate will increase with temperature according to the Arrhenius relation. In the case where gaseous reaction products are formed, the reaction kinetics will also be influenced by the atmosphere. For example, the formation of $BaTiO_3$ according to Eq. (2.3) will depend on the partial pressure of CO_2. The homogeneity of mixing is also important because it influences the diffusion distance and the relative number of contacts between the reactant particles.

Powder preparation by solid-state reactions generally has an advantage in terms of production cost. However, as outlined earlier, the powder quality is also an important consideration for advanced ceramics. The

Synthesis of Powders

powders are normally agglomerated, and a grinding step is almost always required to produce powders with better characteristics. Grinding in ball mills leads to the contamination of the powder with impurities. Incomplete reactions, especially in poorly mixed powders, may produce undesirable phases. We have already outlined the deleterious effects of impurities and undesirable second phases. Furthermore, the particle shape of ground powders is usually difficult to control.

To conclude this section on reactions between solids, we consider the production of SiC powders by the reaction between silica (sand) and carbon (coke):

$$SiO_2 + 3C \rightarrow SiC + 2CO \qquad (2.11)$$

This reaction should occur somewhat above 1500°C but is usually carried out at much higher temperatures so that the SiO_2 is actually a liquid. The production of SiC at high temperatures by the reaction between a mixture of SiO_2 and C, which is carried out on a large scale industrially, is generally referred to as the Acheson process. The mixture is conducting and is heated electrically to temperatures of ≈ 2500°C. The reaction is more complex than that indicated by Eq. (2.11); side reactions such as

$$SiO_2(l) + C(s) \rightarrow SiO(g) + CO(g) \qquad (2.12)$$

occur. The product obtained after several days of reaction consists of an aggregate of black or green crystals. It is crushed, washed, ground, and classified to produce the desired powder sizes. One disadvantage of the Acheson process is that powder quality is often too poor for demanding applications such as the production of high-temperature structural ceramics. We shall see later that smaller scale methods employing gas-phase reactions can produce more desirable powder characteristics.

B. Liquid Solutions

There are two general routes for the production of a powdered material from a solution: (i) evaporation of the liquid and (ii) precipitation by adding a chemical reagent that reacts with the solution. The reader may be familiar with these two routes because they are commonly used in inorganic chemistry laboratories, e.g., in the production of common salt crystals from a solution by evaporation of the liquid or of $Mg(OH)_2$ by the addition of NaOH solution to $MgCl_2$ solution. As outlined earlier and summarized in Table 2.1, ceramic powders must meet certain specified requirements, especially for the fabrication of advanced ceramics. We therefore need to understand the phenomena occurring during the precipitation process in order to produce the desired powder characteristics.

The principles of precipitation from solution were considered more than 40 years ago by LaMer and Dinegar [12], and the main features may be represented in terms of the diagram shown in Fig. 2.11, generally referred to as a *LaMer diagram*. In the cases considered here—evaporation of the liquid and addition of a chemical reagent to the solution—the concentration of the solute to be precipitated, C_x, increases to or above the saturation value, C_s. If the solution is free of dirt particles and the container walls are clean and smooth, then it is possible for C_x to exceed C_s by a large amount to give a supersaturated solution. Eventually a critical supersaturation concentration, C_{ss}, is reached after some time t_1, and homogeneous nucleation and growth of solute particles occurs, leading to a decrease in C_x to a value below C_{ss} after a time t_2. Further growth of the particles occurs by diffusion of solute through the liquid and precipitation onto the particle surfaces. Finally, particle growth stops after a time t_3 when $C_x = C_s$.

It is clear that if we wish to produce particles of a fairly uniform size, then one short burst of nucleation should occur in a short time interval

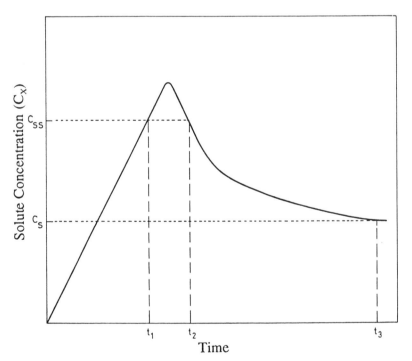

Figure 2.11 Schematic representation of solute concentration versus time in the nucleation and growth of particles from a solution. (From Ref. 12.)

Synthesis of Powders 61

$t_2 - t_1$. One way of achieving this is through the use of a fairly low reactant concentration. Furthermore, uniform growth of the particles requires that the solute be released slowly to allow diffusion to the particles without buildup of the solute concentration and further bursts of nucleation. In practice, this mechanism does not apply to the formation of particles that are agglomerates of primary particles (e.g., particles prepared by the Stober process). Instead, it may apply only to the formation of the primary particles.

We shall now consider in more practical terms the preparation of powders from solutions, starting first with the route based on precipitation.

Precipitation from Solution

The most straightforward use of precipitation is for the preparation of simple oxides. Powders are prepared by the hydrolysis of metal organic compounds (e.g., metal alkoxides) in alcoholic solutions or by the hydrolysis of metal salt solutions. Metal alkoxides have the general formula $M(OR)_z$, where z is an integer equal to the valence of the metal M, and R is an alkyl chain. They can be considered derivatives of either an alcohol, ROH, in which the proton is replaced by the metal M, or of a metal hydroxide, $M(OH)_z$.

Stober et al. [13] carried out a systematic study of the factors that controlled the preparation of fine SiO_2 particles of uniform size by the hydrolysis of silicon alkoxides in the presence of NH_3. The NH_3 served to produce pH values in the basic range. For the hydrolysis of silicon tetraethoxide, $Si(OC_2H_5)_4$, referred to commonly as TEOS, with ethanol as the solvent, the particle size of the powder was dependent on the ratio of the concentration of H_2O to TEOS and on the concentration of NH_3 but not on the TEOS concentration (in the range of 0.02–0.50 mol/dm^3). For a TEOS concentration of 0.28 mol/dm^3, Fig. 2.12 shows the general correlation between particle size and the concentrations of H_2O and NH_3. The particle sizes varied between 0.05 and 0.90 μm and were very uniform, as shown in Fig. 2.13. Different alcoholic solvents or silicon alkoxides were also found to have an effect. The reaction rates were fastest with methanol and slowest with *n*-butanol. Likewise, under comparable conditions, the particle sizes were smallest in methanol and largest in *n*-butanol. Fastest reactions (less than 1 min) and smallest sizes (less than 0.2 μm) were obtained with silicon tetramethoxide, whereas silicon tetrapentoxide reacted slowly ($\approx$24 h) and produced fairly large particles.

The controlled hydrolysis of metal alkoxides has since been applied to the preparation of a wide range of powders. We mentioned in Chapter 1 the work of Barringer and Bowen [14] for the preparation, packing,

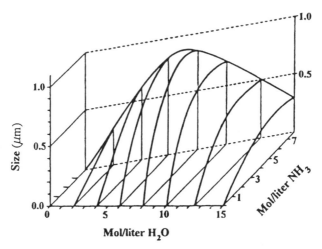

Figure 2.12 Correlation between particle size and the concentrations of water and ammonia in the hydrolysis of a solution of 0.28 mol/dm^3 silicon tetraethoxide in ethanol. (From Ref. 13.)

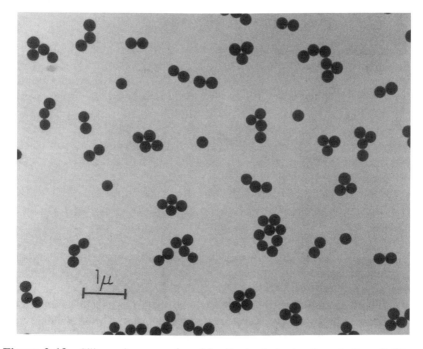

Figure 2.13 Silica spheres produced by the hydrolysis of a solution of silicon tetraethoxide in ethanol. (From Ref. 13.)

Synthesis of Powders

and sintering of monodisperse TiO_2 powders. In a typical experiment, a solution containing 0.30 mol/dm^3 titanium tetraethoxide, $Ti(OC_2H_5)_4$, in anhydrous ethanol was hydrolyzed by a solution containing 1.2 mol/dm^3 deionized water in ethanol. Precipitation of TiO_2 powder occurred in ≈ 5 s at 25.5°C to produce particles with a mean size of ≈ 0.38 μm. Hydrolysis of other titanium alkoxides produced fine particles that were not spherical but equiaxial and agglomerated.

An insight into the mechanism of hydrolysis of $Ti(OC_2H_5)_4$ was provided by Barringer and Bowen [15]. The alkoxide reacts with water to produce a monomeric hydrolysis species according to

$$Ti(OC_2H_5)_4 + 3H_2O \rightleftharpoons Ti(OC_2H_5)(OH)_3 + 3C_2H_5OH \qquad (2.13)$$

However, the presence of dimers and trimers of the hydrolysis species cannot be excluded. Polymerization of the monomer to produce the hydrated oxide is represented by

$$Ti(OC_2H_5)(OH)_3 \rightleftharpoons TiO_2 \cdot xH_2O + (1 - x)H_2O + C_2H_5OH \qquad (2.14)$$

The overall reaction can therefore be represented as

$$Ti(OC_2H_5)_4 + (2 + x)H_2O \rightleftharpoons TiO_2 \cdot xH_2O + 4C_2H_5OH \qquad (2.15)$$

The value of x was found by thermogravimetric analysis to be between 0.5 and 1.

The preparation of powders by the hydrolysis of metal alkoxides is generally carried out above room temperature. The variables that control the precipitation of particles are the concentration of the reactants, the pH of the solution, and the temperature. Oxide or hydrated oxide powders are produced. The precipitated particles can be agglomerates of much finer particles. The hydrolysis of metal salt solutions has the ability to produce a wider range of chemical compositions, including oxides or hydrated oxides, sulfates, carbonates, phosphates, and sulfides. However, the number of parameters that must be controlled can, in some cases, be higher, including the concentration of the metal salts, the chemical composition of the salts used as starting materials, the temperature, the pH of the solution, and the presence of anions and cations that form intermediate complexes. Table 2.4 compares the main process variables for the preparation of powders by hydrolysis of metal alkoxides with those for the hydrolysis of salt solutions.

The hydrolysis of metal salt solutions for the preparation of powders with controlled characteristics has been studied extensively by Matijevic and coworkers [16]. A variety of particle sizes and shapes can be produced (Fig. 2.14). The overall precipitation of hydrated metal oxides can be

Table 2.4 Main Processing Variables for the Precipitation of Particles from Solutions of Metal Alkoxides and Metal Salts

Alkoxide solution (Stober)	Salt solution (Matijevic)
Concentration of reactants	Concentration of reactants
Solution pH	Solution pH
Temperature	Chemical composition of the salt
	Anions and cations in solution
	Temperature

represented as

$$M^{z+}(aq) + zOH^-(aq) \rightleftharpoons M(OH)_z(s) \qquad (2.16)$$

However, the particle characteristics depend on the pH and the nature of the anions present in solution. Frequently, the anions responsible for the particle characteristics are not found in the precipitated powder. This indicates that soluble complexes involving such anions must act as precursors to the nucleation of the particles. The formation of soluble complexes means that the nature of the salt used as the starting material has an important effect on the characteristics of the precipitated particles. As an example, consider the preparation of spherical hydrated aluminum oxide particles with a narrow size distribution [17]. Solutions of $Al_2(SO_4)_3$, $KAl(SO_4)_2$, a mixture of $Al(NO_3)_3$ and $Al_2(SO_4)_3$, or a mixture of $Al_2(SO_4)_3$ and Na_2SO_4 were aged in Pyrex tubes sealed with Teflon-lined caps at $98 \pm 2°C$ for up to 84 h. The pH of the freshly prepared solutions was 4.1, and after aging and cooling to room temperature the pH was 3.1. Particles of uniform size were produced only when the Al concentration was between 2×10^{-4} and 5×10^{-3} mol/dm^3 provided that the $[Al^{3+}]/[SO_4^{2-}]$ molar ratio was between 0.5 and 1. For a constant Al concentration, the particle size increased with increasing sulfate concentration. The temperature of aging was a critical parameter; no particles were produced below 90°C, and the best results were obtained at 98°C. Finally, the particles had reasonably constant chemical composition, which indicates that one or more well-defined aluminum basic sulfate complexes were the precursors to the nucleation of particles.

The conditions for homogeneous precipitation of particles of uniform size can also be met by the slow release of anions in the solution. An example is the precipitation of yttrium basic carbonate particles from a solution of yttrium chloride, YCl_3, and urea, $(NH_2)_2CO$ [18]. Particles of uniform size were produced by aging for 2.5 h at 90°C a solution of 1.5×10^{-2} mol/dm^3 YCl_3 and 0.5 mol/dm^3 urea (Fig. 2.15a). However, solu-

Synthesis of Powders

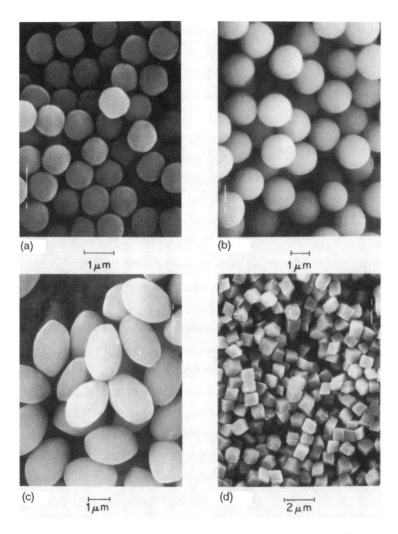

Figure 2.14 Examples of the sizes, shapes, and chemical compositions of powders prepared by precipitation from metal salt solutions, showing particles of (a) hematite (α-Fe$_2$O$_3$), (b) cadmium sulfide, (c) iron(III) oxide, and (d) calcium carbonate. (From Ref. 16, with permission.)

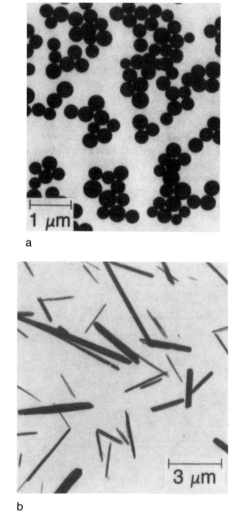

Figure 2.15 Particles obtained (a) by aging for 2.5 h at 90°C a solution of 1.5 × 10^{-2} mol/dm³ YCl₃ and 0.5 mol/dm³ urea and (b) by aging for 18 h at 115°C a solution of 3.0 × 10^{-2} mol/dm³ YCl₃ and 3.3 mol/dm³ urea. (From Ref. 18, with permission.)

tions of YCl₃ containing higher urea concentrations yielded, on aging at 115°C for 18 h, rodlike particles somewhat irregular in size (Fig. 2.15b). At temperatures up to 100°C, aqueous solutions of urea yield ammonium and cyanate ions:

$$(NH_2)_2CO \rightleftharpoons NH_4^+ + OCN^- \tag{2.17}$$

Synthesis of Powders

In acid solutions, cyanate ions react rapidly, according to

$$OCN^- + 2H^+ + H_2O \rightarrow CO_2 + NH_4^+ \quad (2.18)$$

whereas in neutral and basic solutions, carbonate ions and ammonia are formed:

$$OCN^- + OH^- + H_2O \rightarrow NH_3 + CO_3^{2-} \quad (2.19)$$

Yttrium ions are weakly hydrolyzed in water to $YOH(H_2O)_n^{2+}$. The resulting release of hydronium ions accelerates urea decomposition according to Eq. (2.18). The overall reaction for the precipitation of the basic carbonate can therefore be written as

$$YOH(H_2O)_n^{2+} + CO_2 + H_2O \rightarrow Y(OH)CO_3 \cdot H_2O$$
$$+ 2H^+ + (n-1)H_2O \quad (2.20)$$

For the reaction at 115°C, the decomposition of excess urea (>2 mol/dm^3) generates a large number of OH^- ions, which change the medium from acidic to basic (pH 9.7). The reaction of cyanate ions proceeds according to Eq. (2.19). The precipitation of rodlike particles may therefore be represented as

$$2YOH(H_2O)_n^{2+} + NH_3 + 3CO_3^{2-} \rightleftharpoons$$
$$Y_2(CO_3)_3 \cdot NH_3 \cdot 3H_2O + (2n-3)H_2O + 2OH^- \quad (2.21)$$

In addition to an excess of urea and a higher aging temperature, longer reaction times (>12 h) are needed to generate a sufficient amount of free ammonia for reaction (2.21) to dominate.

The largest use of precipitation is the Bayer process for the industrial production of Al_2O_3. The raw material bauxite is first physically beneficiated, then digested in the presence of NaOH at an elevated temperature. During digestion, most of the hydrated alumina goes into solution as sodium aluminate:

$$Al(OH)_3 + NaOH \rightarrow Na^+ + Al(OH)_4^- \quad (2.22)$$

and insoluble impurities are removed by settling and filtration. After cooling, the solution is seeded with fine particles of gibbsite, $Al(OH)_3$. In this case, the gibbsite particles provide nucleating sites for growth of $Al(OH)_3$. The precipitates are continuously classified, washed to reduce the sodium, and then calcined. Powders of α-Al_2O_3 with a range of particle sizes are produced by calcination at 1100–1200°C, followed by grinding and classification (Fig. 2.16). Tabular aluminas are produced by calcination at higher temperatures (≈ 1650°C).

For the preparation of powders of complex oxides (e.g., titanates, ferrites, and aluminates), we have already outlined the main difficulties

a

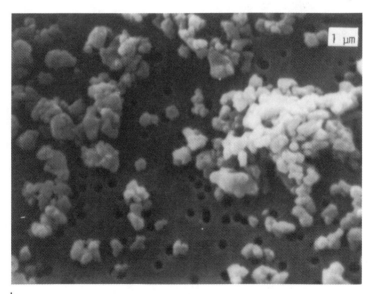

b

Figure 2.16 α-Alumina produced by the Bayer process showing (a) agglomerates after calcination and (b) powders after grinding and classification. (From Ref. 19.)

of solid-state reaction between a mixture of oxide powders when fine, stoichiometric, high-purity powders are required. Some of these difficulties may be alleviated by the use of *coprecipitation* from a solution (sometimes referred to as *cohydrolysis*). A solution of mixed alkoxides, mixed salts, or a combination of salts and alkoxides is generally used. A common problem in coprecipitation is that the different reactants in the solution have different hydrolysis rates, which results in segregation of the precipitated material. Suitable conditions must therefore be found in order to achieve homogeneous precipitation. Consider the preparation of $MgAl_2O_4$ powders [20]. Both Mg and Al are precipitated as hydroxides, but the conditions for their precipitation are quite different. $Al(OH)_3$ is precipitated under slightly basic conditions (pH = 6.5–7.5), is soluble in the presence of excess ammonia, but is only slightly soluble in the presence of NH_4Cl. $Mg(OH)_2$ is completely precipitated only in strongly basic solutions such as NaOH solution. In this case, an intimate mixture of $Al(OH)_3$ and Mg-Al double hydroxide was produced when a solution of $MgCl_2$ and $AlCl_2$ was added to a stirred excess solution of NH_4OH kept at a pH of 9.5–10. Calcination of the precipitated mixture above ≈400°C yielded stoichiometric $MgAl_2O_4$ powder of high purity and fine particle size.

The coprecipitation technique generally produces an intimate mixture of precipitates. In many cases, the mixture has to be calcined at an elevated temperature to produce the desired chemical composition and crystalline phase. One serious consequence is the need for milling the calcined powder, which can introduce impurities into the powder. It is more desirable to produce a precipitate that does not require the use of elevated temperature calcination and subsequent milling. In a few cases, the precipitated powder may have the same cation composition as the desired product. An example is the preparation of $BaTiO_3$ by the hydrolysis of a solution of barium isopropoxide, $Ba(OC_3H_7)_2$, and titanium tertiary amyloxide, $Ti(OC_5H_{11})_4$, by Mazdiyasni et al. [21]. The alkoxides were dissolved in a mutual solvent (e.g., isopropanol) and refluxed for 2 h prior to hydrolysis. While the solution was vigorously stirred, drops of deionized, triply distilled water were slowly added. The reaction was carried out in a CO_2-free atmosphere to prevent the precipitation of barium carbonate. After the precipitate was dried at 50°C for 12 h in a helium atmosphere, a stoichiometric $BaTiO_3$ powder with a purity of more than 99.98% and a particle size of 5–15 nm (with a maximum agglomerate size of <1 μm) was produced.

Hydrothermal preparation involves heating reactants—often metal salts, oxides, hydroxides, or metal powders—as a solution or a suspension, usually in water, at elevated temperature and pressure. Precipitation under these conditions can lead to an anhydrous powder with quite distinctive characteristics, including very fine size (typically ≈10–20 nm), narrow size distribution, single crystal particles, high purity, and good chemical homogeneity.

The hydrothermal method is distinguished from the precipitation process described earlier by the temperature and pressure used in the reactions. Temperatures typically fall between the boiling point and the critical temperature (374°C) of water, and pressures range up to ≈20 MPa. The process is carried out in a hardened steel autoclave, which can be heated to the desired temperature. Normally, the inner surfaces of the vessel are lined with a plastic (e.g., Teflon®) to limit corrosion of the vessel by the solution. Figure 2.17 shows powders of CeO_2 doped with 6 atomic percent (at %) Ca that were produced from a solution of cerium nitrate and calcium nitrate under hydrothermal conditions (4 h at ≈300°C and under a pressure of ≈10 MPa). The particle size of the powder is 13 ± 3 nm. CeO_2 has a cubic crystal structure; the faceted nature of the particles is an indication that they are crystalline. High-resolution transmission electron microscopy also revealed the high crystallinity of the particles (Fig. 2.18).

Another method that employs hydrothermal preparation is *hydrothermal crystallization* [22]. In this case, a poorly ordered precursor material (e.g., amorphous hydrous ZrO_2) is crystallized under hydrothermal condi-

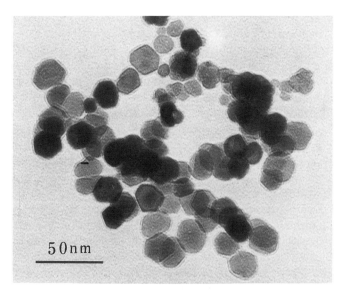

Figure 2.17 Powders of CeO_2 doped with 6 at % Ca prepared by the hydrothermal method.

Synthesis of Powders

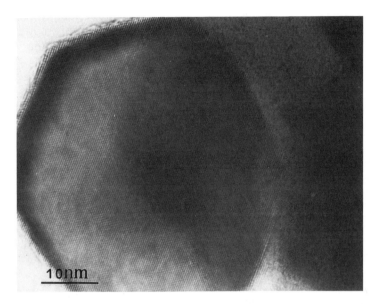

Figure 2.18 High-resolution transmission electron micrograph of a powder described in Fig. 2.17, showing the crystalline nature of the particles.

tions (e.g., 24 h at 300°C under 100 MPa) to produce a crystalline powder with a very fine grain size (e.g., monoclinic ZrO_2 with a particle size of 20–40 nm). The processes occurring in hydrothermal precipitation are essentially dissolution and precipitation.

Hydrothermal preparation has been used on the laboratory scale for the preparation of a variety of simple oxides, doped oxides, and complex oxides, e.g., ZrO_2, Y_2O_3-doped ZrO_2, $BaTiO_3$, and lead lanthanum zirconate titanate. Large-scale commercial exploitation is being investigated. As outlined above, the powders have very desirable characteristics. However, drying the precipitated powder prior to subsequent consolidation may severely retard the sintering rate. Particle adhesion at the contact points results in the gluing together of the particles and the formation of strong agglomerates.

Evaporation of the Liquid
As we outlined earlier, evaporation of the liquid provides another method for bringing a solution to supersaturation and thereby causing the nucleation and growth of particles. As usual, our interest lies in understanding the process for the production of particles with desired characteristics.

The simplest case is for a solution of a single salt. For the production of fine particles, nucleation must be fast and growth slow. This requires that the solution be brought to a state of supersaturation very rapidly so that a large number of nuclei are formed in a short time. One way of doing this is to break the solution up into very small droplets so that the surface area over which evaporation takes place increases enormously. For a solution of two or more salts, a further problem must be considered. Normally the salts will be in different concentrations and will have different solubilities. Evaporation of the liquid will cause different rates of precipitation, leading to segregation of the solids. Here again, the formation of very small droplets will limit the segregation to the droplets, since no mass is transferred between individual droplets. Furthermore, for a given droplet size, the size of the particle becomes smaller for more dilute solutions. This means that we can further reduce the scale of segregation by the use of dilute solutions. We now consider some of the practical ways of producing powders by the evaporation route.

Spray Drying

In spray drying, fine droplets produced by a fluid atomizer are sprayed into a drying chamber and the powder is collected (Fig. 2.19). A variety of atomizers are available, and these are usually categorized according to the manner in which energy is supplied to produce the droplets [23]. A rotary atomizer consists of a spinning disk located at the top of the drying chamber. A jet of solution strikes the disk, accelerates outwards into the chamber and breaks up into droplets. In a pneumatic atomizer, pressure nozzles atomize the solution by accelerating it through a large pressure difference and injecting it into the chamber. Pneumatic atomization occurs when the solution is impacted by a stream of high-speed gas from a nozzle. Ultrasonic atomization involves passing the solution over a piezoelectric device that is vibrating rapidly. Droplet sizes ranging from less than 10 μm to over 100 μm can be produced by these atomizers. In the drying chamber, the flow pattern of the hot air determines the completeness of moisture removal and the maximum temperature that the produced particles will experience. Finally, the particles are carried out in the air stream leaving the chamber and are captured using a bag or cyclone collector.

The variables that must be controlled in spray drying are the size of the droplet, the concentration of the solution, the temperature and flow pattern of the air in the drying chamber, and the design of the chamber. Under suitable conditions, spherical agglomerates with a primary particle size of 0.1 μm or less can be obtained. Spray drying has been found to be useful for the preparation of ferrite powders [24]. For Ni-Zn ferrite, the solution of sulfates was broken up into droplets (10–20 μm) by a rotary

Synthesis of Powders

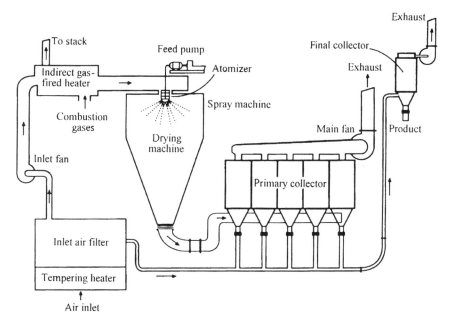

Figure 2.19 Schematic of a spray dryer for the production of powders. (From Ref. 5, with permission.)

atomizer. The powder obtained by spray drying was in the form of hollow spheres having the same size as the original droplets. Calcination at 800–1000°C (required for the decomposition of the sulfates and solid-state reaction of the oxides) produced a fully reacted powder consisting of agglomerates with a primary particle size of ≈0.2 μm. The ground powder (particle size <1 μm) was sintered to almost theoretical density.

By using a higher temperature and a reactive (e.g., oxidizing) atmosphere in the chamber, the salts can be dried and decomposed directly in a single step [25]. This technique is referred to by many terms, including spray pyrolysis, spray roasting, spray reaction, and evaporative decomposition of solutions. In this book we use the term *spray pyrolysis*. The stages in the formation of a dense particle from a droplet are shown schematically in Fig. 2.20. The droplet undergoes evaporation, and the solute concentration in the outer layer increases above the supersaturation limit, leading to the precipitation of fine particles. Precipitation is followed by a drying stage in which the vapor must now diffuse through the pores in the precipitated layer. Decomposition of the precipitated salts produces

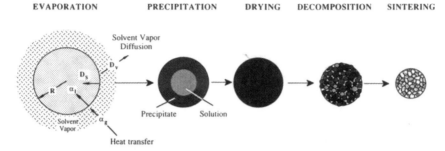

Figure 2.20 Schematic of the stages in the spray pyrolysis process. (From Ref. 25, with permission.)

a porous particle made up of very fine grains, which is finally heated to produce a dense particle.

In addition to the parameters of the solution and the equipment outlined above for spray drying, an additional consideration in spray pyrolysis is the chemical composition of the starting salts. Aqueous solutions are commonly used because of their ease of handling, safety, and low cost combined with the availability of a wide range of soluble metal salts. The solute must also have a high solubility in order to increase the yield of the powder produced. Decomposition, as outlined earlier, is a kinetic process, so for a given set of conditions, salts with relatively high decomposition rates may be decomposed completely whereas others with lower decomposition rates may be incompletely decomposed. For small-scale laboratory equipment in which the reaction times are small, nitrates and acetates are preferable to sulfates because of their lower decomposition temperatures. However, acetates have a low solubility whereas nitrates, acetates, and sulfates can introduce impurities into the powder. Chlorides and oxychlorides are used industrially because of their high solubilities, but the corrosive nature of the gases produced during decomposition and the deleterious effect of residual chlorine on subsequent sintering can be problematic.

A variety of particle morphologies are possible as a result of the drying of a droplet of a salt solution. Some morphologies produced in the spray pyrolysis process are shown in Fig. 2.21. For advanced ceramics, dense particles are preferred over those with highly porous or hollow shell-like morphologies. The major factors that influence the particle morphology are believed to include the solubility of the metal salt and the degree of supersaturation of the solution [25]. The following factors are believed to be necessary for the preparation of solid particles:

Synthesis of Powders

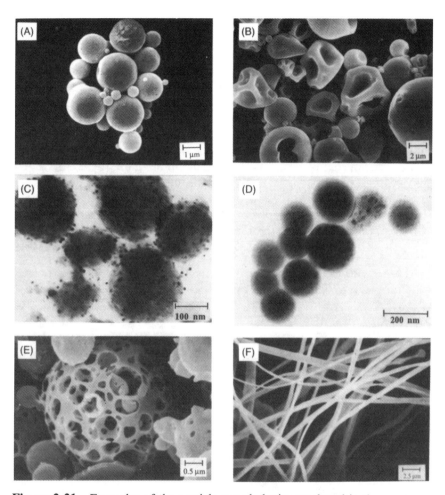

Figure 2.21 Examples of the particle morphologies produced in the spray pyrolysis process. (From Ref. 25, with permission.)

1. There should be a large difference between the critical supersaturation concentration and the saturation concentration of the solute (see Fig. 2.11).
2. The solute should have a high solubility and a positive temperature coefficient of solubility.
3. The precipitated solids should not be thermoplastic or melt during decomposition.

4. The droplet size generated by the atomizer should be small.

Suspensions of fine particles (sometimes referred to as slurries) can also be dried by spray drying. In this case, the liquid is removed in such a way as to limit the agglomeration of the dried powder to a scale equal to or less than the size of the droplet. Limiting the scale of the agglomeration should provide benefits in terms of better overall uniformity of the compacted body, which in turn should also lead to benefits in sintering. An example of a powder produced from a spray-dried slurry is shown in Fig. 2.22 for lead zirconate titanate precipitated from solution [26]. Finally, spray drying is used on a large scale industrially for the drying of slurries (e.g., produced by wet milling) for the production of powders for dry compaction. The process provides good control of the agglomerate size, the flowability of the powder, the moisture content, and the density and strength of the compacted body.

Freeze Drying

In freeze drying the solution is broken up by an atomizer into fine droplets, which are then frozen rapidly by being sprayed into a cold bath of immisci-

Figure 2.22 Scanning electron microscopic (SEM) photomicrograph of a spray-dried lead zirconate titanate powder prepared by spray drying of a slurry. (From Ref. 26.)

Synthesis of Powders

ble liquid such as hexane and dry ice or directly into liquid nitrogen. The frozen droplets are then placed in a cooled vacuum chamber, and the solvent is removed, under the action of a vacuum, by sublimation without any melting. The system may be heated slightly to aid the sublimation. Finally, the dried powder is decomposed at elevated temperatures to produce an oxide. The technique produces spherical agglomerates of fine particles, with the agglomerate being of the same size as the frozen droplets. The size of the particles in the agglomerates depends on the concentration of the solution and the decomposition step.

As we observed for spray drying, the breaking up of the solution into droplets serves to limit the scale of agglomeration or segregation to the size of the droplet. Because the solubility of most salts decreases with temperature, the rapid cooling of the droplets in freeze drying produces very rapidly a state of supersaturation of the droplet solution. Particle nucleation is therefore rapid and growth slow, so the size of the particles in the frozen droplet is very fine. Because the freezing of the droplets can be carried out very rapidly compared to the evaporation of the liquid in spray drying, the surface area of freeze-dried powders can be significantly higher than that for spray-dried powders. Surface areas as high as 60 m^2/g have been reported for freeze-dried powders.

Freeze drying of solutions has generally been used on a laboratory scale for the preparation of ferrite and other oxide powders. In the case of lithium ferrite, $LiFe_5O_8$, powders prepared by freeze drying a solution of oxalates were found to have lower sintering temperatures and to afford better control of the grain size than similar powders prepared by spray drying [27]. Finally, as we outlined for spray drying, the freeze-drying technique is also used for drying slurries. Powders of Al_2O_3 produced from freeze-dried slurries were found to consist of soft agglomerates that could be broken down easily [28]. Pressing of such powders produced fairly homogeneous green bodies. Freeze drying gives softer agglomerates and higher surface area because of the absence of capillary forces (present in conventional drying and spray drying).

Sol-Gel Processing

In Chapter 1, the sol-gel process was outlined as a fabrication route for ceramics. With careful drying, the process can be used for producing monolithic ceramics, but it is best applied to the formation of films and fibers. We shall discuss sol-gel processing in greater detail later in this book (Chapter 5), but at this point we should point out that the process, although expensive, can also be used for the production of powders. The gel formed by hydrolysis, condensation, and gelation of the initial solution (normally with a polymeric structure) is dried and ground to produce a powder. Because the mixing of the constituents in the gel formation is achieved at a molecular level, the powders have a high degree of chemical homogeneity. Controlled drying is not needed in the production of a powder. However, dried gels with lower strength are easier to grind, and the extent of contamination introduced during grinding is lower. Liquid removal under supercritical (or hypercritical) conditions produces almost no shrinkage, so a dried gel with low strength is obtained. Grinding can usually be carried out in plastic media. Powders with the stoichiometric mullite composition ($3Al_2O_3 \cdot 2SiO_2$), produced by supercritical drying of gels have been shown to have fairly high sinterability [29]. The compacted powders sinter to nearly full density below $\approx 1200°C$, which is considerably better than mullite prepared by reaction of mixed powders. For powders produced from dried gels, the benefits in sintering are due to the amorphous structure and high surface area. A crystallization step is necessary for the production of crystalline bodies. If the crystallization occurs prior to full densification, then the sintering benefits may be reduced significantly. Also, a significant volume change during crystallization may cause cracking of the body.

The Pechini and Glycine Nitrate Methods

The *Pechini method* is a process developed by Pechini [30] for the preparation of titanates and niobates for the capacitor industry. With slight modifications, it is also referred to as the citrate gel process or the amorphous citrate process and has been used for the preparation of a wide variety of oxide powders. In this method, metal ions from starting materials such as carbonates, nitrates, and alkoxides are complexed in an aqueous solution with α-carboxylic acids such as citric acid. When heated with a polyhydroxy alcohol, such as ethylene glycol, polyesterification occurs, and on removal of the excess liquid a transparent resin is formed. The resin is then heated to decompose the organic constituents, ground, and finally calcined to produce the powder. Figure 2.23 shows a flow chart for the preparation of powders of strontium titanate, $SrTiO_3$, by the Pechini method [31].

An advantage of the Pechini method is the ability to achieve very good chemical homogeneity. As in the sol-gel process, mixing occurs on the atomic scale during the polymerization process, and this is maintained in the formation of the resin. Provided that none of the constituents are volatilized during the calcination step, the cation composition can be identical to that of the original solution. A disadvantage is that the material formed from the decomposition of the resin is not in the form of a powder but consists of charred lumps. These lumps have to be ground and then calcined to produce a powder. The grinding step may introduce impurities into the powder, and the calcination step generally produces hard agglom-

Synthesis of Powders

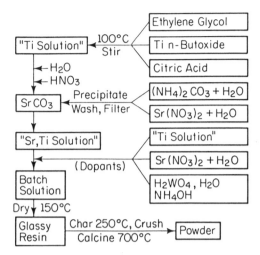

Figure 2.23 Flow chart for the preparation of strontium titanate powder by the Pechini method. (From Ref. 31.)

Figure 2.24 Effect of calcination temperature on the crystallite size of the powder described in Fig. 2.23 for (a) 500°C, (b) 600°C, (c) 700°C, and (d) 800°C (Magnification: 1 cm = 0.25 μm.) (From Ref. 31.)

erates. The primary particle size of the agglomerates also depends on the temperature of calcination, as illustrated in Fig. 2.24 for the SrTiO₃ powder produced according to the process described in Fig. 2.23.

The *glycine nitrate process* is one of a general class of combustion methods for the preparation of ceramic powders. A highly viscous mass formed by evaporation of a solution of metal nitrates and glycine is ignited to produce the powder [32]. The process has certain similarities to the Pechini method. Glycine, an amino acid, forms complexes with the metal ions in solution, which increases the solubility and prevents the precipitation of the metal ions as the water is evaporated. Good chemical homogeneity is therefore achieved, as for the Pechini method. Glycine also serves another important function: it provides a fuel for the ignition step of the process as it is oxidized by the nitrate ions.

The reactions occurring during ignition are highly explosive, and extreme care must be exercised during this step. Normally, only small quan-

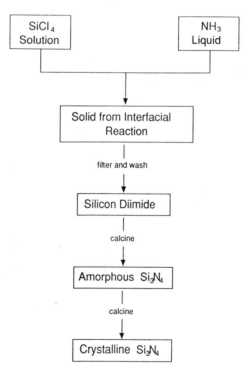

Figure 2.25 Flow chart for the preparation of Si_3N_4 powder by the reaction between liquid $SiCl_4$ and liquid NH_3. (Courtesy UBE, Japan.)

Synthesis of Powders

tities should be ignited at a time. Under well controlled conditions, a loose mass of very fine crystalline powder (particle size less than tens of nanometers) is obtained after ignition, so, in contrast to the Pechini method, no grinding is required. The very fine size and crystalline nature of the powder are believed to be a direct result of the short exposure to high temperatures during the ignition step. The glycine nitrate process therefore offers a relatively inexpensive route to the preparation of very fine, chemically homogeneous powders. It has been used for the preparation of simple oxides as well as complex oxides (e.g., manganites, chromites, ferrites, and oxide superconductors).

Nonaqueous Liquid Reaction

Reactions involving nonaqueous liquids have been used for the preparation of Si_3N_4 powders. The advantage of these methods is the higher purity and finer particle size of the powder produced compared with methods that involve grinding of a solid product. The reaction between liquid $SiCl_4$ and liquid NH_3 has been used on an industrial scale by UBE Industries,

Figure 2.26 Scanning electron micrograph of a commercially available Si_3N_4 powder (UBE-SN-E10) produced by the method described in Fig. 2.25. (Courtesy of UBE, Japan.)

Japan, to produce Si_3N_4 powder. A schematic of the sequence of operations is shown in Fig. 2.25. The products formed by the interfacial reaction between the two liquids were collected and washed with liquid NH_3 and calcined at 1000°C to produce an amorphous Si_3N_4 powder. Subsequent calcination at 1550°C in N_2 yielded a crystalline powder with a particle size of ≈0.2 μm (Fig. 2.26). Metallic impurities (Fe, Ca, and Al) were <0.02 wt %.

C. Vapor-Phase Reactions

Reactions involving the vapor phase have been used extensively for the production of powders of refractory oxides and nonoxides. We will outline these methods, paying particular attention to the preparation of Si_3N_4 and SiC powders. Crystalline Si_3N_4 exists in two different hexagonal polymorphs designated α and β, with the α form having the slightly higher free energy at the formation temperature. Powders of α-Si_3N_4 have a more equiaxial particle shape and sinter more readily than β-Si_3N_4, the particles of which grow in a more elongated shape. The preparation conditions are therefore selected to maximize the amount of α-Si_3N_4 produced. Silicon carbide exists in many polytypes, with the two major forms being designated α and β. The β form is more stable at lower temperatures and transforms irreversibly to the α form at ≈2000°C. Powders produced above ≈2000°C (e.g., in the Acheson process described earlier) therefore consist of α-SiC. Powders of either the α or β form are used in the production of SiC bodies. However, the sintering of β-SiC powders above ≈1800–1900°C results in the transformation to the α phase, which is accompanied by growth of platelike α grains and a deterioration of mechanical properties. The use of β-SiC powder therefore requires very fine powders so that the sintering temperature can be kept below ≈1800°C.

We consider vapor phase preparation methods in the following categories: (i) reactions between a gas and a solid, (ii) reactions between a gas and a liquid, and (iii) reactions between two or more gases.

Gas–Solid Reaction

A widely used method for the preparation of Si_3N_4 powders is direct nitridation, in which Si powder (particle size typically in the range 5–20 μm) is reacted with N_2 at temperatures between 1200 and 1400°C for times in the range 10–30 h. The method is used commercially by many manufacturers; a commercial powder produced by H. C. Starck (Germany) is shown in Fig. 2.27. The reaction and the processing conditions are very similar to those outlined in Chapter 1 for the production of reaction-bonded Si_3N_4 bodies; however, for powder production a loose bed of Si powder is used

Synthesis of Powders

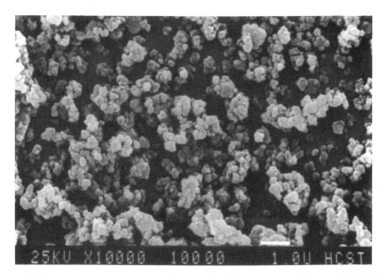

Figure 2.27 Scanning electron micrograph of a commercially available Si_3N_4 powder (LC 12) produced by the nitridation of silicon. (Courtesy of H. C. Starck, Germany.)

rather than the shaped powder form used for the reaction-bonding process. The reaction may be written as

$$3Si + 2N_2 \rightarrow Si_3N_4 \tag{2.23}$$

The thin SiO_2 layer covering the Si particles severely reduces the rate of the reaction. It is found that a nitriding gas of N_2 with $\approx 5\%$ H_2 or a pretreatment in H_2, Ar, or vacuum prior to nitridation to volatilize the SiO_2 layer can improve the nitriding reactivity.

The nitridation of Si has been reviewed by Moulson [33]. For high-purity Si powder, the reaction kinetics at a given temperature between 1250 and 1370°C consist of three regimes: a region obeying linear kinetics followed by one with decreasing rate and finally a region with effectively zero reaction rate. The dominant process is believed to be the volatilization of Si and the vapor-phase reaction with N_2 leading to the formation of α-Si_3N_4 by a chemical vapor deposition process. A minor reaction is the nucleation and growth of β-Si_3N_4 on the solid silicon by surface diffusion of Si to the reaction site. For silicon powder containing impurities (e.g., Fe), solution of N_2 in liquid silicon alloys (formed at the nitriding temperature) occurs with the precipitation of β-Si_3N_4. The nitridation process therefore leads to a mixture of α- and β-Si_3N_4. However, the reaction temperature, the partial pressure of the N_2 gas in the nitriding atmosphere, and the purity of the Si provide some control of the amount of β-Si_3N_4 produced.

Silicon nitride powder is also produced by the carbothermal reduction of SiO_2 in a mixture of fine SiO_2 and C powders followed by nitridation at 1200–1400°C in N_2. This process is used industrially by Toshiba (Japan). The widespread availability of pure fine SiO_2 and C makes this method an attractive alternative to the nitridation of Si. The overall reaction can be written

$$3SiO_2 + 6C + 2N_2 \rightarrow Si_3N_4 + 6CO \tag{2.24}$$

The mechanism is believed to involve the reaction of gaseous silicon monoxide, SiO, as follows:

$$3SiO_2(s) + 3C(s) \rightarrow 3SiO(s) + 3CO \tag{2.25}$$

$$3SiO(s) \rightleftharpoons 3SiO(g) \tag{2.26}$$

$$3SiO(g) + 3C(s) + 2N_2(g) \rightarrow Si_3N_4(s) + 3CO(g) \tag{2.27}$$

Excess carbon is used as an oxygen sink to form gaseous CO and reduce the amount of oxygen on the surface. However, any carbon remaining after the reaction has to be burned out in an oxidizing atmosphere, and this may cause some reoxidation of the Si_3N_4 surfaces.

The two methods described above produce a strongly agglomerated mass of Si_3N_4 that requires comminution, washing, and classification. The impurities introduced into the powder during these steps can cause a significant reduction in the high-temperature mechanical properties of the fabricated material.

Reactions Between a Liquid and a Gas

Mazdiyasni and Cooke [34,35] showed that the reaction between liquid silicon tetrachloride, $SiCl_4$, and NH_3 gas in dry hexane at 0°C can be used to prepare a fine Si_3N_4 powder with very low levels of metallic impurities (<0.03 wt %). The initial reaction may be written as

$$SiCl_4(l) + NH_3(g) \rightarrow Si(NH)_2 + 4NH_4Cl \tag{2.28}$$

but involves a more complex set of reactions. The overall decomposition of silicon diimide to produce Si_3N_4 is

$$3Si(NH)_2 \rightarrow Si_3N_4 + N_2 + 3H_2 \tag{2.29}$$

The powder produced is amorphous to X-rays but crystallizes to α-Si_3N_4 after prolonged heating between 1200 and 1400°C.

Synthesis of Powders

Reaction Between Gases

Reactions between heated gases have been investigated extensively as a route for the preparation of nonoxide powders with controlled characteristics. Particles are formed by homogeneous nucleation in the gas phase. The nucleation rate of spherical particles from a gas may be treated in terms of the same equations used for the nucleation of liquid drops from a supersaturated vapor. The nucleation rate I of liquid drops is [36]

$$I = I_0 \exp\left\{\frac{-16\pi\gamma^3 v^2}{3kT[kT \ln(p/p_0)]^2}\right\} \quad (2.30)$$

where γ is the surface tension, v is the molecular volume, p is the vapor pressure, p_0 is the equilibrium vapor pressure of the liquid at temperature T, k is Boltzmann's constant, and I_0 is given by:

$$I_0 = \alpha \left(\frac{p}{kT}\right)^2 v \left(\frac{2\gamma}{\pi m}\right)^{1/2} \quad (2.31)$$

where α is a constant and m is the molecular mass. Equation (2.31) shows that the nucleation rate is quite sensitive to the supersaturation ratio p/p_0. For a general gas-phase reaction in which a moles of gas A react with b moles of gas B to produce c moles of a solid C and d moles of a gas D, i.e.,

$$aA(g) + bB(g) \rightarrow cC(s) + dD(g) \quad (2.32)$$

the supersaturation ratio is equal to $Kp_A^a p_B^b / p_D^d$, where K is the equilibrium constant for the reaction in Eq. (2.32) and p_A, p_B, and p_D are the partial vapor pressures for the reactants and gaseous product [37]. The initial requirement for the production of a fine powder is the achievement of a high degree of supersaturation to produce a large number of nuclei by homogeneous nucleation. The formation of a powder should therefore be controlled by the value of K. Indeed, it is found that for many oxides and nonoxides the formation of a powder occurs only for relatively large values of K.

In practice, the techniques used for heating the gases include electrically heated tubes, optical means (e.g., lasers), and radio-frequency and microwave plasmas. The following reactions have been used to produce Si_3N_4 and SiC powders:

$$3SiCl_4(g) + 4NH_3(g) \rightarrow Si_3N_4(s) + 12HCl(g) \quad (2.33)$$

$$3SiH_4(g) + 4NH_3(g) \rightarrow Si_3N_4(s) + 12H_2(g) \quad (2.34)$$

$$2SiH_4(g) + C_2H_4(g) \rightarrow 2SiC(s) + 6H_2 \qquad (2.35)$$

The use of silicon tetrachloride leads to highly corrosive HCl as a by-product so that silane, SiH_4, despite being expensive and flammable in air, is generally preferred as the reactant. For Si_3N_4 production, NH_3 is used because N_2 is fairly unreactive.

Prochazka and Greskovich [38] used the reaction between SiH_4 and NH_3 at temperatures between 500 and 900°C in an electrically heated silica tube to produce fine amorphous Si_3N_4 powders. Two main parameters were found to control the reaction: the temperature and the NH_3/SiH_4 molar ratio. For a molar ratio greater than 10 and at temperatures of 500–900°C, nearly stoichiometric powders with a cation purity of >99.99%, a surface area of 10–20 m^2/g, and an oxygen content of <2 wt % were produced. Subsequent calcination above $\approx$1350°C yielded crystalline α-Si_3N_4 powder. The reaction between $SiCl_4$ and NH_3 is used commercially by Toya Soda (Japan) for the production of Si_3N_4 powder.

Our discussions so far have related to four commercially available Si_3N_4 powders (produced by UBE Industries, H. C. Starck, Toshiba, and Toya Soda). Table 2.5 summarizes the properties of these powders [39].

The use of a CO_2 laser as the heat source for the gas phase reactions was employed by Haggerty and coworkers [40] for the preparation of a range of oxide and nonoxide powders. In addition to bringing the reactant gases to the required temperature, the laser heating serves another useful purpose: the frequency of the radiation can be chosen to match one of

Table 2.5 Properties of Commercially Available Silicon Nitride Powders

	Method of preparation			
	Liquid-phase reaction of $SiCl_4/NH_3$	Nitridation of Si in N_2	Carbothermic reduction of SiO_2 in N_2	Vapor-phase reaction of $SiCl_4/NH_3$
Manufacturer	UBE	H. C. Starck	Toshiba	Toya Soda
Grade	SN-E 10	H1	—	TSK TS-7
Metallic impurities (wt %)	0.02	0.1	0.1	0.01
Nonmetallic impurities (wt %)	2.2	1.7	4.1	1.2
α-Si_3N_4 (wt %)	95	92	88	90
β-Si_3N_4 (wt %)	5	4	5	10
SiO_2 (wt %)	2.5	2.4	5.6	—
Surface area (m^2/g)	11	9	5	12
Average particle size (μm)	0.2	0.8	1.0	0.5
Tap density (kg/m^3)	1000	640	430	770

Source: Adapted from Ref. 39.

Synthesis of Powders

the absorption frequencies of one or more of the reactants. A laser can therefore be a very efficient heat source. A laboratory-scale reaction cell is shown in Fig. 2.28. The laser beam enters the cell through a KCl window and intersects the stream of reactant gases, usually diluted with an inert gas such as argon. The powders produced are captured on a filter located between the cell and a vacuum pump. An advantage of the method is that the reactions can be fairly well controlled by manipulation of the process variables such as the cell pressure, the flow rate of the reactant and diluent gases, the intensity of the laser beam, and the reaction flame temperature. The reactions described by Eqs. (2.30) and (2.31) have been used for the production of Si_3N_4 and SiC powders. An advantage of using SiH_4 rather than $SiCl_4$ as a reactant is that it has a strong adsorption band near the wavelength of the laser (10.6 μm).

Table 2.6 summarizes the range of powder characteristics that are generally obtained for Si_3N_4 and SiC powders by this method. An example of an SiC powder produced by laser heating is shown in Fig. 2.29. We see that most of the desirable powder characteristics outlined at the beginning of this chapter, such as very fine particle size, narrow size distribution, equiaxial shape, the absence of strong agglomerates, and high purity are achieved. Although the oxygen content of the powders maintained in

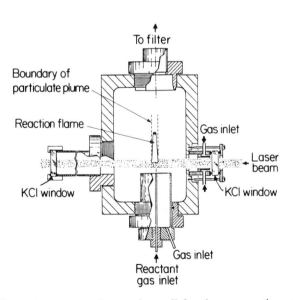

Figure 2.28 Laboratory-scale reaction cell for the preparation of powders by laser heating of gases. (From Ref. 40, with permission.)

Table 2.6 Summary of the Range of Characteristics for Si_3N_4 and SiC Prepared by Laser-Heated Gas-Phase Reactions

Powder characteristic	Si_3N_4	SiC
Mean diameter (nm)	7.5–50	20–50
Standard of deviation of diameters (% of mean)	2.3	≈2.5
Impurities (wt %)		
Oxygen	0.3	0.33–1.3
Total others	<0.01	NA
Major elements	Al, Ca	NA
Stoichiometry (%)	0–60 (excess Si)	0–10 (excess C or Si)
Crystallinity	Amorphous-crystalline	Crystalline Si and SiC
Grain size: mean diameter	≈0.5	0.5–1.0

NA = not applicable
Source: Ref 40.

an inert atmosphere is fairly low, careful handling of the powders is required to prevent excessive exposure to oxygen. As an example, consider a 30 nm diameter SiC particle that has been exposed to an oxidizing atmosphere to produce a typical 3 nm layer of SiO_2; in this case the oxygen content of the powder would have increased to more than 30 wt %. Laser-heated gas-phase reactions are currently used on a laboratory scale mainly for the production of nonoxide powders. However, estimates of the production cost indicate that the method can be very competitive with other methods (e.g., the Acheson process described earlier for SiC), especially for submicrometer powders.

Thermal plasmas have been used for decades as a source of heat for gas-phase reactions. They have been used more in recent years for the production of very fine powders of oxides and to a greater extent for nonoxides such as nitrides and carbides. Typically, the reactant gases (e.g., SiH_4 and CH_4 for the production of SiC powder) are injected into the hot plasma jet, usually an argon plasma, and the resulting powder is collected from the cooled powder–gas mixture. The process parameters that control the powder characteristics are the frequency and powder level of the plasma source, the temperature of the plasma jet, the flow rate of the gases, and the molar ratio of the reactants. Powders with high purity and very fine particle size (e.g., 10–20 nm) can be produced with the thermal plasma method, but a major problem is that the powders are highly

Synthesis of Powders

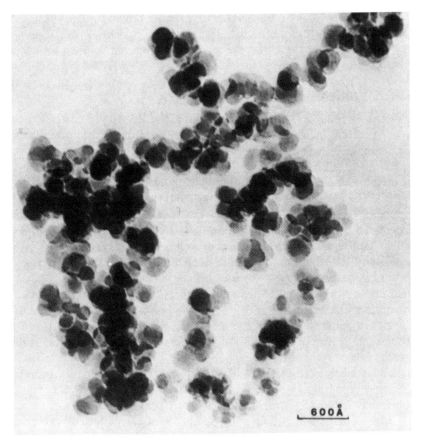

Figure 2.29 Transmission electron micrograph of an SiC powder prepared by laser heating of gases. (From Ref. 40, with permission.)

agglomerated [41]. The method is currently used on a laboratory scale only.

2.6 CONCLUDING REMARKS

In this chapter we have examined the wide range of methods commonly used for the preparation of ceramic powders. It will be recognized that scientifically the methods are based on sound principles of physics and chemistry, which form a framework for understanding how the process variables influence the characteristics of the powder produced. Practically, the methods vary considerably in the quality of the powder produced and in the cost of production. Generally, higher powder quality is associated with higher production cost. For a given application, we therefore need to examine whether the higher production cost is justified by the higher quality of the powder produced.

PROBLEMS

2.1 For a powder with the composition of TiO_2, calculate and plot the surface area of 1 g of powder as a function of the particle size. Use a size range of 5 nm to 100 μm and assume that the particles are spherical. Estimate the percentage of TiO_2 molecules in the surface of the particle for the same size range and plot the results on a graph.

2.2 In an experiment to determine the kinetics of decomposition of magnesium carbonate, a student weighed out 20.00 g of powder and found that the mass of powder remaining in an isothermal experiment was as follows:

Time (min)	Mass of Powder (g)
0	20.00
10	16.91
20	14.73
30	13.19
50	11.35
80	10.16

Determine the order of the reaction.

2.3 Consider the formation of $ZnFe_2O_4$ from spherical particles of ZnO and Fe_2O_3 when the reaction rate is controlled by diffusion through the product layer. What governs the particles on which the product layer forms?

2.4 In the reaction between spherical ZnO and Al_2O_3 powders that have roughly the same size (1 μm in diameter), 20% of the Al_2O_3 was reacted to form zinc aluminate spinel during the first 20 minutes of an isothermal experiment. Assuming that the particle shape remains spherical, estimate how long it will take for all of the Al_2O_3 to be reacted.

2.5 In the preparation of TiO_2 powder by the Stober process, a student starts out with a solution containing 20 vol % of titanium isopropoxide in isopropanol. Assuming that the reaction is stoichiometric, how much water must be added?

2.6 Discuss the factors involved in the design of a continuous process for the preparation of narrow-sized, unagglomerated titania powders. (See Ref. 42.)

2.7 Design a process for preparing zinc oxide powder which is uniformly coated with 1 mol % bismuth oxide.

2.8 A solution containing 0.1 mol/liter of zinc acetate is spray roasted, using a nozzle which produces 40 μm diameter droplets, to produce ZnO particles. If the particles are only 50% dense, estimate their diameter.

2.9 Surface oxidation produces an oxide layer ($\approx$3 nm thick) on silicon carbide. Assuming the composition of the oxide layer to be that of silica, estimate the weight percent of oxygen in silicon carbide powders (assumed to be spherical) with a size of (a) 1 μm and (b) 50 nm.

2.10 Discuss the technical and economic factors which will favor the production of silicon nitride powders by laser heating of gases over the nitridation of silicon. (See Ref. 40.)

REFERENCES

1. D. W. Johnson, Jr., *Am. Ceram. Soc. Bull.*, *60*(2):221 (1981).
2. D. W. Johnson, Jr., in *Advances in Powder Technology* (G. Y. Chin, Ed.), American Society for Metals, Metals Park, OH, 1982, pp. 22–37.
3. R. Polke and R. Stadler, in *Concise Encyclopedia of Advanced Ceramic Materials* (R. J. Brook, Ed.), MIT Press, Cambridge, MA, 1991, pp. 187–193.
4. J. K. Beddow, *Particulate Science and Technology*, Chemical Publishing Co., New York, 1980.
5. J. W. Evans and L. C. De Jonghe, *The Production of Inorganic Materials*, Macmillan, New York, 1991.
6. I. J. Lin and S. Nadiv, *Mater. Sci. Eng.*, *39*:93 (1979).
7. S. N. Patankar, S.-Q. Xiao, J. J. Lewandowski, and A. H. Heuer, *J. Mater. Res.*, *8*(6):1311 (1993).
8. D. A. Young, *Decomposition of Solids*, Pergamon, Oxford, 1966, Chap. 3.
9. O. Kubaschewski, E. L. Evans, and C. B. Alcock, *Metallurgical Thermochemistry*, 5th ed., Pergamon, Oxford, 1979.
10. H. Schmalzried, *Solid State Reactions*, Academic, New York, 1974.
11. R. E. Carter, *J. Chem. Phys.*, *34*:2010 (1961); *35*:1137 (1961).
12. V. K. LaMer and R. H. Dinegar, *J. Am. Chem. Soc.*, *72*:4847 (1950).
13. W. Stober, A. Fink, and E. Bohn, *J. Colloid Interf. Sci.*, *26*:62 (1968).
14. E. A. Barringer and H. K. Bowen, *J. Am. Ceram. Soc.*, *65*(12):C-199 (1982).
15. E. A. Barringer and H. K. Bowen, *Langmuir*, *1*:414 (1985).

16. E. Matijevic, in *Ultrastructure Processing of Ceramics, Glasses, and Composites* (L. L. Hench and D. R. Ulrich, Eds.), Wiley, New York, 1984, pp. 334–352.
17. R. Brace and E. Matijevic, *J. Inorg. Nucl. Chem.*, *35*:3691 (1973).
18. B. Aiken, W. P. Hsu, and E. Matijevic, *J. Am. Ceram. Soc.*, *71*(10):845 (1988).
19. W. D. Kingery, in *Ceramic Powders* (P. Vincenzini, Ed.), Elsevier, New York, pp. 3–18 (1983).
20. R. J. Bratton, *Am. Ceram. Soc. Bull.*, *48*(8):759 (1969).
21. K. S. Mazdiyasni, R. T. Doloff, and J. S. Smith, Jr., *J. Am. Ceram. Soc.*, *52*(10):523 (1969).
22. S. Somiya, in *Concise Encyclopedia of Advanced Ceramic Materials* (R. J. Brook, Ed.), MIT Press, Cambridge, MA, 1991, pp. 375–377.
23. K. Masters, *Spray Drying*, 4th ed., Wiley, New York, 1985.
24. J. G. M. DeLau, *Am. Ceram. Soc. Bull.*, *49*(6):572 (1970).
25. G. L. Messing, S.-C. Zhang, and G. V. Jayanthi, *J. Am. Ceram. Soc.*, *76*(11):2707 (1993).
26. R. Schwartz, D. Eichart, and D. Payne, *Mater. Res. Soc. Symp. Proc.*, *73*:123 (1986).
27. D. W. Johnson, Jr., P. K. Gallagher, D. J. Nitti, and F. Schrey, *Am. Ceram. Soc. Bull.*, *53*(2):167 (1974).
28. M. W. Real, *Proc. Br. Ceram. Soc.*, *38*:59 (1986).
29. M. N. Rahaman, L. C. De Jonghe, S. L. Shinde, and P. H. Tewari, *J. Am. Ceram. Soc.*, *71*(7):C-338 (1988).
30. M. Pechini, U.S. Patent 3,330,697 (1967).
31. K. D. Budd and D. A. Payne, Better ceramics through chemistry, *Mater. Res. Soc. Symp. Proc.*, *32*:239 (1984).
32. L. A. Chick, L. D. Pederson, G. D. Maupin, J. L. Bates, L. E. Thomas, and G. J. Exarhos, *Mater. Lett.*, *10*(12):6 (1990).
33. A. J. Moulson, *J. Mater. Sci.*, *14*:1017 (1979).
34. K. S. Mazdiyasni and C. M. Cooke, *J. Am. Ceram. Soc.*, *56*:628 (1973).
35. K. S. Mazdiyasni and C. M. Cooke, U.S. Patent 3,959,446 (1976).
36. R. F. Strickland-Constable, *Kinetics and Mechanism of Crystallization*, Academic, London, 1968.
37. A. Kato, J. Hojo, and T. Watari, *Mater. Sci. Res.*, *17*:123 (1984).
38. S. Prochazka and C. Greskovich, *Am. Ceram. Soc. Bull.*, *57*(6):579 (1978).
39. D. Segal, *Chemical Synthesis of Advanced Ceramic Materials*, Cambridge Univ. Press, Cambridge, UK, 1989, p. 130.
40. J. S. Haggerty, in *Ultrastructure Processing of Ceramics, Glasses, and Composites* (L. L. Hench and D. R. Ulrich, Eds.), Wiley, New York, 1984, pp. 353–366.
41. C. M. Hollabaugh, D. E. Hull, L. R. Newkirk, and J. J. Petrovic, *J. Mater. Sci.*, *18*:3190 (1983).
42. J. H. Jean, D. M. Goy, and T. A. Ring, *Am Ceram. Soc. Bull.*, *66*:1517 (1987).

Synthesis of Powders

FURTHER READING

The following two conference proceedings, published at regular intervals, contain useful reports on powder preparation.

Advances in Ceramics: Ceramic Powder Science, American Ceramic Society, Westerville, OH.

Materials Research Society Symposium Proceedings: Better Ceramics Through Chemistry, Materials Research Society, Pittsburgh, PA.

3
Powder Characterization

3.1 INTRODUCTION

In Chapter 2 we considered the important methods for the preparation of ceramic powders. You may have observed that the quality of the powder depends on the preparation method. The powder characteristics, as discussed earlier, have a significant influence on the packing uniformity of the consolidated body and on the microstructural evolution of the fired body. Knowledge of the powder characteristics is therefore important. In practice, the extent to which the characterization process is taken depends on the intended application of the fabricated ceramic.

In the case of traditional ceramics, which do not have to meet exacting property requirements, a fairly straightforward observation, with a microscope, of the size, size distribution, and shape of the powders may be sufficient. For advanced ceramics, however, detailed knowledge of the powder characteristics is required for adequate control of the microstructure and properties of the fabricated material. Commercial powders are used in most applications. Normally, the manufacturer has carried out most of the characterization experiments and provides the user with the results, generally referred to as *powder specifications*. The manufacturer's specifications combined with a straightforward observation of the powder with a microscope are sufficient for many applications. For a powder prepared in the laboratory, a detailed set of characterization experiments may have to be carried out.

In the case of advanced ceramics, minor variations in the chemical composition and purity of the powder, as we have seen, can have profound effects on the microstructure and properties of the ceramic. This realization has led to a growing use of analytical techniques that have the capabil-

Powder Characterization

Table 3.1 Powder Characteristics That Have a Significant Influence on Processing

Physical characteristics	Chemical composition	Phase composition	Surface characteristics
Particle size and distribution	Bulk composition	Crystallinity	Surface structure
Particle shape	Minor elements	Phases	Surface chemistry
Degree of agglomeration	Trace impurities		
Surface area			
Density and porosity			

ity of detecting constituents, especially on the surfaces of the particles, at concentrations down to the parts per million level.

The important characteristics of a powder can be categorized into four groups: physical characteristics, chemical composition, phase composition, and surface characteristics. For these four groups, the main powder properties that have a significant influence on processing are summarized in Table 3.1. A bewildering number of techniques have been developed for the characterization of solids [1]. In this chapter we concentrate on those techniques that have broad applications to ceramic powders. The experimental details in performing the characterization will, in general, not be covered, as they are usually discussed at length in the manuals supplied by the manufacturers of the equipment. Instead, we concentrate on the principles of the methods, the range of information that can be obtained with them, and some of their limitations.

3.2 PHYSICAL CHARACTERIZATION

Powders consist of an assemblage of small units with certain distinct physical properties. These small units, loosely referred to as particles, can have a fairly complex structure. A variety of terms have been used to describe them, and this has led to some confusion in the literature. In this book we adopt, with minor modifications, the terminology proposed by Onoda and Hench [2].

A. Types of Particles

Primary Particles
A *primary particle* is a discrete, low-porosity unit and can be either a single crystal, a polycrystalline particle, or a glass. If any pores are present, they

are isolated from each other. A primary particle cannot, for example, be broken down into smaller units by ultrasonic agitation in a liquid. It may be defined as the smallest unit in the powder with a clearly defined surface. For a polycrystalline primary particle, the crystals have been referred to variously as crystallites, grains, or domains. In this book, we use the term *crystal*.

Agglomerates

An *agglomerate* is a cluster of primary particles held together by surface forces, by liquid, or by a solid bridge. Figure 3.1 is a schematic diagram of an agglomerate consisting of dense polycrystalline primary particles. Agglomerates are porous, with the pores being generally interconnected. They are classified into two types: soft agglomerates and hard agglomerates. *Soft agglomerates* are held together by fairly weak surface forces and can be broken down into primary particles by ultrasonic agitation in a liquid. *Hard agglomerates* consist of primary particles that are chemically bonded by solid bridges; they therefore cannot be broken down into primary particles by ultrasonic agitation in a liquid. As outlined in Chapter 2, hard agglomerates generally lead to the formation of microstructural defects in the fabricated ceramic.

Particles

When no distinction is made between primary particles and agglomerates, the term *particles* is used. Particles can be viewed as small units that move as separate entities when the powder is dispersed by agitation and

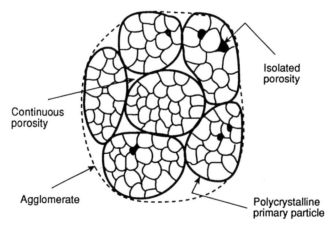

Figure 3.1 An agglomerate of dense polycrystalline primary particles.

Powder Characterization

consist of primary particles, agglomerates, or some combination of the two. Most particle size analysis techniques refer to such particles.

Granules

The term *granules* refers to large agglomerates ($\approx$100–1000 µm in size) that are deliberately formed by the addition of a granulating agent (e.g., a polymer-based binder) to the powder, followed by tumbling or spray drying. These large, nearly spherical agglomerates improve the flowability of the powder during filling and compaction in complex molds.

Flocs

Flocs are clusters of particles in a liquid suspension. The particles are held together weakly by electrostatic forces or by organic polymers and can be redispersed by appropriate modification of the interfacial forces through alteration of the solution chemistry. The formation of flocs is undesirable because it decreases the uniformity of the consolidated body.

Colloids

A *colloid* is any system consisting of a finely divided phase in a fluid. A colloidal suspension or sol consists of fine particles dispersed in a liquid. The particles, referred to as *colloidal particles*, undergo Brownian motion and have a slow (often negligible) sedimentation rate under normal gravity. The size range for colloidal particles is about 1 nm to 1 µm.

Aggregates

An *aggregate* is the coarse constituent in a mixture, which usually also contains a fine constituent called the *bond*. Pebbles in concrete are an example, with the fine cement particles forming the bond.

The approximate size ranges of the various types of particles are summarized in Table 3.2.

Table 3.2 Size Range for Particles in Ceramic Processing

Type of particle	Size range
Powder	
Colloidal particle	1 nm to 1 µm
Coarse particle	1–100 µm
Granule	100 µm to 1 mm
Aggregate	>1 mm

B. Particle Size and Particle Size Distribution

We have seen from Chapter 2 that ceramic powders generally consist of particles of different sizes distributed over a certain range. Some powders may have a very narrow particle size distribution (e.g., those prepared by chemical precipitation under controlled conditions), whereas for others the distribution in sizes may be very broad (e.g., a milled powder that has not been separated into narrow size fractions). Some particles are spherical or equiaxial (i.e., with the same length in each direction), but many are irregular in shape. Often we are required to characterize the particle size and particle size distribution of the powder because, as discussed earlier, these characteristics have an important effect on the consolidation and sintering of the powders.

For a spherical particle, the diameter is taken as the size. However, the size of an irregularly shaped particle is a rather uncertain quantity. We therefore need to define what "particle size" represents. One simple definition of the size of an irregularly shaped particle is that it is equivalent to the diameter of a sphere having the same volume as the particle. This is not much help, because in many cases the volume of the particle is ill defined or difficult to measure. Usually the particle size is defined in a fairly arbitrary manner in terms of a number generated by one of the measuring techniques described later. A particle size measured by one technique may therefore be quite different from that measured by another technique, even when the measuring instruments are operating properly.

Bearing in mind the uncertainty in defining the size of an irregularly shaped particle, we will now attempt to describe the particle size distribution. As a start, let us assume that the powder consists of N particles with sizes $D_1, D_2, D_3, \ldots, D_N$, respectively. We can calculate a mean size $\overline{D}$ and the standard deviation in the mean, S, according to the equations

$$\overline{D} = \sum_{i=1}^{N} \frac{D_i}{N} \tag{3.1}$$

and

$$S = \left(\sum_{i=1}^{N} \frac{(D_i - \overline{D})^2}{N} \right)^{1/2} \tag{3.2}$$

The value of $\overline{D}$ is taken as the particle size of the powder, and S gives a measure of the spread in the particle size distribution. In a random (Gaussian) distribution about two out of every three particles will have their sizes in the range $\overline{D} \pm S$.

Powder Characterization

Most likely the characterization technique will sort the particles into a small number n of size categories, where n is much smaller than N. The technique may also produce a count of the number of particles within each category, so that there would be n_1 particles in a size category centered about D_1, n_2 of size D_2, etc., and n_n of size D_n. Alternatively, the mass or the volume of particles within each size category may be obtained. For convenience, we will consider the number of particles within a size category giving rise to a number-weighted average; the representation of the data in terms of the mass or volume of particles within a size category will follow along similar lines. We can use the data to determine the mean size and the standard deviation according to the equations

$$\overline{D} = \frac{\sum_{i=1}^{n} n_i D_i}{\sum_{i=1}^{n} n_i} \tag{3.3}$$

and

$$S = \left(\frac{\sum_{i=1}^{n} n_i (D_i - \overline{D})^2}{\sum_{i=1}^{n} n_i} \right)^{1/2} \tag{3.4}$$

For comparison, the volume weighted average is

$$\overline{D}_v = \frac{\sum_{i=1}^{n} v_i D_i}{\sum_{i=1}^{n} v_i} = \frac{\sum_{i=1}^{n} n_i D_i^4}{\sum_{i=1}^{n} n_i D_i^3}$$

The reader will realize that the mean particle size determined from Eq. (3.3) is the arithmetic mean. This is not the only mean size that can be defined; others include the geometric mean and the harmonic mean.

The geometric mean, D_g, is the nth root of the product of the diameter of the n particles and is given by

$$\log D_g = \frac{\sum_{i=1}^{n} n_i \log D_i}{\sum_{i=1}^{n} n_i} \tag{3.3a}$$

The harmonic mean, D_h, is the number of particles divided by the sum of the reciprocals of the diameters of the individual particles, i.e.,

$$D_h = \frac{\sum_{i=1}^{n} n_i}{\sum_{i=1}^{n} \frac{n_i}{D_i}} \tag{3.3b}$$

A simple and widely used way to describe the data is in terms of a histogram, showing the percent or fraction of particles within a size category. As an example, consider the data given in Table 3.3. A representation of the data in terms of a histogram is shown in Fig. 3.2. It is also common to plot the cumulative size distribution by summing the percent (or fraction) of particles finer than a given size, defined as the cumulative number percent finer (CNPF), or larger than a given size, defined as the

Table 3.3 Particle Size Distribution Data

Size range (μm)	Number in size range, N_i	Size D_i (μm)	Number fraction	CNPF[a]	$f_N(D)$[b]
<10	35	5	0.005	0.5	0.001
10–12	48	11	0.007	1.2	0.003
12–14	64	13	0.009	2.1	0.004
14–16	84	15	0.012	3.3	0.005
16–18	106	17	0.015	4.8	0.007
18–20	132	19	0.019	6.7	0.009
20–25	468	22.5	0.067	13.4	0.012
25–30	672	27.5	0.096	23.0	0.019
30–35	863	32.5	0.124	35.4	0.025
35–40	981	37.5	0.141	49.5	0.028
40–45	980	42.5	0.141	63.6	0.028
45–50	865	47.5	0.124	76.0	0.025
50–55	675	52.5	0.097	85.7	0.020
55–60	465	57.5	0.067	92.4	0.006
60–70	420	65	0.060	98.4	0.006
70–80	93	75	0.013	99.7	0.001
80–90	13	85	0.002	99.9	0.001
90–100	1	95	0	99.9	—
>100	4	—	0.001	100	—

[a] CNPF = cumulative number percent finer; see text.
[b] $D = 40$ μm [Eq. (3.3)]; $S = 14$ μm [Eq. (3.4)].
Source: Ref. 4.

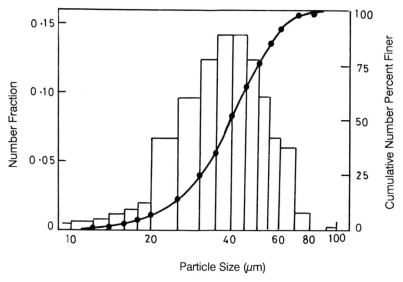

Figure 3.2 Particle size distribution data of Table 3.3 represented in terms of a histogram and the cumulative number percent finer (CNPF) than a given size.

cumulative number percent larger (CNPL). The CNPF is also shown for the data of Table 3.3 in Fig. 3.2, where a smooth curve is drawn through the data points.

In many cases it is necessary to provide a more complete description of the particle size distribution in terms of a mathematical equation, the parameters of which can be used to compare different powders. If the number of particles were fairly large and the size intervals ΔD, were small enough, then we would be able to fit a reasonably smooth curve through the particle size distribution data. Normally, the cumulative size distribution data are used as the starting point, and a smooth curve is fitted through the data. We are now assuming that the cumulative size distribution is some smoothly varying function of D, which we denote as $F_N(D)$. The fractional size distribution function, $f_N(D)$, is then obtained from $F_N(D)$ by taking the derivative:

$$f_N(D) = \frac{d}{dD} F_N(D) \qquad (3.5)$$

where $f_N(D)dD$ is the fraction of particles with sizes between D and $D + dD$.

The function $f_N(D)$ represents the measured size distribution, and this is used for the description of the data in terms of a mathematical equation. Figure 3.3 shows the measured size distribution determined from the data graphed in Fig. 3.2. Usually, the measured size distribution function is fitted in terms of an expected size distribution, such as the normal distribution:

$$f_N(D) = \frac{1}{S\sqrt{2\pi}} \exp\left[-\frac{1}{2}\left(\frac{D - \overline{D}}{S}\right)^2\right] \quad (3.6)$$

where $\overline{D}$ is the mean particle size and S is the standard deviation in the mean particle size. In practice, the values of $\overline{D}$ and S are adjusted in a curve-fitting routine until the best fit to the data is obtained. Since we are assuming that the data are smoothly varying functions, the mean size and standard deviation must now be defined in terms of equations appropriate to the normal distribution:

$$\overline{D} = \frac{\int_{-\infty}^{\infty} D f_N(D) dD}{\int_{-\infty}^{\infty} f_N dD} \quad (3.7)$$

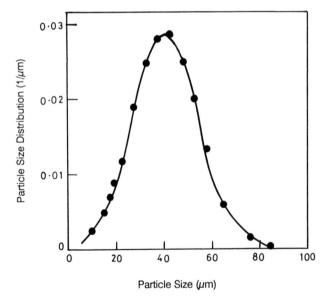

Figure 3.3 The particle size distribution data of Fig. 3.2 fitted to a smooth curve.

Powder Characterization

$$S = \left[\frac{\int_{-\infty}^{\infty} f_N(D)(D - \overline{D})^2 dD}{\int_{-\infty}^{\infty} f_N(D) dD} \right]^{1/2} \tag{3.8}$$

A serious problem with the normal distribution is that it predicts a finite fraction of particles of sizes less than zero. Furthermore, the largest particle size does not have a limit and is infinite. The normal distribution therefore seldom gives a good fit to real particle size data.

A better equation that has been found to give a good approximation to the size distribution of powders prepared by spray drying or by mechanical milling is the log-normal distribution:

$$f_N(\ln D) = \frac{1}{S\sqrt{2\pi}} \exp\left[-\left(\frac{\ln D - \overline{D}}{2S} \right)^2 \right] \tag{3.9}$$

where $\overline{D}$ is now the mean of the natural logarithm of the particle sizes and S is the standard deviation of the natural logarithm of the particle sizes. Unlike the normal distribution, the log-normal distribution accounts for particle sizes greater than zero; however, the largest size is still unlimited.

An empirical function that has been observed to describe the particle size distribution of milled powders is the Rosin–Rammler equation. The equation has undergone a number of modifications; one form is [4]

$$F_M(D) = \frac{\alpha D^{\alpha - 1}}{D_{RR}^\alpha} \exp\left[-\left(\frac{D}{D_{RR}} \right)^\alpha \right] \tag{3.10}$$

where $F_M(D)$ is the mass fraction of particles with sizes between D and $D + dD$, D_{RR} is the Rosin–Rammler modulus, and α is an empirical factor known as the distribution modulus (typically between 0.4 and 1.0). In Eq. (3.10), D_{RR} represents the largest particle size. Another empirical distribution function that has been used to describe the size distribution of milled powders is the Gates–Gaudin–Schuhman equation,

$$F_M(D) = \alpha \frac{D^{\alpha - 1}}{D_{\max}^\alpha} \tag{3.11}$$

where $D_{\max}$ is the maximum particle size. Unlike the log-normal distribution, both the Rosin–Rammler and Gates–Gaudin–Schuhman equations have a finite size for the largest particle.

C. Particle Shape

As outlined earlier, particle shape influences the packing of the powder. It can also influence the flowability of the powder in the filling of dies during compaction. However, except for the fairly simple geometries such as a sphere, cube, or cylinder, the quantitative characterization of particle shape can be fairly complex [5]. Most powder particles have complicated shapes. The shape of a particle is commonly described in terms of the *shape factor*, which provides some measure of the deviation from an idealized geometry such as a sphere or a cube. For elongated particles, the most common way of representing the shape is in terms of the *aspect ratio*, defined as the ratio of the longest dimension to the shortest dimension. For powders, two different definitions have been used for the shape factor. The first uses the sphere as a reference geometry, and the shape factor is defined as

$$\text{Shape factor} = \frac{\text{surface area of the particle}}{\text{surface area of a sphere with the same volume}} \quad (3.12)$$

According to this definition, the shape factor of a sphere is unity; the shape factor for all other shapes is greater than unity [e.g., $(6/\pi)^{1/3}$ or 1.24 for a cube]. The second definition uses two separate shape factors, one for the area and the other for the volume. According to this definition, the area of the particle is

$$A = F_A D^2 \quad (3.13)$$

where F_A is the area shape factor and D is the size of the particle measured by one of the techniques described later. Similarly, the volume of the particle is

$$V = F_V D^3 \quad (3.14)$$

where F_V is the volume shape factor. According to this definition, the area and volume shape factors for the sphere are π and $\pi/6$, respectively, whereas for the cube the area and volume shape factors are 6 and 1, respectively.

In summary, the quantitative description of the shape of an irregularly shaped particle is fairly complex. Although various shape factors can be defined, they have not proved to be useful practically. The trend in ceramic processing is toward the use of spherical or equiaxial particles because they provide better uniformity of the consolidated body. The driving force for the sintering of powders, as we outlined earlier, is the reduction in surface free energy. In practice, the use of spherical or equiaxial particles

Powder Characterization

combined with the direct measurement of the surface area of the particles (using one of the methods described later) provides a far more effective approach than the detailed quantification of the shapes of irregularly shaped particles.

D. Measurement of Particle Size and Size Distribution

Table 3.4 shows the common methods used for the measurement of particle size and particle size distribution. The main features of these methods are described below.

Microscopy

Microscopy is a fairly straightforward technique that offers the advantage of direct measurement of the particle size coupled with simultaneous observation of the particle shape and the extent of agglomeration. It usually forms the first step in the characterization of ceramic powders. Optical microscopes can be used for particle sizes down to ≈ 1 μm, and electron microscopes can extend the range down to ≈ 1 nm. Normally, the sample is prepared by adding a small amount of the powder to a liquid to produce a dilute suspension. After suitable agitation (e.g., with an ultrasonic probe), a drop of the suspension is placed on a glass slide or a microscope stub. Evaporation of the liquid leaves a deposit that is viewed in the microscope. Good separation of the particles must be achieved. For powders with a fairly wide distribution of sizes, care must also be taken to ensure that the deposit is representative of the original powder batch.

Particle size measurements are usually made from micrographs. The method is extremely tedious if it is done manually. Normally a large num-

Table 3.4 Common Methods for the Measurement of Particle Size and Their Approximate Range of Applicability

Method	Range (μm)
Microscopy	
Optical	>1
Scanning electron	>0.1
Transmission electron	>0.001
Sieving	5–1000
Sedimentation	0.1–100
Light scattering	0.1–1000
Coulter counter	0.5–400
X-ray line broadening	<0.1

ber of particles (a few hundred) and two diameters of each particle (in orthogonal directions) must be measured. Automatic image analysis of micrographs or an electronic display can reduce the amount of work considerably. As would be apparent, microscopy produces a particle size distribution based on the number of particles within an appropriate size range.

Sieving

The use of sieves for separating particles into fractions with various size ranges is the oldest of the classification methods. The particles are classified in terms of their ability or inability to pass through an aperture having a controlled size. Sieves with openings greater than 37 μm are constructed with wire mesh and are identified in terms of a mesh size and a corresponding aperture size. The mesh size is equal to the number of wires per linear inch of the sieve screen, for example, a 325 mesh sieve has 325 wires per inch. The screen has square apertures, the size of which depends on both the number of wires per linear dimension and the size of the wire. Special metal sieves can have openings as small as 1 μm.

Sieving is most often used for powders with a significant fraction of particles greater than ≈40 μm. Typically, several sieves are stacked together, with the coarsest mesh aperture at the top and the smallest at the bottom. The powder is placed on the top screen, and the stack is shaken for several minutes to produce separation of the various fractions. Agglomeration of the powder and clogging of the screens during sieving of a dry powder can lead to significant problems below ≈40 μm. The use of pulsed jets of air to reduce clogging or wet sieving in which the particles are dispersed in a liquid can alleviate the problems.

Adequate separation of the particles into their true size fractions requires a fairly long time. Since it is normally too time consuming to sieve for such a long time, the particle size distribution obtained from most sieving operations is only approximate. However, this approximate size characterization can be quite useful for the selection or verification of raw materials in the traditional ceramic industry. The various fractions produced by sieving are weighed so that a particle distribution based on mass is obtained. For elongated particles, sieving generally favors measurement of the longer particle dimension. Sieving is not feasible for powders of advanced ceramics, where the particle size is normally less than 1 μm. Furthermore, it should not be used for clean powders because of the expected contamination with metallic impurities from the sieves.

Sedimentation

A spherical particle falling in a viscous liquid with a sufficiently small velocity soon reaches a constant velocity called the *terminal velocity*,

Powder Characterization

where the effective weight of the particle is balanced by the frictional force exerted on it by the liquid. The frictional force, F, on the particle is given by Stokes' law:

$$F = 6\pi\eta a v \tag{3.15}$$

where η is the viscosity of the liquid, a is the radius of the particle, and v is the terminal velocity. Equating F to the effective weight of the particle gives

$$v = \frac{D(\rho_s - \rho_l)g}{18\eta} \tag{3.16}$$

where D is the diameter (size) of the sphere, g is the acceleration due to gravity, and ρ_s and ρ_l are the densities of the particle and the liquid, respectively. Equation (3.16) is normally referred to as Stokes' equation, which is not to be confused with Stokes' law, Eq. (3.15). Measurement of the sedimentation rate can therefore be used to determine the particle size from Stokes' equation. When particle shapes other than spheres are tested, the measured particle size is referred to as the Stokes diameter or the equivalent spherical diameter.

Determination of particle size from Eq. (3.16) has a limited range of validity. Stokes' law (Eq. 3.15) is valid for laminar or streamline flow (i.e., where there is no turbulent flow) and assumes that there are no collisions or interactions between the particles. The transition from turbulent to laminar flow is governed by the Reynolds number (Re)

$$\text{Re} = \frac{v\rho_l D}{\eta} \begin{cases} >0.2 \text{ turbulent flow} \\ <0.2 \text{ laminar flow} \end{cases} \tag{3.17}$$

Laminar flow is found to be restricted to Reynolds numbers of less than 0.2. For Al_2O_3 particles dispersed in water at room temperature, the restriction to laminar flow means that the method is reliable for particle sizes below ≈ 100 μm. For sufficiently small particles undergoing sedimentation under gravity, Brownian motion resulting from collisions with the molecules of the liquid may displace the particle by a measurable amount. This effect puts a lower limit to the use of gravitational settling in water at ≈ 1 μm. However, faster settling in a centrifuge can extend the size range down to ≈ 0.1 μm.

Sedimentation methods can be divided into two categories: (i) those that measure the cumulative sedimentation through a fixed distance as a function of time and (ii) those that measure, also as a function of time, the concentration of particles in a plane at some fixed height in a dilute suspension. In the first category of methods, a dilute suspension is carefully placed on a column of settling liquid and the fraction of particles that settle to a fixed depth is measured as a function of time. The second category of methods is sometimes referred to as turbidimetry. The concentration of the particles at a fixed height in the liquid is found by measuring the turbidity of the liquid, usually by an optical or X-ray technique. A light beam or an X-ray beam penetrates a glass cell containing a suspension of the powder (usually in water). As the powder settles, the intensity of the transmitted beam increases according to the equation

$$\frac{I}{I_0} = \exp(-KACx) \qquad (3.18)$$

where I_0 and I are the intensities of the incident and transmitted beams, respectively; K is a constant called the extinction coefficient; A is the projected area per unit mass of particles; C is the concentration by mass of the particles; and x is the length of the light path through the suspension. The fraction of particles by mass is found by measuring the intensity ratio and using Eqs. (3.16) and (3.18).

Light Scattering

The scattering of light from a monochromatic source can be used to determine the particle size and particle size distribution using a principle very different from that described above for turbidimetry. When a beam of light strikes an assembly of particles, some of it is transmitted, some is absorbed, and some is scattered. As shown schematically in Fig. 3.4, for

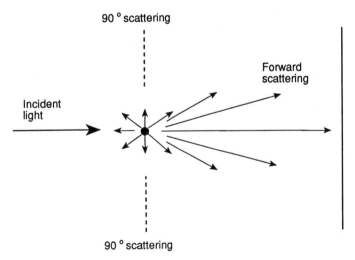

Figure 3.4 Forward scattering of light by small particles.

Powder Characterization

sufficiently small particle sizes, a greater intensity of the scattered light occurs at low angles, so most scattering is in the forward direction. Particle size data can be found by measuring the variation in the intensity of the forward-scattered light. Relating the particle size data to the measurement of the intensity of the scattered light, however, is governed by rigorous diffraction theory. Usually, three limiting cases, stated in terms of the ratio of the particle size D to the wavelength λ of the incident light are considered:

1. $D \ll \lambda$ Rayleigh scattering theory
2. $D \approx \lambda$ Mie theory
3. $D \gg \lambda$ Fraunhofer diffraction theory

Most instruments for measurement of particle size by light scattering are based on the Fraunhofer diffraction theory. According to this theory, the intensity of the light scattered by particles is proportional to the particle size. However, the size of the diffraction pattern (the scattering angle θ) is inversely proportional to the particle size:

$$\sin \theta = \frac{1.22\lambda}{D} \tag{3.19}$$

Figure 3.5 shows examples of the light scattered from two particles of different sizes. Smaller particles scatter a small, definite amount of light through a fixed but larger angle. Conversely, larger particles scatter a greater amount of light but through a smaller angle.

Most light-scattering instruments for particle size analysis employ measuring techniques that give the particle size based on the volume or

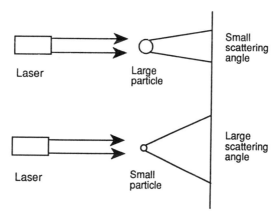

Figure 3.5 Scattering of light by small and large particles.

projected area of the particle. The light source is usually a laser, e.g., a helium-neon laser emitting light with a wavelength of 0.63 μm. For instruments based on the Fraunhofer diffraction theory, a reliable size range is ≈2–100 μm. However, the use of special light collection systems and the Mie theory can extend the range to ≈0.1–1000 μm.

In practice, the powder can be analyzed in the dry state or, more usually, as a dilute suspension in a liquid. Light-scattering methods have the advantages that the measurements can be made accurately and fairly rapidly, only a small amount of the sample is required, and the results can be recorded automatically.

Electrical Sensing Zone Techniques (The Coulter Counter)

In the Coulter counter, the number and size of particles suspended in an electrolyte are measured by causing the particles to flow through a narrow orifice on either side of which an electrode is immersed (Fig. 3.6). As a particle passes through the orifice, it displaces an equivalent volume of the electrolyte and causes a change in the electrical resistance, the magnitude of which is proportional to the volume of the particle. The changes in resistance are converted to voltage pulses which are amplified, sized, and counted to produce data for the particle size distribution of the suspended particles.

As outlined above, the particle size data are based on the volume of the particle. However, the effect of particle shape on the data is subject

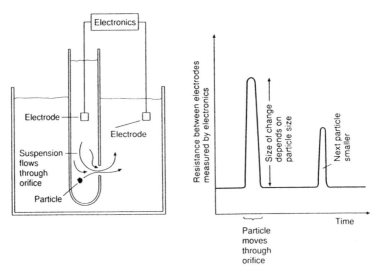

Figure 3.6 Electrical sensing zone technique (Coulter counter) for particle size determination. (From Ref. 4, with permission.)

Powder Characterization

to some doubt. It is generally believed that the parameter that is measured corresponds to the overall envelope of the particle. The method is suitable for particle sizes in the range of 1–100 μm, but with care and the use of multiple orifices the range can be extended to 0.5–400 μm. The method is most sensitive when the particle size is close to the diameter of the orifice. However, blocking of the orifice by larger particles can be a tedious problem. Originally applied to the counting of blood cells, the Coulter counter technique has become fairly popular for particle size analysis of ceramic powders.

X-Ray Line Broadening

The broadening of X-ray diffraction peaks provides a convenient method for measuring particle sizes below ≈0.1 μm. The origins of the X-ray line broadening effect, like many other diffraction effects of crystals, is best understood in terms of the reciprocal lattice. However, a discussion of the reciprocal lattice and its use in visualizing diffraction effects is beyond the scope of this book. We therefore refer the reader to the text by Cullity [6] for a discussion. Essentially, the width of the diffraction peak (or the size of the diffraction spot) is inversely proportional to a parameter that is a measure of the size of the crystal. As the size of the crystal increases, the width of the peak decreases. If the crystal gets too large—greater than ≈0.1 μm in average diameter—the peaks are so narrow that their width cannot be distinguished from the broadening inherent in the X-ray diffraction instrument. In addition to size, other factors such as lattice strain can produce broadening of the X-ray peaks. However, for particle sizes below ≈20 nm, the broadening due to size dominates.

Figure 3.7 shows the X-ray diffraction profiles for the (111) peak of a ZrO_2 powder doped with 3 mol % Y_2O_3. The powder was prepared by hydrothermal synthesis, and the particle size, as determined by TEM, was 10–15 nm. Heating the powder to higher temperatures leads to coarsening and a corresponding decrease in the peak width. Eventually, the particles become large enough that any change in the peak width is too small to be measured. The size of the crystal, D, is obtained from the Scherrer equation,

$$D = \frac{C' \lambda}{\beta \cos \theta} \tag{3.20}$$

where C' is a constant (≈0.9), λ is the wavelength of the X-rays, β is the width of the peak at half-maximum corrected for instrumental broadening, and θ is the Bragg angle of reflection. Normally, β is obtained from the equation

$$\beta^2 = \beta_m^2 - \beta_s^2 \tag{3.21}$$

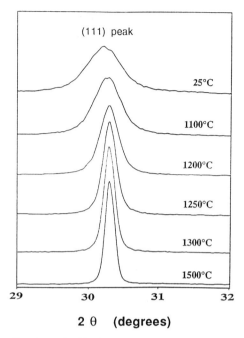

Figure 3.7 X-ray line broadening for a $ZrO_2/3$ mol % Y_2O_3 powder prepared by hydrothermal synthesis (particle size 10–15 nm) after heating to the temperatures indicated.

where β_m is the measured half-width and β_s is the half-width of a standard sample of the same material having a crystal size greater than ≈ 1 μm. Most modern diffractometers have computer software for extracting the crystal size from the measured half-widths of the peaks. For the powder shown in Fig. 3.7, the average particle size of the synthesized powder as determined by X-ray line broadening was 13 nm, which agrees very well with the value obtained by TEM.

The reader may have noticed that we have used the term *crystal size* to describe the size obtained by X-ray line broadening. This is because the technique gives a measure of the size of the crystals regardless of whether the particles consist of single crystals, polycrystals, or agglomerates. If primary particles consist of single crystals, as in the Y_2O_3-doped ZrO_2 powder described earlier, then the crystal size determined by X-ray line broadening will be comparable to the particle size determined by other methods such as electron microscopy. For polycrystalline primary

Powder Characterization

particles or agglomerates, it will be much smaller than the size determined by other methods.

Particle size can be obtained as an average value from measurement of the surface area of the powder when some assumptions are made concerning the particle shape and the presence or absence of pores. The surface area of fine powders is important in its own right, so we consider this property separately below.

E. Surface Area

Techniques for the determination of the surface area of a powder are based on the phenomenon of gas adsorption. Adsorption processes are generally classified into two categories: physical adsorption and chemical adsorption. A physically adsorbed gas is bonded to the surface primarily by weak van der Waals forces. Chemically adsorbed gases, on the other hand, will have developed a strong chemical bond with the surface.

In the determination of the surface area, the amount of gas adsorbed is measured at a fixed temperature for different gas pressures P. The most common adsorbate is liquid nitrogen at its boiling point (77K). Normally, a known volume of gas is contacted with the powder and the gas laws are used to determine the volume V (or mass) of gas removed from the system due to adsorption. A graph of V versus P, referred to as the adsorption isotherm, is plotted. If the gas is at a pressure below its saturation vapor pressure P_0, then the relative pressure P/P_0 is used instead of P. As outlined below, further manipulation of the adsorption data gives the monolayer capacity V_m, defined as the amount of gas required to cover the surface of the powder with a monolayer, from which the surface area is obtained.

The first equation relating the volume of adsorbed gas to the equilibrium gas pressure was developed by Langmuir. The Langmuir equation is more widely applicable to chemisorption of a monolayer and is inappropriate for multilayer adsorption characteristic of physical adsorption. According to the Langmuir model, the rate of condensation $\dot{C}$ on the surface is proportional to the number of unoccupied sites, S_0, and to the gas pressure P; i.e.,

$$\dot{C} = K_c P S_0 \tag{3.22}$$

where K_c is a constant. The rate of evaporation of the gas from the surface, $\dot{E}$, is proportional only to the number of occupied surface sites S_1 of a total number S; i.e.,

$$\dot{E} = K_e S_1 \tag{3.23}$$

where K_e is a constant. At equilibrium, the rates $\dot{C}$ and $\dot{E}$ are equal. Equating Eqs. (3.22) and (3.23) and defining Θ as the fraction of surface covered with adsorbed molecules, i.e., $\Theta = S_1/S$, then

$$\Theta = \frac{bP}{1 + bP} \tag{3.24}$$

where b, equal to K_c/K_e, is known as the Langmuir constant. Equation (3.24) is known as the Langmuir adsorption isotherm. Alternatively, Θ may be replaced by V/V_m, where V is the volume of gas adsorbed at a gas pressure P and V_m is the volume of gas required to form one monolayer, so that

$$\frac{V}{V_m} = \frac{bP}{1 + bP} \tag{3.25}$$

Equation (3.25) is usually written in the form

$$\frac{P}{V} = \frac{1}{bV_m} + \frac{P}{V_m} \tag{3.26}$$

A plot of P/V versus P yields the monolayer capacity V_m.

To relate V_m to the surface area, it is necessary to know the area occupied by an adsorbed gas molecule, σ, which is usually taken from the density of the liquefied gas. The surface area can then be calculated from

$$S_w = \frac{N_A \sigma V_m}{V_0} \tag{3.27}$$

where the symbols, expressed in their most commonly used units, are defined as follows:

S_w = the specific surface area (m²/g)
N_A = the Avogadro number (6.023×10^{23} mol^{-1})
σ = area of an adsorbed gas molecule (16.2×10^{-20} m² for nitrogen)
V_m = monolayer volume (cm³/g)
V_0 = the volume of 1 mol of the gas at STP (22,400 cm³/mol)

Substituting these values into Eq. (3.27) gives, for nitrogen adsorption,

$$S_w = 4.35 V_m \tag{3.28}$$

As discussed earlier, the Langmuir model is more appropriate to monolayer chemical adsorption. A major advance was made in 1938 by S. Brunauer, P. H. Emmett, and E. Teller, who analyzed the physical adsorption of multilayers of gases on solid surfaces. The method, referred to as the BET method, is now the basis for surface area analysis of powders. The theory and practice of the BET method is considered in many

Powder Characterization

books [3]. We will not repeat the discussion; instead, only a general consideration is provided.

The important equation for the BET adsorption isotherm is usually expressed in the form

$$\frac{1}{V(P/P_0 - 1)} = \frac{1}{V_m c} + \frac{c-1}{V_m c}\left(\frac{P}{P_0}\right) \quad (3.29)$$

where c is a constant and the other terms are as defined earlier. As outlined earlier, V is measured at a constant temperature (usually the boiling point of liquid nitrogen) for various values of P/P_0 (usually in the range 0.05–0.3). A plot of the left-hand side of Eq. (3.29) versus P/P_0 should yield a straight line with a slope s and intercept i given by

$$s = \frac{c-1}{V_m c} \quad (3.30)$$

and

$$i = \frac{1}{V_m c} \quad (3.31)$$

The monolayer volume V_m can therefore be obtained from

$$V_m = \frac{1}{s+i} \quad (3.32)$$

Finally, the specific surface area S_w is obtained from Eq. (3.27) or, for nitrogen adsorption, from Eq. (3.28).

Although any gas may be used as the adsorbate, the standard technique employs the adsorption of nitrogen at the boiling point of liquid nitrogen. This works well for powders with a specific surface area greater than a few square meters per gram. For powders with somewhat lower surface areas, the use of krypton at the boiling point of liquid nitrogen is recommended. Although there is some disagreement as to the correct value for the area occupied by a krypton molecule, most investigations assume a value of 18.5×10^{-20} m^2.

As discussed earlier, the average particle size can be obtained from the surface area data. Assuming that the powder is unagglomerated and that the particles are spherical and dense, the particle size D can be estimated from the equation

$$D = \frac{6}{S_w \rho_s} \quad (3.33)$$

where ρ_s is the density of the solid. Large differences between the particle size value determined from the surface area and that measured by other techniques may be due to a number of factors, including agglomeration of the powder, the presence of porosity in the particles, and a large deviation from spherical shape.

F. Porosity of Particles

Many powders consist of agglomerates that are highly porous. Sometimes it is necessary to characterize, quantitatively, the porosity and pore size distribution of the agglomerates. For pores that are accessible (i.e., not isolated from the external surface of the sample), two methods have been used: gas adsorption and, more commonly, mercury intrusion porosimetry (referred to simply as mercury porosimetry). These two methods are also used to characterize the porosity and pore size distribution of compacted powders, and we shall refer to them in Chapter 6 when we consider powder consolidation methods. For isolated pores in particles, liquid or gas pycnometry is used.

Gas Adsorption

As noted above in discussing the measurement of surface area, at low gas pressures the adsorbed gas is present as an incomplete monolayer. At higher pressures, multilayer coverage, which forms the basis for the BET method, occurs. At still higher pressures, the gas may condense to a liquid. This condensation at higher gas pressures can be used to determine the pore size and pore size distribution in a porous material.

The adsorption of a gas onto a porous solid is favored in the small capillaries, resulting in the condensation of the gas to a liquid. If condensation occurs in cylindrical capillaries with a radius up to a value r, then the gas pressure P in the capillaries of radius r is governed by the Kelvin equation

$$\ln \frac{P}{P_0} = \frac{-2\gamma_{lv} V_L \cos \theta}{RTr} \tag{3.34}$$

where P_0 is the saturation gas pressure of the liquid over a plane surface, γ_{lv} is the surface tension of the liquid–vapor interface, V_L is the molar volume of the liquid, θ is the contact angle between the liquid and the pore wall, R is the gas constant [8.314 J/(K·mol)], and T is the absolute temperature. For nitrogen at the boiling point of liquid nitrogen (77 K), $\gamma_{lv} = 8.72 \times 10^{-3}$ N/m^2, $V_L = 34.68 \times 10^{-6}$ m^3/mol, and θ is assumed to be zero, so that

$$r = \frac{4.05 \times 10^{-4}}{\log(P/P_0)} \, \mu\text{m} \tag{3.35}$$

Typical values for P/P_0 may range from 0.6 to 0.99, indicating that gas adsorption is appropriate for the pore size range of ≈2–100 nm.

Pore volume and pore size distribution can be determined from the adsorption or desorption isotherm in which the volume of gas adsorbed or desorbed is measured as a function of P/P_0. If the amount of gas adsorbed on the external surface is small compared to that adsorbed in the pores, the total pore volume is the condensed volume adsorbed at the saturation pressure. With many adsorbents, a hysteresis loop occurs between the adsorption and desorption isotherms (Fig. 3.8). This has been explained in terms of capillary condensation augmenting multilayer adsorption at the pressures at which hysteresis is present [3].

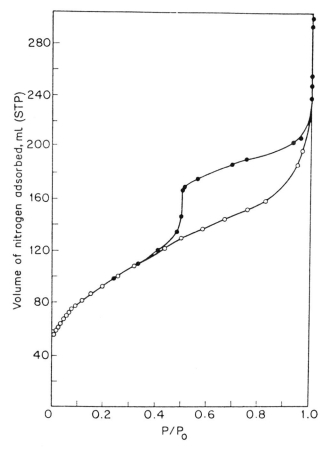

Figure 3.8 Nitrogen adsorption and desorption data used in the determination of pore size distribution. (From Ref. 3.)

The simplest procedure for determining the pore size distribution starts with the adsorption or desorption isotherm. In practice, the desorption isotherm is usually used. At a given value of P/P_0, if the volume of gas desorbed, in cm^3 per gram at standard temperature and pressure (STP) is V_d, then the condensed volume of the desorbed gas is $V_c = MV_d/(\rho V_L)$ where M is the molecular weight of the adsorbate, ρ is the density of the liquified gas at its saturated vapor pressure and V_L is the molar gas volume at STP. For nitrogen M = 28 g and ρ = 0.808 g/cm^3. Since V_L = 22.4 liters, $V_c = 1.547 \times 10^{-3} V_d$. The radius of the pore, r_c, at a given value of P/P_0 can be found from Eq. (3.35). In this way, for each value of P/P_0 both V_c and r_c can be determined from the desorption isotherm. The pore size distribution curve, i.e., $v(r)$, versus r_c, is found by differentiating the distribution for V_c as a function of r_c.

Mercury Porosimetry

Compared to the gas adsorption method described above, mercury porosimetry is a more widely used technique for the characterization of porosity in a sample. It provides a more direct method for the measurement of pore size and pore size distribution. Mercury porosimetry is based on the phenomenon of capillary rise (Fig. 3.9). A liquid that wets the walls of a narrow capillary (contact angle $\theta < 90°$) will climb up the walls of the capillary. However, the level of a liquid that does not wet the walls of a capillary ($\theta > 90°$) will be depressed. For the case of a nonwetting liquid, a pressure must be applied to cause the liquid to flow up the capillary to the level of the reservoir. For a capillary with principal radii of curvature

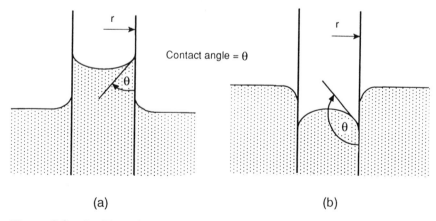

Figure 3.9 Capillary rise phenomena for (a) a wetting liquid (contact angle <90°), and for (b) a nonwetting liquid (contact angle >90°).

Powder Characterization

r_1 and r_2 in two orthogonal directions, this pressure is given by the equation of Young and Laplace:

$$P = -\gamma_{lv} \left(\frac{1}{r_1} + \frac{1}{r_2} \right) \cos \theta \tag{3.36}$$

where γ_{lv} is the surface tension of the liquid–vapor interface. For a cylindrical capillary of radius r, Eq. (3.36) becomes

$$P = \frac{-2\gamma_{lv} \cos \theta}{r} \tag{3.37}$$

For mercury, γ_{lv} is assumed to be 0.48 N/m and θ is assumed to be in the range 130–140°. Using $\theta = 130°$, then for P expressed in megapascals (MPa) and r in micrometers (μm),

$$r = 0.436/P \tag{3.38}$$

In mercury porosimetry, a sample holder is partially filled with the powder, evacuated, and then filled with mercury. Since mercury does not wet most ceramic materials, pressure must be applied to force it into the pores. In the experiment, the volume of mercury intruded into the sample, V_i, is measured as a function of the applied pressure P. Figure 3.10 shows data taken from Ref. 3. Assuming that the pores are cylindrical, the pressure required to fill pores larger than a radius r is given by Eq. (3.38). Using this relation between P and r, the data are replotted to produce a graph of V_i as a function of r (Fig. 3.11), sometimes referred to as the cumulative pore size distribution by volume. It is more common to express the results as a relative pore size distribution by volume, normally referred to simply as the pore size distribution. If $v(r)dr$ is the volume of pores with radius between r and $r + dr$, then

$$v(r) = \frac{dV}{dr} \tag{3.39}$$

that is, pore size distribution $v(r)$ is obtained by differentiating the cumulative pore size distribution (Fig. 3.11). The function $v(r)$ can also be obtained directly from the data for V_i versus P, using

$$v(r) = \frac{dV}{dP} \frac{P}{r} \tag{3.40}$$

A derivation of Eq. (3.40) from first principles is given in Ref. 4.

The pressures available in mercury porosimeters are such that pore sizes in the range of ≈2 nm to 200 μm can be measured. It should be

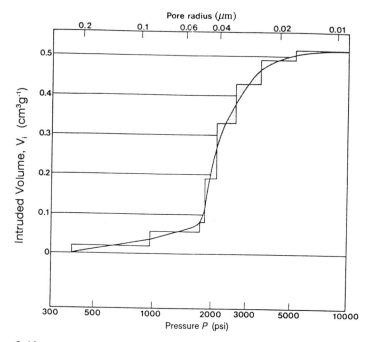

Figure 3.10 Mercury porosimetry data showing the intruded volume V_i versus pressure. Steps indicate changes in the intruded volume for finite changes in the pressure. It is preferable to fit a smooth curve through the data. (From Ref. 3.)

remembered that Eq. (3.37) assumes that the pores are circular in cross section, which may hardly be the case in practice. Furthermore, for ink-bottle pores, i.e., pores with constricted necks opening to large volumes, the use of Eq. (3.37) gives a measure of the neck size that is not truly indicative of the actual pore size. These types of pores can also give rise to hysteresis because they fill at a pressure characteristic of the neck size whereas they empty at a pressure characteristic of the larger volume of the pore. Finally, the compressibilities of the mercury and the sample are assumed to be independent of the applied pressure, which may not be the case at fairly high pressure.

Pycnometry

As mentioned earlier, pycnometry can be used to determine the isolated porosity in particles. In practice, the apparent density of the particles, ρ_a, defined as the overall density of the solid plus the isolated porosity, is measured. Then, from a knowledge of the theoretical density of the parti-

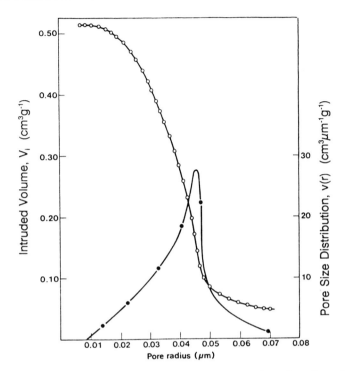

Figure 3.11 The mercury porosimetry data of Fig. 3.10 replotted to show the intruded volume V_i (open circles) and the pore size distribution $v(r)$ (filled circles) versus pore radius r. (From Ref. 3.)

cles, ρ_t, defined as the density of the pore-free solid, the amount of isolated porosity is found as $1 - \rho_a/\rho_t$. For a crystalline material, the theoretical density may be obtained from the crystal structure and the dimensions of the unit cell determined by X-ray diffraction.

For coarser powders (particle size $> \approx 10$ μm), a liquid pycnometer is used. A calibrated bottle is weighed (mass m_0), and the powder is then added (total mass m_1). A liquid of known density ρ_l is then added (total mass m_2). In a separate run, the pycnometer bottle containing the liquid only (i.e., no powder) is weighed (mass m_3). The apparent density of the particles ρ_a is found from

$$\rho_a = \frac{m_1 - m_0}{(m_3 - m_0) - (m_2 - m_1)} (\rho_l - \rho_{air}) + \rho_{air} \qquad (3.41)$$

where ρ_{air} is the density of air. Care must be taken to ensure good wetting of the particles by the liquid and to remove trapped air (by boiling the liquid).

Helium gas pycnometry is usually used for powders finer than ≈ 10 μm. The small size of the helium molecule enables it to penetrate into very fine pores. The volume occupied by the solid is measured from the volume of gas displaced. The apparent density is then calculated from the mass of the powder used and its measured volume.

3.3 CHEMICAL COMPOSITION

As shown earlier in Table 3.1, the chemical characteristics that have the greatest influence on the processing and properties of ceramics are the bulk composition, minor elements, and trace impurity elements. The bulk composition controls the intrinsic properties of the material. Usually, small concentrations of elements, e.g., $\approx 0.1-10$ at %, are added deliberately to modify the structural and chemical characteristics of the powder or may be retained in the powder from the preparation step. Trace impurity elements at concentrations less than a few hundred parts per million are invariably present even in the cleanest powders; they result from impurities present in the starting materials and from contamination during the preparation step. The minor elements and trace impurities, as discussed in Chapter 1, can have a profound effect on the microstructure of the fabricated body. The importance of determining the chemical composition of ceramic powders is therefore widely recognized. Table 3.5 summarizes the more common methods used to analyze the chemical composition of ceramic powders.

Table 3.5 Techniques for the Analysis of the Chemical Composition of Ceramic Powders

Technique	Detection limit	State of sample	Application
Optical atomic spectroscopy	1–1000 ppm	Liquid	Bulk composition; minor elements; trace impurities
X-ray fluorescence	>0.1%	Solid	Bulk composition; minor elements
Secondary ion mass spectrometry	<1 ppm	Solid	Trace impurities

A. Bulk Composition

Optical Atomic Spectroscopy

The common optical atomic spectroscopies used for the characterization of ceramic powders are atomic absorption spectroscopy, atomic emission spectroscopy, and atomic fluorescence spectroscopy. They are all based on transitions between energy levels in atoms. A portion of the powder must therefore be converted into individual atoms. In practice, this is achieved by dissolving the solid in a liquid to form a solution, which is then broken into droplets and vaporized into individual atoms by heating.

For a multielectron atom, there are many energy levels in the ground state, i.e., the state of lowest energy. There are also a large number of unoccupied energy levels into which electrons may be excited. When an atom is in an excited state, some of its electrons move to higher energy levels. In the simplest case, we shall consider two energy levels only, one in the ground state of the atom with energy E_g, and one in the excited state with energy E_e (Fig. 3.12). We are therefore making the very simplified assumption that only one type of electron undergoes transitions. The frequency of radiation absorbed or emitted during the transition is given by Planck's law,

$$\nu = \frac{E_e - E_g}{h} \tag{3.42}$$

where h is Planck's constant. The transitions that are useful in optical atomic spectroscopy involve fairly low energy transitions such that the wavelength of the radiation given by Eq. (3.42) occurs in the range of ≈ 10 nm to 400 μm. Typically, these transitions occur between the outer energy levels of the atom.

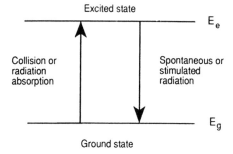

Figure 3.12 Schematic showing the transitions in atomic spectroscopy.

For atoms at thermal equilibrium at a temperature T, the number of atoms in the excited state, n_e, is related to the number in the ground state, n_g, by the Boltzmann equation,

$$\frac{n_e}{n_g} = \exp \frac{-(E_g - E_e)}{kT} \tag{3.43}$$

where k is the Boltzmann constant. The number of atoms in the ground state will be less than the total number n due to the excitation of n_e atoms. However, unless the temperature is very high, n_g will be approximately equal to n. From Eq. (3.43), the number of excited atoms, n_e, will be proportional to n. The intensity of the absorbed or emitted radiation, which is proportional to n_e, will therefore be proportional to the total number of atoms, n. While the relation between intensity of the absorbed or emitted radiation and the concentration of the atoms can be rigorously derived [7], it is rarely used practically in quantitative analysis. Instead, standard samples of known concentrations are made up, and the intensities are determined for these standards. The concentration of the unknown sample, C_u, is determined from a calibration curve or more simply from the relation

$$C_u = C_s \frac{I_u}{I_s} \tag{3.44}$$

where C_s is the concentration of a standard and I_u and I_s are the intensities of the unknown and standard samples, respectively.

Atomic absorption spectroscopy is based on the absorption of radiation by an electron in the ground state of an atom to produce a transition to a higher (excited) state. Radiation from an external light source is passed through the atomized vapor, and the absorption is measured by standard techniques. Vaporization is achieved by introducing droplets of the sample solution into a flame (a technique referred to as flame atomic absorption spectrometry) or into a graphite tube furnace (electrothermal atomic absorption spectroscopy). A schematic of the technique is shown in Fig. 3.13a.

Atomic emission spectroscopy is based on the spontaneous emission of radiation as an electron that was previously excited to a higher energy level in the atom undergoes a transition to the electronic ground state. Unlike the case of atomic absorption spectroscopy, no external light source is required, because the method used to vaporize the solution droplets is also used to thermally excite the electrons to higher energy levels (Fig. 3.13b). In flame atomic emission spectroscopy, the solution droplets are sprayed into a flame, which vaporizes the solution and thermally excites the electrons. Other common methods of vaporization and excitation

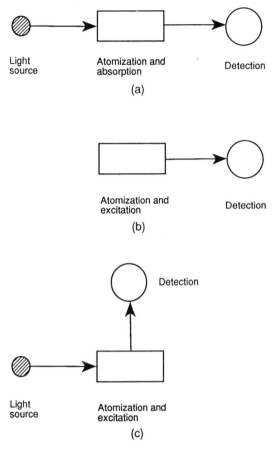

Figure 3.13 Schematic showing (a) atomic absorption spectroscopy, (b) atomic emission spectroscopy, and (c) atomic fluorescence spectroscopy.

that do not involve the use of a flame include an inductively coupled plasma, a direct current plasma, a direct current arc, an alternating current spark, and carbon furnaces.

Atomic fluorescence spectroscopy combines aspects of atomic absorption spectroscopy and atomic emission spectroscopy. The solution is vaporized in the same way as for atomic absorption spectroscopy and atomic emission spectroscopy. As in atomic absorption spectroscopy, the atoms are excited by irradiation with an external light source. Finally, as in atomic emission spectroscopy, the spontaneously emitted radiation is used to determine the concentration (Fig. 3.13c).

The three optical spectroscopic techniques described above are complementary. Depending on a variety of factors such as the element, the method of vaporization, and the method of excitation, the detection limit varies from a fraction of a part per million to a few thousand parts per million. The techniques appear to be significantly more sensitive when vaporization methods that do not rely on a flame are used. In general, atomic fluorescence spectroscopy has a greater sensitivity than atomic absorption spectroscopy or atomic emission spectroscopy. As outlined earlier, the sample must be dissolved to produce a solution for analysis by these techniques. The low solubility of many ceramics may present difficulties in their analysis by optical atomic spectroscopy.

X-Ray Fluorescence Spectroscopy

Atomic X-ray fluorescence spectroscopy, referred to simply as X-ray fluorescence (XRF), relies on the transitions of electrons from the outer energy levels to the inner subshells of the atom. The energy associated with these transitions (≈ 0.6–60 keV) is significantly greater than that associated with the transitions used in optical atomic spectroscopy ($\approx$ a few electronvolts). XRF yields a spectrum of lines in which each line results from a transition between two energy levels. For the energy range used in XRF, spectra consisting of the K series of lines for the elements with atomic number Z between 9 and 72 and the L series of lines for Z between 25 and 92 are obtained.

The X-ray spectrum does not depend on the way in which the atomic excitation is produced, but generally highly energetic X-rays are used. The sample is bombarded with the X-rays, and the intensity of the fluorescent radiation emitted is analyzed with respect to its wavelength (wavelength-dispersive technique) or its energy (energy-dispersive technique). Unknown elements in a sample are identified by comparing the spectrum with reference spectra. Quantitative analysis is normally performed by comparing the line intensities of the sample with those for standards. Alternatively, the elemental concentrations can be found from the measured line intensities using theoretical equations [8].

The detection limit of XRF is $\approx 0.1\%$, which means that the technique is not as sensitive as the optical spectroscopic techniques described earlier. However, XRF analysis can be performed with the sample in the solid form. For ceramic powders, many of which have a low solubility, the use of the solid powder can be a distinct advantage.

B. Minor Elements and Trace Impurities

Minor elements, typically in concentrations of 0.1–10%, can also be analyzed by all of the methods outlined earlier for bulk chemical analysis.

Trace impurities in concentrations of <0.1% down to a few parts per million can be analyzed by the optical atomic spectroscopies described earlier for bulk chemical analysis. For these optical atomic spectroscopies, the methods that do not use a flame for vaporization of the sample solution (e.g., inductively coupled plasma) can provide detection down to the ppm level. A very powerful technique for trace analysis is secondary ion mass spectrometry (SIMS). The sensitivity of SIMS is at the ppm level or lower. Although it can be used for bulk analysis, it is more appropriate for surface analysis. We will therefore describe the SIMS technique later, in the section devoted to surface characterization (Section 3.5).

3.4 PHASE COMPOSITION

The presence of crystallinity in a powder can be ascertained by X-ray diffraction. For amorphous powders, it is often useful to know the temperature range in which crystallization occurs. One way of determining the crystallization range is to heat the powder to known temperatures and check for the presence of crystallization by X-ray diffraction. However, a more convenient way is by differential thermal analysis (DTA). In DTA, the test sample under investigation and a reference sample are heated (or cooled) at a uniform rate and the difference in temperature between the two is plotted as a function of the temperature of the reference sample or as a function of the furnace temperature. Chemical reactions and phase changes are accompanied by heat changes, and these are observed as exothermic or endothermic peaks.

For crystalline powders, X-ray diffraction is the main method for the analysis of phase composition. The technique is described in detail in a number of texts, including those of Cullity [6] and Klug and Alexander [9]. The diffraction of X-rays by crystals is governed by Bragg's law,

$$2d \sin \theta = n\lambda \tag{3.45}$$

where d is the spacing between the lattice planes in the crystal, θ is the angle of diffraction, n is an integer (sometimes referred to as the order of diffraction), and λ is the wavelength of the X-rays (0.154 nm for Cu K_α radiation). The identification of a phase is achieved by determining the d spacings and the relative intensities of the peaks and comparing the values with reference data for known materials. In recent years, computer automation has greatly decreased the amount of work required for the identification of phases and the quantitative determination of the phases present. The intensities and diffraction angles of the diffraction spectrum of the sample can be readily compared with those of known materials by simple data retrieval.

The detection limit of most modern diffractometers is about 0.5–1 wt %, which means that phases with concentrations lower than this value generally go undetected. For the quantitative determination of phases, some care should be taken in the preparation of the sample if accurate and reliable results are to be obtained. The effects of preferred orientation, texturing, particle size broadening, and other factors must be minimized. Quantitative analysis should also be performed by calculating the integrated line intensities (i.e., the area under the peaks) instead of using the heights of the peaks. However, this is not usually done because of the additional effort required or overlap of the major peaks.

Two methods have generally been used in quantitative X-ray analysis. The simplest and more widely used method employs an internal standard. To illustrate this method, consider a powder consisting of a mixture of two phases, α and β. Quantitative determination of the weight percentages of the α and β phases in the powder can be best performed by first constructing a calibration curve. In the construction of the calibration curve, the diffraction patterns of mixtures containing known amounts of the α and β phases and a fixed fraction of an internal standard (e.g., 25 wt % MgO powder) are obtained. From these patterns, a prominent peak for each of the phases is selected, and its peak height relative to that for a prominent peak of internal standard is measured. Quantitative determination of the weight percent of the α and β phases in the powder under investigation involves (i) mixing the powder with the same fraction of internal standard, (ii) obtaining the diffraction pattern, (iii) measuring the appropriate peak height ratios, and (iv) using the peak height ratios to read off the phase contents from the calibration curve. If the analysis is performed carefully, the error in the measured phase contents is less than about ±3%.

The second method for quantitative X-ray analysis does not rely on the availability of standard samples of the phases for constructing a calibration curve. Instead, the peak intensities are computed mathematically using available atomic parameters. For a powder mixture containing j crystalline phases ($\alpha, \beta, \gamma, \ldots$), the intensity I of a given diffraction peak from phase α of the mixture is given by the expression [6]

$$I_\alpha = \frac{C_{\alpha v} K_0 |F|^2 m (LP)}{2\mu V^2} \exp(-2M) \tag{3.46}$$

where $C_{\alpha v}$ is the volume fraction of the α phase, K_0 is a constant, F is the structure factor, m is the multiplicity, LP is the Lorentz polarization factor, μ is the linear absorption coefficient, V is the volume of the unit cell, and $\exp(-2M)$ is the Debye–Waller temperature factor. Assuming that the Debye–Waller temperature factor is constant, Eq. (3.46) can be

Powder Characterization

simplified to

$$I_\alpha = C_{\alpha v} K'_0 L \tag{3.47}$$

where K'_0 is a constant and L is given by

$$L = \frac{|F|^2 m(LP)}{V^2} \tag{3.48}$$

The value of L must be calculated for each reflection of the different phases. In practice, only two or three major reflections are considered for each phase. Assuming that n_α is the number of reflections considered for the α phase,

$$C_{\alpha v}\overline{K'_0} = \frac{1}{n_\alpha} \sum_\alpha \left(\frac{I_\alpha}{L_\alpha}\right) \tag{3.49}$$

where the summation is over the n_α reflections. If $C_{\alpha w}$ is the weight fraction of the α phase, then

$$C_{\alpha w} = C_{\alpha v} \frac{\rho_\alpha}{\rho} \tag{3.50}$$

where ρ_α is the density of the α phase and ρ is the overall density of the sample. Since the sum of the weight fractions of the phases must be equal to 1, the weight fraction of the α phase may be written as

$$C_{\alpha w} = \frac{(\rho_\alpha/n_\alpha) \sum_\alpha (I_\alpha/L_\alpha)}{\sum_j (\rho_j/n_j) [\sum_k (I_k/L_k)]} \tag{3.51}$$

Table 3.6 Calculated L Values for the Major X-Ray Diffraction Peaks in Si, α-Si$_3$N$_4$, and β-Si$_3$N$_4$

Phase	Peak	2θ (Cu K_α)	L
Si	(111)	28.48	26.69
	(220)	47.37	17.56
α-Si$_3$N$_4$	(201)	30.99	7.44
	(102)	34.54	6.66
	(210)	35.31	6.79
β-Si$_3$N$_4$	(101)	33.66	10.90
	(210)	36.07	11.21

Source: Ref. 10.

where the summations in the denominator are over the n_j phases of the mixture and over the n_k reflections for phase k.

The values of L for the major reflections of Si, α-Si$_3$N$_4$, and β-Si$_3$N$_4$ are given in Table 3.6. They are extremely useful in the determination of the phase content of Si$_3$N$_4$ powders when standards for the two phases are not available [10]. Furthermore, the method is not nearly as time-consuming as the internal standard method, which requires the construction of a calibration curve. Using the calculated L values, the method gave the measured composition, in weight percent, of a Si$_3$N$_4$ powder to $\pm 2\%$ of the composition of a standard [10].

3.5 SURFACE CHARACTERIZATION

As the particle size of a powder decreases below 1 μm, the surface area of the powder and the volume fraction of the outermost layer of ions on the surface increase quite significantly with decreasing particle size. We would therefore expect the surface characteristics to play an increasingly important role in the processing of fine powders as the particle size decreases. The surface area and surface chemistry are the characteristics that have the most profound influence on processing. Another characteristic, the surface structure, while having only a limited role in processing, may nevertheless have an important influence on surface phenomena (e.g., vaporization, corrosion, and heterogeneous catalysis). Having considered the surface area and its characterization earlier, we now turn to the characterization of the surface chemistry and the surface structure.

Although the characterization of the powder surface area and its role in processing have received considerable attention for decades, it is only within recent years that the importance of surface chemistry in ceramic processing has begun to be recognized. As we have outlined in Chapter 1, the consolidation of fine ceramic powders from liquid suspensions to produce more uniform green bodies has been shown to produce significant benefits in fabrication. However, the quality of the microstructure of the consolidated body is controlled by the dispersion behavior of the powder and the interaction between the particles in the liquid, which in turn are controlled by the chemistry of the powder surfaces. The surface chemistry of the powders can also have a direct influence on densification and microstructural evolution during the firing process, regardless of the powder consolidation method. During the firing stage the surfaces of the particles become interfaces and grain boundaries that act as sources and sinks for the diffusion of matter. Densification and coarsening will be controlled by the structure and composition of the interfaces and grain boundaries. Impurities on the surfaces of the powder, for example, will alter the grain boundary composition and, in turn, the microstructure of the fabricated body. In many ceramics, the control of the structure and chemistry of the grain boundaries through manipulation of the surface chemistry of the powders forms one of the most important considerations in processing. A classic example is the case of Si_3N_4 materials for high-temperature structural applications for which the grain boundary structure and chemistry control the microstructural evolution during firing and the properties of the fabricated body.

With the enormous and increasing importance of surfaces to modern technological processes, a large number of techniques have been developed. We consider only those techniques that have emerged to be most widely applicable to the characterization of ceramic powders. Readers interested in the more specialized techniques are referred to texts devoted to surface analysis such as Ref. 11. In general, the surface characterization techniques rely on the interaction between atomic particles (e.g., ions, electrons, and neutrons) or radiation (e.g., X-rays and ultraviolet rays) with the sample. The interaction produces various emissions that can be used to analyze the sample. Figure 3.14 shows the principal emissions produced by the interaction between an electron beam and a sample. Depending on the thickness of the sample and the energy of the electron beam, a certain fraction of electrons will be scattered in the forward direction, another fraction will be absorbed, and the remaining fraction will be scattered in the backward direction. The forward-scattered electrons consist of electrons that have undergone elastic scattering (i.e., interactions with the atoms of the sample that result in a change in direction but virtually no loss in energy) and electrons that have undergone inelastic scattering (i.e., interactions that result in both a change in direction and a reduction in the energy of the incident electrons). The elastically scattered electrons are much greater in number than the inelastically scattered electrons. They are used in the transmission electron microscope to produce diffraction effects in the determination of the structure of the sample. The backscattered electrons are highly energetic electrons that have been scattered in the backward direction. The majority of these will also have undergone elastic collisions with the atoms of the sample.

The incident electron beam can also generate secondary effects in the sample. One of these effects is that the incident electrons can knock electrons out of their orbits around an atom. If the ejected electrons are near the surface of the sample (within ≈ 20 nm), they may have enough energy to escape from the sample and become what are called secondary electrons. These secondary electrons are used in the scanning electron microscope to produce an image. A second type of effect occurs when an electron fills a vacant site in one of the orbitals of an excited atom.

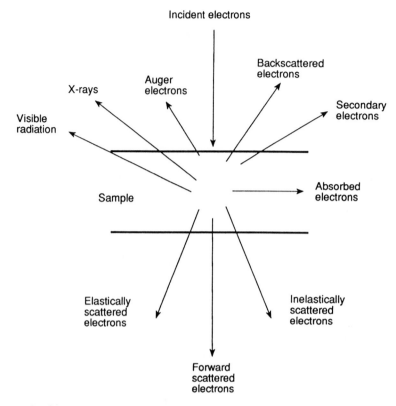

Figure 3.14 Emissions produced by the interaction of an electron beam with a solid sample.

As we observed earlier, radiation in the form of X-rays or light can be produced in the transition and used to obtain information about the chemical composition of the sample. Furthermore, some of the X-rays can be absorbed by electrons in the orbitals around the atoms. If these electrons are very close to the surface of the sample, they may escape. These electrons, called Auger electrons, can also provide chemical information about the sample.

The depth to which the incident electron beam penetrates into the sample depends on the energy of the beam and the nature of the sample. For electron energies of 1 MeV, the penetration depth is ≈ 1 μm. However, for energies in the range of 20–1000 eV, the penetration depth is only ≈ 1 nm. An incident electron beam with energy below ≈ 1 keV can therefore be used to probe surface layers one to two monolayers thick. Ions and

Powder Characterization

X-rays can also be used to produce diffraction effects and to eject electrons. They too can be used as incident beams to probe the surfaces of materials.

A. Surface Structure

When viewed on a microscopic or submicroscopic scale, the surface of a particle is not smooth and homogeneous. Various types of irregularities are present. Most ceramic particles, as we observed in Chapter 2, are agglomerates of finer primary particles. Features such as porosity and boundaries between the primary particles will be present on the surface. For nonoxide particles, a thin oxidation layer will cover the surfaces of the particles. Even for single-crystal particles, the surface can be fairly complex. Figure 3.15 shows a schematic model of a solid surface that has been used to develop the theory of various surface phenomena [12]. The surface has several atomic irregularities that are distinguishable by the number of nearest neighbors or coordination number. The atoms in terraces are surrounded by the largest numbers of neighbors. Atoms located in steps or in kinks in steps have fewer neighbors than atoms in the terraces. Finally, adatoms located on top of the terraces have the smallest number of neighbors.

The most widely applicable techniques for the characterization of the surface structure of solids include microscopy and electron diffraction. Two of the most prominent techniques, scanning tunneling microscopy (STM) and low-energy electron diffraction (LEED), have limitations for the characterization of powder surfaces because they require a fairly flat surface with sufficient coherency over a relatively large area. Readers

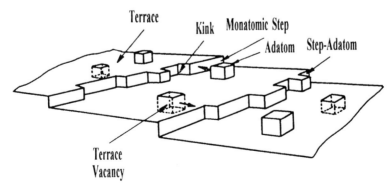

Figure 3.15 Model of a heterogeneous solid surface showing different surface sites. (From Ref. 12.)

interested in these two techniques are referred to excellent treatments elsewhere [1,12].

Modern scanning electron microscopes provide a very simple method for observing the powder structure with a point-to-point resolution of ≈ 2 nm. Finer details of the structure can be observed by TEM. Figure 3.16a shows a high-resolution TEM micrograph of an Si_3N_4 powder having an amorphous layer (3–5 nm thick) on the surface [13]. This surface layer, with the composition of a silicon oxynitride, has a significant effect on the sintering and high-temperature mechanical properties of the fabricated body. A micrograph of another Si_3N_4 powder observed in a high-voltage TEM (1.5 MeV) shows that the particles are connected by a strong amorphous bridge, presumably a silicon oxynitride (Fig. 3.16b). Such hard agglomerates, as we observed earlier, can have a detrimental effect on the packing of the powder.

B. Surface Chemistry

Electron, ion, and photon emissions from the outermost layers of the surface can be used to provide qualitative or quantitative information about the chemical composition of the surface. The most widely applicable techniques for the characterization of the surface chemistry of ceramic powders are Auger electron spectroscopy (AES), X-ray photoelectron spectroscopy (XPS), which is also referred to as electron spectroscopy for chemical analysis (ESCA), and secondary ion mass spectrometry (SIMS). Table 3.7 provides a summary of the main measurement parameters for these three techniques [14]. In practice, all surface analytical techniques (with the exception of Rutherford backscattering) require the use of an ultrahigh-vacuum environment. The analysis conditions are therefore rather limited and do not correspond to those normally used in ceramic processing, where the powder experiences fairly prolonged exposure to the atmosphere.

Although the techniques described here derive their usefulness from being truly surface sensitive, they can also be used to provide information from much deeper layers. This is normally done by sequential (or simultaneous) removal of surface layers by ion beam sputtering and analysis. This mode of analysis in which the composition can be found layer by layer is referred to as composition-depth profiling. It is one of the most important modes of surface analysis because the composition of the surface is usually different from that of the bulk. The ability to perform composition-depth profiling, therefore, is often a measure of the effectiveness of a technique.

Powder Characterization

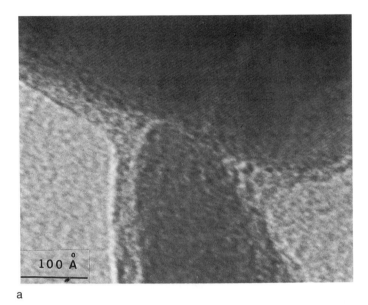

a

b

Figure 3.16 Surface structure of an Si_3N_4 powder as revealed by (a) high-resolution TEM and (b) high-voltage TEM.

Table 3.7 Methods for the Surface Chemical Analysis of Ceramic Powders

Measurement parameter	Auger electron spectroscopy (AES)	X-ray photoelectron spectroscopy (XPS)	Secondary ion mass spectrometry (SIMS)
Incident particle	Electrons (1–20 keV)	X-rays (1254 and 1487 eV)	Ions (Ar^+, Xe^+) (100 eV to 30 keV)
Emitted particle	Auger electrons (20–2000 eV)	Photoelectrons (20–2000 eV)	Sputtered ions
Element range	>Li ($Z = 3$)	>Li ($Z = 3$)	>H ($Z = 1$)
Detection limit	10^{-3}	10^{-3}	10^{-6}–10^{-9}
Depth of analysis	2 nm	2 nm	1 nm
Lateral resolution	>20 nm	>15 nm	50 nm to 10 mm

Source: Ref. 14, with permission.

Auger Electron Spectroscopy (AES)

In AES, a beam of electrons is used to excite Auger electrons from the surface. A schematic of the overall process is shown in Fig. 3.17a. When an electron is ejected from an inner shell of an atom through collision with an incident electron, the resultant vacancy is soon filled by an electron from one of the outer shells. The energy released in the transition may appear as an X-ray photon (used, e.g., for compositional analysis in X-ray fluorescence as described earlier) or be transferred to another electron, which is ejected from the atom with an energy E_A given by

$$E_A = E_1 - E_2 - E^* \qquad (3.52)$$

where E_1 and E_2 are the binding energies of the atom in the singly ionized state and E^* is the binding energy for the doubly ionized state. The Auger electron moves through the solid and soon loses its energy through inelastic collisions with bound electrons. However, if the Auger electron is emitted sufficiently close to the surface, it may escape from the surface and be detected by an electron spectrometer. The energy spectrum of these electrons is recorded and analyzed. Since each type of atom has its own characteristic electron energy levels, the peaks in the observed Auger spectrum can be used for determining the elemental composition by comparison with standard Auger spectra for the elements.

The technique is fairly rapid and provides highly reproducible results. When calibration standards are used, quantitative elemental composition can be obtained to better than ±10%. Although AES can, in principle, provide information on chemical composition, it is largely used for ele-

Powder Characterization

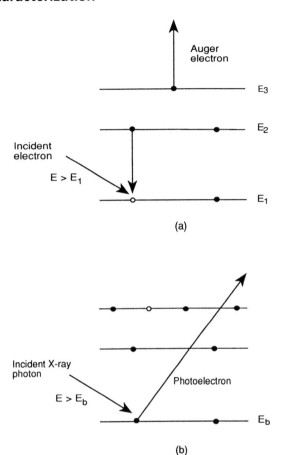

Figure 3.17 Schematic showing the electron emissions in (a) Auger electron spectroscopy and (b) X-ray photoelectron spectroscopy.

mental analysis only. For ceramic powders, which are mostly insulating, severe electrostatic charging of the surface may occur, leading to a reduction in the effectiveness of the technique.

X-Ray Photoelectron Spectroscopy (XPS)

In XPS, the sample is irradiated by a source of low-energy X-rays that cause photoionization of the atoms. The photoelectrons produced as a result of the process have a kinetic energy E_k given by the Einstein relation

$$E_k = h\nu - E_b \tag{3.53}$$

where h is Planck's constant, ν is the frequency of the incident X-rays, and E_b is the binding energy of the photoelectron. If the photoelectrons have sufficient kinetic energy to overcome the work function of the sample, photoemission occurs (Fig. 3.17b). The photoelectrons emitted from the surface are analyzed in an electron spectrometer to produce an energy spectrum. In addition to the valence electrons, which take part in chemical bonding, each atom (except hydrogen) possesses core electrons, which do not take part in bonding. We are generally concerned with the core electrons in XPS.

X-ray photoelectron spectroscopy is a fairly versatile technique that can be used for qualitative and quantitative analysis of the surface composition as well as for determining the chemical bonding (or oxidation state) of the surface atoms. For qualitative analysis, a survey scan is performed over a wide energy range (typically 0–1000 eV) to determine the elements present. As outlined for AES, the position of each peak is compared with a standard for qualitative analysis. A survey scan for a commercial Si_3N_4 powder prepared by a gas-phase reaction between $SiCl_4$ and NH_3 is shown in Fig. 3.18. For quantitative analysis, the principal peak for each element is selected, and its intensity (i.e., peak area after removal of the background) is measured. The fractional atomic concentration of an element A is given by

$$C_A = \frac{I_A/S_A}{\sum_i I_i/S_i} \tag{3.54}$$

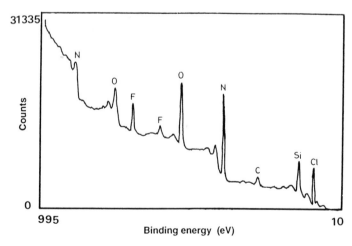

Figure 3.18 A survey scan of the surface chemistry of a commercial Si_3N_4 powder by X-ray photoelectron spectroscopy.

Powder Characterization

where I_i is the measured peak intensity of element i and S_i is the atomic sensitivity factor for that peak. The atomic sensitivity factors, which can be calculated theoretically or derived empirically, are usually provided in the reference manuals supplied by the manufacturer of the instrument. The accuracy of the quantitative analysis ($\pm 10\%$) is similar to that for AES.

Information about the chemical bonding of the surface atoms can be determined from the chemical shifts in the peak positions in the XPS spectrum. The binding energies of the core electrons are sufficiently affected by their chemical environment to produce a measurable shift (of a few electronvolts) in the measured electron energy. It is necessary to scan over a fairly narrow energy range (≈ 25 eV) for each peak to improve the accuracy in measuring the chemical shift. Figure 3.19 illustrates the nature of the chemical shift for the $2p$ peak in Si and in SiO_2. In the practical determination of the chemical bonding, the measured peak position is compared with reference values for the corresponding peak in a number of compounds.

Unlike AES, in which, especially for insulating samples, electron beam damage may occur, the damage to the sample in XPS is minimal. Furthermore, electrostatic charging of the sample in XPS is minimal. As discussed earlier, AES, XPS, and SIMS are complementary techniques, and it is not uncommon to find instruments that incorporate more than one of them. Figure 3.20 is a sketch of the sample chamber and the electronic detection system for a combined AES/XPS instrument.

Secondary Ion Mass Spectrometry (SIMS)

In SIMS, a beam of low-energy ions is used to sputter off surface atoms into a vacuum, in which ionized species are detected directly using a mass spectrometer. There are two distinct modes of analysis. In static SIMS, an ion beam of low current density is used so the analysis is confined to the outermost layers. In dynamic SIMS, an ion beam of high current density is used to erode successive atomic layers at a relatively fast rate. Here we limit our attention to static SIMS. Figure 3.21 shows the positive and negative secondary ion mass spectra of the same Si_3N_4 referred to earlier in Figure 3.18.

The mechanism or mechanisms involved in the sputtering process using a beam of low-energy electrons are far from understood. In one model, referred to as the molecular model, secondary ions are thought to result from dissociation of sputtered neutral molecular species some distance from the surface (Fig. 3.22). Three steps are involved in the process: (i) energy transfer in the collision cascade in the solid, (ii) collision energy transfer between the atoms of the molecule leaving the surface, and (iii) charge exchange during dissociation of the molecule.

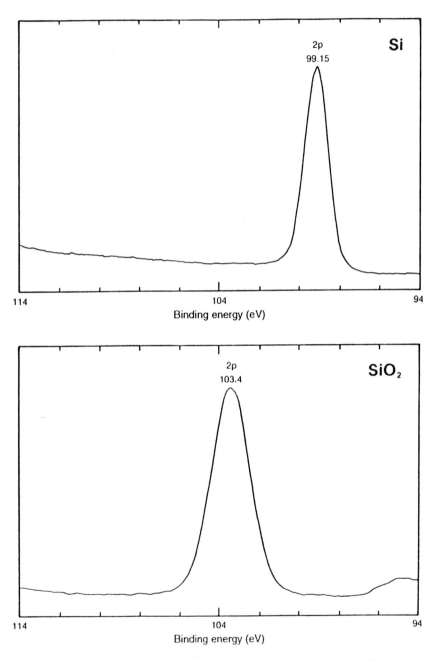

Figure 3.19 Peak positions for the silicon 2p peak in Si and in SiO_2, showing the chemical shift produced by the formation of SiO_2. (Courtesy of Perkin-Elmer Corporation, Eden Prairie, MN.)

Powder Characterization

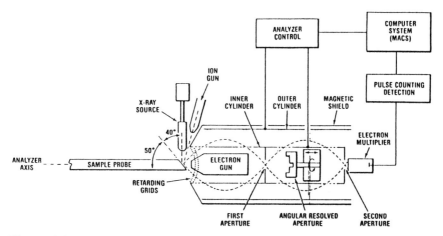

Figure 3.20 Schematic of the combined AES/XPS instrument. (Courtesy of Perkin-Elmer Corporation, Eden Prairie, MN.)

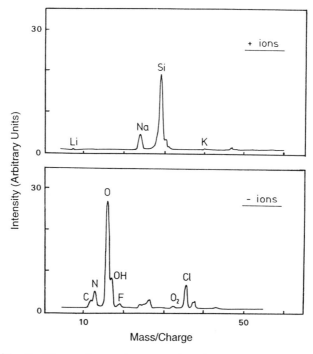

Figure 3.21 Positive and negative secondary ion mass spectra of the Si_3N_4 powder referred to in Fig. 3.18.

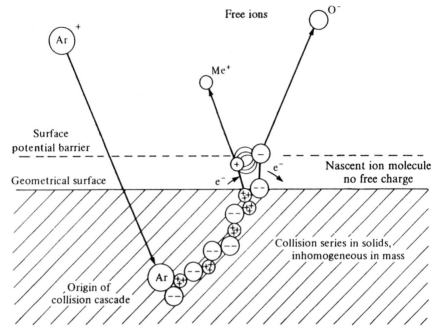

Figure 3.22 Model for the emission of secondary ions in the sputtering process. (From Ref. 14, with permission.)

Certain key advantages of SIMS are well known. It combines high spatial resolution, high sensitivity for qualitative elemental analysis, and the ability to provide a detailed analysis of the chemical composition of the surface. However, at present, quantitative analysis can be performed only with a certain degree of difficulty. Furthermore, SIMS has the disadvantage of essentially destroying the surface during analysis.

3.6 CONCLUDING REMARKS

In this chapter we have examined a variety of techniques with broad application to the characterization of ceramic powders. In addition to their use in ceramic processing, powders are also important in a number of other technologies (e.g., powder metallurgy and pollution control), so the techniques described here have broad applicability. Some of the techniques (e.g., gas adsorption and mercury porosimetry) are also important for characterizing porous compacts, whereas others (e.g., the techniques for phase analysis, chemical composition, and surface analysis) are also

Powder Characterization

very important for solids. We shall therefore encounter the use of these techniques again when we consider powder consolidation and the microstructural characterization of solids produced by firing. Finally, as stated earlier in this book, when powders form the starting materials for our fabrication route, a proper characterization is essential for understanding the effects accompanying the consolidation and firing stages.

PROBLEMS

3.1 The particle size distribution of a powder is as follows:

Mean size (μm):   4   6   9   12   18   25
Number of particles: 10 16 20 25   15   7

Calculate the arithmetic, geometric and harmonic mean sizes.

3.2 Discuss what methods you would use to measure the particle sizes of the following powders:
  (a) A powder with particles less than 50 μm.
  (b) A nanoscale powder with particles less than 20 nm.
  (c) A sub-micrometer powder.
  (d) A powder with particles in the range 20–85 μm which must be analyzed dry.
  (e) A powder with particles less than 50 μm where the surface area is the important property.

3.3 (a) Plot the shape factor for a cylinder as a function of the length to diameter ratio.
  (b) Calculate the shape factor for a tetrahedron, an octahedron and a dodecahedron.
  (c) An agglomerate consists of n spherical particles in point contact. Calculate and plot the shape factor as a function of n, for n in the range of 2–100.

3.4 In a nitrogen adsorption experiment at the boiling point of liquid nitrogen, the volume (V) of gas adsorbed at a pressure (P) were measured as follows:

P (mm Hg):   80   100   125   140   200
V (cm^3/g): 0.420 0.439 0.464 0.476 0.534

Determine the surface area of the powder.

3.5 The following data were obtained in a liquid pycnometry experiment performed at 20°C: mass of pycnometer = 35.827 g; mass of pycnometer and powder = 46.402 g; mass of pycnometer and water = 81.364 g; mass of pycnometer, powder and water = 89.894 g. If the theoretical density of the solid is 5.605 g/cm^3, determine the amount of closed porosity in the powder.

3.6 (a) How would you determine the solid solubility limit of calcium in cerium oxide?

(b) How would you determine the effect of calcium concentration on the theoretical density of calcium-doped cerium oxide?

3.7 In the reaction between ZnO and Fe_2O_3, the amount of $ZnFe_2O_4$ formed at any temperature depends on the heating rate. How would you measure this for heating rates of 1 and 10°C/min?

3.8 In an X-ray photoelectron spectrum, would you expect the 1s peak for carbon to be at a higher or lower binding energy compared to that for the oxygen 1s peak? Would the fluorine 1s peak occur at higher or lower binding energy compared to that for the oxygen 1s peak? Explain your answer.

3.9 The chemical composition of the oxide layer on silicon nitride particles varies as a function of the thickness of the layer. How would you confirm this? What methods would you use to measure the composition as a function of the thickness of the oxide layer?

REFERENCES

1. J. B. Wachtman, *Characterization of Materials*, Butterworth-Heinemann, Boston, MA, 1991.
2. G. Y. Onoda, Jr., and L. L. Hench, *Ceramic Processing Before Firing*, Wiley, New York, 1978, pp. 35–37.
3. T. Allen, *Particle Size Measurement*, 3rd ed., Chapman and Hall, New York, 1981.
4. J. W. Evans and L. C. De Jonghe, *The Production of Inorganic Materials*, Macmillan, New York, 1991, p. 131.
5. J. K. Beddow, *Particulate Science and Technology*, Chemical Publishing Co., New York, 1980.
6. B. D. Cullity, *Elements of X-Ray Diffraction*, 2nd ed., Addison-Wesley, Reading, MA, 1978.
7. J. D. Winefordner and M. S. Epstein, *Physical Methods of Chemistry*, Vol. IIIA, *Determination of Chemical Composition and Molecular Structure*, Wiley, New York, 1987, pp. 193–327.
8. C. J. Sparks, *Advances in X-Ray Analysis*, Vol. 19, Plenum, New York, 1976.
9. H. P. Klug and L. E. Alexander, *X-Ray Diffraction Procedures*, 2nd ed., Wiley, New York, 1974.
10. C. P. Gazzara and D. R. Messier, *Am. Ceram. Soc. Bull.*, 56(9):777 (1977).
11. D. P. Woodruff and T. A. Delchar, *Modern Techniques of Surface Analysis*, Second Edition, Cambridge Univ. Press, Cambridge, U.K., 1994.

12. G. A. Somorjai and M. Salmeron, in *Ceramic Microstructures '76* (R. M. Fulrath and J. A. Pask, Eds.), Westview Press, Boulder, CO, 1977, pp. 101–128.
13. M. N. Rahaman, Y. Boiteux, and L. C. De Jonghe, *Am. Ceram. Soc. Bull.*, *65*(8):1171 (1986).
14. J. M. Walls, *Methods of Surface Analysis*, Cambridge Univ. Press, Cambridge, U.K., 1989.

4
Science of Colloidal Processing

4.1 INTRODUCTION

A colloid consists of two distinct phases: a continuous phase (referred to as the dispersion medium) that is either liquid or gas and a fine dispersed particulate phase that is either solid or liquid. The dispersed particles generally have dimensions ranging between 1 and 1000 nm, sometimes referred to as the colloidal size range. Colloids can therefore consist of solid particles in a liquid (referred to as a sol, colloidal suspension, or simply a suspension, e.g., paint), liquid particles in a liquid (an emulsion, e.g., milk), liquid particles in a gas (an aerosol, e.g., fog), and other combinations.

In the processing of ceramics, colloidal suspensions and colloidal solutions are of particular interest. Colloidal suspensions are being used increasingly in the consolidation of ceramic powders to produce the green body. Compared to powder consolidation in the dry or semidry state (e.g., pressing in a die), colloidal methods can lead to better packing uniformity in the green body, which in turn leads to better microstructural control during firing. Colloidal solutions consist of polymer molecules dissolved in a liquid. The size of the polymer molecules in solution falls in the colloidal size range, so these systems are considered part of colloid science. Polymer solutions are relevant to the fabrication of ceramics by the solution sol-gel route, which we shall consider in the next chapter. For the present chapter, we concentrate on colloidal suspensions.

A basic problem with which we shall be concerned is the stability of colloidal suspensions. Clearly the particles must not be too large; otherwise gravity will produce rapid sedimentation. The other important factor is the attractive force between the particles. Attractive van der Waals forces exist between the particles regardless of whether other forces may be involved. If the attractive force is large enough, the particles will collide and stick together, leading to rapid sedimentation of particle clusters (i.e., to flocculation or coagulation). Although in principle the reduction in the attractive force can also be used, the techniques employed to prevent flocculation rely on the introduction of repulsive forces. Repulsion between electrostatic charges (referred to as electrostatic stabilization), repulsion between polymer molecules (polymeric stabilization), or some combination of the two (electrosteric stabilization) are the basic techniques used.

Because it controls the packing uniformity of the consolidated powder, the stability of colloidal suspensions must be understood at a basic level. A stable colloidal suspension may be consolidated into a densely packed structure; however, an unstable suspension may lead to a loosely packed structure or, under certain conditions, to a particulate gel with a fairly high volume.

Colloidal suspensions exhibit a number of interesting properties. For electrostatically stabilized suspensions, the motion of the particles in an electric field (referred to as electrophoresis) can be a source of valuable information in the characterization of the charges associated with the particles and the stability of the suspensions. Colloidal suspensions, especially concentrated suspensions, also have remarkable deformation and flow properties (i.e., rheological properties). A good paint must flow easily when applied to the surface to be painted but must then become rigid enough to prevent it from flowing off a vertical surface. In the forming of ceramic powders, the shaped powder form is often produced from the colloidal suspension by conventional casting methods (e.g., slip casting or tape casting). The rheological behavior of the suspension and its characterization for the optimization of the consolidation process are important factors.

4.2 TYPES OF COLLOIDS

Colloids consisting of particles dispersed in a liquid are generally divided into two broad classes: lyophilic colloids and lyophobic colloids. When the liquid is specifically water, the colloids are described as hydrophilic and hydrophobic. Lyophilic (i.e., liquid-loving) colloids are those in which there is a strong affinity between the dispersed particle and the liquid. The liquid is strongly adsorbed onto the surfaces of the particle, so the interface between the particle and the liquid is very similar to the interface between the liquid molecules. This system will be intrinsically stable. Polymer solutions and soap solutions are good examples of lyophilic colloids.

Lyophobic (liquid-hating) colloids are those in which the liquid does not show affinity for the particle. If attractive forces exist between the particles, there will be a strong tendency for them to stick together when they come into contact. This system will be unstable, and flocculation will result. Suspensions of insoluble particles in a liquid (e.g., most ceramic particles dispersed in a liquid) are well-known examples of lyophobic colloids. We therefore need to understand the attractive forces that lead to flocculation and how they can be overcome to produce colloids with the desired stability.

4.3 ATTRACTIVE SURFACE FORCES

Attractive surface forces, generally referred to as van der Waals forces, exist between all atoms and molecules regardless of what other forces may be involved. We shall first examine the origins of the van der Waals forces between atoms and molecules and later consider the attractive forces between macroscopic bodies such as particles.

A. Van der Waals Forces Between Atoms and Molecules

The van der Waals forces between atoms and molecules can be divided into three groups [1]:

1. *Dipole–dipole forces* (*Keesom forces*). A molecule of HCl, because of its largely ionic bonding, has a positive charge and a negative charge separated by a small distance (≈ 0.1 nm); that is, it consists of a minute electric dipole. A polar HCl molecule will interact with another polar HCl molecule and produce a net attractive force.
2. *Dipole–induced dipole forces* (*Debye forces*). A polar molecule such as HCl can induce a dipole in a nonpolar atom or molecule such as an argon atom, causing an attractive force between them.
3. *Dispersion forces* (*London forces*). This type of force exists between nonpolar atoms or molecules, for example, between argon atoms. To understand the origin of this type of force, consider an argon atom. Although the argon atom has a symmetrical distribution of electrons around the nucleus so that it has no dipole, this situation is only true as an average over time. At any instant, the electron cloud undergoes local fluctuations in charge density, leading to an instantaneous dipole. This instantaneous dipole produces an electric field that polarizes the electron distribution in a neighboring argon atom, so the neighboring atom itself acquires a dipole. The interaction between the two dipoles leads to an attractive force.

The attractive dispersion force can be calculated as follows [2]. Suppose the first atom has a dipole moment μ at a given instant in time. The electric field E at a distance x along the axis of the dipole is

$$E = \frac{1}{4\pi\epsilon_0}\left(\frac{2\mu}{x^3}\right) \qquad (4.1)$$

where ϵ_0 is the permittivity of free space and x is much greater than the length of the dipole. If there is another atom at this point (Fig. 4.1), it becomes polarized and acquires a dipole moment μ', given by

$$\mu' = \alpha E \qquad (4.2)$$

where α is the polarizability of the atom. A dipole μ' in a field E has a potential energy V given by

$$V = -\mu' E \qquad (4.3)$$

Substituting for μ' and E gives

$$V = -\left(\frac{1}{4\pi\epsilon_0}\right)^2\left(\frac{4\alpha\mu^2}{x^6}\right) \qquad (4.4)$$

A rigorous derivation by London in 1930 using quantum mechanics gives

$$V = -\frac{3}{4}\left(\frac{1}{4\pi\epsilon_0}\right)^2\left(\frac{\alpha^2 h\nu}{x^6}\right) \qquad (4.5)$$

where h is Planck's constant and ν is the frequency of the polarized orbital. The force between the atoms has a magnitude

$$F = \frac{\partial V}{\partial x} = \frac{\text{constant}}{x^7} \qquad (4.6)$$

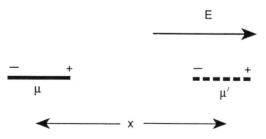

Figure 4.1 Polarization of a molecule by the field E due to a dipole, giving rise to mutual attraction.

The Keesom and Debye forces are also proportional to x^{-7} and lead to a potential energy proportional to x^{-6}. For dispersion forces, the dependence of V on x^{-6} is strictly applicable to separation distances less than a few tens of nanometers. For larger separations, when the forces are described as retarded forces, V varies as x^{-7}. However, unless otherwise stated, we shall assume that only unretarded dispersion forces operate and assume an x^{-6} dependence for the remainder of this chapter.

B. Van der Waals Forces Between Macroscopic Bodies

To determine the van der Waals forces between macroscopic bodies (e.g., two particles), we assume that the interaction between one molecule and a macroscopic body is simply the sum of the interactions with all the molecules in the body. We are therefore assuming simple additivity of the van der Waals forces and ignore how the induced fields are affected by the intervening molecules.

In the computation due to Hamaker, individual atoms are replaced by a smeared-out uniform density of matter (Fig. 4.2). An infinitesimally small volume Δv_1 in the first body exerts an attractive potential on an infinitesimally small volume Δv_2 according to the equation

$$V' = -\frac{C}{x^6} \rho_1 \Delta v_1 \rho_2 \Delta v_2 \qquad (4.7)$$

where C is a constant and ρ_1 and ρ_2 are the numbers of molecules per unit volume in the two bodies. Of course, $\rho_1 = \rho_2$ if the two bodies consist of the same material. The additivity of the forces now becomes an integration, with the total potential energy being

$$V_A = \int -\frac{C}{x^6} \rho_1 \rho_2 \, dv_1 \, dv_2 \qquad (4.8)$$

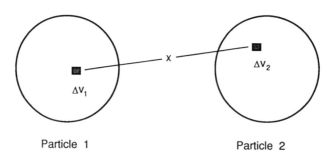

Particle 1 Particle 2

Figure 4.2 Additivity of the van der Waals forces between macroscopic bodies. The total interaction is taken as the sum of the interactions between infinitesimally small elements in the two bodies.

Science of Colloidal Processing

Hamaker showed that Eq. (4.8) can be expressed as

$$V_A = -\frac{A}{\pi^2} \int \frac{dv_1 \, dv_2}{x^6} \tag{4.9}$$

where A, called the Hamaker constant, is equal to $C\pi^2 \rho_1 \rho_2$. For a large number of solids, the Hamaker constant has values in a fairly narrow range between 10^{-20} and 10^{-19} J ($\approx 2.5kT$–$25kT$, where k is the Boltzmann constant and T is room temperature in degrees Kelvin).

For two semi-infinite parallel surfaces, the potential energy is infinite. We need to restrict the calculation to unit area. In this case (Fig. 4.3a), the attractive potential per unit area between two semi-infinite parallel surfaces that are a distance h apart is given by

$$V_A = -A/12\pi h^2 \tag{4.10}$$

For a sphere of radius a at a distance h from a flat semi-infinite solid (Fig.

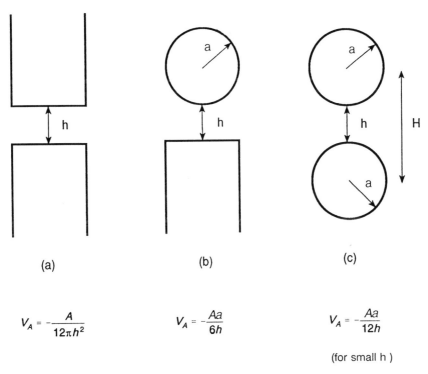

(a) $V_A = -\dfrac{A}{12\pi h^2}$ (b) $V_A = -\dfrac{Aa}{6h}$ (c) $V_A = -\dfrac{Aa}{12h}$

(for small h)

Figure 4.3 The potential energy of attraction between two bodies with specified geometries. (a) Two semi-infinite parallel surfaces; (b) a sphere and a semi-infinite surface; (c) two spheres.

4.3b), V_A is given by

$$V_A = -Aa/6h \qquad (4.11)$$

Finally, the potential energy between two spheres of radius a that are separated by a distance h (Fig. 4.3c) is given by

$$V_A = -\frac{A}{6}\left[\frac{2a^2}{H^2 - 4a^2} + \frac{2a^2}{H^2} + \ln\left(\frac{H^2 - 4a^2}{H^2}\right)\right] \qquad (4.12)$$

where the distance H between the centers of the spheres is equal to $h + 2a$. For small separations, where $h \ll a$, Eq. (4.12) gives

$$V_A = -Aa/12h \qquad (4.13)$$

showing that V_A is proportional to h^{-1}. The equations for the three cases considered are summarized in Fig. 4.3.

C. Effect of the Intervening Medium

The method of Hamaker for the computation of the van der Waals force of attraction between macroscopic bodies assumes that the intervening medium is air or vacuum. There is no simple way to modify it to account for the presence of a dielectric medium between the bodies. A different approach used by Lifshitz treats the interactions in terms of the dielectric properties of the materials. The Lifshitz approach is beyond the scope of this book, and the reader is referred to suitable texts and review articles on colloid science for a discussion [1, 3]. For particles with the same dielectric constant ϵ_s separated by a medium with a dielectric constant ϵ_l, the theory shows that the Hamaker constant is proportional to $(\epsilon_s - \epsilon_l)^2/(\epsilon_s + \epsilon_l)^2$, i.e.,

$$A = c\left(\frac{\epsilon_s - \epsilon_l}{\epsilon_s + \epsilon_l}\right)^2 \qquad (4.14)$$

where c is a constant of proportionality. We see from Eq. (4.14) that for the same solid A is always positive so the van der Waals force is always attractive. However, the intervening medium always leads to a reduction in the attractive force compared to the force when the medium is air or vacuum.

4.4 LYOPHOBIC COLLOIDS

Consider a system of colloidal particles dispersed in a liquid. The particles undergo Brownian motion (similar to the motion of pollen in a liquid) and will eventually collide. The potential energy of attraction, V_A, between two particles of radius a at a distance h from each other is given by Eq. (4.13). Assuming that this equation is valid down to contact between the particles, then we can calculate V_A by putting $h \approx 0.3$ nm. Taking the Hamaker constant as 10^{-20} J, for particles with a radius $a = 0.2$ μm, V_A has a value of $\approx 5 \times 10^{-19}$ J. The thermal energy of the particles is kT, where k is the Boltzmann constant and T is the absolute temperature. Near room temperature ($T \approx 300$ K), kT is on the order of 4×10^{-21} J, and this value is much smaller than V_A for the particles in contact. Thus the particles will stick together on collision because the thermal energy is insufficient to overcome the attractive potential energy. Flocculation will therefore occur unless some method is found to produce a repulsion between the particles that is sufficiently strong to overcome the attractive force. There are two basic ways for achieving this:

1. *Electrostatic stabilization* in which the repulsion between the particles is based on electrostatic charges on the particles
2. *Polymeric stabilization* in which the repulsion is produced by polymer molecules adsorbed (or chemically attached) onto the particle surfaces or existing freely in solution

Stabilization may also be achieved as a result of both electrostatic and steric repulsion. In this case, it is referred to as *electrosteric stabilization*. This type of stabilization is produced by ionic polymers (referred to as polyelectrolytes) adsorbed onto the surfaces of the particles.

4.5 ELECTROSTATIC STABILIZATION

As outlined above, electrostatic stabilization is said to occur when the repulsion between the particles is achieved by electrostatic charges on the particles. The repulsion is not, however, a simple case of repulsion between charged particles. A diffuse electrical double layer of charge is produced between the particles, and the repulsion occurs as a result of the interaction of the double layers. By way of an introduction to the analysis of the electrical double layer, we consider how particles acquire an electrostatic charge in a liquid and the general principle of the double layer.

A. Charges on Particles in a Liquid

The main processes by which particles dispersed in a liquid can acquire a surface charge are (i) preferential adsorption of ions, (ii) dissociation of surface groups, (iii) isomorphic substitution, (iv) adsorption of polyelectrolytes, and (v) accumulation of electrolytes [4]. Of these, preferential adsorption of ions from solution is the most common process for ceramic particles, whereas isomorphic substitution is commonly found in clays.

Adsorption of Ions from Solution

In the process of preferential adsorption of ions from solution, an electrolyte such as an acid, a base, or a metal salt is added to the solution. Ions preferentially adsorb onto the surface of the particles, leading to a charge on the particle surface. Because the system consisting of the particle and the electrolyte must be electrically neutral, an equal and opposite countercharge exists in the solution. Most oxide surfaces are hydrated; for an oxide of a metal M, there will be MOH groups on the surface, as illustrated in Fig. 4.4 for SiO_2. In acid solutions, adsorption of H^+ ions (or hydronium ions, H_3O^+) produces a positively charged surface, whereas in basic solutions, adsorption of OH^- ions (or the dissociation of H^+ ions) leads to a negatively charged surface. Thus oxide surfaces are positively charged at low pH and negatively charged at high pH. At some intermediate pH,

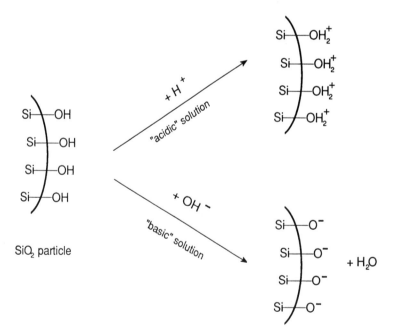

Figure 4.4 The production of surface charges on SiO_2 particles by adsorption of ions from acidic or basic solutions.

referred to as the point of zero charge (PZC), the adsorption of H^+ ions will balance that of the OH^- ions, and the particle surface will be effectively neutral.

For oxides, the PZC is related to the acidity or basicity of the surface groups. Acidic oxides such as SiO_2 have a low PZC, whereas basic oxides such as Al_2O_3 have a high PZC. For an oxide of a metal M, the adsorption of H^+ and OH^- ions can be written

$$MOH + H^+ \rightleftharpoons MOH_2^+ \tag{4.15a}$$

$$MOH + OH^- \rightleftharpoons MO^- + H_2O \tag{4.15b}$$

The equilibrium constants K_a and K_b for the reactions described by Eqs. (4.15a) and (4.15b) can be written as

$$K_a = \frac{[MOH_2^+]}{[MOH][H^+]}, \quad K_b = \frac{[MO^-][H_2O]}{[MOH][OH^-]} \tag{4.16}$$

where the square brackets are used to denote concentration. At the PZC, $[MO^-] = [MOH_2^+]$, so

$$[H^+]_{PZC} = \left(\frac{K_b K_w}{K_a}\right)^{1/2} \tag{4.17}$$

where K_w is the equilibrium constant for the dissociation of H_2O, normally referred to as the ionic product of water. Since pH is defined as the negative logarithm of $[H^+]$, the PZC can be written

$$PZC = \frac{1}{2}(pK_a - pK_b - pK_w) \tag{4.18}$$

where pK is the negative logarithm of the appropriate equilibrium constant K. As Eq. (4.18) shows, the IEP is closely related to pK_a and pK_b, so for a given oxide its value will be influenced by the surface chemistry of the powder.

As discussed later (see Section 4.5G: Electrokinetic Phenomena), it is more convenient to measure the potential (or charge) at a short distance from the particle surface (at the surface of the Stern layer). The potential is called the ζ (zeta) potential. The pH where the ζ potential is zero is called the isoelectric point (IEP). Thus, the PZC is distinguished from the IEP as follows

PZC = pH where the particle surface is neutral
IEP = pH where the charge at the Stern layer is zero (ζ-potential = 0).

Table 4.1 gives the approximate IEP values for several ceramic oxides [5].

Table 4.1 Nominal Isoelectric Points (IEPs) of Some Oxides

Material	Nominal composition	IEP
Quartz	SiO_2	2
Soda lime silica glass	$1.0Na_2O \cdot 0.6CaO \cdot 3.7SiO_2$	2–3
Potassium feldspar	$K_2O \cdot Al_2O_3 \cdot 6SiO_2$	3–5
Zirconia	ZrO_2	4–6
Apatite	$10CaO \cdot 6PO_2 \cdot 2H_2O$	4–6
Tin oxide	SnO_2	4–5
Titania	TiO_2	4–6
Kaolin	$Al_2O_3 \cdot SiO_2 \cdot 2H_2O$	5–7
Mullite	$3Al_2O_3 \cdot 2SiO_2$	6–8
Chromium oxide	Cr_2O_3	6–7
Hematite	Fe_2O_3	8–9
Zinc oxide	ZnO	9
Alumina (Bayer process)	Al_2O_3	8–9
Calcium carbonate	$CaCO_3$	9–10
Magnesia	MgO	12

Source: Ref. 5, with permission.

Isomorphic Substitution

Isomorphic substitution is commonly found in clays. In the crystal lattice, some of the cations are replaced by other cations of lower valence without altering the crystal structure; for example, Si^{4+} ions are replaced by Al^{3+} or Mg^{2+} ions, and Al^{3+} ions by Mg^{2+} ions. This process leads to a deficit of positive charges, which is balanced by other positive ions (e.g., Na^+, K^+, or Ca^{2+}) adsorbed on the surface of the clay particles. In the case of the clay mineral pyrophyllite, $Al_2(Si_2O_5)_2(OH)_2$, isomorphic substitution in which Mg^{2+} replaces some of the Al^{3+} ions in the lattice leads to montmorillonite, $Na_{0.33}(Al_{1.67}Mg_{0.33})(Si_2O_5)_2(OH)_2$, another mineral with the same crystal structure in which the charge deficit is balanced by Na^+ ions on the particle surfaces. When the clay mineral montmorillonite is dispersed in water, the Na^+ ions pass freely into solution, leaving negatively charged particles (Fig. 4.5).

The extent of isomorphic substitution is dictated by the nature of the clay, and this is expressed by the cation exchange capacity (CEC). The CEC of a clay is the number of charges on the clay (expressed in coulombs per kilogram) that can be replaced in solution. It is typically in the range of 10^3–10^5 C/kg, and for a given clay it is not sensitive to variables such as pH or concentration of the electrolyte in solution.

Science of Colloidal Processing

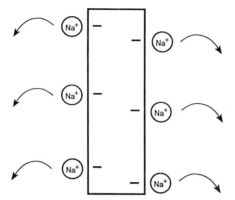

Figure 4.5 The production of surface charges on clay particles (montmorillonite) in aqueous liquids.

B. Origins of the Electrical Double Layer

For electrostatically stabilized colloidal suspensions, the charges, as we have seen, consist of a surface charge on the particles and an equal and opposite countercharge in the solution. Suppose the particle has a positive surface charge. In the complete absence of thermal motion, an equal num-

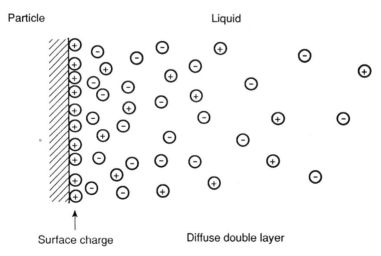

Figure 4.6 The distribution of positive and negative charges in the electrical double layer associated with a charged surface in a liquid.

ber of negative ions (counterions) would adsorb onto the positive charge and neutralize it. However, such a compact double layer does not form because of thermal motion. Instead, the counterions are spread out in the liquid to form a diffuse double layer as shown in Fig. 4.6. There is a fairly rapid change in the concentration of the positive and negative ions as we move away from the surface (Fig. 4.7a). As a result, the electric potential also falls off rapidly with distance from the surface (Fig. 4.7b).

As two particles approach one another in the liquid, the diffuse double layers will start to overlap. It is the interaction between the double layers that gives rise to the repulsion between the particles. If the repulsion is strong enough, it can overcome the van der Waals attractive force, thereby producing a stable colloidal suspension. We therefore need to examine the interactions between double layers and the factors that influence the repulsion between interacting double layers. In the analysis, we will start with an isolated double layer associated with a single particle. Following this, we will examine the interactions between two double layers.

C. Isolated Double Layer

The electrical properties of an isolated double layer can be described in terms of the variation of the electric charge or the electric potential (see Fig. 4.7). The two are related through the capacitance, so if one is known, the other can be found. We shall consider the potential mainly. We shall also consider the analysis in one dimension (the x direction) and make two simplifying assumptions: (i) The particle surface will be taken to be flat, and (ii) the electrolyte will be considered to be symmetrical (e.g., HCl), such that the valence of the positive ions is equal in magnitude to that of the negative ions.

The variation in the electric potential ϕ with distance x is governed by Poisson's equation,

$$\nabla^2 \phi(x) = -\frac{\rho(x)}{\epsilon \epsilon_0} \tag{4.19}$$

where $\rho(x)$ is the charge density at a distance x from the surface, ϵ is the dielectric constant of the liquid medium, and ϵ_0 is the pemittivity of vacuum. The charge density is obtained from the sum of the contributions of the individual ions:

$$\rho(x) = \sum_i c_i z_i F \tag{4.20}$$

where c_i is the concentration and z_i is the valence of the i ions and the summation is over the positive and negative ions in the solution. The concentration c_i is given by the Boltzmann distribution [6]:

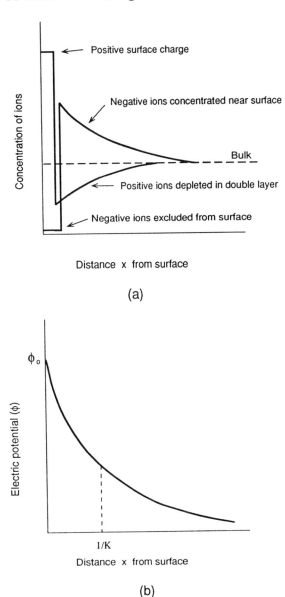

Figure 4.7 (a) Concentration of positive and negative ions as a function of distance from the particle surface. (b) The electric potential ϕ as a function of distance from the particle surface. The distance equal to $1/K$ is the Debye length.

$$c_i = c_{\infty i} \exp\left(\frac{-z_i F\phi(x)}{RT}\right) \quad (4.21)$$

where $c_{\infty i}$ is the concentration of the i ions very far from the particle surface, F is the Faraday constant, R is the gas constant, and T is the absolute temperature. Equations (4.19)–(4.21) give

$$\nabla^2\phi(x) = -\frac{1}{\epsilon\epsilon_0}\sum_i c_{\infty i} z_i F \exp\left(\frac{-z_i F\phi(x)}{RT}\right) \quad (4.22)$$

Summing over the positive (+) and negative (−) ions and putting $z_+ = -z_- = z$ (symmetrical electrolyte) and $c_{\infty+} = c_{\infty-} = c_{\infty}$ (the concentrations of the positive and negative ions far from the surface are equal), Eq. (4.22) reduces to

$$\nabla^2\phi(x) = \frac{2czF}{\epsilon\epsilon_0}\sinh\left(\frac{zF\phi(x)}{RT}\right) \quad (4.23)$$

For low surface potential, referred to as the Debye–Hückel approximation, such that

$$zF\phi \ll RT \quad \text{(Debye–Hückel approximation)} \quad (4.24)$$

and making the substitution

$$K^2 = \frac{2cz^2F^2}{\epsilon\epsilon_0 RT} \quad (4.25)$$

Eq. (4.23) becomes

$$\nabla^2\phi(x) = K^2\phi(x) \quad (4.26)$$

With the appropriate boundary conditions, i.e.,

$$\phi = \phi_0 \text{ at } x = 0; \quad \phi = 0 \text{ at } x = \infty \quad (4.27)$$

where ϕ_0 is the potential at the surface of the particle, the solution of Eq. (4.26) is

$$\phi = \phi_0 \exp(-Kx) \quad (4.28)$$

The term K occurs frequently in the analysis of the electrical double layer, and 1/K has the dimensions of length. At a distance $x = 1/K$, the potential ϕ has fallen to $1/e$ of its value at the surface of the particle, and beyond this the change in ϕ is small. Thus 1/K may be considered the thickness of the double layer and is usually referred to as the Debye length. According to Eq. (4.25), the thickness of the double layer depends on a number of experimental parameters. As we shall see later, these parameters can be varied to control the magnitude of the repulsion between two double layers and thereby the stability of the suspension.

Equation (4.23) can be integrated twice to give [7]

$$\tanh\left(\frac{zF\phi(x)}{4RT}\right) = \tanh\left(\frac{zF\phi_0}{4RT}\right)\exp(-Kx) \qquad (4.29)$$

Equation (4.29), referred to as the Gouy–Chapman equation, is valid for any value of the surface potential ϕ_0. However, for ϕ_0 less than ≈ 50 mV, the difference in the ϕ values found from the Debye–Hückel approximation and the Gouy–Chapman equation is insignificant (Fig. 4.8).

A more rigorous analysis shows that the electrical double layer consists of a compact layer (about a few molecular diameters thick), referred to as the Stern layer, and a more diffuse layer referred to as the Gouy–Chapman layer. The electric potential in the Gouy–Chapman layer decreases exponentially with distance according to Eq. (4.28) or (4.29) but decreases more steeply in the Stern layer (Fig. 4.9). However, this refinement in the double layer theory will not be considered any further in this book.

D. Surface Charge

For electroneutrality, the surface charge density (charge per unit area) must be equal to the integrated charge density in the solution; that is,

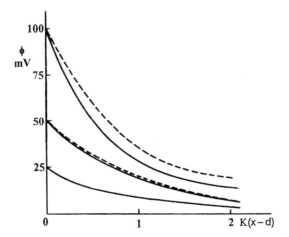

Figure 4.8 Comparison of the predictions of the Debye–Hückel and Gouy–Chapman equations for the potential ϕ in the electrical double layer. The Debye–Hückel equation can be used with insignificant error up to ≈ 50 mV. (From Ref. 4, with permission.)

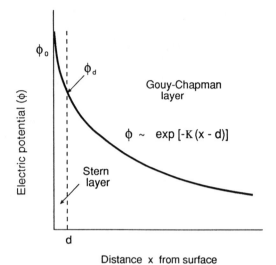

Figure 4.9 The electrical double layer consists of a compact Stern layer and a more diffuse Gouy–Chapman layer.

$$\sigma_0 = -\int_0^\infty \rho \, dx \tag{4.30}$$

Substituting for ρ from Eq. (4.19), then,

$$\sigma_0 = \int_0^\infty \epsilon\epsilon_0 \nabla^2 \phi(x) \, dx = \left[\epsilon\epsilon_0 \frac{d\phi}{dx}\right]_0^\infty = -\epsilon\epsilon_0 \left(\frac{d\phi}{dx}\right)_{x=0} \tag{4.31}$$

Equation (4.23) can be integrated once to give

$$\left(\frac{d\phi}{dx}\right)_{x=0} = -\left(\frac{8RTc}{\epsilon\epsilon_0}\right)^{1/2} \sinh\left(\frac{zF\phi_0}{2RT}\right) \tag{4.32}$$

Substituting in Eq. (4.31) gives:

$$\sigma_0 = (8\epsilon\epsilon_0 RTc)^{1/2} \sinh\left(\frac{zF\phi_0}{2RT}\right) \tag{4.33}$$

For low potential (Debye–Hückel approximation), Eq. (4.33) becomes

$$\sigma_0 = \epsilon\epsilon_0 K\phi_0 \tag{4.34}$$

The capacitance is defined as the charge divided by the voltage, and for

Science of Colloidal Processing

a parallel-plate capacitor the capacitance is equal to $\epsilon\epsilon_0/t$, where t is the distance between the plates. We can therefore see from Eq. (4.34) that the electrical double layer can be treated as a parallel plate capacitor with a thickness of $1/K$.

E. Repulsion Between Two Double Layers

Two colloid particles will begin to interact as soon as their double layers overlap. We consider the simplest case of two parallel surfaces at a distance h apart in an electrolyte. The electric potential within the double layers consists of two symmetrical curves resembling that shown in Fig. 4.7b, and the net effect is roughly additive (Fig. 4.10). As the distance h decreases, the overlap of the double layers causes the potential to increase. This increase in the potential implies a repulsive force.

The general theory of the interaction between electrical double layers is known as the DLVO theory (after Derjaguin, Landau, Verwey, and Overbeek). It applies to the diffuse Gouy–Chapman layer. The DLVO theory is beyond the scope of this book and the reader is referred to texts on colloid science for a discussion (e.g., Ref. 7). Instead, we shall consider a fairly simple method, referred to as the Langmuir force method, for determining the repulsion.

The interaction between the particles can be analyzed in terms of a repulsive force or a repulsive potential energy (i.e., the work done in bringing the particles from infinity to the desired distance apart). Normally, the repulsive potential energy is used. Once again, we shall consider the particles to be two parallel surfaces. According to the Langmuir force method, the ionic concentration at a point midway between the two charged surfaces (e.g., at A in Fig. 4.11) will be greater than that far away from the surfaces (e.g., at B). This gives rise to an excess osmotic pressure Π that acts to push the surfaces apart. The excess osmotic pressure at A is given by [8]

$$\Pi = 2RTc \left[\cosh\left(\frac{zF\phi_m}{RT}\right) - 1 \right] \qquad (4.35)$$

where ϕ_m is the net electric potential at A arising from the overlap of the double layers. For large separations h, the net potential is roughly additive, so

$$\phi_m = 2\phi \qquad (4.36)$$

where ϕ is the potential of an isolated double layer at A. When ϕ_m is small, that is, $zF\phi_m \ll RT$, Eq. (4.35) reduces to

$$\Pi = cz^2F^2\phi_m^2/RT \qquad (4.37)$$

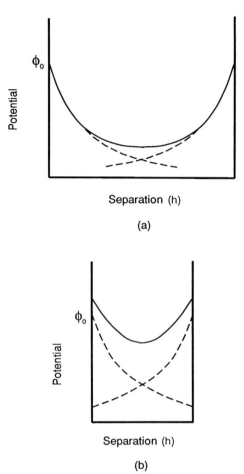

Figure 4.10 The electric potential in the electrolyte solution between two parallel surfaces for (a) large separation and (b) small separation. As the separation is decreased, the potential is increased, implying a repulsion.

For low surface potential ϕ_0, the Debye–Hückel approximation can be used and the potential ϕ obeys Eq. (4.28), that is,

$$\phi = \phi_0 \exp\left(\frac{-Kh}{2}\right) \tag{4.38}$$

The repulsive potential energy V_R between the charged surfaces at a separation h is the work done in bringing the surfaces from infinity to a distance

Science of Colloidal Processing

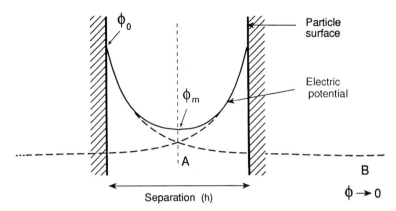

Figure 4.11 Overlap of the electrical double layers associated with two parallel surfaces leads to an increase in the ionic concentration. The osmotic pressure that results from the increased concentration acts to push the surfaces apart.

h apart, so that

$$V_R = \int_h^\infty \Pi \, dh \tag{4.39}$$

Using Eqs. (4.36)–(4.38) and integrating,

$$V_R = 2\epsilon\epsilon_0 K \phi_0^2 \exp(-Kh) \tag{4.40}$$

It should be remembered that Eq. (4.40) is valid for low ϕ_0 and for large separations where ϕ_m is small. For large ϕ_0, the Gouy–Chapman equation, Eq. (4.29), must be used. Using Eq. (4.37) (valid for low ϕ_m) and integrating [8], we obtain

$$V_R = 2\epsilon\epsilon_0 K \left(\frac{4RT\gamma}{zF}\right)^2 \exp(-Kh) \tag{4.41}$$

where γ is defined by

$$\gamma = \tanh\left(\frac{zF\phi_0}{4RT}\right) \tag{4.42}$$

For two spherical particles of radius a at a distance h apart (see Fig. 4.3c), V_R is given by [7]

$$V_R = 2\pi a \epsilon\epsilon_0 \left(\frac{4RT\gamma}{zF}\right)^2 \exp(-Kh) \tag{4.43}$$

When ϕ_0 is low, Eq. (4.43) reduces to

$$V_R = 2\pi a \epsilon \epsilon_0 \phi_0^2 \exp(-Kh) \tag{4.43a}$$

This equation should be compared with the result obtained by the DLVO theory

$$V_R = 2\pi a \epsilon \epsilon_0 \phi_0^2 \ln[1 + \exp(-Kh)] \tag{4.43b}$$

F. Stability of Electrostatically Stabilized Colloids

We now consider what happens when two colloid particles approach one another. As the double layers start to overlap, the double layer repulsion opposes the attraction from van der Waals interactions. The overall effect, as noted earlier, is normally described in terms of the potential energy rather than the force. Using the convention that repulsive potentials are positive and attractive potentials are negative, Fig. 4.12 shows an example of the potential energies for the van der Waals attraction and the double layer repulsion. The resultant curve for the total potential energy V_T shows a deep minimum at M_1 corresponding approximately to contact between the particles and a secondary minimum at M_2. For two particles initially far apart approaching one another to a separation M_2, if the ther-

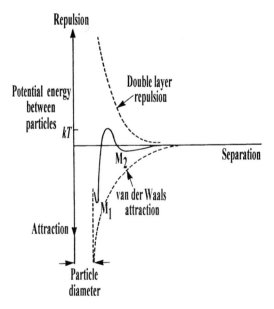

Figure 4.12 The potential energy between two particles in a liquid resulting from the effects of the van der Waals attraction and the double layer repulsion.

Science of Colloidal Processing

mal energy (kT) of the particles is small compared to the depth of M_2, the particles will not be able to escape from one another. Flocculation will result, leading to a sediment of loosely packed particles. However, the colloid can be restabilized by heating (increasing kT), by changing the electrolyte concentration to increase the double layer repulsion, or by a combination of the two. This process of restabilization of a flocculated suspension is referred to as peptization.

The curve for V_T in Fig. 4.12 depends on the relative magnitudes of V_A and V_R. For a given sol (e.g., a given suspension of Al_2O_3 particles in water), V_A is approximately constant. However, as seen from Eq. (4.40) or (4.41), the value of the double layer repulsion can be changed significantly by changes in the concentration and in the valence of the ions in solution. These two experimental parameters provide a useful way to control the stability of the colloid. Figure 4.13 summarizes four main types

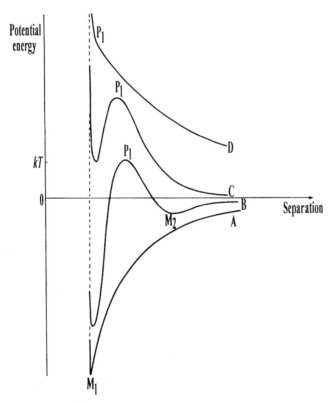

Figure 4.13 Four examples of the resultant potential energy between two particles. The repulsion between the particles increases in going from curve A to curve D.

of curves for the total potential energy. In curve A, the repulsion is weak and V_T is not significantly different from the curve for the van der Waals curve. The particles attract one another and reach equilibrium at the primary minimum M_1. A sediment consisting of clusters of particles almost in contact is produced. This is known as coagulation, in contrast to flocculation, in which, as stated above, the particles are loosely held together at relatively large separations.

In curve B, the repulsion is increased, and this leads to a secondary minimum at M_2 that is shallow compared to kT. There is therefore little chance for flocculation. However, the height of the maximum at P_1 is comparable to kT, so particles may be able to surmount the energy at P_1 and fall into the primary minimum at M_1. Irreversible coagulation will result because the depth of M_1 is much greater than kT.

The double layer repulsion has become so large in curve C that there is no secondary minimum at M_2. Furthermore, the height of the maximum P_1 is much greater than kT, so the particles have almost no chance of surmounting the energy barrier. Curve C therefore represents the situation for a stable colloid.

Finally, in curve D, the attraction is much smaller than the repulsion over all separations, so there is no minimum in the curve. An extremely stable colloid will be produced under these conditions. However, the strong repulsion limits the concentration of particles in suspension. Addition of a salt can lead to a compression of the double layer and a reduction in the viscosity.

G. Electrokinetic Phenomena

Electrokinetic phenomena involve the combined effects of motion and an electric field. When an electric field is applied to a colloidal suspension, the particles move with a velocity that is proportional to the applied field strength. The motion is called *electrophoresis*. It is a valuable source of information on the sign and magnitude of the charge and on the potential associated with the double layer. The measured potential, called the ζ (zeta) potential, is an important guide to the stability of lyophobic colloids.

In a particle electrophoresis apparatus, the suspension is placed in a cell and a dc voltage V is applied to two electrodes at a fixed distance l apart in the cell (Fig. 4.14). The sign of the particle charge is obtained directly, because it is opposite to that of the electrode toward which the particle is migrating. The particle velocity is measured by using a microscope, and the velocity per unit field strength, called the electrophoretic mobility, is used in the determination of the ζ potential and the charge.

Science of Colloidal Processing

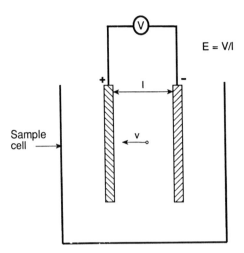

Figure 4.14   Schematic of a particle electrophoresis cell.

Conversion of the Electrophoretic Mobilities to ζ-Potentials
A particle with a charge q in an electric field E experiences a force directed toward the oppositely charged electrode given by

$$F = qE \tag{4.44}$$

In a viscous medium, the terminal velocity v is reached rapidly, so Stokes' law can be used to account for the force on the particle; that is,

$$F = 6\pi\eta a v \tag{4.45}$$

where η is the viscosity of the medium and a is the radius of the particle. From these two equations, the electrophoretic mobility is obtained:

$$u = \frac{v}{E} = \frac{q}{6\pi\eta a} \tag{4.46}$$

For particles in a dilute electrolyte solution (i.e., for $Ka < 0.1$), the potential on the surface of the particle can be taken as that of an isolated particle; that is,

$$\zeta = q/4\pi\epsilon\epsilon_0 a \tag{4.47}$$

Substituting for q in Eqs. (4.46) and (4.47) gives

$$u = 2\epsilon\epsilon_0\zeta/3\eta \tag{4.48}$$

For concentrated electrolyte solutions ($Ka > 200$), the Helmholtz–Smoluchowski equation can be used:

$$u = \epsilon\epsilon_0\zeta/\eta \tag{4.49}$$

For values of Ka between 0.1 and 200, Henry's equation applies:

$$u = \frac{2\epsilon\epsilon_0\zeta}{3\eta}[1 + f(Ka)] \tag{4.50}$$

where $f(Ka)$ has the values

Ka	0	0.1	1	5	10	50	100	∞
$f(Ka)$	0	0.0001	0.027	0.16	0.239	0.424	0.458	0.5

Significance of the ζ Potential

The ζ potential determined from the measured electrophoretic mobility is the potential at the surface of the electrokinetic unit moving through the solution (Fig. 4.15). It is not the potential at the surface of the particle but must correspond to a surface removed from the particle surface by at least one hydrated radius of the counterion (sometimes referred to as the shear surface). It is believed that the value of the ζ potential is close to that of the Stern potential (see Fig. 4.9). The double layer on the solution side of the ζ potential will be diffuse, so the ζ potential is the appropriate

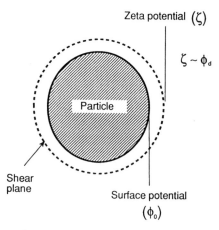

Figure 4.15 The particle moves with an attached layer of liquid so that a shear plane develops between the moving particle and the liquid. The ζ potential is taken to be the electric potential at the plane of shear between the particle and the electrolyte solution. It is approximately equal to the potential at the surface of the Stern layer.

potential to be used in effects that depend on the diffuse double layer (e.g., the double layer repulsion discussed earlier) provided that a reasonable guess can be made as to the distance of the ζ potential from the surface of the particle.

The charge calculated from the ζ potential is the charge within the shear surface. If the particle surface is assumed to be planar, the surface charge density σ_d (i.e., the charge density at the surface of the Stern layer) can be determined approximately from Eq. (4.33) or (4.34). Since the ζ potential and σ_d depend only on the surface properties and the charge distribution in the electrical double layer, they are independent of the particle size.

Typical data are shown in Fig. 4.16 for the electrophoretic mobility and charge density at the Stern plane for TiO_2 (rutile) as a function of pH in aqueous solutions of potassium nitrate [9]. The ζ potential will show the same variation with pH because it is proportional to the mobility. For many oxides, ζ potential values determined from the mobility generally fall within the range of $+100$ to -100 mV.

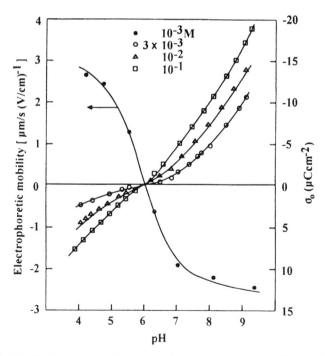

Figure 4.16 Surface charge density and electrophoretic mobility as functions of pH for rutile in aqueous solutions of potassium nitrate. (From Ref. 9.)

The ζ potential plays an important role in that it is widely used as a measure of the stability of colloidal suspensions. It must be remembered, however, that it controls only the repulsive part of the particle interaction. Suspensions prepared at pH values close to the isoelectric point (IEP) may flocculate fairly rapidly because the repulsion may not be sufficient to overcome the van der Waals attraction. Farther away from the IEP, we should expect the rate of flocculation to be slower. In practice, for good stability, suspensions are often prepared at pH values comparable to those of the plateau regions of the ζ potential or electrophoretic mobility curve, e.g., <5 or >7 for the data shown in Fig. 4.16.

4.6 POLYMERIC STABILIZATION

Polymeric stabilization is a generic term used to describe all aspects of the stabilization of colloidal particles by organic (nonionic) polymer molecules. The principle has been exploited for over 4500 years, since the Egyptians applied it empirically in the production of ink for writing on papyrus. Polymeric stabilization is today exploited for a wide range of industrial products such as paints, glues, inks, food emulsions, pharmaceuticals, detergents, and lubricants and plays a part in many biological systems. In the processing of ceramics, it is more widely used than electrostatic stabilization for the production of stable suspensions in the consolidation of ceramic powders by casting methods such as slip casting and tape casting.

Polymeric stabilization may be accomplished in two distinct ways:

1. Steric stabilization, the more common of the two, in which the stability is achieved by polymer molecules adsorbed or attached to the surfaces of the colloidal particles
2. Depletion stabilization, which is achieved by polymer molecules in free solution

A. Steric Stabilization

For good stability, the polymer molecule must consist of two parts; one part is nominally insoluble and anchors (chemically or physically) onto the particle, and the other part is soluble in the liquid. Figure 4.17 is a schematic of a polymer molecule in which one end anchors chemically onto a hydrated silica surface. The role of the anchor is to prevent desorption of the polymer under the shearing stress induced by the close approach of a second particle. The part of the molecule in solution waggles around and takes up numerous configurations available to a flexible polymer chain. As we shall see below, it is the overlap of the flexible chains

Science of Colloidal Processing

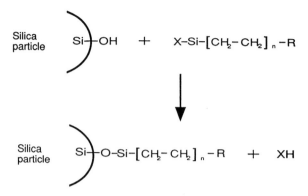

Figure 4.17 Chemical attachment of a polymer chain to the surface of a silica particle.

associated with neighboring particles that produces a repulsion between the particles. Homopolymers consisting of a single type of species in the chain are generally ineffective because of the conflicting requirements that the liquid be a poor solvent to ensure strong adsorption but a good solvent to impart good stabilization.

B. Range of Steric Repulsion

Steric repulsion starts to become significant when the polymer molecules on the surfaces of the particles start to interact. It is therefore important to understand the geometrical configuration that a polymer molecule takes up, because this will determine the range of the steric repulsion. Polymer molecules in solution take on the configuration of a coil rather than that of an extended chain (Fig. 4.18). The diameter of the coil is a difficult parameter to calculate. A more accessible parameter that has been found to be extremely useful for the description of many properties of polymers is the end-to-end distance of the coil. For any polymer in solution, the chains have a distribution of end-to-end distances. Usually, the root mean square (rms) end-to-end distance, denoted $\langle r^2 \rangle^{1/2}$, is the parameter considered to provide a measure of the size of the polymer molecule.

The statistics of polymer chain molecules is fairly complex [10]. However, making a number of simplifying assumptions (e.g., each segment of the polymer chain can rotate freely and independently of the neighboring segment), then

$$\langle r^2 \rangle^{1/2} = l\sqrt{N} \tag{4.52}$$

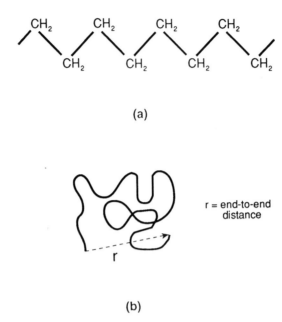

Figure 4.18 Two configurations of a polymer chain: (a) an extended chain in a highly extended polymer; (b) a coiled chain in a solution. The end-to-end distance of the coil is r.

where l is the length of a segment in the polymer chain and N is the number of segments in the chain (i.e., the degree of polymerization of the molecule). We see that for $l = 0.15$ nm and $N = 10^4$, $\langle r^2 \rangle^{1/2} = 15$ nm.

C. Origins of Steric Stabilization

Let us consider what happens when two particles covered with polymer molecules come within range of one another (Fig. 4.19a). The polymer chains will start to interact when the distance between the particles is equal to $\approx 2L$, where L is the size of the polymer molecule taken as $\langle r^2 \rangle^{1/2}$. On further approach such that the interparticle distance is between L and $2L$ (Fig. 4.19b), the polymer chains may interpenetrate. The concentration of the polymer is increased in the interpenetration region, and in certain solvents this can lead to a repulsion. The repulsion is said to arise from a mixing (or rather a demixing) effect due to the need of the polymer chains to avoid other chains in the interpenetration region of increased concentration [11]. The repulsion can also be thought of as corresponding to an osmotic pressure and is sometimes described in these terms. In some

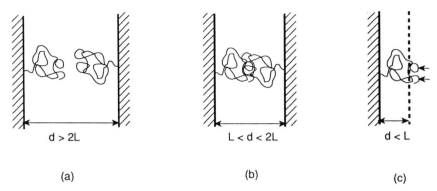

Figure 4.19 The range of steric repulsion in polymeric stabilization. (a) There is no interpenetration for large separations. (b) When the particles become close enough, interpenetration of the coils leads to a repulsion in good solvents or an attraction in poor solvents. (c) For small separations, compression of the coils always leads to a repulsion.

solvents, as we shall discuss below, the interpenetration of the coils can lead to an attraction, thereby causing flocculation. As the distance of approach between the particles decreases further to less than L (Fig. 4.19c), not only does interpenetration occur but also the polymer chain on one particle may be compressed by the rigid surface of the other particle. This compression generates an elastic contribution to the stabilization that always opposes flocculation [11]. The elastic contribution to the repulsion is sometimes described as an entropic effect or a volume restriction effect. At these separations, there are regions of space that are no longer accessible to a given chain. Conformations that would otherwise have been accessible are excluded, so there is a loss of configurational entropy. The loss in entropy is given by the Boltzmann equation,

$$\Delta S = k \ln \Omega \tag{4.53}$$

where Ω is the loss in the number of conformations of the polymer chain. The free-energy change due to the loss in entropy is

$$\Delta G_{en} = -T \Delta S \tag{4.54}$$

where T is the temperature. The term ΔG_{en} is always positive (because ΔS is negative), and this is equivalent to a repulsion.

To summarize at this stage, the interaction between the polymer chains on the particles may be separated into (i) a mixing effect that produces either a repulsion or an attraction and (ii) an elastic effect that is always repulsive (Table 4.2). The mixing effect is also described as an

Table 4.2 Interactions Between Polymer Chains in Steric Stabilization

Interaction	Nature of force
Mixing effect (osmotic effect)	Repulsive (good solvent)
	Attractive (poor solvent)
Elastic effect (entropic effect or volume restriction effect)	Repulsive (always)

osmotic effect, whereas the elastic effect is described as an entropic effect or a volume restriction effect. The free energy of the polymeric interaction can be written

$$\Delta G_{\text{steric}} = \Delta G_{\text{mix}} + \Delta G_{\text{elastic}} \tag{4.55}$$

where ΔG_{mix} and $\Delta G_{\text{elastic}}$ are the free-energy change due to the mixing and elastic effects, respectively.

D. Effect of Temperature: The Theta Temperature

When compared to electrostatic stabilization, temperature has a significantly greater effect on the stability of sterically stabilized suspensions. As we saw above (Fig. 4.19b), interpenetration of the polymer chains gives rise to a mixing effect. At a certain temperature, referred to as the Θ (theta) temperature, the interpenetration of the coils does not lead to a change in the free energy of mixing ($\Delta G_{\text{mix}} = 0$), and a system of the polymer dissolved in the solvent behaves like an ideal solution. The solvent is referred to as a Θ solvent.

At temperatures greater than the Θ temperature, the coils may repel one another, and this repulsion leads to stability of the suspension ($\Delta G_{\text{mix}} > 0$). The solvent is referred to now as a good solvent (or a "better than Θ" solvent). However, at temperatures less than the Θ temperature, the coils may attract each other ($\Delta G_{\text{mix}} < 0$), leading to flocculation of the suspension. In this case, the solvent is referred to as a poor solvent (or a "worse than Θ" solvent). As shown schematically in Fig. 4.20, we would therefore expect the stability of the suspension to change very markedly within a few degrees of the Θ temperature of the polymer solution. Indeed, for many sterically stabilized suspensions, the temperature at which the stability of the suspension changes dramatically (referred to as the critical flocculation temperature) correlates very well with the Θ temperature of the solution [11].

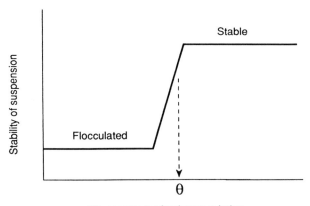

Figure 4.20 The stability of the suspension as a function of the temperature of the polymer solution. The stability changes dramatically within a few degrees of the Θ temperature of the solution.

E. Stability of Sterically Stabilized Suspensions

As outlined earlier for electrostatically stabilized suspensions, the total energy of interaction, V_T, must be considered. For sterically stabilized suspensions, V_T is the sum of the free energy of the steric interaction, ΔG_{steric}, given by Eq. (4.55) and the van der Waals energy of attraction, V_A, discussed earlier in this chapter; that is,

$$V_T = \Delta G_{\text{steric}} + V_A \tag{4.56}$$

The variation in V_T can be illustrated with the results shown in Fig. 4.21 for the interaction between mica surfaces in the presence of a solution of polystyrene in cyclopentane [12]. The degree of polymerization of the polymer (i.e., the number of segments in the chain) is 2×10^4, which gives an $\langle r^2 \rangle^{1/2}$ of ≈ 20 nm. In the absence of polymer molecules, there is simply a van der Waals attraction. At moderate coverage of the polymer chains on the mica surfaces, the repulsion overcomes the attraction at a separation of ≈ 20 nm, whereas for high coverage the interaction is repulsive at all separations.

The results shown in Fig. 4.21 serve to illustrate the influence of two important issues in steric stabilization: the relative unimportance of van der Waals attraction in steric stabilization, especially at fairly high coverage, and the effect of the surface coverage of polymers. In certain cases,

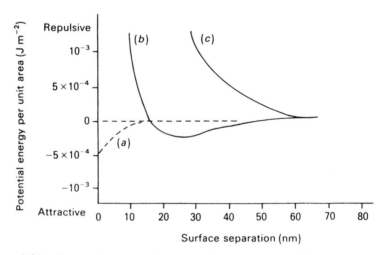

Figure 4.21 Interaction energy between mica surfaces in the presence of a solution of polystyrene in cyclopentane. (a) In the absence of polymer molecules, there is a simple van der Waals attraction. (b) At moderate surface coverage, the steric repulsion begins to counteract the attraction at separations less than 20 nm. (c) At high surface coverage, the interaction is repulsive at all separations. (From Ref. 12.)

low coverage may actually lead to flocculation. This happens when the particle has usable anchoring sites that are not occupied. Polymers from a neighboring particle can then attach themselves and form bridges, leading to flocculation. This type of flocculation is referred to as bridging flocculation.

As discussed earlier, other factors are also important:

1. The nature of the attachment of the polymers to the particle surfaces. Irreversible adsorption or chemical attachment is required for stability.
2. The character of the solvent. Solvents with good solubility for the polymer molecules are required for stability.
3. The temperature of the system. Temperatures above the Θ temperature (required for a good solvent) must be used.
4. The size of the polymer chain in solution. The size, taken as $\langle r^2 \rangle^{1/2}$, increases as the square root of the degree of polymerization and controls the range of the steric repulsion.

Science of Colloidal Processing

F. Depletion Stabilization

Depletion stabilization, as mentioned earlier, refers to stabilization of colloidal suspensions by polymers in free solution. To gain an understanding of the stability in such systems, let us consider the approach of two particles from a large separation distance (Fig. 4.22). Closer approach of the

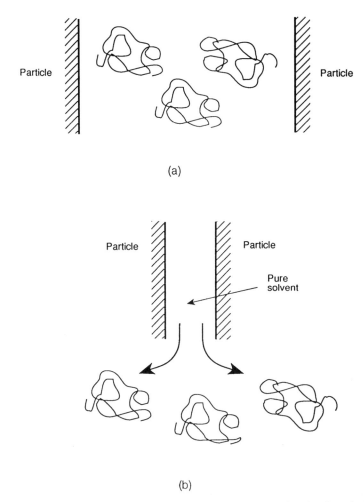

Figure 4.22 Polymeric stabilization by free polymer molecules in solution. (a) Depletion stabilization at separations greater than the diameter of the polymer coil. (b) Depletion flocculation at small separations.

particles must be accompanied by demixing of the polymer molecules and the solvent in the interparticle region. Consequently, work must be done to make the polymer molecules leave the interparticle region. This corresponds to a repulsion between the particles that, if high enough, can lead to stabilization of the suspension. This type of stabilization is referred to as depletion stabilization.

If the repulsion is not high enough, the particles can approach closer, and when the separation becomes smaller than the size of the polymer chain in solution (i.e., approximately $\langle r^2 \rangle^{1/2}$), all of the polymer chains will have been excluded from the interparticle region (Fig. 4.22b). Closer approach of the particles will be favored because we have now created a type of reverse osmotic effect. The pure solvent between the particles will diffuse into the surrounding region in an attempt to lower the concentration of the polymer. This reverse osmotic effect is equivalent to an attraction between the particles, and flocculation occurs. This type of flocculation is referred to as depletion flocculation.

4.7 STRUCTURE OF CONSOLIDATED COLLOIDS

So far in this chapter we have been concerned mainly with the principles of colloid stability. The factors that influence the stability of the suspensions and how these suspensions can be manipulated to alter their stability have been discussed. We now proceed to an issue of great practical importance in ceramic processing: the relation between the nature of the suspension and the structure of the consolidated colloid. The structure of the consolidated powder form, as we found earlier, has a significant influence on the quality of the fired microstructure. We therefore need to understand how the properties of the suspension can be manipulated to produce the desired structure in the consolidated colloid.

An initial step in the colloidal processing of most ceramic powders may consist of the removal of hard agglomerates and large particles, if present, from the suspension by sedimentation and removal of the supernate. The next step is then the consolidation of the suspension. This is achieved by a variety of methods, including gravitational settling, centrifuging, and filtration for colloids with a low concentration of particles and casting (i.e., pressure, slip, or tape casting) for concentrated suspensions. Stable suspensions of submicrometer particles are known to settle very slowly under normal gravity, so the settling rate is normally increased by centrifuging or filtration. The consolidated colloid has a densely packed structure (Fig. 4.23a). Because the particles flocculate while settling, unstable suspensions settle faster and produce a loosely packed structure (Fig. 4.23b).

Science of Colloidal Processing

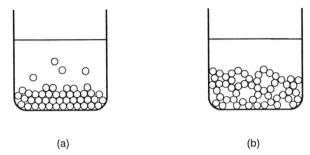

Figure 4.23 Schematic of a consolidated colloid formed from (a) a stable suspension and (b) a flocculated suspension.

The structure of the consolidated colloid formed from the stable suspension and the flocculating suspension is somewhat more complex than the arrangements illustrated in Fig. 4.23. The microstructures of nearly monodisperse SiO_2 particles shown in Fig. 4.24 indicate that the difference in the structures formed from the two systems is not controlled by the way in which the individual particles arrange themselves in the sediment but rather the way in which they group themselves to form densely packed multiparticle units called domains [13]. As Fig. 4.24c shows that, for a consolidated colloid formed from a stable suspension of spherical, nearly monodisperse powders, the particles within the domains are arranged in a periodic pattern not unlike that found in crystalline grains. Domains are also formed if the particles are irregular in shape, but the periodic arrangement of the domains is lost.

The microstructures of Fig. 4.24 also show that the size of the domains can be altered through manipulation of the interparticle forces. The size of the domains increases for higher double layer repulsion; furthermore, the efficiency of the packing of the domains increases, leading to high packing density in the consolidated colloid (Fig. 4.24c). As the repulsion decreases (or equivalently, as the attraction increases), the domain size decreases and the interdomain pore size increases. The interdomain porosity is therefore the main cause of the low packing density of flocculated colloids.

We can attempt to describe the structure of the consolidated colloid in terms of the particle interactions. For highly attractive forces, the particles will "stick" on contact with another particle, leading to the formation of a highly disordered system (Fig. 4.24a). As the repulsion increases, the particle will be able to undergo a certain amount of rearrangement into low-energy positions, leading to an increase in the domain size and a

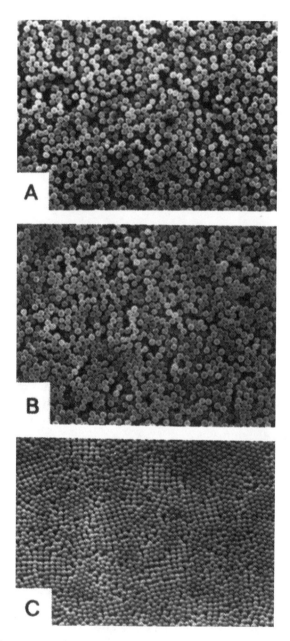

Figure 4.24 Microstructures of particle domains formed by centrifugal consolidation of SiO$_2$ colloidal suspensions at (A) $\zeta = 0$ mV, (B) $\zeta = 68$ mV, and (C) $\zeta = 110$ mV. The average particle diameter is 0.7 µm. (From Ref. 13, with permission.)

Science of Colloidal Processing

reduction of the interdomain pore size (Fig. 4.24b). Finally, when the repulsion becomes dominant, each additional particle can interact with the others to produce larger domains with small interdomain pores. (Fig. 4.24c). However, it should be noted that the domain boundaries constitute defects that interfere with sintering.

We have discussed the structure of consolidated colloids formed from electrostatically stabilized suspensions, but the same principles apply to sterically stabilized suspensions. Studies with polystyrene spheres stabilized with polymers have yielded structures similar to those shown in Fig. 4.24.

4.8 RHEOLOGY OF COLLOIDAL SUSPENSIONS

Rheology plays an important part in the processing of ceramics from colloidal suspensions. Rheological measurements can be used to characterize the state of dispersion and for optimizing, especially in the case of concentrated systems, the flow behavior of the suspension. The stability of the dispersion, as discussed before, has a significant influence on the structure of the consolidated colloid. When the suspension is consolidated by casting methods (e.g., slip or tape casting), we require, on the one hand, that the suspensions contain the highest possible fraction of particles. A concentrated suspension serves to reduce the shrinkage during drying of the cast and to produce a consolidated powder form with high packing density. On the other hand, we also want the suspension to have a low enough viscosity that it can be poured. Rheological measurements provide an important means of optimizing these requirements. We start by considering the general types of rheological properties exhibited by colloidal suspensions.

A. Rheological Properties

A characteristic property that is used to describe the flow of a liquid is the viscosity η, defined by

$$\eta = \tau/\dot{\gamma} \tag{4.57}$$

where τ is the shear stress and $\dot{\gamma}$ is the shear rate. If η is independent of the shear rate (or the shear stress), the liquid is said to be Newtonian. Many simple liquids show Newtonian behavior. For more complex systems such as polymer solutions and colloidal suspensions, η is not independent of the shear rate, and the behavior is said to be non-Newtonian. We must now write

$$\eta(\dot{\gamma}) = \frac{d\tau}{d\dot{\gamma}} \tag{4.58}$$

where $\eta(\dot{\gamma})$, the viscosity at a given strain rate, is found from the slope of the curve of shear stress versus strain rate. Figure 4.25a shows the curves for Newtonian behavior and types of non-Newtonian behavior found with colloidal suspensions. When the viscosity increases with increasing shear rate, the behavior is described as shear thickening. Suspensions containing a high concentration of nearly equiaxial particles may show shear thickening and eventually dilatancy at high strain rates. Moderately concentrated suspensions of small elongated particles may show shear thinning or pseudoplasticity, where the viscosity decreases with increasing shear rate. Shear thickening or shear thinning behavior is sometimes described by an empirical relation of the form

$$\tau = K\dot{\gamma}^n \tag{4.59}$$

where K is called the consistency index and n is an exponent that indicates the deviation from Newtonian behavior. If $n = 1$, the suspension is Newtonian; when $n < 1$, the suspension shows shear thinning or pseudoplastic flow; and when n is > 1, the flow is described as shear thickening or dilatant.

For systems in which the viscosity decreases with increasing shear rate after an initial threshold stress called the yield stress, τ_y, the flow behavior is described as plastic. Concentrated suspensions of clay particles commonly exhibit plastic behavior. A material obeying the equation

$$\tau - \tau_y = \eta\dot{\gamma} \tag{4.60}$$

where η is independent of the shear rate, is said to show Bingham-type behavior. The viscosity of most plastic materials, however, is dependent on the shear rate, and the flow behavior is often described by a power

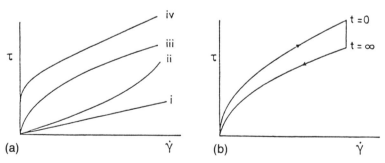

Figure 4.25 Typical rheological behavior of colloidal suspensions. (a) (i) Newtonian, (ii) shear thickening, (iii) shear thinning or pseudoplastic, (iv) plastic. (b) Thixotropic. (From Ref. 14, with permission.)

Science of Colloidal Processing

law relation,

$$\tau - \tau_y = K\dot{\gamma}^n \tag{4.61}$$

where K and n were defined earlier in Eq. (4.59).

When the viscosity depends not only on the shear rate but also on the time, we have thixotropic behavior (Fig. 4.25b). This type of behavior is more often observed in flocculated suspensions and colloidal gels. When the suspension is sheared, the flocs are broken down, leading to a distribution of floc sizes. Often the regeneration of the flocs is slow, which causes the resistance to flow to decrease.

For the range of values encountered with colloidal suspensions, a concentric cylinder rotating viscometer is often used to measure the viscosity. Less often used is the cone-and-plate viscometer. For liquids and solutions, a capillary viscometer is often used. These instruments are available commercially and are well described by the manufacturers.

B. Viscosity of Colloidal Suspensions

The rheology of the suspension depends on the concentration of particles and on the interparticle forces between the particles [14]. The effects of these two factors are considered in the following sections.

Suspensions of Hard Spheres

To understand the effect of the concentration of particles, we consider the particles to be hard spheres and neglect the presence of charged electrical double layers or adsorbed polymers. For dilute suspensions, for which all interactions between the particles are neglected, the viscosity is given by the Einstein equation,

$$\frac{\eta}{\eta_0} = 1 + 2.5f \tag{4.62}$$

where η is the viscosity of the suspension, η_0 is the viscosity of the liquid medium, and f is the volume fraction of the particles. Equation (4.62) can be used with negligible error for $f < 0.01$.

When interactions are considered, the viscosity at low shear rates when the particle distribution is only slightly altered by the flow is given by

$$\frac{\eta(0)}{\eta_0} = 1 + 2.5f + 6.2f^2 \tag{4.63}$$

where $\eta(0)$ is the viscosity of the suspension at low shear rates. Equation (4.63) has been shown to provide an adequate fit to experimental data for $f < 0.15$.

For $f > 0.15$, no rigorous theory exists, and empirical equations have to be used. One of these is

$$\frac{\eta}{\eta_0} = (1 - k_p f)^{-[\eta]/k_p} \tag{4.64}$$

where $[\eta]$ is the intrinsic viscosity of the suspension. When η is expressed relative to η_0 as in Eq. (4.62), $[\eta]$ is defined by

$$[\eta] = \frac{1}{\eta_0} \left(\frac{d\eta}{df} \right)_{f \to 0} \tag{4.65}$$

We can see from Eq. (4.62) that $[\eta] = 2.5$ for a suspension of hard spheres. The term k_p in Eq. (4.64) is defined as the reciprocal of the volume fraction of particles, $1/f$, at which the viscosity becomes practically infinite. This limiting volume fraction is ≈ 0.65 for hard spheres but may be as low as 0.40–0.50 for suspensions, for which interactions between the particles are significant.

The shape of the particles also has an effect on the viscosity [14]. The effect of shape would be important, for example, in the processing, by colloidal methods, of ceramic composites reinforced with whiskers (short single-crystal fibers) or platelets. Figure 4.26 shows the results of theoretical calculations for the intrinsic viscosity $[\eta]$ as a function of the axial ratio for particles with the shape of prolate ellipsoids. For an axial

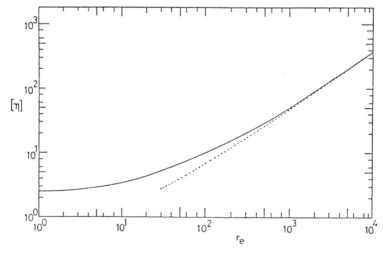

Figure 4.26 Intrinsic viscosity as a function of axial ratio for prolate ellipsoids. (From Ref. 14, with permission.)

ratio of 15–20, which may be relevant to the use of whiskers or platelets in ceramic composites, the results show that [η] has a value of ≈4.5 compared to 2.5 for spherical particles.

Effect of Particle Charge

For colloidal suspensions stabilized by electrostatic repulsion, the charges, as we discussed earlier, consist of a surface charge and a more or less diffuse double layer of charges. When the suspension is sheared, interactions between the charges provide additional mechanisms for energy dissipation that lead to an increase in the viscosity. In the case of concentrated suspensions, the increase in the viscosity can be quite significant compared to the case of a suspension with the same concentration of hard spheres.

Three different mechanisms have been identified to account for the effects of electrostatic charges. They are often referred to as the primary, secondary, and tertiary electroviscous effects. The primary electroviscous effect arises from the deformation of the diffuse double layer due to the shearing forces of the flow. For weak flows, which cause only a small distortion of the double layer, and for small ζ potentials (<25 mV), theoretical analysis shows that [14]

$$[\eta] = \frac{5}{2}\left(1 + \frac{(2\epsilon\epsilon_0\zeta)^2 \, Ka \, (1 + Ka)^2 \, F(Ka)}{\sigma\eta_0 a^2}\right) \quad (4.66)$$

where ϵ is the dielectric constant of the liquid, ϵ_0 is the permittivity of free space, a is the radius of the particle, K^{-1} is the Debye length given by Eq. (4.25), σ is the specific conductivity of the liquid, and $F(Ka)$ is a power series with limiting values of

$$F(Ka) = \frac{1}{200\pi Ka} + \frac{11Ka}{3200\pi} \quad \text{for small } Ka \quad (4.67a)$$

and

$$F(Ka) = \frac{3}{2\pi(Ka)^4} \quad \text{for large } Ka \quad (4.67b)$$

For a given value of K, we see that [η] is predicted to increase as ζ^2. This increase in [η] may be explained in terms of an increase in the hydrodynamic radius of the particle due to its higher ζ potential. Equations (4.66) and (4.67) also predict that [η] is independent of K at small values of Ka but is inversely proportional to K for large Ka. To summarize, although the effect of the ζ potential is at least qualitatively predictable, it must be realized that other important effects such as shear thinning may occur and that these are difficult to treat theoretically.

The secondary electroviscous effect is due to the electrostatic repulsion between the particles when they pass close to one another in the shear field. This repulsion provides a mechanism for energy dissipation because, compared to hard spheres, additional work has to be done against the repulsion between the particles. The third electroviscous effect occurs in suspensions stabilized by charged polymers (polyelectrolytes). It is due to the stretching of the polymers caused by the mutual repulsion of the charges. The phenomena associated with the electroviscous effects are complex, and the reader is referred to Ref. 14 for a discussion.

Effect of Adsorbed Polymer Layers

The stabilization of suspensions, as discussed earlier, can also be produced by polymer molecules adsorbed (or chemically bonded) onto the surfaces of the particles. The polymers are either organic, giving rise to steric stabilization, or ionizable (i.e., polyelectrolytes), in which case the stabilization is produced by steric and electrostatic repulsion (i.e., electrosteric repulsion). They may produce dramatic changes in the viscosity of the suspension; often, polymer additions of less than 0.5% of the weight of the particles allow the incorporation of high solids content (e.g., 50–60 vol % of the suspension) while keeping the viscosity fairly low.

Poly(vinyl alcohol) (PVA) is an organic polymer with the general formula ($-CH_2-CHOH-$)$_n$, where n is the degree of polymerization. Its effect on the viscosity of suspensions of SiO_2 particles dispersed in water has been investigated by Sacks et al. [15]. The SiO_2 particles were spherical and nearly monodisperse, and the particle concentration in the suspension was 20 vol %. The results serve to illustrate some of the phenomena described earlier. The adsorption of the PVA onto the SiO_2 particle surfaces is shown in Fig. 4.27 for a suspension prepared at a pH equal to 3.7. Initially, all of the PVA added to the suspension is adsorbed onto the particle surfaces. As the PVA concentration increases, the particle surfaces become saturated, and a plateau region occurs in the adsorption isotherm. This first plateau is believed to be due to the development of monolayer coverage of PVA on the particle surfaces. Further increases in the PVA concentration lead to an increase in the adsorption and the occurrence of a second plateau region; it is believed that this is an indication of the development of a denser packing of the adsorbed polymer molecules or the occurrence of multilayer adsorption.

Figure 4.28 shows the effect of the adsorbed PVA on the relative viscosity of the suspension, η/η_0, for suspensions prepared at a pH of 3.7. Also shown are the data for an electrostatically stabilized suspension prepared at pH 7.0 without any PVA. For the suspensions containing no PVA, we see that the stabilized suspension with a pH of 7.0 has a fairly

Science of Colloidal Processing

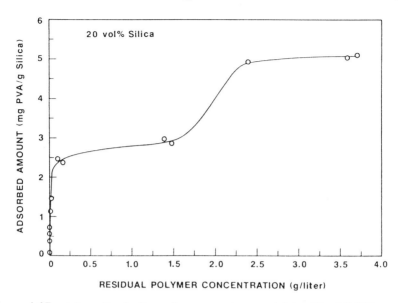

Figure 4.27 Adsorption isotherm for suspensions containing 20 vol % SiO$_2$ particles in solutions of poly(vinyl alcohol). (From Ref. 15, with permission.)

low viscosity and shows Newtonian behavior. The viscosity of the unstabilized suspension with a pH of 3.7, however, is much higher, and a high degree of shear thinning occurs. Following our earlier discussion, we see clearly the benefits of this type of rheological data for the characterization of the state of dispersion of the suspension.

Considering now the effect of the PVA, we see that initially the viscosity increases as the amount of adsorbed PVA increases from 0 to 1.1 mg per gram of SiO$_2$ and a high degree of shear thinning occurs. The observed effect is most likely due to bridging flocculation discussed earlier (see Section 4.6E).This is also supported by the yield stress data shown in Fig. 4.29. However, when the amount of adsorbed PVA becomes greater than 1.1 mg/g SiO$_2$, the trend is reversed, and the viscosity decreases with increasing amounts of adsorbed PVA. The yield stress of the suspensions also decreases.The PVA coverage on the particles has now become sufficient to cause a transition from bridging flocculation to steric stabilization. For an amount of adsorbed PVA equivalent to that of the first plateau in the adsorption isotherm of Fig. 4.27 ($\approx$2.9 mg/g of SiO$_2$), we see that the viscosity is fairly low and the behavior is almost Newtonian. The viscosity decreases slightly for PVA concentrations above the first plateau of the adsorption isotherm, but it is still somewhat higher

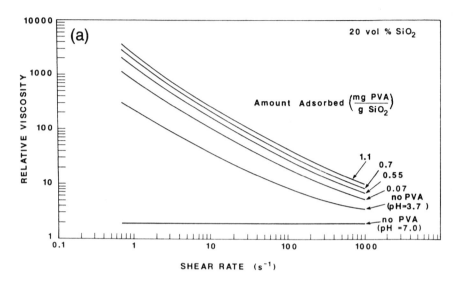

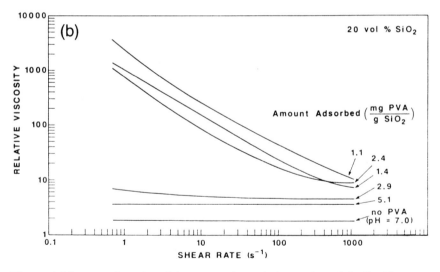

Figure 4.28 The viscosity of the suspension relative to that of the liquid versus shear rate for suspensions containing 20 vol % SiO$_2$ particles. The suspensions prepared at pH 3.7 (the isoelectric point of the SiO$_2$) and pH 7.0 contained no poly(vinyl alcohol). All of the suspensions containing poly(vinyl alcohol) were prepared at pH 3.7. (From Ref. 15, with permission.)

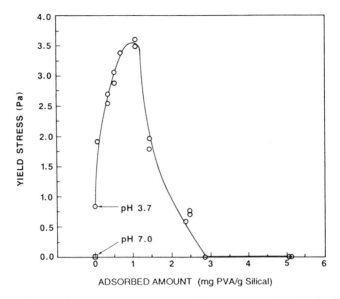

Figure 4.29 Yield stress as a function of the amount of poly(vinyl alcohol) adsorbed on to the SiO_2 for suspensions prepared at pH 3.7 with 20 vol % SiO_2. (From Ref. 15, with permission.)

than that for the electrostatically stabilized suspension prepared at a pH of 7.0.

4.9 STABILIZATION OF SUSPENSIONS WITH POLYELECTROLYTES: ELECTROSTERIC STABILIZATION

As outlined earlier, suspensions can also be stabilized by electrosteric repulsion, involving a combination of electrostatic repulsion and steric repulsion. Electrosteric stabilization requires the presence of adsorbed polymers and a significant double layer repulsion. An important way of achieving electrosteric stabilization is through the use of polyelectrolytes, i.e., polymers that have ionizable groups that dissociate to produce charged polymers. Polyelectrolytes are widely used industrially in the preparation of highly concentrated ceramic suspensions (>50 vol% particles) which are subsequently consolidated and fired to produce dense ceramic articles.

Because of the combination of electrostatic effects and steric effects, the role of polyelectrolytes in the processing of suspensions is not clear.

However, a detailed study of the stabilization of Al_2O_3 suspensions with polyelectrolytes was made recently by Aksay and coworkers [16,17]. Here, we describe the basic features of the polyelectrolyte role based on their work.

A. Charging of Polyelectrolytes and Particles

Figure 4.30 shows schematically the structure of two common polyelectrolytes: poly(methacrylic acid), denoted as PMAA, and poly(acrylic acid), PAA. The polyelectrolytes are normally used in the form of solutions of the sodium salt (e.g., the sodium salt of poly(methacrylic acid), denoted as PMAA-Na) or the ammonium salt (e.g., PMAA-NH_4). However, the PMAA (or PAA) portion of the polyelectrolyte is of primary importance in adsorption and stabilization. The degree of polymerization, n (or the molecular weight) can vary between $\approx 1,000$ and $\approx 50,000$. The functional groups of PMAA and PAA consist of carboxylic acid (COOH) groups, which can exist as COOH or dissociated to COO^-. The dissociation can be written in a general form as

$$A\text{—}COOH + H_2O \leftrightarrow A\text{—}COO^- + H_3O^+ \quad (4.68)$$

Depending on the conditions of the solvent (e.g., the pH and the ionic concentration), the fraction of the functional groups which are dissociated (i.e., COO^-) and those which are not dissociated (i.e., COOH) will vary. As the fraction dissociated, α, increases from ≈ 0 to ≈ 1, the charge on the polymer varies from neutral to highly negative. The behavior of the polyelectrolyte is therefore dependent on the solvent conditions. Figure 4.31 shows the fraction of dissociated acid groups as a function of pH and NaCl salt concentration for the sodium salt of PMAA. As the pH and salt concentration increases, the extent of dissociation and the negative charge of the polymer increase. At a pH greater than ≈ 8.5, the polymer is highly negative with $\alpha \approx 1$. Under these conditions, the free polymer is in the

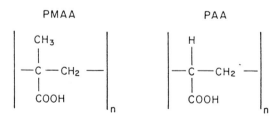

Figure 4.30 Schematic diagram showing polymer segments of poly(methacrylic acid), PMAA, and poly(acrylic acid), PAA. (From Ref. 17, with permission.)

Science of Colloidal Processing

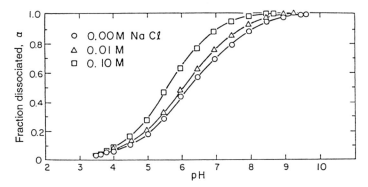

Figure 4.31 Fraction of acid groups dissociated versus pH for the sodium salt of poly(methacrylic acid), PMAA-Na, for three different salt concentrations. (The molecular weight of the polyelectrolyte is 15,000.) (From Ref. 16, with permission.)

form of relatively large expanded random coils (diameter ≈ 10 nm). As the pH decreases, the number of negative charges decreases, with the polyelectrolyte being effectively neutral at a pH of ≈3.4. In the neutral condition, the polymer forms small coils (diameter ≈ 3 nm) or clumps.

As discussed earlier, the surface charge of the Al_2O_3 particles also depends on the conditions of the solution. Figure 4.32 shows the surface

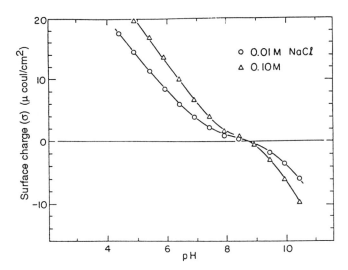

Figure 4.32 Surface charge density versus pH for a suspension of α-Al_2O_3 powder (Sumitomo AKP-30.) (From Ref. 16, with permission.)

charge density, σ_0, of the Al_2O_3 particles as a function of pH for two NaCl salt concentrations. At any pH, there is a large number of positive, negative and neutral charges on the particle surface; σ_0 gives the net charge. At the point of zero charge PZC, the surface is neutral, i.e., the number of positive charges is equal to the number of negative charges. From Fig. 4.32, the PZC of the Al_2O_3 powder is at a pH of ≈ 8.7.

Figure 4.31 and 4.32 show that considerable electrostatic attraction should occur between the negatively charged polyelectrolyte and the positively charged Al_2O_3, particularly in the pH range of ≈ 3.5 to 8.7. This attraction will influence the adsorption of the polymer and, hence, the stability of the suspension.

B. Polyelectrolyte Adsorption and Suspension Stability

Figure 4.33 shows the adsorption of the polyelectrolyte at various pH values. The results are plotted in the form of milligrams of PMAA adsorbed per square meter of the Al_2O_3 surface versus the initial amount of PMAA-Na added (as a percent of the dry weight of Al_2O_3, i.e., on a dry weight basis, dwb). The solid diagonal line represents the adsorption behavior that would occur if 100% of the PMAA added were to adsorb. It is clear that the amount of PMAA adsorbed increases with decreasing pH. For pH values greater than the PZC, $\alpha \approx 1$ and the negative sites on

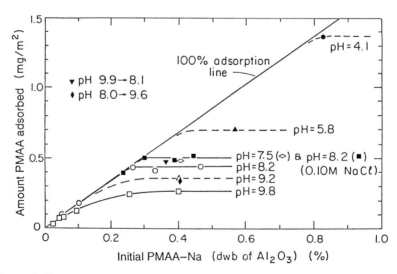

Figure 4.33 Amount of PMAA salt adsorbed on α-Al_2O_3 powder as a function of the initial amount of PMAA-Na added to the suspension. (From Ref. 16, with permission.)

Science of Colloidal Processing

the polyelectrolyte tend to repel each other. The repulsion suppresses the formation of loops in the conformation of the polyelectrolyte adsorbed on to the particle surface. The polyelectrolyte adsorbs in a flat conformation with each chain apparently covering a relatively large amount of surface area. As the pH decreases and α approaches zero, the polyelectrolyte chains become uncharged and the formation of loops in the adsorbed polymer conformation is, in principle, enhanced. Thus the projected area per adsorbed chain is relatively small and more adsorbed chains are required to establish a monolayer.

By studying the sedimentation behavior of the Al_2O_3 suspension, the conditions for stability can be determined as a function of pH and polyelectrolyte concentration. The observations can be conveniently plotted on a map, called a colloid stability map (Fig. 4.34). The map shows, for a given concentration of Al_2O_3 particles (20 vol%), the regions of stability and instability (tendency towards flocculation). Regions below the curve are unstable. Regions near and slightly above the curve are stable and dispersed; stabilization results from electrosteric repulsion due to the adsorbed polyelectrolyte. Regions further above the curve have appreciable amounts of free polymer in solution. Below a pH value of ≈ 3.3, the polyelectrolyte is essentially neutral and stabilization is believed to be produced

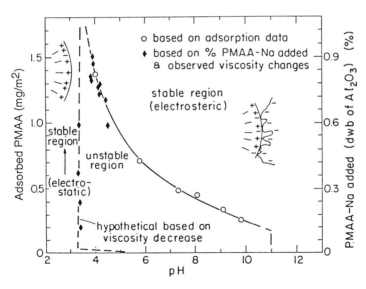

Figure 4.34 Colloid stability map of α-Al_2O_3 powder (20 vol%) suspensions as a function of adsorbed PMAA and pH. (From Ref. 16, with permission.)

predominantly by electrostatic repulsion. The role of the neutral polyelectrolyte is not clear. It may be sufficiently desorbed or the amount adsorbed may be sufficiently low that double layer repulsion due to the particle charge can still occur.

C. Rheological Behavior and Suspension Processing

Figure 4.35 shows the viscosity of Al_2O_3 suspensions stabilized with PMAA (molecular weight $\approx 15{,}000$) as a function of pH for several particle concentrations. All of the suspensions were in the stable region of Fig. 4.34 and fairly close to the stability curve (i.e., the amount of polyelectrolyte was close to that required for saturation of the particle surface). For the suspension with 20 vol% Al_2O_3, the viscosity is fairly low and pH has little effect on the viscosity. However, as the particle concentration becomes higher, the viscosity increases and pH has a significant effect. The viscosity goes through a minimum at a pH close to 8.8. This value coincides with the PZC of the Al_2O_3. Based on our earlier discussion of electrostatic stabilization, the viscosity would be expected to be a maximum at this pH. The results of Fig. 4.35 indicate that, upon introduction of the polyelectrolyte, the particle interactions are governed by the adsorbed polyelectrolyte and not by the bare Al_2O_3 surface.

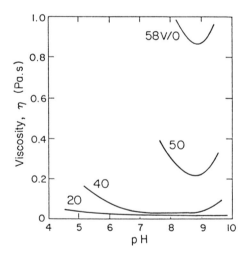

Figure 4.35 Viscosity (at a shear rate of 9.3 s^{-1}) versus pH for α-Al_2O_3 suspensions with particle concentrations of 20 to 58 vol% and stabilized with PMAA. The amounts of PMAA added were sufficient to provide complete surface adsorption. (From Ref. 17, with permission.)

Science of Colloidal Processing

The amount of polyelectrolyte added and the molecular weight also have an effect. Figure 4.36 shows the viscosity of a suspension containing 50 vol% Al_2O_3 versus the amount of polyelectrolyte (PAA) added for three different molecular weights. For each molecular weight, there is a critical amount of polyelectrolyte that must be added before stabilization and low viscosities can be achieved. This critical amount corresponds to the amount required to saturate the surface. Further additions of polyelectrolyte past the saturation limit only serve to produce excess polyelectrolyte in solution. The effects of excess polyelectrolyte are more drastic for higher molecular weight.

To summarize, by understanding the chemistries of the polyelectrolyte and the particle surface, the polyelectrolyte adsorption behavior and the polyelectrolyte rheological behavior, the viscosity of the suspension can be controlled for a given concentration of particles. In this way, stable suspensions with high particle concentration (>60 vol%) and relatively low viscosity can be prepared and consolidated to produce homogeneous powder compacts with high green density (>65% of the theoretical). Compacts prepared from such highly concentrated suspensions provide considerable benefits for sintering, including a lowering of the firing temperature, high density and fine grain size [17].

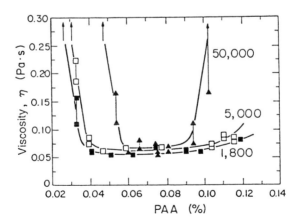

Figure 4.36 Viscosity versus the amount of poly(acrylic acid), PAA, added (as a percent of the dry weight of Al_2O_3) for various PAA molecular weights. The suspensions, prepared at a pH of 9, contained 50 vol% α-Al_2O_3 particles (Sumitomo AKP-20). The unfilled data points are the initial values and the filled points are the values after 10 min. (From Ref. 17, with permission.)

4.10 CONCLUDING REMARKS

In this chapter we outlined the basic principles of electrostatic and steric stabilization of colloidal suspensions. As we have remarked at various instances, and especially for the case of steric stabilization, a considerable gap still exists in our understanding of many aspects of colloid stability. Nevertheless, the principles outlined in this chapter provide a basis for the manipulation of the experimental parameters to produce a suspension of the desired stability. Colloidal methods have been used for a long time in the forming of traditional, clay-based ceramics, and they are now being used increasingly in the advanced ceramics sector. As discussed in Chapter 1, the use of colloidal methods for improving the packing uniformity of the consolidated powder form has yielded clear benefits in the control of the fired microstructure. Finally, many of the principles outlined in this chapter will be encountered again in Chapter 6 when we consider the methods used to consolidate ceramic powders.

PROBLEMS

4.1 Calculate the van der Walls potential energy, V_A, and the gravitational potential energy, V_G, between two colloidal particles of radius $a = 0.2$ μm separated by a distance $h = 10$ nm in a dilute hydrochloric acid solution at 27°C, given that the Hamaker constant $A = 10^{-20}$ J, the surface potential of the particles $\phi_o = 25$ mV, the dielectric constant of the solution $\epsilon = 80$, the concentration of the hydrochloric acid = 5 millimoles/liter and the density of the particles d = 4 g/cm^3.

4.2 For the conditions given in Problem 1, determine the electrostatic repulsive potential energy, V_R, between the particles and the total potential energy, V_T.

4.3 The clay mineral montmorillonite has the formula

$$Na_{0.33}(Al_{1.67}Mg_{0.33})(Si_2O_5)_2(OH)_2$$

Assuming that all of the Na$^+$ ions pass into solution, determine the ion exchange capacity of montmorillonite (in coulombs per kilogram).

4.4 Stable suspensions of alumina particles and silica particles are prepared separately at pH = 6. Discuss the stability of the suspension formed by mixing the two suspensions.

4.5 Derive Eq. (4.29). Hence show that for a high surface potential, the double layer potential at a large distance from a flat surface decays as

Science of Colloidal Processing

$$\phi = \frac{4RT}{zF} \exp(-Kx)$$

4.6 Calculate the free energy of repulsion, V_R, between two double layers, each having an area of 1 cm² at a surface potential $\phi_0 = 4RT/F$, in a monovalent electrolyte giving a Debye length, $1/K$, equal to 10^{-6} cm and a distance of separation equal to 10 nm.

4.7 The surface potential of a flat surface in dilute hydrochloric acid solution at 25°C is 50 mV. If the acid concentration is 5 millimoles/liter, calculate the surface charge density. The dielectric constant of the solution can be assumed to be equal to that of water, i.e., 78.

4.8 (a) How does the double layer repulsion change if the medium water is replaced by methanol, assuming that the surface potential and the ionic concentration remain the same?
(b) If a monovalent electrolyte is replaced by a bivalent electrolyte but the Debye length, $1/K$, remains the same, how would the concentration of the electrolyte have to change?

4.9 In a particle electrophoresis apparatus, the particles in a dilute suspension were observed to move with a velocity of 0.2 mm/sec when a potential difference of 100 V was applied to the electrodes that were 10 mm apart. Determine the zeta-potential of the particles.

4.10 Derive Eq. (4.52) for the root mean square end-to-end distance, $\langle r^2 \rangle^{1/2}$, of a polymer chain. Derive expressions for (i) the average end-to-end distance $\langle r \rangle$ and (ii) the most probable end-to-end distance, r_{mp}, in terms of $\langle r^2 \rangle^{1/2}$.

REFERENCES

1. D. Tabor, in *Colloidal Dispersions* (J. W. Goodwin, Ed.) The Royal Society of Chemistry, London, 1982, Chap. 2.
2. D. Tabor, *Gases, Liquids and Solids*, 3rd ed., Cambridge Univ. Press, Cambridge, U.K., 1991, Chap. 12.
3. J. Mahanty and B. W. Ninham, *Dispersion Forces*, Academic, New York, 1976.
4. J. Lyklema, in *Colloidal Dispersions* (J. W. Goodwin, Ed.), The Royal Society of Chemistry, London, 1982, Chap. 3.
5. J. S. Reed, *Introduction to the Principles of Ceramic Processing*, Wiley, New York, 1988, p. 134.
6. J. W. Evans and L. C. De Jonghe, *The Production of Inorganic Materials*, Macmillan, New York, 1991.
7. D. J. Shaw, *Introduction to Colloid and Surface Chemistry*, 3rd Ed., Butterworths, London, 1980.

8. J. Th. G. Overbeek, *Colloid and Surface Science: A Self-Study Course,* Part II, *Lyophobic Colloids,* Massachusetts Institute of Technology, Cambridge, MA, 1972.
9. H. M. Jang and D. W. Fuerstenau, *Colloids Surf., 21*:238 (1986).
10. P. J. Flory, *Statistical Mechanics of Chain Molecules,* Hanser, New York, 1969.
11. D. H. Napper, in *Colloidal Dispersions* (J. W. Goodwin, Ed.), The Royal Society of Chemistry, London, 1982, Chap. 5.
12. J. Klein, *Physics World,* June 1989, pp. 35–38.
13. I. A. Aksay, in *Advances in Ceramics,* Vol. 9, *Forming of Ceramics* (J. A. Mangels and G. L. Messing, Eds.), The American Ceramic Society, Columbus, OH, 1984, pp. 94–104.
14. J. W. Goodwin, in *Colloidal Dispersions* (J. W. Goodwin, Ed.), The Royal Society of Chemistry, London, 1982, Chap. 8.
15. M. D. Sacks, C. S. Khadilkar, G. W. Scheiffele, A. V. Shenoy, J. H. Dow, and R. S. Sheu, in *Advances in Ceramics,* Vol. 21, *Ceramic Powder Science* (G. L. Messing, K. S. Mazdiyasni, J. W. McCauley, and R. A. Haber, Eds.), The American Ceramic Society, Westerville, OH, 1987, pp. 495–515.
16. J. Cesarano III, I. A. Aksay, and A. Bleier, *J. Am. Ceram. Soc., 71*(4):250 (1988).
17. J Cesarano III and I. A. Aksay, *J. Am. Ceram. Soc., 71*(12):1062 (1988).

FURTHER READING

J. Th. G. Overbeek, *Colloid and Surface Chemistry: A Self-Study Course,* Parts 1-4; Massachusetts Institute of Technology, Cambridge, MA, 1972.

J. Israelachvili, *Intermolecular and Surface Forces,* Academic Press, Orlando, FL, 1985.

D. H. Napper, *Polymeric Stabilization of Colloidal Dispersions,* Academic Press, New York, 1983.

R. J. Hunter, *Zeta Potential in Colloid Science-Principles and Applications,* Academic Press, New York, 1981.

5
Sol-Gel Processing

5.1 INTRODUCTION

The term *sol-gel* is used broadly to describe several processes in the areas of chemistry and the chemical synthesis of inorganic materials such as ceramics and glasses. A *sol* is a suspension of colloidal particles in a liquid or a solution of polymer molecules. The term *gel* refers to the semirigid mass formed when the colloidal particles are linked by surface forces to form a network or when the polymer molecules are cross-linked or interlinked. In materials synthesis, two different sol-gel processing routes can therefore be distinguished, depending on whether the gel consists of a network of colloidal particles (in which case the gel is sometimes referred to as a colloidal or particulate gel) or polymer chains (a polymeric gel). In many cases, however, the distinction between a colloidal gel and a polymeric gel may not be very clear.

The basic steps in the production of ceramics by the two sol-gel processing routes were outlined in Chapter 1 when we surveyed the common methods used for the fabrication of ceramics (see Fig. 1.5). The present chapter provides a more detailed examination of the science and practice of sol-gel processing in the fabrication of ceramics and glasses. The subject is covered in great depth in Ref. 1. We will recall that the polymeric gel route, sometimes referred to as the solution sol-gel process, provides a fabrication route that does not involve the use of solid particles as the starting materials. The colloidal gel route is more widely used industrially in applications such as nuclear fuels, catalysts and gas condensation columns. However, the research and development interest in the polymeric gel route has been growing rapidly. We will therefore focus more closely on the polymeric gel route.

The starting material for the fabrication of ceramics and glasses by the polymeric gel route is typically a solution of metal-organic compounds, especially metal alkoxides. The chemical reactions that occur during the conversion of the solution to the gel have a significant influence on the structure and chemical homogeneity of the gel. A basic problem, therefore, is to understand how the rates of the chemical reactions are controlled by the processing variables such as chemical composition of the metal alkoxide, concentration of reactants, pH of the solution, and temperature. The problem becomes more complex when a solution of two or more alkoxides is used in the fabrication of multicomponent gels (i.e., gels containing more than one metal cation). There will be a loss of chemical homogeneity if steps are not taken to control the reaction.

After preparation, the gel contains a large amount of liquid existing in fine interconnected channels. It must be dried prior to firing and conversion to a useful material. During drying, the large capillary stresses associated with the evaporation of the liquid can cause severe cracking and warping. The problem is especially critical for polymeric gels, in which the pores are much finer than those in colloidal gels. Considerable care must be taken to control the drying process in normal evaporation of the liquid. Two general approaches have been used to circumvent the problems of drying. The use of chemicals added to the solution prior to gelation, referred to as drying control chemical additives, or DCCAs, is sometimes claimed to permit relatively rapid drying. However, the mechanism by which these chemicals operate is not very clear. The removal of the liquid under supercritical conditions eliminates the liquid–vapor interface and thereby prevents the development of capillary stresses. The gels undergo relatively little shrinkage during supercritical drying. The dried gels are therefore fragile and may shrink considerably during firing.

The dried polymeric gel is typically amorphous in structure and contains fine pores, which, in many cases, have a fairly uniform size. These characteristics are very favorable for good densification, so reduced firing temperatures are a characteristic advantage of the sol-gel process. However, in some cases, crystallization prior to full densification may limit the sintering rate for compositions that are crystallizable.

The polymeric gel route can provide substantial benefits in fabrication when compared to the conventional methods such as sintering of powders or melting and casting of glasses. These benefits include good chemical homogeneity, high purity, and low processing temperatures. However, the disadvantages are also real. Many metal alkoxides are fairly expensive, and most are very sensitive to moisture, so they must be handled in a dry environment (e.g., an inert atmosphere glove box). The large shrinkage of the gel during the drying and firing steps makes dimensional control of

Sol-Gel Processing

large articles difficult. The rate of production of ceramics and glasses is severely limited when the drying process is carried out by normal evaporation of the liquid. It is difficult to dry monolithic gels thicker than ≈ 1 mm or films thicker than ≈1 μm. The sol-gel process is therefore seldom used for the production of large articles. Instead, it has seen considerable use for the production of small or thin articles such as films, fibers, and powders, and its use in this area is expected to grow substantially.

5.2 TYPES OF GELS

As mentioned above, two sol-gel routes can be distinguished depending on whether the gel structure consists of colloidal particles or polymer molecules. The structure of the gel has important implications for the fabrication of ceramics and glasses by the sol-gel route [1,2]. We therefore need to consider the practical as well as the fundamental distinction between colloidal gels and polymeric gels.

A. Colloidal Gels

Colloidal gels consist of a skeletal network of essentially anhydrous particles held together by surface forces (see Fig. 1.6). The structure of the particles normally corresponds to that of the bulk solid of the same composition. For example, colloidal SiO_2 particles have the same structure as bulk silica glass produced by melting. Furthermore, hydroxyl groups are present only on the surface of the particles. The pores in colloidal gels are much larger than in polymeric gels. Because of the larger pores, the capillary stress developed during the removal of the liquid from the pores (drying) is lower so that less shrinkage occurs. Because of the larger pores, the permeability of colloidal gels is higher; this combined with the lower capillary stress means that colloidal gels are less likely to crack during drying. The structure of the dried gel is characterized by a relatively high porosity (≈70–80%) and pores that are relatively large compared with the size of the particles (i.e., the average pore size is typically 1–5 times the particle size).

The structure of the gel has important consequences for the firing of the dried gel in its conversion to the final article. When the particles are amorphous (glassy), densification of the gel occurs by viscous flow. Densification occurs when the firing temperature is well above the glass transition temperature so that the viscosity is low enough for appreciable flow to occur. For colloidal SiO_2 gels (particle size ≈50 nm), sintering is performed in the range of ≈1200–1500°C. During sintering, the viscous glass flows under the action of the capillary forces of the pores to fill up the porosity. The driving force for the process comes primarily from a reduction in the surface free energy of the gel.

The colloidal gel route is not used, in general, for the processing of crystalline particles. For the same number of identical particles, the volume of the gel is greater than that for a consolidated colloid formed from a stable suspension or from a flocculated suspension (Fig. 5.1). The low packing density and the large pore size relative to the particle size present in colloidal gels are microstructural characteristics that generally lead to severe problems in sintering. As noted earlier, the trend is toward the use of fine particles that are consolidated uniformly from stable colloidal suspensions.

Colloidal gels are generally prepared by reducing the volume of a stable colloidal suspension (e.g., by evaporating some of the liquid) or by reducing the stability of the suspension slightly through the addition of an acid or base. The colloidal gel route has been used on a limited basis for the production of glasses such as fused silica and high silica glasses [3,4]. For the formation of multicomponent gels that contain more than one metal cation (e.g., SiO_2 doped with TiO_2), the colloidal gel route typically involves mixing two or more sols (e.g., sols of SiO_2 and TiO_2) in the desired proportion followed by gelling. The range of chemical homogeneity in the gelled material can, at best, be on the order of the particle size. When the dopant concentration is small (i.e., less than a few percent), the production of good chemical homogeneity becomes difficult.

B. Polymeric Gels

Polymeric gels, as we will recall from Chapter 1, are prepared by chemical reactions involving the hydrolysis, condensation, and polymerization of

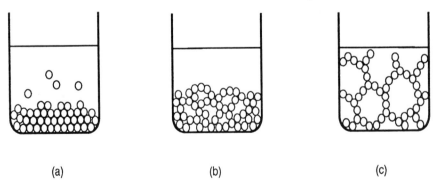

Figure 5.1 Sediments formed from (a) a stable colloid and (b) a flocculated colloid compared with (c) a colloidal gel.

metal alkoxides in solution. The gels are formed by the entanglement and cross-linking of the growing polymer chains or polymer clusters. As we shall see later in this chapter, the polymeric structure of the gels can vary considerably depending on the conditions used in the preparation of the gel. After preparation, polymeric gels consist of a weak amorphous solid structure containing an interconnected network of very fine pores filled with liquid. The volume of the pores is very high, typically 90–95% of the total volume, and the diameters of the pore channels are typically on the order of 2–10 nm.

Gelation in polymeric gels is accompanied by a sharp increase in the viscosity that essentially freezes in the polymer structure at the point of gelling. However, the frozen-in structure can change appreciably during subsequent aging of the gel and during removal of the liquid from the gel (drying). Under certain conditions, the aging gel can shrink considerably while expelling liquid. Because of the very fine pores, huge capillary stresses are developed during normal evaporation of the liquid. The capillary stresses are equivalent to the application of external compressive stresses on the gel. Removal of the liquid by evaporation therefore collapses the weak polymer network and results in additional cross-linking of the polymer structure. Cross-linking and collapse of the gel continues until the structure can withstand the compressive action of the capillary stresses. Depending on the structure of the gel, the porosity of the dried gel can be anywhere between $\approx$30 and 70%.

As we observed earlier for colloidal gels, the structure of the dried gel has a significant effect on the firing process. Indeed, compared to the colloidal gel case, structural effects on sintering are more pronounced as well as more complex. On a macroscopic scale, we can consider the polymeric gel to be an amorphous solid containing fine, interconnected porosity. The average pore size of polymeric gels is usually much finer than that of colloidal gels. For a fixed value of the porosity, the surface area of the solid/vapor interface increases inversely as the radius of the pore. Therefore, the driving force for sintering due to reduction in surface area is usually much higher for polymeric gels. The result is a lowering of the sintering temperature for polymeric gels. For example, in the case of SiO_2, viscous sintering of polymeric gels occurs generally between 800 and 1000°C, compared with 1200–1500°C for colloidal gels.

The local chemical structure also has an important influence on the sintering of polymeric gels. There is some collapse of the gel structure during drying, and the solid skeletal phase that makes up the dried gel is not identical to the corresponding bulk glass produced, for example, by melting. Evidence suggests that the gel structure contains fewer cross-links and more free volume than the melted glass [2]. This means that during the firing process the gel structure will change to become more highly cross-linked, with a corresponding reduction in its free volume and its surface area.

C. The Polymeric Gel Route: Advantages and Disadvantages

The polymeric gel route has seen considerable use during the last 15–20 years or so in the fabrication of ceramics and glasses. The use of liquid starting materials (e.g., solutions of metal alkoxides) provides distinct advantages for the preparation of pure, chemically homogeneous materials. In the preparation of the gel, the reactions occur in such a way that the mixing may be achieved on a molecular scale. Exceptionally good chemical homogeneity may therefore be achieved. Furthermore, the preparation of compounds with a small amount of dopants is relatively easy because the dopant is added in the form of a solution. Because liquids can be purified easily by distillation, materials of high purity can be produced by the polymeric gel route. Another distinct advantage is the low processing temperature, which, as noted earlier, is a result of the high driving force for densification. For example, the firing temperature of polymeric SiO_2 gel is typically in the range of 800–1000°C compared to 1200–1500°C for a colloidal gel of SiO_2 particles. The more traditional fabrication method for glasses involving melting and casting requires a melting temperature of >1800°C and a working temperature of >2000°C for SiO_2.

The use of metal alkoxides in the polymeric gel route is, however, not without its problems. Many metal alkoxides are very expensive, so the production cost can be high. Although commercial availability is steadily increasing, many metal alkoxides are still difficult to obtain. As we shall see later, metal alkoxides are very sensitive to moisture, and special handling in a dry, inert atmosphere (e.g., in a glove box) is usually required. The capillary stresses accompanying the removal of the liquid during the drying process can be very high. Conventional drying by evaporation of the liquid must therefore be carried out very slowly if cracking and warping of the dried gel are to be avoided. For large articles, the drying step can take up to a few weeks. The total shrinkage in the drying and sintering stages is very high, and this makes dimensional control of the fabricated article very difficult. Because of the drying problems combined with the large fabrication shrinkage, the polymeric gel route is best suited for the production of small articles such as films, fibers, and powders. In the firing stage, crystallization prior to full densification can reduce the rate of sintering. Higher fabrication temperatures are required for the achievement of high densities, so the advantage of low-temperature fabrication is limited. When full densification is achieved prior to crystallization, a

Table 5.1 Advantages and Disadvantages of the Polymeric Gel Route Compared with Conventional Fabrication Methods of Ceramics and Glasses

Advantages
1. High purity
2. High chemical homogeneity with multicomponent systems
3. Low temperature of preparation
4. Preparation of ceramics and glasses with novel compositions
5. Ease of fabrication for special products such as films and fibers

Disadvantages
1. Expensive raw materials
2. Large shrinkage during fabrication
3. Drying step leads to long fabrication times
4. Limited to the fabrication of small articles
5. Special handling of raw materials usually required

---

controlled heating step is necessary for the production of crystalline materials. If the dimensional changes accompanying the amorphous to crystalline phase change are relatively large, then cracking may result.

To summarize, the polymeric gel route has been used for the production (on a laboratory scale) of a wide variety of ceramics and glasses, and its use is expected to increase substantially in the future. We have seen that the route has distinct advantages as well as disadvantages compared with conventional methods for the preparation of ceramics and glasses. These are summarized in Table 5.1.

5.3 METAL ALKOXIDES (METAL-ORGANIC COMPOUNDS)

As discussed earlier, the starting materials for the production of ceramics and glasses by the polymeric gel route are, in most cases, metal alkoxides, which are also referred to as metal-organic compounds. Metal alkoxides are also frequently referred to in the literature as organometallic compounds. However, it is not correct to call alkoxides organometallic. An organometallic has a metal-carbon (M—C) bond (unlike most alkoxides) whereas a metal-organic need not have M and C directly bonded. We recall that these compounds are also used in the preparation of fine oxide powders (e.g., the Stober process discussed in Chapter 2). Because of their importance, especially in sol-gel processing, we need to have a basic understanding of the chemistry and properties of metal alkoxides in order to prepare materials with the desired structure and chemical homogeneity. The synthesis and properties of metal alkoxides are considered in detail in the book by Bradley et al. [5]. In addition, the synthesis of metal alkoxides is described in a review article by Okamura and Bowen [6]. Here we consider only a few of the main features relevant to the production of ceramics and glasses by the sol-gel process.

Metal alkoxides have the general formula $M(OR)_z$, where M is a metal of valence z and R is an alkyl group. They can be considered derivatives of either an alcohol in which the hydroxylic hydrogen is replaced by the metal or of a metal hydroxide in which the hydrogen is replaced by an alkyl group. Accordingly, the chemistry of the metal alkoxides involves the metal–oxygen–carbon bond system.

A. Preparation of Metal Alkoxides

The method used for the preparation of a metal alkoxide depends, in general, on the electronegativity of the metal. The main methods can be divided into two groups: (i) reactions between metals and alcohols for the more electropositive metals (i.e., those with relatively low electronegativity values) and (ii) reactions involving metal chlorides for the less electropositive metals or electronegative elements (i.e., those with relatively high electronegativity values). In addition, there are miscellaneous methods that are useful for the synthesis of some alkoxides. These include alcohol interchange or alcoholysis reactions, transesterification reactions between alkoxides and esters, and esterification reactions between oxides or hydroxides and alcohols [5].

Reactions Between Metals and Alcohols

The alkoxides of the more electropositive metals with valences up to 3 can be prepared by the simple reaction

$$M + zROH \rightarrow M(OR)_z + (z/2)H_2 \uparrow \qquad (5.1)$$

The nature of the alcohol also has a significant effect on these reactions. For example, sodium reacts vigorously with methanol and ethanol, but the reaction rate is considerably slower with isopropanol and extremely slow in *tert*-butanol. For the alkali metals (e.g., Li, Na, K) and the alkali earth metals (e.g., Ca, Sr, Ba), the reaction proceeds without the use of a catalyst. However, beryllium, magnesium, aluminum, the lanthanides, and yttrium require catalysts to cause them to react with alcohols. Depending on the metal, various catalysts have been used, including I_2, $HgCl_2$, and $BeCl_2$. Although the mechanism of these catalytic reactions is not well understood, one suggestion is that the catalyst aids in the removal of the surface oxide layer that acts as a passivating layer on the metal.

Sol-Gel Processing

Reactions Involving Metal Chlorides

For less electropositive metals or electronegative elements, the alkoxides are obtained via the anhydrous metal chloride. For some highly electronegative elements such as boron, silicon and phosphorus, the direct reaction between metal chlorides and alcohols is effective:

$$MCl_z + zROH \rightarrow M(OR)_z + zHCl \uparrow \qquad (5.2)$$

However, for most metals the reaction must be forced to completion by using bases such as ammonia:

$$MCl_z + zROH + NH_3 \rightarrow M(OR)_z + zNH_4Cl \downarrow \qquad (5.3)$$

The reaction described by Eq. (5.3) forms the most useful procedure for the preparation of many alkoxides (including those of Zr, Hf, Si, Ti, Fe, Nb, Ge, V, Ta, Th, Sb, U, and Pu) and is used widely for commercial production. The reaction between an anhydrous metal chloride and sodium alkoxide in the presence of excess alcohol and an inert solvent such as benzene or toluene is also a useful method:

$$MCl_z + zNaOR \rightarrow M(OR) + zNaCl \downarrow \qquad (5.4)$$

The nature of the alcohol has a significant influence on the preparation of alkoxides by the reaction involving metal chlorides and alcohols. For the lower straight chain alcohols such as methanol and ethanol, the alcohol undergoes a relatively straightforward reaction with the metal chloride and the base. However, for other alcohols, side reactions may assume a dominant role, so the yield of the alkoxide product is usually low.

Miscellaneous Methods

The alkoxides of the alkali metals can also be prepared by dissolving the metal hydroxide in the alcohol. For example, sodium ethoxide can be produced from sodium hydroxide and ethanol:

$$NaOH + C_2H_5OH \rightleftharpoons NaOC_2H_5 + H_2O \qquad (5.5)$$

Alkoxides of some highly electronegative elements (e.g., B, Si, Ge, Sn, Pb, As, Se, V, and Hg) can be prepared by an esterification reaction involving an oxide and an alcohol:

$$MO_{z/2} + zROH \rightleftharpoons M(OR)_z + (z/2)H_2O \qquad (5.6)$$

The reactions described by Eqs. (5.5) and (5.6) are reversible, so the water produced must be continually removed. In practice, this is usually done by using solvents such as benzene or xylene, which form azeotropes with the water. The azeotropic mixture behaves like a single substance in that the vapor produced by distillation has the same composition as the liquid. It can be easily fractionated out by distillation.

Metal alkoxides have the ability to exchange alkoxide groups with alcohols, and this has been used in the preparation of new alkoxides for a variety of metals, including Zn, Be, B, Al, Si, Sn, Ti, Zr, Ce, Nb, Nd, Y, and Yb. The reaction is called alcoholic interchange or alcoholysis. The general reaction can be written as

$$M(OR)_z + zR'OH \rightleftharpoons M(OR')_z + zROH \qquad (5.7)$$

To complete the reaction, the alcohol ROH produced in the reaction is removed by fractional distillation. Benzene or xylene, which form azeotropes with alcohol, are used to aid in the removal of the alcohol by fractional distillation. As an example, the alcoholysis of aluminum isopropoxide with n-butanol can be used to prepare aluminum n-butoxide:

$$Al(O\text{—}i\text{-}C_3H_7)_3 + 3n\text{-}C_4H_9OH \rightleftharpoons Al(O\text{—}n\text{-}C_4H_9)_3 + 3\ i\text{-}C_3H_7OH \qquad (5.8)$$

where i and n refer to the secondary (or iso) and normal alkyl chains, respectively.

Metal alkoxides undergo transesterification with carboxylic esters, and this affords a method of conversion from one alkoxide to another. The reaction is reversible and can be written

$$M(OR)_z + zCH_3COOR' \rightleftharpoons M(OR')_z + zCH_3COOR \qquad (5.9)$$

Fractional distillation of the more volatile ester CH_3COOR in Eq. (5.9) is required to complete the reaction. Transesterification reactions have been used for the preparation of alkoxides of various metals including Zr, Ti, Ta, Nb, Al, La, Fe, Ga, and V. For example, zirconium *tert*-butoxide has been prepared from the reaction between zirconium isopropoxide and *tert*-butyl ester followed by distillation of isopropyl acetate:

$$Zr(O\text{—}i\text{-}C_3H_7)_4 + 4CH_3COO\text{—}t\text{-}C_4H_9 \qquad (5.10)$$
$$\rightarrow Zr(O\text{—}t\text{-}C_4H_9)_4 + 4CH_3COO\text{—}i\text{-}C_3H_7$$

Double alkoxides are metal alkoxides with two different metal cations within each molecule or molecular species. Bradley et al. [5] provide an excellent summary of the methods used to synthesize double alkoxides. One of these involves dissolving each alkoxide in a mutual solvent, mixing the solutions, and refluxing at an elevated temperature. As an example, the synthesis of the double alkoxide $NaAl(OC_2H_5)_4$ by the reaction between sodium ethoxide and aluminum ethoxide may be written

$$NaOC_2H_5 + Al(OC_2H_5)_3 \rightarrow NaAl(OC_2H_5)_4 \qquad (5.11)$$

Sol-Gel Processing

The stability of double alkoxides varies considerably, depending on the nature of the two metals and the alkoxide group. Currently, a few double alkoxides are available commercially.

B. Basic Properties of Metal Alkoxides

In this section, we consider some of the basic physical and chemical properties of metal alkoxides that are important for the production of ceramics and glasses by sol-gel processing.

Physical Properties of Metal Alkoxides

The physical properties of metal alkoxides depend primarily on the characteristics of the metal (e.g., electronegativity, valence, atomic radius, and coordination number) and secondarily on the characteristics of the alkyl group (e.g., size and shape). There is a change from the solid nonvolatile ionic alkoxides of some of the alkali metals to the volatile covalent liquids of elements of valence 3, 4, 5, or 6 (e.g., Al, Si, Ti, Zr, Sb, and Te), whereas alkoxides of metals with intermediate electronegativities, such as La and Y, are mainly solids (Table 5.2).

The alkyl group has a striking effect on the volatility of metal alkoxides. Many metal methoxides are solid nonvolatile compounds (e.g., sodium methoxide). However, as the number of methyl groups increases and the size of the metal atom decreases, methoxides become sublimable solids or fairly volatile liquids (e.g., silicon tetramethoxide). Many metal alkoxides are strongly associated by intermolecular forces that depend on the size and shape of the alkyl group. The degree of association of metal alkoxides is sometimes described by the term *molecular complexity*,

Table 5.2 Physical State of the Alkoxides of Some Metals with Different Electronegativities

Alkoxide	State
$Na(OC_2H_5)$	Solid (decomposes above $\approx$530 K)
$Ba(O\text{—}i\text{-}C_3H_7)_2$	Solid (decomposes above $\approx$400 K)
$Al(O\text{—}i\text{-}C_3H_7)_3$	Liquid (bp 408 K at 1.3 kPa)
$Si(OC_2H_5)_4$	Liquid (bp 442 K at atmospheric pressure)
$Ti(O\text{—}i\text{-}C_3H_7)_4$	Liquid (bp 364.3 K at 0.65 kPa)
$Zr(O\text{—}i\text{-}C_3H_7)_4$	Liquid (bp 476 K at 0.65 kPa)
$Sb(OC_2H_5)_3$	Liquid (bp 367 K at 1.3 kPa)
$Te(OC_2H_5)_4$	Liquid (bp 363 K at 0.26 kPa)
$Y(O\text{—}i\text{-}C_3H_7)_3$	Solid (sublimes at $\approx$475 K)

which refers to the average number of empirical units in a complex. Figure 5.2 shows a schematic of a coordination complex of aluminum isopropoxide with a molecular complexity of 3. The physical properties of the alkoxides are determined by two opposing tendencies. One tendency is for the metal to increase its coordination number by using the bridging property of the alkoxo groups. The opposite tendency is the screening or steric effect of the alkyl groups, which interfere with the coordination process. The degree of screening depends on the size and shape of the alkyl group. The result is that alkoxides with a wide variety of properties, ranging from nonvolatile polymeric solids to volatile monomeric liquids, can be achieved depending on the nature of the alkyl group. Table 5.3 shows the boiling points and molecular complexity of some titanium alkoxides. In spite of the large increase in molecular weight, the boiling point decreases dramatically from 138.3°C for the ethoxide to 93.8°C for the *tert*-butoxide. For the same molecular weight, there is also a significant reduction in the boiling point for the branched isopropoxide compared to the linear *n*-propoxide.

Chemical Properties of Metal Alkoxides

Metal alkoxides are characterized by the ease with which they undergo hydrolysis. In many cases the alkoxides are so sensitive to traces of moisture that special precautions must be taken in their handling and storage. The use of an inert, dry atmosphere (e.g., normally available in a glove box) and dehydrated solvents is essential in most experiments. The mechanism of hydrolysis of metal alkoxides is fairly complex and depends on the experimental conditions. However, in the initial step a water molecule interacts with the alkoxide and, following an electronic rearrangement, a molecule of alcohol is expelled:

$$M(OR)_z + H_2O \rightarrow M(OH)(OR)_{z-1} + ROH \tag{5.12}$$

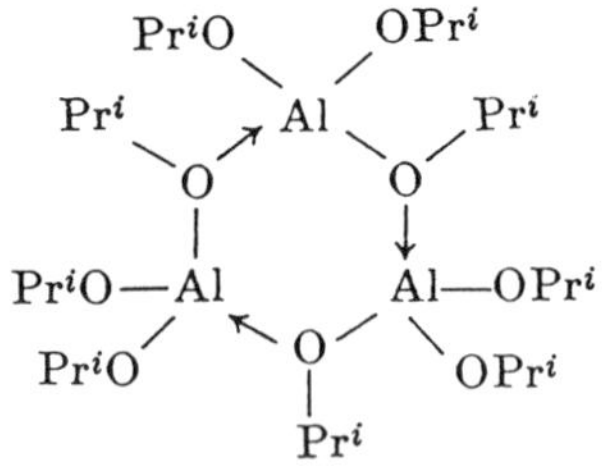

Figure 5.2 Coordination complex of aluminum isopropoxide consisting of three molecules.

Sol-Gel Processing

Table 5.3 Boiling Point and Molecular Complexity of Some Titanium Alkoxides

Alkoxide	Molecular weight	Boiling point (K) at 0.65 kPa	Molecular complexity
$Ti(OCH_3)_4$	172	(Solid)	—
$Ti(OC_2H_5)_4$	228	411.3	2.4
$Ti(O\text{—}n\text{-}C_3H_7)_4$	284.3	410	(unknown)
$Ti(O\text{—}i\text{-}C_3H_7)_4$	284.3	364.3	1.4
$Ti(O\text{—}t\text{-}C_4H_9)_4$	340.3	366.8	1.0

The hydroxy metal alkoxide product in Eq. (5.12) may react further by condensation to form polymerizable species:

$$M(OH)(OR)_{z-1} + M(OR)_z \rightarrow (RO)_{z-1}MOM(OR)_{z-1} + ROH \tag{5.13}$$

$$2M(OH)(OR)_{z-1} \rightarrow (RO)_{z-1}MOM(OR)_{z-1} + H_2O \tag{5.14}$$

The reactions described by Eqs. (5.12)–(5.14) must, however, be considered somewhat simplistic in light of the evidence that most metal alkoxides containing lower aliphatic alkyl groups are actually coordinated complexes and not single molecules.

The rate of hydrolysis of a metal alkoxide depends on the characteristics of the metal and those of the alkyl group. In general, silicon alkoxides are among the slowest to hydrolyze, and for a given metal alkoxide the hydrolysis rate increases as the length of the alkyl group decreases.

In excess water, metal alkoxides form insoluble hydroxides or hydrated metal oxides, depending on the metal [7]. Aluminum alkoxides, for example, initially form the monohydroxide (boehmite), which later may convert to the trihydroxide (bayerite):

$$Al(OR)_3 + 2H_2O \rightarrow AlO(OH)\downarrow + 3ROH \tag{5.15}$$

$$AlO(OH) + H_2O \rightarrow Al(OH)_3\downarrow \tag{5.16}$$

Boron alkoxides form the oxide or boric acid when reacted with excess water. Oxide formation can be written as

$$2B(OR)_3 + 3H_2O \rightarrow B_2O_3 + 6ROH \tag{5.17}$$

As we shall see later, the formation of insoluble precipitates as represented by Eqs. (5.15)–(5.17) makes it impossible for polymerization to occur.

For the formation of ceramics and glasses by the polymeric gel route, we therefore want to prevent the formation of insoluble precipitates.

Silicon alkoxides show a different type of reaction. They form soluble "silanols," (i.e., compounds with Si—OH group), in excess water rather than an insoluble oxide or hydroxide. For example, the compound $Si(OR)_{4-x}(OH)_x$ contains x silanol groups. The hydrolysis reaction may be written

$$Si(OR)_4 + xH_2O \rightarrow Si(OR)_{4-x}(OH)_x + xROH \qquad (5.18)$$

Complete hydrolysis leading to the formation of the silicic acid monomer $Si(OH)_4$ generally does not occur except at low pH and high water concentration. Condensation of the silanol groups occurs prior to the replacement of all the OR groups by hydroxyl groups, leading to the formation of polymeric species, as in Eqs. (5.13) and (5.14), for example.

Metal alkoxides are soluble in their corresponding alcohols. In practice, dissolution of solid alkoxides or dilution of liquid alkoxides is normally performed in the corresponding alcohol. As discussed earlier, metal alkoxides have the ability to exchange alkoxide groups with alcohols [Eq. (5.7)].

5.4 THE SOL-GEL PROCESS FOR METAL ALKOXIDES

The sol-gel process, as noted in Chapter 1, can be divided into three stages: (i) gel formation by hydrolysis and condensation reactions, (ii) drying of the synthesized gel, and (iii) conversion of the dried gel into the final product by firing. Here we consider the processes occurring in these three stages in more detail. The experimental parameters employed in the hydrolysis and condensation reactions (e.g., concentration of alkoxide, reaction medium, concentration of catalyst, and temperature) have a significant influence on the structure of the gel, which in turn affects the behavior of the gel during drying and ultimately the kinetics of densification during firing. An important issue with which we shall be concerned is the relationship between gel formation and the conversion of the gel to the final article.

A. Hydrolysis, Condensation, and Gelation

Silica has been used as a model system in many studies on the mechanism of gel formation by hydrolysis and condensation of alkoxides [8]. The structural evolution of silica gels has been characterized far better than that of other materials. Our discussion of the mechanism of gel formation will therefore focus on the silica system.

Sol-Gel Processing

A commonly used alkoxide for the sol-gel processing of silica is tetraethylorthosilicate, $Si(OC_2H_5)_4$, abbreviated TEOS, which is also referred to as silicon tetraethoxide. Since TEOS is immiscible with water, it is normally dissolved in the corresponding alcohol (ethanol), which provides a mutual solvent for the hydrolysis reaction. The hydrolysis and condensation of TEOS can be described by Eqs. (5.12)–(5.14). However, the amount of water added in the hydrolysis reaction can vary over a wide range (e.g., 1–20 mol H_2O to 1 mol TEOS), and acids or bases are normally used to catalyze the hydrolysis and condensation reactions. Depending on the parameters used in the synthesis, the gelation process may change considerably. The structure of the gel can range from that of a polymeric gel to that of a colloidal gel. It is determined by the relative rates of the hydrolysis and condensation reactions.

The formation of silica gel is usually divided into acid catalyzed conditions (pH < 2.5) and base catalyzed conditions (pH > 2.5). This is unusual terminology, since "basic" normally refers to pH > 7. In this case, the dividing line between acidic and basic behavior is near the point of zero charge (PZC) because the mechanism changes with the charge on the silanol. Under acid-catalyzed conditions and low additions of water (less than ≈4 mol H_2O to 1 mol TEOS), the rate of hydrolysis is much faster than that of condensation. In the simplest description, the reaction occurs by electrophilic attack on the alkoxide group (Fig. 5.3). The oxygen atom of the alkoxide group with its two pairs of unshared electrons is an attractive site for coordination with H^+ ions from the acid. Further hydrolysis and condensation lead to the growth of primarily linear polymers or randomly branched polymers. Gelation occurs when the growing polymers entangle and form occasional cross-links, leading to a rapid increase in viscosity (Fig. 5.4).

In acid-catalyzed conditions and higher additions of water (greater than ≈20 mol H_2O to 1 mol TEOS) or in base catalyzed conditions with

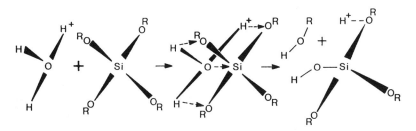

Figure 5.3 Schematic representation of the reaction mechanism for acid-catalyzed hydrolysis of TEOS by electrophilic attack. (From Ref. 8.)

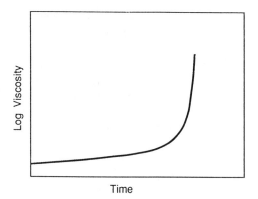

Figure 5.4 Schematic of viscosity versus aging time for the hydrolysis of TEOS. Gelation is accompanied by a rapid increase in the viscosity.

moderately high pH, the rate of hydrolysis becomes less than that of condensation. The reaction occurs by a mechanism involving nucleophilic attack on the silicon (Fig. 5.5). The silicon atom bonded to the more electronegative oxygen and surrounded by the relatively small ethoxide groups is a favorable site for attack by the negative OH^- ions. Highly branched polymeric clusters are formed that do not interpenetrate prior to gelation. In this case, gelation occurs by the linking together of the clusters in a manner similar to the formation of colloidal gels. As outlined above, gelation is accompanied by a rapid increase in viscosity (Fig. 5.4). It is only in conditions of very high pH and a large excess of water that colloidal silica particles are formed. An example of such conditions was described in Chapter 2 for the preparation of nearly monodisperse SiO_2

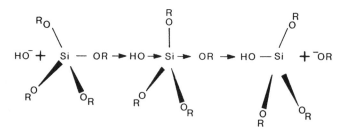

Figure 5.5 Schematic representation of the reaction mechanism for base-catalyzed hydrolysis of TEOS by nucleophilic attack. (From Ref. 8.)

Sol-Gel Processing

particles by Stober, Fink, and Bohn. Gel formation in acid-catalyzed and base-catalyzed reactions as well as for particles is summarized schematically in Fig. 5.6.

The description of the acid and base catalyzed reactions given above, while useful for visualizing the differences in gel structure, is however too simple. Reaction-limited aggregation clusters form under both acid and base catalysis but the cross-linking is much higher at high pH and high H_2O/TEOS ratio. The clusters grow and link together. Gelation occurs by a percolation process when the cluster spans the vessel holding the sol.

B. Aging of Polymeric Gels

The condensation reactions that lead to gelation do not stop when gelation occurs. If the gel is aged in the original pore liquid, small clusters continue to diffuse and attach to the main network. As these new links form, the network becomes stiffer and stronger. Many gels exhibit the phenomenon shown in Fig. 5.7 where the gel network shrinks and expels the liquid [10]. This phenomenon is known as syneresis. Studies of syneresis in silica gels show that the rate depends on the processing conditions (e.g., water concentration and pH) in much the same way as the condensation reactions leading to gelation. The shear modulus of the gel increases with aging time and the rate of increase of the modulus is larger at higher temperatures. However, when compared to drying where evaporation of the liquid is allowed to take place, the shrinkage and the shear modulus of the gel during syneresis increase much less rapidly with time.

C. Drying of Gels

After preparation, polymeric gels typically consist consist of a weak amorphous solid structure containing an interconnected network of very fine pores filled with liquid. Usually an excess volume of alcohol is used as a common solvent in the preparation of the gel so that the liquid composition is predominantly an alcohol. The gel is sometimes referred to as an alcogel. Colloidal gels consist of a particulate network in which the pores are filled with an aqueous liquid and in this case the gel is sometimes referred to as an aquagel (or a hydrogel). The gel must be dried to remove the liquid prior to conversion to the final article. The simplest method, sometimes referred to as conventional drying, is to remove the liquid by evaporation in air or in a common drying chamber such as an oven. However, drying must be carried out slowly and under carefully controlled conditions, especially in the case of polymeric gels, if monolithic crack-free bodies are to be obtained. The gel produced by conventional drying is referred to as a xerogel. Alternatively, the gel may be dried by removal of the liquid under

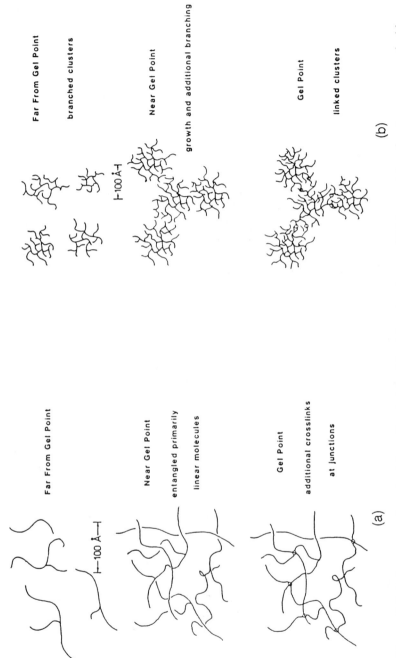

Figure 5.6 Polymer growth and gel formation in (a) acid-catalyzed systems and (b) base-catalyzed systems compared with (c) gel formation in colloidal systems. (From Ref. 2.)

Sol-Gel Processing

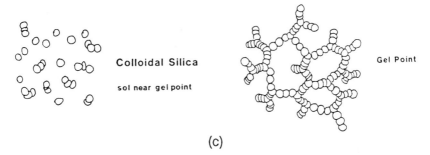

Figure 5.6 Continued.

supercritical conditions. In the ideal case, no shrinkage occurs during supercritical drying so that the dried gel is highly porous (e.g., typically ≈90–95% porosity in polymeric gels). The gel produced by supercritical drying is referred to as an aerogel.

In practice, the drying stage presents one of the major difficulties of the sol-gel process. It is often difficult to dry polymeric gels thicker than 1 mm or films thicker than 1 μm. We will examine some of the main factors which control the drying process so that a better understanding of the problems and their solution may be achieved.

D. Conventional Drying

Drying involves the interaction of three independent processes: (i) evaporation, (ii) shrinkage, and (iii) fluid flow in the pores. It is therefore a complex process. As in most complex physical phenomena, an insight

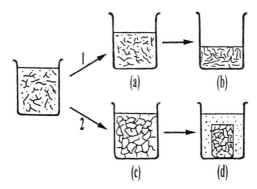

Figure 5.7 Two modes of shrinkage of a wet gel: shrinkage produced when the liquid is allowed to evaporate (route 1) and expulsion of liquid from the gel giving rise to shrinkage without evaporation (syneresis) (route 2). (From Ref. 10.)

into the problem is best achieved through a combination of theoretical modeling and experimental investigations. A detailed analysis of the drying process has been carried out recently by Scherer [9]. While special attention is given to the drying of gels, the theory is sufficiently general that it can also be applied to the drying of ceramics formed by other methods (e.g., slip casting and extrusion). Here we consider the basic features of Scherer's theory and its application to the drying of gels. In the next chapter, we shall consider the application of some of these principles to the drying of ceramics formed by other methods (e.g., slip casting and extrusion).

Stages in Drying

As sketched in Fig. 5.8, if the pore liquid contains a pure liquid, the drying process can be divided into two major stages: (i) a constant rate period (abbreviated as CRP) where the evaporation rate is nearly constant and (ii) a falling rate period (FRP) where the evaporation rate decreases with time. In some materials it is possible to further separate the FRP into two parts. In the first part, called the first falling rate period (FRP1), liquid flows to the exterior surface along a continuous film on the pore walls and the evaporation rate decreases approximately linearly with time. In the second part, called the second falling rate period (FRP2), the continuous film breaks down, so evaporation occurs inside the pores, and the rate decreases more rapidly. The evaporation rate is defined as the volume (or mass) of liquid evaporating per unit area of the drying surface per unit time. Drying curves are plotted as a function of time (Fig. 5.8) or, more commonly in clay-based ceramics, as a function of the moisture content. In the ceramic literature, the moisture content is defined in two ways, called the dry basis and the wet basis

$$\text{Moisture Content (Dry Basis)} = \frac{\text{Wet Mass} - \text{Dry Mass}}{\text{Dry Mass}} \quad (5.19a)$$

$$\text{Moisture Content (Wet Basis)} = \frac{\text{Wet Mass} - \text{Dry Mass}}{\text{Wet Mass}} \quad (5.19b)$$

However, the dry basis is commonly used. At the end of the CRP, the time is called the critical point while the moisture content is called the critical moisture content.

The stages in the drying of a gel are illustrated schematically in Fig. 5.9. During the CRP, the liquid/vapor meniscus remains at the surface of the gel. Evaporation occurs at a rate close to that of a free liquid surface (e.g., an open dish of liquid). For every unit volume of liquid that evaporates, the volume of the gel decreases by one unit volume. This stage of constant evaporation rate accompanied by shrinkage lasts until the critical

Sol-Gel Processing

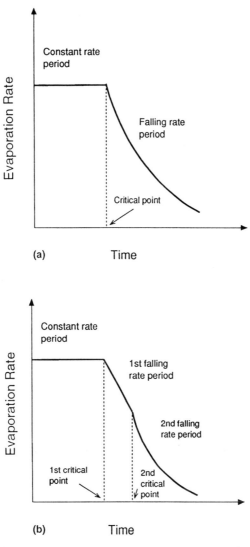

Figure 5.8 Schematic drying curves for the evaporation rate showing (a) one falling rate period and (b) the first and second falling rate periods.

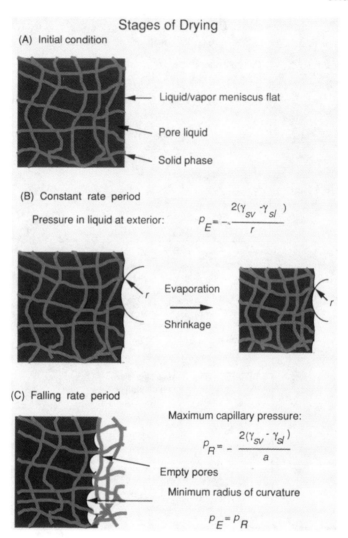

Figure 5.9 Schematic illustration of the drying process: red network represents solid phase and shadowed area is liquid filled pores. (A) before evaporation occurs, the meniscus is flat. (B) Capillary tension develops in liquid as it "stretches" to prevent exposure of the solid phase, and network is drawn back into liquid. The network is initially so compliant that little stress is needed to keep it submerged, so the tension in the liquid is low and the radius of the meniscus is large. As the network stiffens, the tension rises and, at the critical point (end of the constant rate period), the radius of the meniscus drops to equal the pore radius. (C) During the falling rate period, the liquid recedes into the gel. (From Ref. 9, with permission.)

Sol-Gel Processing

point when shrinkage stops and the FRP begins. At the end of the CRP, the gel may shrink to as little as one-tenth of its original volume. During the FRP, the liquid receds into the gel.

Driving Force for Shrinkage: Decrease in Interfacial Energy

Consider a tube of radius held vertically in a reservoir of liquid which wets it (Fig. 5.10). If the contact angle is θ, the negative pressure under the liquid/vapor meniscus in the capillary is:

$$p = -\frac{2\gamma_{lv}\cos\theta}{a} \tag{5.20}$$

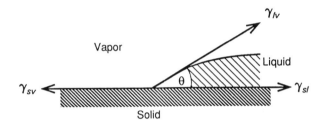

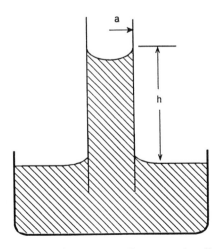

Figure 5.10 Capillary rise phenomenon for a wetting liquid with a contact angle θ.

Assuming for simplicity that $\theta = 0$, the liquid is drawn up the tube to a height h, given by

$$h = \frac{2\gamma_{lv}}{a\rho_L g} \tag{5.21}$$

where γ_{lv} is the surface energy of the liquid-vapor interface, ρ_L is the density of the liquid and g is the acceleration due to gravity. The potential energy gained by the liquid is equivalent to raising a mass of liquid $\pi a^2 h \rho_L$ through a height $h/2$. Hence,

$$PE = \frac{1}{2} \pi a^2 h^2 \rho_L g \tag{5.22}$$

This energy comes from the wetting of the walls of the tube by the liquid. We may describe the process in the following way. The surface energy of a solid arises from the asymmetric forces at the free surface. If we cover the surface with another material, for example, a liquid, we reduce this asymmetry and hence reduce the surface energy. The energy liberated in the process is available for pulling the liquid up the tube. The energy liberated can also be thought of as giving rise to a "capillary pressure" or "capillary force" which acts on the liquid to pull it up the tube. Quantitatively, if the liquid rises up the tube to cover 1 m² of the surface, then we destroy 1 m² of solid-vapor interface and create 1 m² of solid/liquid interface. The energy given up by the system is then

$$\Delta E = \gamma_{sv} - \gamma_{sl} \tag{5.23}$$

where γ_{sv} and γ_{sl} are the surface energies of the solid/vapor and solid/liquid interfaces, respectively.

Consider now a wet gel with pores that are assumed, for simplicity, to be cylindrical. If evaporation occurs to expose the solid phase, a solid/liquid interface is replaced by a solid/vapor interface. If the liquid wets the solid (i.e., the contact angle $\theta < 90°$), then, as seen from Fig. 5.10, $\gamma_{sv} > \gamma_{sl}$. The exposure of the solid phase would lead to an increase in the energy of the system. To prevent this, liquid tends to spread from the interior of the gel to cover the solid/vapor interface. (This is analogous to the example of liquid flow up a capillary tube discussed earlier.) Since the volume of the liquid has been reduced by evaporation, the meniscus must become curved, as indicated in Fig. 5.11. The hydrostatic tension in the liquid is related to the radius of curvature, r, of the meniscus by

$$p = -\frac{2\gamma_{lv}}{r} \tag{5.24}$$

where γ_{lv} is the specific energy (or surface tension) of the liquid/vapor

Sol-Gel Processing

Figure 5.11 To prevent exposure of the solid phase (A), the liquid must adopt a curved liquid/vapor interface (B). Compressive forces on the solid phase cause shrinkage. (From Ref. 9, with permission.)

interface. The negative sign in this equation arises from the sign convention for stress and pressure. The stress in the liquid is positive when the liquid is in tension; the pressure follows the opposite sign convention, so tension is negative pressure.

The maximum capillary pressure, p_R, in the liquid occurs when the radius of the meniscus is small enough to fit into the pore. For liquid in a cylindrical pore of radius a, the minimum radius of the meniscus is

$$r = -\frac{a}{\cos\theta} \tag{5.25}$$

where θ is the contact angle. The maximum tension is

$$p_R = -\frac{2(\gamma_{sv} - \gamma_{sl})}{a} = -\frac{2\gamma_{lv}\cos\theta}{a} \tag{5.26}$$

It is equal to the tension under the liquid/vapor meniscus [Eq. (5.20)]. The capillary tension in the liquid imposes a compressive stress on the solid phase, causing contraction of the gel. However, as outlined later, the capillary tension is smaller than the maximum value during most of the drying process.

It is observed that for silica gels, shrinkage proceeds faster when evaporation is allowed to take place, indicating that capillary pressure is the dominant factor driving the shrinkage. However, other factors can make a contribution to the driving force and these can be significant in other systems.

Osmotic Pressure. Osmotic pressure (Π) is produced by a concentration gradient. A common example is the diffusion of pure water through a semi-permeable membrane to dilute a salt solution on the other side. As indicated in Fig. 5.12, a pressure Π would have to be exerted on the salt

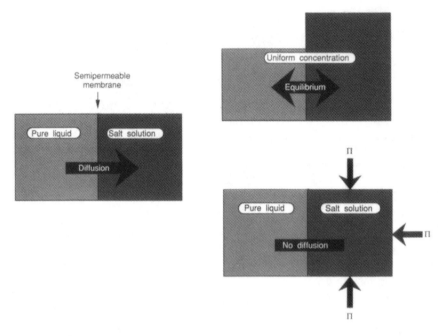

Figure 5.12 Water diffuses into the salt solution to equilibrate the concentration on either side of the impermeable membrane; pressure Π would have to be exerted on the solution to prevent the influx of water. (From Ref. 9, with permission.)

solution or a tension $-\Pi$ must be exerted on the pure water to prevent it from diffusing into the salt solution. As discussed earlier, gels prepared by the hydrolysis of metal alkoxides contain a solution of liquids (e.g., water and alcohol) that differ in volatility. Evaporation creates a composition gradient and liquid diffuses from the interior to reduce the gradient. If the pores are large, a counterflow of liquid to the interior occurs and no stress is developed. However, if the pores are small enough to inhibit flow, diffusion away from the interior can produce a tension in the liquid. The balancing compression on the solid phase (which, in principle can approach the value of Π) can cause shrinkage of the gel.

Disjoining Forces. Disjoining forces are short range forces resulting from the presence of solid/liquid interfaces. An example of such forces is the repulsion between electrostatically charged double layers discussed in Chapter 4. Short range forces in the vicinity of a solid surface can also induce some degree of structure in the adjacent liquid. The molecules in the more ordered regions adjacent to the solid surface have reduced mobil-

Sol-Gel Processing

ity compared with those in the bulk of the liquid. Disjoining forces are important in layers that are within ≈ 1 nm of the solid surface. They would be expected to be important in gels with very fine pores. For gels with relatively large pore sizes, disjoining forces will be important only in the later stages of drying.

Transport of Liquid

Transport of liquid during drying can occur by (i) flow, if a pressure gradient existes in the the liquid and (ii) diffusion, if a concentration gradient exists. According to Fick's first law (see Chapter 7), the flux, J, caused by a concentration gradient, ∇C, is given by

$$J = -D\nabla C \tag{5.27}$$

where D is the diffusion coeffeicient. The flux J is defined as the number of atoms (molecules or ions) diffusing across unit area per second down the concentration gradient and, in one dimension (e.g., the x-direction), $\nabla C = dC/dx$. As discussed earlier, diffusion may be important during drying if the liquid in the pores consists of a solution and a concentration gradient develops by preferential evaporation of one component of the solution. In general, however, diffusion is expected to be less important than flow.

As in most porous media, liquid (or fluid) flow obeys Darcy's law

$$J = -\frac{K\nabla p}{\eta_L} \tag{5.28}$$

where J is the flux of liquid (the volume flowing across unit area per unit time down the pressure gradient), ∇p is the pressure gradient (equal to dp/dx in one dimension), η_L is the viscosity of the liquid and K is the permeability of the porous medium. As in Fick's law where the physics of the diffusion process is subsumed in the diffusion coefficient, in Darcy's law the parameters of the porous medium that control the liquid flow are accounted for in terms of the permeability. A variety of models, based on the representation of pores by arrays of tubes, has been put forward for the permeability of porous media. One of the most popular, based on its simplicity and accuracy, is the Carman-Kozeny equation

$$K = \frac{P^3}{5(1-P)^2 S^2 \rho_s^2} \tag{5.29}$$

where P is the porosity, S is the specific surface area (i.e., per unit mass of the solid phase) and ρ_s is the density of the solid phase. For a silica gel prepared from an alkoxide, $S \approx 400\text{--}800$ m^2/g, $P \approx 0.9$ and $\rho_s \approx 1.5\text{--}1.8$ g/cm^3. Substituting these values in Eq. (5.29) gives K in the range of 10^{-13}–10^{-14} cm^2 (i.e., very low). While Eq. (5.29) is fairly successful for many types of porous materials, it also fails often; the equation must therefore be used with caution.

Darcy's law, as outlined earlier, is obeyed by many materials, including some with very fine pores ($\leq$ 10 nm in size). Figure 5.13 shows the data for silica gel where the flux is indeed proportional to the pressure gradient [11]. The experiment was performed by casting a sheet of gel ($\approx$ 2 mm thick) onto a Teflon® filter, then imposing a pressure drop across it and measuring the rate of flow through it.

The Physical Process of Drying

Colloidal gels and polymeric gels show the same general behavior in drying: a constant rate period follwed by a falling rate period (Fig. 5.8). However, because of the profound difference in structure, the gels respond differently to the compressive stresses imposed by capillary forces. Initially, colloidal gels show a very small elastic contraction but this is insignificant compared to the total drying shrinkage (typically 15–30 vol %) that occurs predominantly by rearrangement of the particles. Sliding of the particles over one another leads to denser packing and an increase in the stiffness of the gel. Eventually the network is stiff enough to resist the comprssive stresses and shrinkage stops. Rearrangement processes are difficult to analyze; however, like clays and soils, colloidal gels are expected to have elastic/plastic behavior.

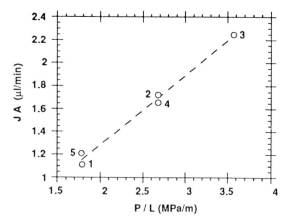

Figure 5.13 Flow rate JA (where J is the flux and A is the area of sheet of gel) versus pressure gradient in a silica gel is linear, in accordance with Darcy's law, Eq. (5.28); numbers next to points indicate order of experiments. (From Ref. 11.)

Sol-Gel Processing

Polymeric gels are highly deformable. As for other materials with a polymeric structure, the response to a stress is is viscoelastic, i.e., a combination of an instantaneous elastic deformation and a time-dependent viscous deformation. The elastic deformation arises from stretching and bending of the polymer chains while the viscous deformation arises from reorientation and relaxation of polymer chains (or clusters) into lower energy configurations. By assuming the gel to be a continuum, constitutive equations relating the strain to the imposed stress have been developed. As a result of such analysis, considerable insight has been gained into the stress development in the gel and the shrinkage during drying. In the following discussion, particular attention will therefore be paid to the drying of polymeric gels.

The Constant Rate Period. The first stage of drying, as outlined earlier, is called the constant rate period (CRP) because the rate of evaporation is independent of time (or moisture content). When evaporation starts, the temperature at the surface of the gel drops because of a loss of heat due to the latent heat of vaporization of the liquid. However, heat flow to the surface from the atmosphere quickly establishes thermal equilibrium where transfer of heat to the surface balances the heat loss due to the latent heat of vaporization. The temperature at the surface becomes steady and is called the wet-bulb temperature (T_W). The surface of the gel is therefore at the wet-bulb temperature during the CRP. The rate of evaporation, $\dot{V}_E$, is proportional to the difference between the vapor pressure of the liquid at the surface, p_W, and the ambient vapor pressure, p_A

$$\dot{V}_E = H(p_W - p_A) \tag{5.30}$$

where H is a factor that depends on the temperature, the velocity of the drying atmosphere and the geometry of the system. Since $\dot{V}_E$ increases as p_A decreases, T_W decreases with a decrease in ambient humidity. For polymeric gels, the ambient vapor pressure must be kept high to avoid rapid drying so that the temperature of the sample remains near the ambient.

Let us consider a wet gel in which some liquid suddenly evaporates. As outlined earlier, the liquid in the pores stretches to cover the dry region and a tension develops in the liquid. The tension is balanced by compressive stresses on the solid phase of the gel. Since the network is compliant, the compressive forces cause it to contract into the liquid and the liquid surface remains at the exterior surface of the gel (Fig. 5.9b). In a polymeric gel, it does not take much force to submerge the solid phase, so that initially the capillary tension of the liquid is low and the radius of the meniscus is large. As drying proceeds, the network becomes stiffer because new bonds are forming (e.g., by condensation reactions) and the porosity is decreasing. The menicus also deepens (i.e., the radius decreases) and the tension in the liquid increases [Eq. (5.24)]. When the radius of the meniscus becomes equal to the pore radius in the gel, the liquid exerts the maximum possible stress [Eq. (5.26)]. This point marks the end of the CRP; beyond this the tension in the liquid cannot overcome the further stiffening of the network. The liquid menicus recedes into the pores and this marks the start of the FRP (Fig. 5.9c). Thus, the characteristic features of the CRP are (i) the shrinkage of the gel is equal to the rate of evaporation, (ii) the liquid meniscus remains at the surface, and (iii) the radius of the liquid meniscus decreases.

At the end of the CRP (i.e., at the critical point or the critical moisture content), shrinkage virtually stops. According to Eq. (5.26), for an alkoxide gel with $\gamma_{lv}\cos\theta \approx 0.02$–$0.07$ J/m^2 and $a \approx 1$–10 nm, the capillary tension, p_R, at the critical point is ≈ 4–150 MPa. This shows that the gel can be subjected to enormous pressures at the critical point. The amount of shrinkage that precedes the critical point depends on p_R, which, according to Eq. (5.26) increases with the interfacial energy, γ_{lv}, and with decreasing pore size, a. If additives (e.g., surfactants) are added to the liquid to reduce γ_{lv}, then p_R decreases; as a result, less shrinkage occurs and the porosity of the dried gel increases.

The Falling Rate Period. When shrinkage stops, further evaporation forces the liquid meniscus into the pores and the evaporation rate decreases (Fig. 5.9c). This stage is called the falling rate period (FRP). As outlined earlier, the FRP can be divided into two parts. In the first falling rate period (FRP1), most of the evaporation is still occurring at the exterior surface. The liquid in the pores near the surface exists in channels that are continuous with the rest of the liquid. (The liquid is said to be in the funicular state.) These continuous channels provide pathways for liquid flow to the surface (Fig. 5.14a). At the same time, some liquid evaporates in the pores and the vapor diffuses to the surface. In this stage of drying, as air enters the pores, the surface of the gel may loose its transparency.

As the distance between the liquid-vapor interface (the drying front) and the surface increases, the pressure gradient decreases and the flux of liquid also decreases. If the gel is thich enough, eventually a stage is reached where the flux becomes so slow that the liquid near the surface is in isolated pockets. (The liquid is now said to be in the pendular state.) Flow to the surface stops and the liquid is removed from the gel by diffusion of the vapor. This marks the start of the second falling rate period (FRP2), where evaporation occurs inside the gel (Fig. 5.14b).

Drying From One Surface. In many cases the wet gel is supported so that liquid evaporates from one surface only (Fig. 5.15). As evaporation occurs, capillary tension develops first on the drying surface. This tension

Sol-Gel Processing

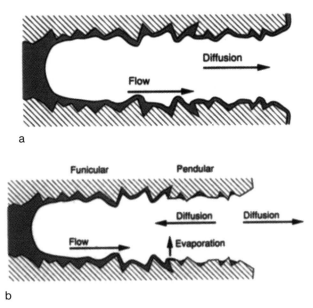

Figure 5.14 Schematic diagram illustrating fluid transport during the falling rate period. After the critical point, the liquid/vapor meniscus retreats into the pores of the body. (a) in the first falling rate period, liquid is in the funicular state, so transport by liquid flow is possible. There is also some diffusion in the vapor phase. (b) During the second falling rate period evaporation occurs inside the body, at the boundary between the funicular (continuous liquid) and pendular (isolated pockets of liquid) regions. Transport in the pendular region occurs by diffusion of vapor. (From Ref. 9, with permission.)

draws liquid from the interior to produce a uniform hydrostatic pressure. If the permeability of the gel is high, liquid flow is produced by only a small pressure gradient. However, if the permeability is low (or the gel is fairly thick), a significant pressure gradient is developed. The solid network is therefore subjected to a greater compression on the drying surface. This causes the gel to warp upwards (Fig. 5.15a). Later in the drying process, as outlined earlier, the liquid-vapor interface moves into the interior of the gel and the pores are filled with air. The gel network surrounding the air-filled pores is relieved of any compressive stress. However, the lower part of the gel still contains liquid so that it is subjected to compression due to capillary forces. This causes the gel to warp in the opposite direction (Fig. 5.15b).

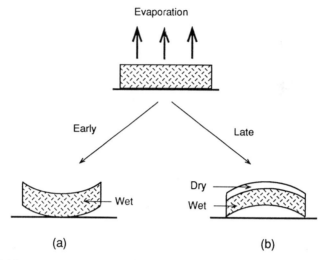

Figure 5.15 Warping of plate of gel dried by evaporation from upper surface. (a) Plate warps upward initially (b) then reverses curvature after top surface becomes dry. (Adapted from Ref. 9.)

Drying Stresses

During the CRP, the pores remain full of incompressible liquid. The change in liquid content must be equal to the change in pore volume, which is related to the volumetric strain rate, $\dot{V}$. Equating these changes, we obtain

$$\dot{V} = -\nabla \cdot J = -\nabla \cdot \left(\frac{K}{\eta_L}\nabla p\right) \tag{5.31}$$

The flux of liquid to the surface also matches the evaporation rate (i.e., $J = \dot{V}_E$), so that Eq. (5.28) requires that

$$\dot{V}_E = \frac{K}{\eta_L}\nabla p\,|_{surface} \tag{5.32}$$

To calculate the drying stresses, Eq. (5.31) must be solved using Eq. (5.32) as a boundary condition. The methods for solving the equation are discussed in Ref. 9. For a viscoelastic flat plate (including one that is purely elastic or viscous), the stress in the solid phase of the gel in the plane of the plate is given by

$$\sigma_x = \sigma_y \approx \langle p \rangle - p \tag{5.33}$$

Sol-Gel Processing

where p is the negative pressure (tension) in the liquid and $\langle p \rangle$ is the average pressure in the liquid. According to this equation, if the tension in the liquid is uniform, $p = \langle p \rangle$ and there is no stress on the solid phase. However, when p varies through the thickness, the network tends to contract more where p is high and this differential strain causes warping or cracking. The situation is analogous to the stress produced by a temperature gradient: cooler regions contract relative to warmer regions and the differential shrinkage causes the development of stresses.

If the evaporation rate is high, p can approach its maximum value, given by Eq. (5.26), while $\langle p \rangle$ is still small so that the total stress at the surface of the plate is

$$\sigma_x \approx \frac{2\gamma_{lv} \cos\theta}{a} \qquad (5.34)$$

For slow evaporation, the stress at the drying surface of the plate is

$$\sigma_x \approx \frac{L\eta_L \dot{V}_E}{3K} \qquad (5.35)$$

where the half-thickness of the plate is L and evaporation occurs from both faces of the plate.

Cracking During Drying

It is commonly observed that cracking is more likely to occur if the drying rate is high or the gel is thick. It is also observed that cracks generally appear at the critical point when shrinkage stops and the liquid/vapor meniscus moves into the body of the gel. In Scherer's theory, cracking is attributed to stresses produced by a pressure gradient in the liquid. However, the stress that causes failure is not the macroscopic stress σ_x acting on the network. If the drying surface contains flaws, such as that sketched in Fig. 5.16, the stress is concentrated at the tip of the flaw. The stress at the tip of a flaw of length c is proportional to

$$\sigma_c = \sigma_x (\pi c)^{1/2} \qquad (5.36)$$

Fracture occurs when $\sigma_c > K_{Ic}$, where K_{Ic} is the critical stress intensity of the gel. Assuming that the flaw size distribution is independent of the size and drying rate of the gel, the tendency to fracture would be expected to increase with σ_x. According to Eq. (5.35), σ_x increases with the thickness, L, of the sample and the drying rate, $\dot{V}_E$. Scherer's analysis therefore provides a qualitative explanation for the dependence of cracking on L and $\dot{V}_E$.

In another explanation, it has been proposed that cracking in gels occurs as a result of local stresses produced by a distribution of pores sizes

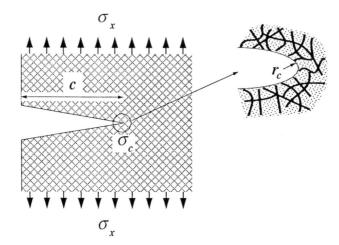

Figure 5.16 Surface flaw of length c and tip radius r_c acted upon by far-field stress σ_x creates amplified stress σ_c at the crack tip. Inset indicates that r_c is on the scale of the mesh size of the network. (From Ref. 11.)

[22]. As sketched in Fig. 5.17, after the critical point, liquid is removed first from the largest pores. It has been suggested that the tension in the neighboring small pores deforms the pore wall and causes cracking. As discussed in Ref. 9, this explanation does not appear to be valid. As outlined earlier, the drying stresses in Scherer's theory are macroscopic in that the pressure gradient extends through the thickness of the gel, i.e., the stresses are not localized. If the stresses were localized on the scale of the pores, the gel would be expected to crumble to dust as the drying front advanced. Instead, gels crack into only a few pieces. However, flaws that lead to failure may be created by local stresses (resulting, for example, from non-uniform pore sizes) and then propagated by the macroscopic stresses.

Control of Cracking

Scherer's theory of drying provides useful guidelines for the control of stress development and cracking. As outlined earlier, cracking is attributed to macroscopic stresses produced by a pressure gradient, ∇p, in the liquid. Fast evaporation leads to a high ∇p and hence to large differential strain. To avoid cracking the gel must be dried slowly. However, the "safe" drying rates are so slow that gels thicker than ≈ 1 cm require uneconomically long drying times. However, a number of procedures that increase the safe drying rates can be used.

Sol-Gel Processing

Constant rate period

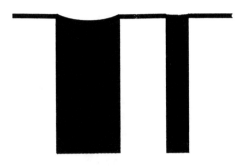

Largest pores empty first

Greater tension in smaller pores causes cracking

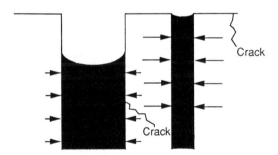

Figure 5.17 Illustration of microscopic model: during the constant rate period, meniscus has same radius of curvature for pores of all sizes; after the critical point, the largest pores are emptied first. The capillary tension compressing the smaller pores causes local stresses that crack the network. (After Ref. 22.)

The boundary condition at the surface of the gel [Eq. (5.32)] shows that high $\dot{V}_E$ leads to high ∇p. Furthermore, low permeability (low K) leads to high ∇p for a given $\dot{V}_E$. Using Eq. (5.29), it can be shown that K increases roughly as the square of the pore size. An obvious strategy is to increase the pore size of the gel. One approach that has been used for silica gels involves mixing colloidal silica particles with the alkoxide (TEOS). In addition to producing gels with large pores, the silica particles also strengthen the gel. Another approach is to use colloidal particles to produce gels that are easier to dry (e.g., colloidal silica gel). However, as outlined later, gels with larger pores require higher firing temperatures for densification so that some compromise between ease of drying and ease of firing is made.

The capillary pressure sets a limit on the magnitude of the drying stresses, i.e., $\sigma_x \leq p_R$, and is probably responsible for the creation of critical flaws. The probability of fracture can therefore be reduced by reducing the capillary pressure through (i) increasing the pore size (discussed above) and (ii) decreasing the liquid/vapor interfacial tension, [see Eq. (5.26)]. The interfacial tension can be reduced by using a solvent with a lower volatility than water that, in addition, has a low γ_{lv}. The surface tension of a liquid can also be reduced by raising the temperature. Beyond the critical temperature and pressure, there is no tension. Since $p = 0$, ∇p must also be zero so that no drying stresses can be produced. This approach forms the basis of supercritical drying described in the next section.

Another approach is to strengthen the gel so that it is more able to withstand the drying stresses. For example, as discussed earlier, aging the gel in the pore liquid at slightly elevated temperatures stiffens the gel network and also reduces the amount of shrinkage in the drying stage.

For gels prepared from metal alkoxides, certain organic compounds added to the alkoxide solution have been claimed to speed up the drying process considerably while avoiding cracking of the gel. It has been reported that, with the use of these compounds, gels thicker than 1 cm can be dried in ≈ 1 day. These compounds, referred to as drying control chemical additives (DCCA), include formamide (NH_2CHO), glycerol ($C_3H_8O_3$), and oxalic acid ($C_2H_2O_4$). While the role of these compounds during drying is not clear, it is known that they increase the hardness (and presumably the strength) of the wet gel. However, they also cause serious problems during firing because they are difficult to burn off. Decomposition leading to bloating of the gel and chemical reaction (e.g., the formation of carbonates) severely limits the effectiveness of these compounds.

E. Supercritical (Hypercritical) Drying

As mentioned earlier, supercritical drying (sometimes referred to as hypercritical drying) can be used to avoid the problems associated with the capillary stresses in conventional drying. In supercritical drying, the liquid in the pores is removed above the critical temperature, T_c, and the critical pressure, p_c, of the liquid. Under these conditions, there is no distinction between the liquid and the vapor states. The densities of the liquid and vapor are the same, there is no liquid/vapor meniscus and no capillary pressure.

In supercritical drying, the wet gel is placed in an autoclave and heated fairly slowly (less than ≈0.5°C/min) to a temperature and pressure above the critical point. A sketch of the temperature/pressure path during the drying process is sketched in Fig. 5.18. The temperature and pressure are increased in such a way that the liquid/vapor phase boundary is not crossed. After equilibration above the critical point, the fluid is released slowly.

The critical temperature and pressure for ethanol are 243°C and 6.4 MPa (≈63 atmospheres). The need to cycle an autoclave through such high temperature and pressure makes the process time consuming, expensive and somewhat dangerous. The total drying time can be as long as 2–3 days for a large gel. A convenient alternative is to replace the liquid in the pores with a substance having a much lower critical point. Table 5.4 shows the critical points of some common fluids. Carbon dioxide has a T_c of 31°C and a p_c of 7.4 MPa. The low T_c means that the drying process can be carried out near ambient temperatures. Carbon dioxide is also relatively cheap. In supercritical drying with CO_2, the liquid (e.g., alcohol) in the pores of the gel must first be replaced. This is accomplished by placing the gel in the autoclave and flowing liquid CO_2 through the system until no trace of alcohol can be detected. Following this, the temperature and pressure of the autoclave are raised slowly to ≈40°C and ≈8.5 MPa. After maintaining these conditions for ≈30 min, the CO_2 is slowly released. The step where the liquid in the pores is replaced with CO_2 can

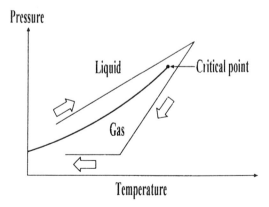

Figure 5.18 Schematic illustration of the pressure–temperature schedule during hypercritical drying.

Table 5.4 Critical Points of Some Fluids

Fluid	Formula	Critical temperature (K)	Critical pressure (MPa)
Carbon dioxide	CO_2	304.1	7.36
Nitrous oxide	N_2O	309.8	7.24
Freon-13	$CClF_3$	301.9	3.86
Freon-23	CHF_3	298.9	4.82
Freon-113	$CCl_2F-CClF_2$	487	3.40
Freon-116	CF_3-CF_3	292.7	2.97
Methanol	CH_3OH	513	7.93
Ethanol	C_2H_5OH	516	6.36
Water	H_2O	647	22.0

take several hours so that supercritical drying with CO_2 is not much faster overall.

With supercritical drying, monolithic gels as large as the autoclave can be produced. Very little shrinkage (<1% linear shrinkage) occurs during supercritical drying so that the aerogel is very porous and fragile. The solid phase occupies less than 10% of the aerogel. Because of the low density, the sintering of aerogels to produce dense polycrystalline articles is impractical. Considerable shrinkage (>50% linear shrinkage) would have to take place and this makes control of the dimensions of the article difficult. However, as discussed in Chapter 11, a common problem is the crystallization of the gel prior to the attainment of high density which severely limits the final density. Because of these problems, supercritical drying is better suited to the fabrication of porous materials and powders. Monolithic aerogels are of interest because of their exceptionally low thermal conductivity. Powders with good chemical homogeneity can be easily produced by grinding the fragile aerogel.

F. Structural Changes During Drying

We saw earlier that there are fundamental differences in structure between colloidal gels and polymeric gels and, within polymeric gels, between gels prepared by acid-catalyzed and base-catalyzed reactions. During conventional drying, we would expect the structure of these gels to evolve differently under the compressive action of the capillary stresses. Polymeric gels will gradually collapse and cross-link as unreacted hydroxyl and alkoxy groups come into contact with each other. When the structure be-

Sol-Gel Processing

comes stiff enough to resist the capillary stresses, residual porosity will be formed. For gels produced by acid-catalyzed reactions, in which the polymer chains are weakly cross-linked, the structure can be highly compacted before it is sufficiently cross-linked to produce residual porosity (Fig. 5.19a). The xerogel will be characterized by a relatively high density and very fine pores. In contrast, after drying, base-catalyzed gels will

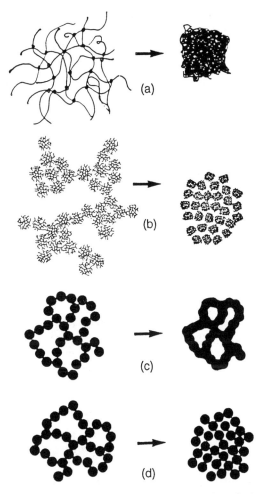

Figure 5.19 Schematic representation of the structural evolution during drying for (a) acid-catalyzed gels, (b) base-catalyzed gels, (c) colloidal gel aged under conditions of high silica solubility, and (d) colloidal gel composed of weakly bonded particles. (From Ref. 2.)

have a relatively low density and larger pores because the structure of polymeric clusters will be more difficult to collapse (Fig. 5.19b). Silica gels prepared by acid-catalyzed reactions can lead to xerogels with porosities as low as 35–40% compared to values as high as 60–70% for similar xerogels produced by the base-catalyzed route.

For colloidal gels, the structure will fold or crumple under the action of the capillary stresses. Neck formation or an increase in the coordination number of the gel will continue until the structure is strong enough to resist the capillary stresses, at which point residual porosity is formed. Because the particulate structure of the gel can better withstand the smaller capillary stresses, the shrinkage for colloidal gels will be significantly smaller than that of polymer gels. The structure of the xerogel will be a somewhat contracted and distorted version of the original structure (Figs. 5.19c and d). The porosity is typically in the range of 70–80%.

G. Gel Densification During Firing

Colloidal gels, as discussed above, contract much less than polymeric gels during drying. Since the solid phase is composed of a fully cross-linked network that is similar in structure to that of the corresponding melted glass, the drying process does not alter the structure of the solid skeletal phase of the gel. Polymeric gels undergo significant contraction and further polymerization by condensation reactions during drying. However, the resulting polymer structure that makes up the solid skeletal phase of the xerogel is still less highly cross-linked than the corresponding melted glass. For example, the number of nonbridging oxygen atoms (i.e., those ending in OH and OR groups) has been estimated in the range of 0.33–1.48 for every silicon atom in silica xerogels with a polymeric structure compared to a value of ≈ 0.003 for a melted silica glass with 0.05 wt % water. One effect of the lower cross-link density is that the solid skeletal phase of the gel has a lower density than that of the corresponding melted glass. This is equivalent to saying that the skeletal phase has extra "free volume" compared to the corresponding glass produced by melting.

The structure of the xerogel has important implications for densification. Compared to the corresponding dense glass prepared by melting, xerogels have a high free energy, which will act as a powerful driving force for densification during firing. Three physicochemical characteristics contribute to this high free energy [2]. The surface area of the solid/vapor interface makes the largest contribution, estimated at 30–300 J/g (which corresponds to 100–1000 m^2/g of interfacial area). The reduction of the surface area provides a high driving force for densification by viscous flow. The two other contributions to the high free energy result from the reduced cross-link density of the polymer chains compared to the corresponding melted glass. Polymerization reactions can occur according to

$$\text{Si(OH)}_4 \rightarrow \text{SiO}_2 + 2\text{H}_2\text{O}, \quad \Delta G_{f(298\text{K})} = -14.9 \text{ kJ/mol} \quad (5.37)$$

More weakly cross-linked polymers containing more nonbridging oxygen atoms will therefore make a greater contribution to the free energy. For silica gels containing 0.33–1.48 OH groups per silicon atom, the contribution to the free energy resulting from the reaction described by Eq. (5.37) is estimated to range from ≈20 to 100 J/g. Finally, the free volume of the solid skeletal phase is also expected to contribute to the free energy. Figure 5.20 summarizes the free energy versus temperature relations for polymeric and colloidal gels, glass, and an ideal supercooled liquid of the same oxide composition.

Scherer and coworkers have made a detailed study of the densification of gels [12,13]. Figure 5.21 illustrates data for the linear shrinkage of three silica gels during firing at a constant heating rate of 2°C/min [14]. The colloidal gel (curve C) shrinks only at elevated temperatures

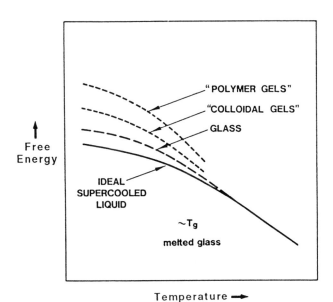

Figure 5.20 Schematic representation of free energy–temperature relations between dried polymeric gels, dried colloidal gels, glass, and an ideal supercooled liquid of the same oxide composition. (From Ref. 2.)

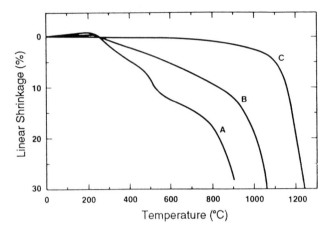

Figure 5.21 Linear shrinkage versus temperature during constant heating rate sintering at 2°C/min for an acid-catalyzed silica gel (A), a base-catalyzed silica gel (B), and a colloidal silica gel (C). The acid- and base-catalyzed gels were prepared with a water concentration of 4 mol H_2O per mole of TEOS. (From Ref. 14.)

($\approx$1200°C), and its densification behavior can be accurately described by models for viscous sintering of porous melted glass. We shall examine these models later in this book when we consider viscous sintering. The shrinkage of the polymeric gels differs markedly from that of the colloidal gel. Starting from fairly low temperatures, continuous shrinkage occurs, and the extent of the shrinkage depends on the method used to prepare the gel. At any temperature, the shrinkage of the acid-catalyzed gel (curve A) is significantly greater than that of the base-catalyzed gel (curve B). The results show that the sintering of polymeric gels is not as simple as that of colloidal gels. Furthermore, the densification behavior of polymeric gels cannot be explained on the basis of viscous sintering alone.

For polymeric gels, Scherer and coworkers proposed that four mechanisms operate during the conversion of the dried gel to a dense glass: (i) capillary contraction of the gel, (ii) condensation polymerization leading to an increase in the cross-link density, (iii) structural relaxation by which the structure approaches that of a supercooled liquid, and (iv) viscous sintering. The temperature range in which each of these mechanisms contributes to the densification depends on the structure of both the porous and solid phases of the gel (e.g., pore size and skeletal density) as well as the rate of heating and the previous thermal history. A good illustration of the temperature range in which each of these mechanisms operates is provided in Fig. 5.22, which shows the shrinkage and weight loss of a

Sol-Gel Processing

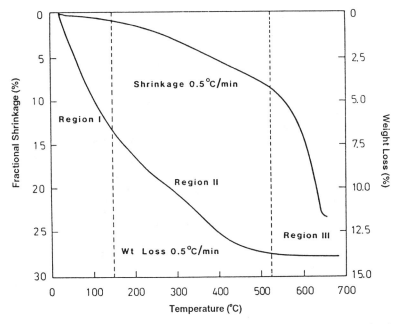

Figure 5.22 Linear shrinkage and weight loss for a borosilicate gel during heating at a constant rate of 0.5°C/min. (From Ref. 12.)

borosilicate gel during heating at a constant rate of 0.5°C/min. Three regions have been identified corresponding to the following trends:

Region I: weight loss without shrinkage (25–150°C)
Region II: weight loss with concurrent shrinkage (150–525°C)
Region III: shrinkage without weight loss (>525°C)

Region I. The weight loss in region I is, according to differential thermal analysis (DTA), due to the desorption of physically adsorbed water and alcohol. A very small shrinkage is observed, and this occurs as a result of the increase in the surface energy due to the desorption process. The surface energy increases from ≈ 0.03 J/m^2 for a surface saturated with water and alcohol to ≈ 0.15 J/m^2 for a pure silanol surface (i.e., produced after desorption). This increase in surface energy is equivalent to an increase in the capillary pressure.

Region II. Weight loss in this region is attributed to two processes. Water is removed as a by-product of polycondensation reactions:

$$\equiv\text{Si}-\text{OH} + \text{HO}-\text{Si}\equiv \rightarrow \equiv\text{Si}-\text{O}-\text{Si}\equiv + \text{H}_2\text{O} \qquad (5.38)$$

The reaction described by Eq. (5.38) may occur within the skeletal phase or on the surface of the skeletal phase. The second process is the oxidation of carbonaceous residues present originally as unhydrolyzed alkoxy groups. Differential thermal analysis indicates that the removal of carbon is essentially complete by 400°C but that hydrogen is continuously removed over the entire temperature range (150–525°C).

The shrinkage in region II can be attributed solely to densification of the solid skeletal phase of the gel. Alternatively, the removal of very fine pores by viscous sintering may be postulated, but the pore size would have to be less than ≈0.4 nm. Pores of this size would, however, be indistinguishable from free volume in the skeletal phase. Two mechanisms have been identified for the densification of the skeletal phase: (i) polymerization reactions [e.g., Eq. (5.38)], which lead to a higher cross-link density of the polymer chain, and (ii) structural relaxation of the polymeric network as the structure approaches the configuration of the melt-prepared glass. Structural relaxation occurs by diffusive motion of the polymer network without the expulsion of water and other products. The relative contribution of each mechanism to the overall densification in region II is difficult to quantify.

We would expect skeletal densification to be insignificant for colloidal gels in which the particulate network has the same structure as the corresponding melt-prepared glass and to make the greatest contribution to shrinkage for weakly cross-linked gels. This is fully borne out by the results of Fig. 5.21.

Region III. The large, fairly fast shrinkage in region III is consistent with a mechanism of densification by viscous flow. The pores in the gel that were formed after the removal of the liquid in the drying stage are removed in this process. The driving force for densification is the reduction in the solid/vapor interfacial area associated with the pores. Although the weight loss in this region is low, further studies indicate that polymerization reactions and structural relaxation can have a strong influence on the kinetics of densification.

Table 5.5 summarizes the dominant shrinkage mechanisms and their relative contributions to the overall shrinkage for the borosilicate gel described in Fig. 5.22. The significant contribution due to skeletal densification by condensation polymerization and structural relaxation should not go unnoticed.

In concluding this section on densification of gels, a few additional points should be made. First, we have discussed the conversion of the xerogel to the dense body under the influence of the available driving forces. However, densification involves transport of matter, so it is limited by the kinetics of the process. The kinetics must also be considered be-

Sol-Gel Processing

Table 5.5 Dominant Mechanisms of Shrinkage During the Firing of a Borosilicate Gel and Their Relative Contributions to the Total Shrinkage

Region	Temperature range (°C)	Mechanism	Relative contribution (%)
I	25–150	Capillary contraction	3
II	150–525	Condensation polymerization and structural relaxation	33
III	>525	Viscous flow	63

Source: Ref. 13.

cause it determines how long the process will take. The kinetics of densification will depend on the physical and chemical structure of the gel as well as on the time–temperature schedule of heating. The rate of heating, for example, may have a significant effect on the densification [2]. Increased heating rate reduces the time spent at each increment of temperature and therefore reduces the amount of viscous flow that can occur in a given temperature interval. However, an increase in heating rate also reduces the amount of cross-linking and structural relaxation that can occur over the same temperature interval, so the viscosity of the gel at each temperature is reduced. Under certain conditions, the rate of viscous sintering can increase at sufficiently high heating rates, as observed by Scherer and coworkers [12].

Second, even after drying, the gel structure normally contains a fairly large amount of unhydrolyzed alkoxy groups. As outlined above, these are oxidized and removed as gases below ≈400°C. However, two major problems may arise if the firing process is not carried out carefully. One is that trapped gases in the pores can lead to bloating of the gel at higher temperatures. This may occur, for example, if the heating rate is so fast that the pores become isolated prior to complete oxidation of the alkoxy groups. Further heating to higher temperatures causes an increase in the pressure of the gas in the pores, resulting in enlargement of the pores and bloating. The other problem is the presence of carbon residues in the gel due to incomplete oxidation. In some cases, the gel may turn black around 400°C. In general, for this part of the firing process the temperature must be high enough to oxidize the alkoxy groups but at the same time low enough to prevent the pores from sealing off.

Third, we have assumed that a dense glass is produced at the end of the firing process. This is not always so. For crystallizable compositions such as ceramics, crystallization may occur prior to full densification. A crystalline body is generally more difficult to densify than the corresponding amorphous body. Crystallization prior to full densification will, in general, hinder densification. If full densification is achieved prior to crystallization, a controlled heating step is necessary to nucleate and grow the crystalline grains. We shall examine this point in more detail later in this book when we consider viscous sintering with crystallization (see Chapter 11).

Fourth, we have seen that in the production of a glass by the polymeric gel route the network structure evolves very differently from that of the corresponding glass prepared by conventional melting and casting. An important question is whether the structure and properties of the fabricated sol-gel-derived glass are any different from those of the corresponding melt-prepared glass. Current indications are that for gels densified above the glass transition temperature the structure and properties can be indistinguishable from those of the corresponding melt-prepared glass.

5.5 SOL-GEL PREPARATION TECHNIQUES

Having considered some of the basic physics and chemistry of the sol-gel method, we come now to some of the practical issues involved in the preparation of gels. For the production of simple oxides (e.g., SiO_2), the techniques used in the preparation of the gel are fairly straightforward. Further considerations must be taken into account for the production of complex oxides, for we must ensure that the desired composition and uniformity of mixing are achieved during the gelation stage. In sol-gel processing, gel compositions with one type of metal cation (such as silica or alumina gel), which, on pyrolysis, yield simple oxides, are referred to as single-component gels. Multicomponent gels have more than one type of metal cation and yield complex oxides on pyrolysis.

A wide range of ceramic and glass compositions have been prepared by sol-gel processing, and details of the experimental procedure can be found in the literature. We will not repeat the procedure for the preparation of specific gel compositions. Instead, the common techniques will be outlined with the aid of a few specific examples. To keep the discussion at a suitable level, we shall limit the consideration of multicomponent gels to the case of gels containing two metal cations only. Since the preparation procedures are quite different for colloidal gels and polymeric gels, we shall consider them separately.

A. Preparation of Colloidal Gels

The colloidal gel route was outlined in Chapter 1 (Fig. 1.5b). For single-component gels, colloidal particles are dispersed in water and peptized with acid or base to produce a sol. Two main methods can be employed to achieve gelation: (i) removing of water from the sol by evaporation to reduce its volume or (ii) changing the pH to slightly reduce the stability of the sol.

The silica system is a good example of the sol-gel process for the preparation of colloidal gels. The preparation of alumina sols from alkoxides has also been well characterized [15]. Aluminum alkoxides such as aluminum *sec*-butoxide and aluminum isopropoxide are readily hydrolyzed by water to form hydroxides. However, which hydroxide is formed depends on the conditions employed in the hydrolysis. The initial hydrolysis reaction of aluminum alkoxides can be written

$$Al(OR)_3 + H_2O \rightarrow Al(OR)_2(OH) \tag{5.39}$$

The reaction proceeds rapidly with further hydrolysis and condensation:

$$2Al(OR)_2(OH) + H_2O \rightarrow (RO)(HO)Al\!-\!O\!-\!Al(OH)(OR) + 2ROH \tag{5.40}$$

Assuming the formation of polymers that are not too highly crosslinked, the incorporation of n aluminum ions into the chain is given approximately by the formula $Al_nO_{n-1}(OH)_{(n+2)-x}(OR)_x$. As the reaction proceeds, the number of OR groups, i.e., x, relative to n should decrease to a value that depends on the hydrolysis temperature and the concentration of OR groups in the solvent. [The reader should keep in mind, as we mentioned earlier, that Eqs. (5.39) and (5.40) are not exact formulas but merely represent simplifications.] Hydrolysis by cold water (20°C) results in the formation of a monohydroxide that is predominantly amorphous. The structure contains a relatively high concentration of OR groups. It is believed that the presence of the OR groups is directly related to the structural disorder in the amorphous phase, because their removal (e.g., by aging in the solvent) inevitably leads to conversion of the amorphous hydroxide to a crystalline hydroxide, boehmite (AlO(OH)) or bayerite (Al(OH)$_3$). Aging at room temperature leads to the formation of bayerite by a process involving the dissolution of the amorphous hydroxide and subsequent precipitation as the crystalline phase. Aging of the amorphous hydroxide above 80°C leads to rapid conversion to boehmite. Since the conversion of the amorphous hydroxide to boehmite or bayerite is accompanied by the liberation of OR groups, the rate of conversion is inhibited by the presence of alcohol in the solvent used for the aging process. Hydrolysis of aluminum alkoxides by hot water (80°C) results in the formation of boehmite, which is relatively unaffected by aging.

Using aluminum alkoxides as the starting material, the production of alumina by the colloidal gel route involves the following main steps: (i) hydrolysis of the alkoxide to precipitate a hydroxide; (ii) peptization of the precipitated hydroxide, e.g., by the addition of acids, to form a clear sol; (iii) gel formation by, for example, evaporation of solvent; (iv) drying of the gel; and (v) firing of the dried gel. In this case, the formation of the sol can be a critical part of the process. While boehmite and the amorphous hydroxide prepared by cold water hydrolysis can be peptized to a clear sol, bayerite will not form a sol. The formation of bayerite during hydrolysis should therefore be avoided. In addition, the nature of the acid has a significant effect on the peptization step. Table 5.6 shows the peptizing effect of various acids on the precipitate formed by hydrolysis of aluminum *sec*-butoxide. The results are similar when aluminum isopropoxide is used. It appears that only strong or fairly strong acids, which do not form chemical complexes (or form only very weak complexes) with aluminum ions, are effective for achieving peptization. For these acids, the concentration of the acid also has an effect. Peptization requires the addition of at least 0.03 mol of acid per mole of hydroxide (followed by heating at ≈80°C for a sufficient time).

The amount of acid used in the peptization step also has a significant influence on the gelation of the sol and on the properties of the fabricated

Table 5.6 Peptizing Effect of Various Acids on the Precipitate Formed by the Hydrolysis of Aluminum *sec*-Butoxide

Acid	Formula	Condition of precipitate[a]
Nitric	HNO_3	Clear sol
Hydrochloric	HCl	Clear sol
Perchloric	$HClO_4$	Clear sol
Hydrofluoric	HF	Unpeptized
Iodic	HIO_4	Unpeptized
Sulfuric	H_2SO_4	Unpeptized
Phosphoric	H_3PO_4	Unpeptized
Boric	H_3BO_3	Unpeptized
Acetic	CH_3COOH	Clear sol
Trichloroacetic	CCl_3COOH	Clear sol
Monochloroacetic	$CH_2ClCOOH$	Clear to cloudy
Formic	$HCOOH$	Clear to cloudy
Oxalic	$H_2C_2O_4 \cdot H_2O$	Unpeptized
Phthalic	$C_8H_4O_3$	Unpeptized
Citric	$H_3C_6H_5O_7 \cdot H_2O$	Unpeptized
Carbolic	C_6H_5OH	Unpeptized

[a] After 7 days at 95°C.
Source: Ref. 15, reprinted with permission.

Sol-Gel Processing

aluminum oxide. There is a critical acid concentration at which the volume of the gel is a minimum. For nitric acid, this critical concentration is ≈0.07 mol per mole of hydroxide. At this minimum volume, the gel contains an equivalent of 25 wt % Al_2O_3. Deviation from the critical acid concentration, to either higher or lower values, causes a sharp increase in the volume of the gel (Fig. 5.23). At higher acid concentrations, the gels may contain an equivalent of only 2–3 wt % Al_2O_3. Finally, because of the large shrinkages that occur, gels containing an equivalent of less than ≈4 wt % Al_2O_3 do not retain their integrity after drying and firing.

In the case of multicomponent colloidal gels, a primary concern is to prevent segregation of the individual components so that uniform mixing can be achieved. Various routes have been used for their preparation, including (i) coprecipitation of mixed oxides or hydroxides, (ii) mixing of sols of different oxides or hydroxides, and (iii) mixing of sols and solutions. In the coprecipitation technique, the general approach is to mix

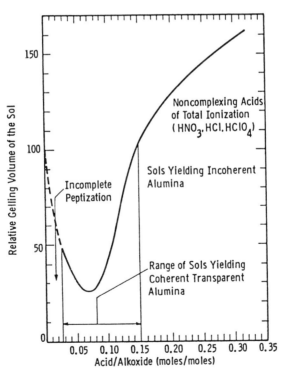

Figure 5.23 Effect of acid concentration on the volume of the gel formed from a peptized boehmite sol. (From Ref.10.)

different salt solutions or alkoxide solutions to give the required composition, followed by hydrolysis with water. The precipitated material is usually referred to as a gel, but, unlike the gels produced from dispersions of colloidal particles, it is not normally dispersible in water. The success of the method depends on controlling the concentration of the reactants and the pH and temperature of the solution to produce mixed products with the desired chemical homogeneity. We considered the coprecipitation technique earlier, in Chapter 2, for the preparation of powders of complex oxides. An example considered was the work of Bratton (Ref. 20, Chap. 2) involving the coprecipitation of a mixture of the double hydroxide $Mg(OH)_2 \cdot Al(OH)_3$ and $Al(OH)_3$ from a solution of magnesium chloride and aluminum chloride.

When gels are formed by mixing sols of different oxides or hydroxides, the uniformity of mixing is, at best, on the order of the size of the colloid particles. The chemical homogeneity will therefore be worse than that obtained in the coprecipitation method assuming ideal coprecipitation (i.e., without aggregation). As an example of this method, consider the preparation of an aluminosilicate gel, for example, one with the mullite composition, $(3Al_2O_3 \cdot 2SiO_2)$, by the mixing of boehmite sol and silica sol. Although these sols can be prepared in the laboratory by, for example, hydrolysis of alkoxides, they are also available commercially. In the pH range of $\approx 2.5-8$, the surfaces of the boehmite particles are positively charged whereas those of the silica particles are negatively charged. If the mixture of the two sols is gelled within this pH range, then a fairly homogeneous colloidal gel can be obtained because of the attraction and intimate contact between the oppositely charged boehmite and silica particles.

The aluminosilicate system can also be used to illustrate the third method of mixing sols and solutions. In one case, boehmite sol is mixed with a solution of TEOS in ethanol. Gelling is achieved by heating the mixture to evaporate some solvent. Alternatively, silica sol is mixed with a solution of aluminum nitrate, and the mixture is gelled by heating.

B. Preparation of Polymeric Gels

For single-component gels such as silica gel, we have considered in detail the conditions that lead to the formation of polymeric gels. Turning now to the preparation of multicomponent gels, further considerations must be taken into account. As an example, consider the formation of silica-titania glasses by the polymeric gel route. A convenient starting point for the formation of the gel is the hydrolysis and condensation of a mixed solution of a silicon alkoxide (e.g., TEOS) and a titanium alkoxide (e.g.,

Sol-Gel *Processing*

titanium tetraethoxide). However, from our earlier discussion of the properties of alkoxides, we would expect the hydrolysis of the titanium alkoxide to be much faster than that of the silicon alkoxide. Uncontrolled additions of water to the mixture of the two alkoxides would lead to vigorous hydrolysis of the titanium alkoxide and the formation of precipitates that are useless for polymerization. The problem of mismatched hydrolysis rates must be considered seriously if good chemical homogeneity is to be achieved. In general, four different approaches are used to overcome the problem: (i) use of double alkoxides, (ii) partial hydrolysis of the slowest reacting alkoxide, (iii) slow addition of small amounts of water, (iv) matching hydrolysis rates of the individual alkoxides by changing the length of the alkyl group or using chelating agents. Of these, methods (ii) and (iii) are used most often.

(i) Use of Double Alkoxides. The use of double alkoxides as the starting material in the gelation process eliminates the problem of mismatched hydrolysis rates, because each molecule or molecular species of the double alkoxide already contains the metal cations mixed in the desired ratio. The double alkoxide is hydrolyzed and polymerized in the way described earlier for simple alkoxides containing one metal cation, and the gel formed has the same ratio of the two metals as the double alkoxide. The homogeneity of mixing therefore extends to the atomic level. The high degree of homogeneity obtained by this method is illustrated by the work of Dislich [16], who prepared magnesium aluminum spinel, $MgAl_2O_4$, from a double alkoxide formed by reacting magnesium methoxide with aluminum *sec*-butoxide (Fig. 5.24). On heating the dried gel, the major X-ray

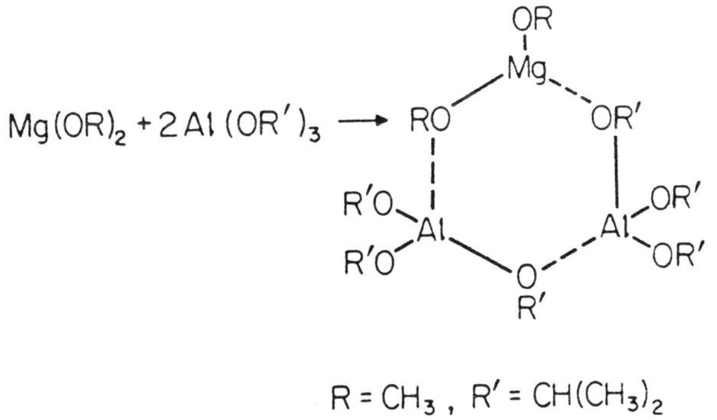

$$Mg(OR)_2 + 2 Al (OR')_3 \longrightarrow$$

$R = CH_3$, $R' = CH(CH_3)_2$

Figure 5.24. Magnesium aluminum double alkoxide formed by the reaction of a solution containing 1 mol magnesium methoxide to 2 mol aluminum *sec*-butoxide in alcohol. (From Ref.16.)

reflections of spinel started to appear at ≈250°C, and crystallization was completed by ≈400°C. In comparison, as outlined above, the colloidal gels of the same material prepared by Bratton using a coprecipitation technique still contained gibbsite, Al(OH)$_3$, at temperatures below ≈400°C. Despite the high degree of chemical homogeneity obtainable by the double alkoxide route, the method is used only on a limited basis for the formation of multicomponent gels, because many double alkoxides are difficult to synthesize or are unstable. Furthermore, for a given two-component gel, changing the chemical composition of the gel by changing the atomic ratios of the two metal cations becomes a tedious task because a new double alkoxide has to be synthesized in each case.

(ii) Partial Hydrolysis of the Slowest Reacting Alkoxide. This method, involving partial hydrolysis of the slowest reacting alkoxide prior to the addition of the other alkoxide, is best illustrated by the work of Yoldas [7] on the silica-alumina and silica-titania systems. For the silica-titania system, the starting materials are TEOS and titanium tetraethoxide, Ti(OEt)$_4$. The hydrolysis rate of TEOS, we recall, is much slower than that for the titanium ethoxide. If the TEOS is diluted with ethanol (e.g., 1 mol of ethanol to 1 mol of TEOS) and partial hydrolysis of the solution is carried out with the addition of 1 mol of water per mole of TEOS, then the majority of the silanol species will contain one OH group:

$$Si(OR)_4 + H_2O \rightarrow (RO)_3Si\text{—}OH + ROH \qquad (5.41)$$

As noted earlier, these soluble silanols may polymerize on aging. However, for the present situation where the number of OH groups in the silanol is limited to about one per molecule and the system is diluted, the polymerization rate is slow during the first few hours if the temperature is reasonably low (e.g., room temperature):

$$(RO)_3Si\text{—}OH + RO\text{—}Si(OR)_2(OH)$$
$$\xrightarrow{\text{slow}} (RO)_3Si\text{—}O\text{—}Si\text{—}(OR)_2(OH) + ROH \qquad (5.42)$$

If titanium ethoxide is introduced into the cooled solution with vigorous stirring, the following reaction occurs:

$$(RO)_3Si\text{—}OH + RO\text{—}Ti(OR)_3 \xrightarrow{\text{fast}} (RO)_3Si\text{—}O\text{—}Ti(OR)_3 + ROH \qquad (5.43)$$

Further polymerization with, for example, other silanols occurs with aging:

Sol-Gel Processing

$$(RO)_3Si\text{—}O\text{—}Ti(OR)_3 + HO\text{—}Si(OR)_3$$

$$\rightarrow (RO)_3Si\text{—}O\underset{|}{\overset{OR}{\text{—}Ti\text{—}}}O\text{—}Si(OR)_3 + ROH \quad (5.44)$$
$$\phantom{\rightarrow (RO)_3Si\text{—}O\text{—}}OR$$

Since the reactions described by Eqs. (5.43) and (5.44) occur at a faster rate than the self-condensation reaction described by Eq. (5.42), dissimilar constituents (i.e., molecular species of silicon and titanium rather than those of silicon alone) tend to become neighbors so that the mixing is at a molecular level. Furthermore, the product remains in solution because there are too few hydroxyl groups to cause precipitation. After the silicon and titanium have been incorporated into the polymeric network, the addition of more water completes the hydrolysis, condensation, and polymerization reactions to produce a single-phase gel.

(iii) Slow Addition of Small Amounts of Water. As illustrated by Eqs. (5.15)–(5.17), on hydrolysis with excess water, most metal alkoxides form insoluble oxide or hydroxide precipitates that are useless for further polymerization reactions. However, if small amounts of water are added slowly to a sufficiently dilute solution, it is possible to form polymerizable molecular species from these alkoxides also. For example, when boron alkoxide is exposed to water, a number of transient molecular species such as $B(OR)_2(OH)$ and $B(OR)(OH)_2$ form initially, e.g.,

$$B(OR)_3 + H_2O \rightarrow B(OR)_2(OH) + ROH \quad (5.45)$$

These transient molecular species represent various degrees of hydrolysis of the boron alkoxide. They are soluble in solution and can undergo condensation reactions leading to the formation of a polymer network.

For the SiO_2–B_2O_3 system, Yoldas [7] investigated the effect of water additions on the homogeneity of a solution of TEOS and boron methoxide, $B(OCH_3)_3$. His results are shown in Fig. 5.25. For a given solution, if the water content exceeds a certain value, solution homogeneity is lost due to precipitation. Boron methoxide hydrolyzes much faster than TEOS and precipitates as B_2O_3. However, for lower water content, the partially hydrolyzed molecular species are soluble, so a clear solution is obtained. Condensation and polymerization reactions between the partially hydrolyzed species lead to the production of a homogeneous gel. The line separating the clear solution from that containing precipitates can be represented by

$$M_{\text{water}} = AM_{\text{BE}} + BM_{\text{TEOS}} \quad (5.46)$$

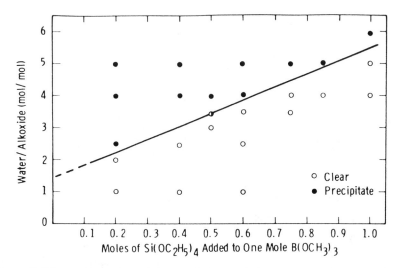

Figure 5.25 Regions of precipitate and clear solution formation for the hydrolysis of a solution containing various ratios of TEOS and boron methoxide. (From Ref. 7.)

where M_{water}, M_{BE}, and M_{TEOS} are the number of moles of water, boron ethoxide, and TEOS, respectively. For $M_{\text{BE}} = 1$, the results of Fig. 5.25 give $A \approx 1.5$ and $B \approx 4$. These values indicate that to cause precipitation by the slow addition of water, ≈ 1.5 mol of water is required for 1 mol of boron ethoxide and an additional 4 mol of water is required for each mole of TEOS added to the system. The requirement of ≈ 1.5 mol of water to cause precipitation in the boron alkoxide agrees with the reaction described by Eq. (5.17).

(iv) Matched Hydrolysis Rates. We noted earlier that for a given metal M, the hydrolysis rate of the alkoxide $M(OR)_x$ generally decreases with increasing length of the alkyl group R. This suggests that the hydrolysis rate of alkoxides of different metals can be matched by careful choice of the alkoxide group. As an example, consider the SiO_2–TiO_2 system. Silicon alkoxides are among the slowest to hydrolyze, but, of these, tetramethylorthosilicate (TMOS) has the fastest hydrolysis rate. Therefore, by selecting a titanium alkoxide with a sufficiently long alkyl group, it may be possible to match the hydrolysis rates of the two alkoxides. For this system, it was found by Yamane and coworkers [17] that the hydrolysis of a mixed solution of TMOS and titanium *tert*-amyloxide, $Ti[OC(CH_3)_2C_2H_5]_4$, produced a homogeneous gel. In general, however, this method is rarely used because of the difficulties in matching the hydrolysis rates closely or in obtaining alkoxides with the desired hydrolysis rates.

A more promising approach has been described by Livage et al. [18] who showed that chelating agents such as acetylacetone can react with alkoxides at a molecular level giving rise to new molecular precursors. The whole hydrolysis-condensation process can therefore be modified. For example, with the use of chelating agents, the hydrolysis of transition metal alkoxides can be slowed so that better chemical homogeneity can be achieved in multicomponent gels.

Comparison of the Preparation Methods. To compare the homogeneity of multicomponent gels prepared by different methods, Yamane and co-workers [17] produced gels by methods (ii), (iii), and (iv) outlined above for a composition corresponding to 93.75 mol % SiO_2 and 6.25 mol % TiO_2. Each gel was melted to form a glass, and the optical transmission of the glass was used as a measure of the homogeneity of the gel. In this way, it was found that the homogeneity of the glasses prepared by methods (ii), (iii), and (iv) did not differ significantly. However, the method based on matching the hydrolysis rates of the alkoxides, method (iv), produced a glass with a slightly better homogeneity than methods (ii), partial hydrolysis, and (iii), slow additions of water. The homogeneity of these three glasses was also compared with that of a glass produced from a mechanically mixed gel. For the mixed gel, SiO_2 and TiO_2 gels were prepared separately by hydrolysis of the individual alkoxides, after which they were mixed mechanically. The three gels prepared from mixtures of alkoxides gave glasses with a homogeneity that was far better than that of the glass produced from the mechanically mixed gel.

Multicomponent gels can also be prepared for systems in which the starting materials are not all alkoxides. In this method, a solution of alkoxides and metal salts is normally used. This is particularly useful when the alkoxides are not available, difficult to prepare or use, or too expensive. The metal salts used are citrates, acetates, nitrates, chlorides, and sulfates. The salt should be soluble in alcohol so that it can be mixed with alkoxide solutions without premature hydrolysis of the alkoxide. Nitrates are possibly the most widely used salts because of their solubility and the ease with which the anion can be removed by thermal decomposition. However, they are highly oxidizing, and care should be taken during heating of the gel. Acetates are also useful but do not decompose as easily as nitrates. Chlorides and sulfates are less widely used because the anions are fairly difficult to remove by thermal decomposition. One problem encountered with this method is that the salt often crystallizes during the drying process, which results in loss of homogeneity. Finally, little is known about the reaction mechanism by which the polymerizing alkoxide incorporates the ions of the metal salt into the gel structure.

5.6 APPLICATIONS OF SOL-GEL PROCESSING

The applications of ceramics and glasses produced by sol-gel processing are widespread, and their number is expected to grow substantially [19,20]. It is not possible to consider all of these applications in the limited space remaining in this chapter. Instead, we consider a few general areas in which the sol-gel process has had, and will continue to have, the greatest impact. These areas are thin films or coatings, fibers, and monolithic ceramics and glasses with novel compositions.

A. Thin Films and Coatings

Thin films and coatings form the most common application of sol-gel processing because of the difficulties in making gel layers thicker than ≈ 1 μm. The starting material for the coating process is generally a solution of metal alkoxides but the use of solutions of metal salts or sols (fine particles dispersed in a liquid) has also been reported. Using a solution of metal alkoxides, the hydrolysis and condensation reactions discussed earlier play an important role in the development of film properties.

Two main techniques have been used for producing the film: (i) dip coating, where the object to be coated is lowered into the solution and withdrawn at a suitable speed, and (ii) spin coating, where the solution is dropped onto the object, which is spinning at a high speed. Of the two techniques, dip coating is currently the more widely used. One requirement is that the contact angle between the solution and the surface of the object be low so that the solution wets and spreads over the surface. Usually water is present in the sol but moisture from the atmosphere is also sufficient to cause hydrolysis and condensation reactions. The final film is obtained after firing. There are various advantages and disadvantages inherent in each coating technique. Dip coating does not require any specialized apparatus (Fig. 5.26). Both internal and external surfaces wetted by the solution are coated. The solution must not be too sensitive to moisture, because it undergoes some exposure to the atmosphere during the coating process. Spin coating leads to coating one side of the object only, and the coating solution can be kept away from moisture prior to being dropped onto the object. Edge effects may occur for objects that are not axisymmetric. Furthermore, the spinning of large objects is impractical.

For dip coating, the thickness of the liquid film depends on the viscosity of the solution and the speed of withdrawal of the object. The idealized

Sol-Gel Processing

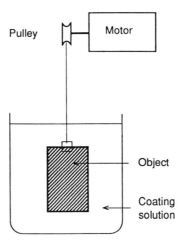

Figure 5.26 Schematic illustration of a dip-coating apparatus.

case of an infinite plate being withdrawn vertically from a sufficiently large vessel to allow boundary effects to be ignored was analyzed by Landau and Levich in 1942. For low capillary numbers defined by

$$N_{ca} = \eta U/\gamma_{lv} \tag{5.47}$$

where U is the withdrawal speed, η is the dynamic viscosity, and γ_{lv} is the surface tension of the solution, the thickness of the liquid film can be expressed in the form

$$t = 0.944 \, N_{ca}^{1/6} \left(\frac{\eta U}{\rho g}\right)^{1/2} \tag{5.48}$$

where ρ is the density of the solution and g is the acceleration due to gravity. From Eqs. (5.47) and (5.48), the thickness of the film is predicted to vary as $(U\eta)^{2/3}$. If the final solid film is densified to full density, ρ_f, the thickness of the final film is

$$t_f = 0.944 \, \frac{\rho - \rho_s}{\rho_f - \rho_s} N_{ca}^{1/6} \left(\frac{\eta U}{\rho g}\right)^{1/2} \tag{5.49}$$

where ρ_s is the density of the solvent. In the case of SiO_2 films prepared from a solution of TEOS in ethanol, $\rho_f = 2.2$ g/cm^3 and $\rho_s = 0.8$ g/cm^3. Empirical modifications have been proposed, but the relatively simple expression of Landau and Levich, Eq. (5.48) or (5.49), gives a reasonably good fit to most of the reported experimental data.

The physics of spin coating is different from that of dip coating. Analyses show that the thickness of the film prepared by spin coating varies inversely as $\omega^{2/3}$, where ω is the angular velocity of the spin coater, and as $\eta^{1/3}$.

Table 5.7 summarizes some possible applications of films and coatings prepared by sol-gel processing of metal alkoxide solutions [21].

B. Fibers

As discussed earlier, the viscosity of metal alkoxide solutions increases with time as the hydrolysis and condensation reactions proceed (Fig. 5.4). At viscosities greater than ≈ 1 Pa·s, the solution becomes sticky and fibers can be drawn. At this stage the solution is said to be spinnable. At longer times, as gelation is approached, the viscosity becomes too high, and the spinnability decreases dramatically. However, the viscosity is not the only parameter that controls the spinnability. Another important factor is the nature of the polymeric species in the solution. It is found that only solutions that contain relatively linear polymers can be fairly easily spun into fibers.

In the case of TEOS, we recall that linear polymers will be formed for acid-catalyzed reactions (pH <2.5) and low water content (less than ≈ 4 mol H_2O per mole of TEOS). In practice, it has been found that such solutions can be readily spun into fibers. An example of continuously drawn silica fibers from TEOS solutions is shown in Fig. 5.27. Further-

Table 5.7 Applications of Films Prepared from Metal Alkoxide Solutions

Application	Example	Composition
Mechanical	Protection	SiO_2
Chemical	Protection	SiO_2
Optical	Absorbing	TiO_2—SiO_2; SiO_2—R_mO_n; oxides of Fe, Cr, and Co
	Reflecting	In_2O_3—SnO_2
	Antireflecting	Na_2O—B_2O_3—SiO_2
Electrical	Ferroelectric	$BaTiO_3$; $KTaO_3$; PLZT
	Electronic conductor	In_2O_3—SnO_2; SnO_2—CdO
	Ionic conductor	β-Alumina
Catalytic	Photocatalyst	TiO_2
	Catalyst carrier	SiO_2; TiO_2; Al_2O_3

Source: Ref. 21, reprinted with permission.

Sol-Gel Processing

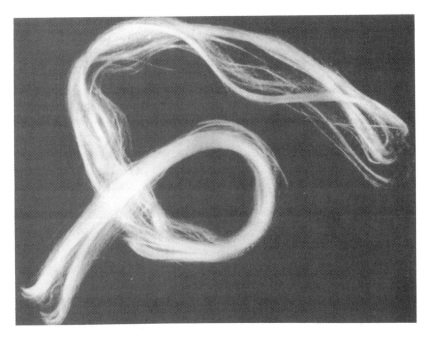

Figure 5.27 Silica fibers drawn from a hydrolyzed solution of TEOS. (From Ref. 20, with permission.)

more, acid-catalyzed reactions with high water content (e.g., 20 mol H_2O to 1 mol TEOS) or base-catalyzed reactions produced solutions that were not spinnable. For TEOS solutions that are spinnable, another observation is that the uniformity of cross section of the fiber depends on the composition of the starting solution. Starting solutions that lead to a lower shrinkage in the gelation of the drawn fiber generally lead to uniform circular cross sections. In other cases, the cross-sectional geometry can be fairly irregular [21]. For acid-catalyzed conditions, the effect of the composition of the starting solution on the spinnability and the fiber cross section is summarized in Fig. 5.28.

In addition to SiO_2, other fiber compositions that have been made by sol-gel processing include SiO_2-TiO_2 (10–50 mol % TiO_2), SiO_2-Al_2O_3 (10–30 mol % Al_2O_3), SiO_2-ZrO_2 (10–33 mol % ZrO_2), and SiO_2-Na_2O-ZrO_2 (25 mol % ZrO_2). The advantages of low-temperature processing and good chemical homogeneity provided by the sol-gel route are clearly evident from these compositions, which are highly refractory and difficult to produce by conventional methods.

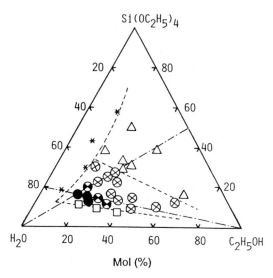

Figure 5.28 Relation between fiber drawing behavior and composition of TEOS–water–ethanol solutions after hydrolysis at 80°C with an acid concentration of 0.01 mol HCl per mole of TEOS: (∗) immiscible, (□) not spinnable, (△) no gel formation, (●) circular cross section, (⊗) non-circular cross section, (◓) circular and noncircular cross section. (Reprinted by permission of the American Ceramic Society.)

C. Monolithic Ceramics and Glasses

Monolithic ceramics and glasses can be prepared by three methods that involve sol-gel processing: (i) firing of xerogels or aerogels, (ii) compaction and firing of gel-derived powders (i.e., powders prepared by sol-gel processing), and (iii) melting of gel-derived powders. As outlined earlier, the sol-gel process is unlikely to be an attractive process for the production of monolithic ceramics and glasses that can normally be obtained by conventional fabrication methods. Its main advantage lies in the possibility of preparing new ceramics and glasses that cannot normally be prepared by those methods. Examples include the highly refractory fiber compositions discussed above. Additional but less important advantages include the achievement of high chemical homogeneity and lower firing temperatures. The high chemical homogeneity of gel-derived powders leads to lower melting temperatures and good uniformity in the compostion of the fabricated glasses, as demonstrated by the work of Yamane et al. [17] on SiO_2-TiO_2 glasses discussed in Section 5.6B. Gel-derived powders are normally amorphous and, provided there is no crystallization prior to the

Sol-Gel Processing

achievement of full densification, sinter at lower temperatures than crystalline powders of the same composition. For example, gel-derived mullite powders can be sintered to full density below ≈1300°C, whereas the sintering temperature is ≈1600°C for crystalline mullite powders available commercially.

5.7 CONCLUDING REMARKS

In this chapter we have examined the basic physics and chemistry of sol-gel processing and the ways in which these relate to the practical fabrication of ceramics and glasses. It is apparent that sol-gel processing offers considerable opportunities in both science and technology. An understanding of the scientific issues provides the most useful basis for successful application of the method. Sol-gel processing offers considerable advantages as a fabrication method, but its disadvantages are also very real. Successful applications (e.g., films and fibers) are those in which the advantages in processing are used and the disadvantages are minimized. The applications of sol-gel processing are widespread, and certain applications such as coatings on window glass are being used on a commercial basis. Although the number of applications is expected to grow substantially, the method cannot compete with the more conventional fabrication methods used in the mass production of ceramics and glasses.

PROBLEMS

5.1 You wish to prepare approximately 50 g of silica by the sol-gel process under the following conditions: You start with a solution of 50 mol % TEOS in ethanol and you add 10 mol of water per mole of alkoxide. Determine the volume of each starting material required for the process, assuming that the reaction goes to completion.

5.2 An alcogel contains 5 vol % of solid. Careful drying leads to the production of a xerogel that is approximately 50% porous. Estimate the linear shrinkage during drying. On firing, a fully dense solid is obtained. If the weight loss during firing is approximately 20%, estimate the linear shrinkage during the firing stage.

5.3 The alcogel in Problem 2 is dried supercritically (where the shrinkage negligible) and the aerogel is fired to full density. Estimated the linear shrinkage during firing.

5.4 Estimate the surface free energy of 1 g of a silica gel that is 50% porous with pores (assumed spherical) of 10 nm in diameter. Compare your answer with the surface free energy of 1 g of silica glass spheres with a diameter of 10 μm. The specific suface energy of both materials can be assumed to be 0.25 J/m².

5.5 Describe three methods for the preparation of $BaTiO_3$ by sol-gel processing and comment on the chemical homogeneity of the gel prepared by each method.

5.6 Assuming Eq (5.48), derive Eq. (5.49) for the thickness of the fully densified film produced by dip coating.

REFERENCES

1. C. J. Brinker and G. W. Scherer, *Sol-Gel Science*, Academic, New York, 1990.
2. C. J. Brinker and G. W. Scherer, *J. Non-Cryst. Solids, 70*:301 (1985).
3. E. M. Rabinovich, D. W. Johnson, Jr., J. B. MacChesney, and E. M. Vogel, *J. Am. Ceram. Soc., 66*:683 (1983).
4. G. W. Scherer and J. C. Luong, *J. Non-Cryst. Solids, 63*:163 (1984).
5. D. C. Bradley, R. C. Mehrotra, and D. P. Gaur, *Metal Alkoxides*, Academic, London, 1978.
6. H. Okamura and H. K. Bowen, *Ceram. Int., 12*:161 (1986).
7. B. E. Yoldas, *J. Mater. Sci., 14*:1843 (1979).
8. K. D. Keefer, *Mater. Res. Soc. Symp. Proc., 32*:15 (1984).
9. G. W. Scherer, *J. Non-Cryst. Solids, 109*:171 (1989); *J. Am. Ceram. Soc. 73*:[1]3 (1990).
10. D. P. Partlow and B. E. Yoldas, *J. Non-Cryst. Solids, 46*:153 (1981).
11. G. W. Sherer, in *Drying '92* (A. S. Mujumdar, Ed.), Elsevier, New York, 1992, pp. 92–113.
12. C. J. Brinker, G. W. Scherer, and E. P. Roth, *J. Non-Cryst. Solids, 72*:345 (1985).
13. G. W. Scherer, C. J. Brinker, and E. P. Roth, *J. Non-Cryst. Solids, 72*:369 (1985).
14. C. J. Brinker, E. P. Roth, D. R. Tallant, and G. W. Scherer, in *Science of Ceramic Chemical Processing* (L. L. Hench and D. R. Ulrich, Eds.), Wiley, New York, 1986, Chap. 3.
15. B. E. Yoldas, *Am. Ceram. Soc. Bull., 54*:289 (1975).
16. H. Dislich, *Angew. Chem., Int. Ed., 10*:363 (1971).
17. M. Yamane, S. Inoue, and N. Keiichi, *J. Non-Cryst. Solids, 48*:153 (1982).
18. C. Sanchez, J. Livage, M. Henry, and F. Babonneau, *J. Non-Cryst. Solids, 100*:65 (1988).
19. B. J. J. Zelinski and D. R. Uhlmann, *J. Phys. Chem. Solids, 45*:1069 (1984).
20. R. W. Jones, *Fundamental Principles of Sol-Gel Technology*, Institute of Metals, London, 1989.
21. S. Sakka, *Am. Ceram. Soc. Bull., 64*:1463 (1985).
22. J. Zarzycki, M. Prassas, and J. Phalippou, *J. Mater. Sci., 17*:3371 (1982).

Sol-Gel Processing

FURTHER READING

Reference 1 describes in detail the physics and chemistry of sol-gel processing.

Technological applications are reviewed in: L. C. Klein (Ed.), *Sol-Gel Technology for Thin Films, Fibers, Preforms, Electronics and Specialty Shapes*, Noyes, Park Ridge, NJ (1988).

The conference proceedings, published at regular intervals, contain useful reports on sol-gel processing: Materials Research Society Symposium Proceedings: Better Ceramics Through Chemistry, Materials Research Society, Pittsburgh, PA.

6
Powder Consolidation and Forming of Ceramics

6.1 INTRODUCTION

The commonly used methods for the consolidation of ceramic powders are described in this chapter. We recall that the microstructure of the consolidated powder form (i.e., the green body) has a significant effect on the subsequent firing stage. If severe variations in packing density occur in the green body, the fired body will, in general, contain heterogeneities that, in turn will have strong consequences for properties. The uniform packing of particles in the green body is the goal of the consolidation step. Since the packing density controls the amount of shrinkage during firing, the achievement of high packing densities is also desirable. Geometrical particle packing concepts provide a useful basis for understanding how the structure of the consolidated powder comes about. An important practical consideration is the extent to which the parameters of the consolidation process can be manipulated to control the packing uniformity and packing density of the green body.

The colloidal techniques described in Chapter 4 provide considerable benefits for the control of the packing uniformity of the consolidated powder form. The production of green bodies with uniform microstructure from a fully stabilized colloidal suspension of spherical, fine, monodisperse particles has not been incorporated into industrial applications, where mass production is desired and fabrication cost is a serious consideration. Colloidal techniques, however, play an important role in two low-cost consolidation methods that involve the casting of ceramic bodies from particulate slurries: slip casting and tape casting. A drawback of the casting methods, however, is that they are limited to thin articles or thin sheets.

Mechanical compaction of dry or semidry powders in a die is one of the most widely used forming operations in the ceramic industry. In gen-

Powder Consolidation

eral the applied pressure is not transmitted uniformly because of friction between the particles themselves and between the particles and the walls of the die. The stress variations lead to density variations in the compacted powder, thereby placing considerable limits on the degree of packing uniformity that can be achieved. Although the density variations can be reduced significantly by isostatic pressing, mechanical compaction provides far less control in the manipulation of the microstructure of the green body than the casting methods.

Plastic forming methods in which a mixture of the ceramic powder and additives is deformed plastically through a nozzle or in a die provide a convenient route for the mass production of ceramic green bodies. These methods are similar to those used in the polymer (plastics) industry. Extrusion is used extensively in the traditional ceramics industry and to a lesser extent in the advanced ceramics sector. Injection molding has been the subject of intense investigation in recent years. However, it has not yet made any significant inroads in the forming of ceramics for industrial applications.

Polymeric additives (e.g., binders) play an important role in the production of the consolidated powder form. In many cases, the selection of the additives can be vital to the success of the forming process. The additives must, however, be removed from the green body prior to firing. For consolidation techniques such as tape casting and injection molding that use a considerable amount of additives, binder removal can be one of the limiting steps in the overall fabrication process.

6.2 PACKING OF PARTICLES

The packing of particles is generally divided into two types: (i) regular packing and (ii) random packing. While various parameters can be used to characterize the packing arrangement, two of the most widely used are (i) the packing density, defined as

$$\text{Packing density} = \frac{\text{volume of solids}}{\text{total volume of the arrangement (solids + voids)}} \quad (6.1)$$

and (ii) the coordination number, which is the number of particles in contact with any given particle. The packing density is sometimes referred to as the packing fraction or the fractional solids content. As a first step toward understanding the packing of real powder particles, we consider the regular packing of spheres of the same size (monosize spheres). The particles are assumed to be rigid and inert.

A. Regular Packing of Monosize Spheres

The reader is probably familiar with the packing of atoms in crystalline solids to produce regular, repeating, three-dimensional patterns such as the simple cubic, body-centered cubic, face-centered cubic, and hexagonal close-packed structures. The packing densities and coordination numbers for these crystal structures are listed in Table 6.1.

In order to build up a three-dimensional packing pattern of particles, we can begin, conceptually by (i) packing spheres in two dimensions to form layers, and then (ii) stacking the layers on top of one another. Two types of layers are shown in Fig. 6.1, where the angle of intersection between the rows has limiting values of 90° (referred to a square layer) and 60° (simple rhombic or triangular layer). Although other types of layers that have angles of intersection between these two values are possible, only the square layer and the simple rhombic layer will be considered here. There are three geometrically simple ways of stacking each type of layer on top of one another, giving rise to six packing arrangements altogether. However, examination of the arrangements will show that, neglecting the difference in orientation in space, two of the ways of stacking the square layers are identical to two of the ways of stacking the simple rhombic layers [1]. Therefore there are only four different regular packing arrangements shown in Fig. 6.2. The packing densities and coordination numbers of these arrangements are summarized in Table 6.2.

The rhombohedral packing arrangement, which has the highest packing density, is the most stable packing arrangement. Even for monosized spherical powders, such dense packing arrangements have been achieved over only very small regions (called domains) of the consolidated powder form when rather special consolidation procedures have been used. An example is the slow sedimentation of monosize SiO_2 particles from a stable suspension shown earlier in Fig. 4.24. The domains are separated from one another by boundaries of disorder much like the granular microstructure of

Table 6.1 Packing Density and Coordination Number of Some Common Crystal Structures for a Pure Metal

Crystal structure	Packing density	Coordination number
Simple cubic	0.524	6
Body-centered cubic	0.680	8
Face-centered cubic	0.740	12
Hexagonal close-packed	0.740	12

Powder Consolidation

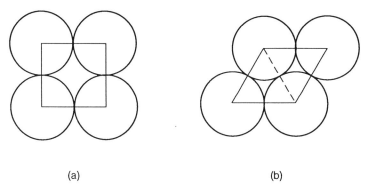

Figure 6.1 Two types of layers for the regular packing of monosize spheres. (a) Square; (b) rhombic or triangular.

polycrystallne materials. Figure 6.3 shows one problem that can arise when the consolidated powder is fired: cracklike voids open at the domain boundaries [2]. In general, the commonly used ceramic forming methods produce more random packing arrangements in the consolidated powder form.

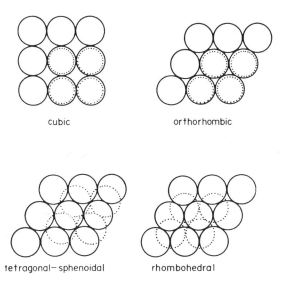

Figure 6.2 The four packing arrangements produced by stacking square and triangular layers of monosize spheres.

Table 6.2 Packing Density and Coordination Number of the Four Regular Packing Arrangements of Monosize Spheres

Packing arrangement	Packing density	Coordination number
Cubic	0.524	6
Orthorhombic	0.605	8
Tetragonal-sphenoidal	0.698	10
Rhombohedral	0.740	12

B. Random Packing of Spherical Particles

Starting with the simplest case of monosize spherical particles, two different states of random packing have been distinguished. If the particles are poured into a container that is then vibrated to settle the assembly of particles, the resulting packing arrangement reaches a state of minimum porosity referred to as *dense random packing*. On the other hand, if the particles are simply poured into the container so that they are not allowed to rearrange and settle into as favorable a position as possible, the resulting packing arrangement is referred to as *loose random packing*. An infinite number of packing arrangements may exist between these two limits.

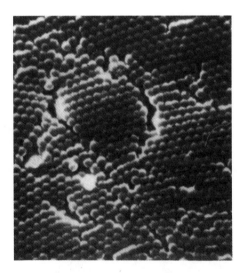

Figure 6.3 Partial densification of a periodically packed, multilayered arrangement of polymer spheres. Note the opening displacements at the domain boundaries. (From Ref. 2, with permission.)

Powder Consolidation

Dense random packing of monosized spheres has been studied experimentally by shaking hard spheres in a container. The upper limit of the packing density consistently ranges between 0.635 and 0.640. Computer simulations give a value of 0.637. A good value for the maximum packing density of a randomly packed structure of monosize spheres may be taken as 0.64. The maximum packing density for random packing of monosize spheres is predicted to be independent of the sphere size, and this prediction has been verified experimentally. For loose random packing, theoretical simulations as well as experiments give values in the range of 0.57–0.61 for the packing density. For dense random packing of monosize spheres, calculations show that fluctuations in the packing density become weak beyond a distance of three sphere diameters from the center of any given sphere. For density fluctuations existing over such a small scale, uniform sintering may be achieved during firing of the consolidated powder form. Therefore, the production of regular, crystal-like particle packing, achievable at present only over very small domains, may after all be unnecessary from the point of view of fabrication.

The interstitial holes between the spheres in dense random packing can be filled with spheres that are smaller than those of the original structure, thereby leading to an increase in the packing density (Fig. 6.4). For this type of random packing of a binary mixture of spheres, the packing

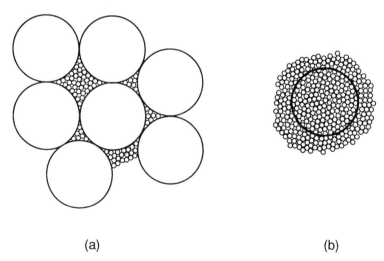

(a) (b)

Figure 6.4 Increase in packing density achieved by (a) filling the interstices between large spheres with small spheres and (b) replacing small spheres and their interstitial porosity by large spheres.

density is a function of (i) the ratio of the sphere diameters and (ii) the fraction of the large (or small) spheres in the mixture. By filling the interstitial holes with a large number of very fine spheres, we can maximize the packing density of the binary mixture. Starting with an aggregate of large (coarse) spheres in dense random packing, as we add fine spheres the packing density of the mixture increases along the line CR in Fig. 6.5. A stage will be reached when the interstitial holes between the large spheres are filled with fine spheres in dense random packing. Further additions of fine spheres will only serve to expand the arrangement of large spheres, leading to a reduction in the packing density. Assuming a packing density of 0.64 for dense random packing, the volume fraction of interstitial holes

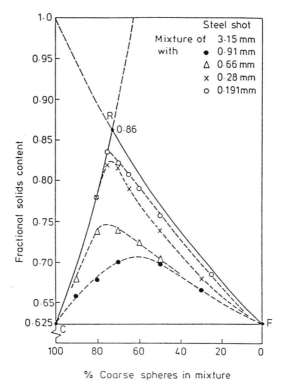

Figure 6.5 Binary packing of spheres showing the packing density (fractional solids content) as a function of the composition of the mixture. CRF represents the theoretical predictions for dense random packing when the ratio of the large sphere diameter to the small sphere diameter approaches infinity. The data of McGeary for the mechanical packing of steel shot are also shown. (From Ref. 3, with permission.)

Powder Consolidation

in the original aggregate of large spheres is 1 − 0.64, or 0.36. At the maximum packing density of the binary mixture, the interstitial holes are filled with a large number of fine spheres in dense random packing. The maximum packing density is therefore 0.64 + 0.36 × 0.64 = 0.87. The fractional volumes occupied by the large spheres and fine spheres are 0.64 and 0.87 − 0.64, respectively. The fraction of large spheres in the binary mixture is therefore 0.64/0.87, or 0.735. Alternatively, we can increase the packing density of an aggregate of fine spheres in dense random packing by replacing some of them and their interstitial holes by large spheres. In this case, the packing density of the mixture will increase along the line FR in Fig. 6.5. The intersection at R of the two curves CR and FR represents the state of optimum packing. Figure 6.5 also shows the experimental data of McGeary [3] for binary mixtures of spherical steel particles with size ratios. (A packing density of 0.625 was assumed for dense random packing.) It is seen that as the ratio of the size of the large sphere to the small sphere increases, the data move closer to the theoretical curve.

The packing of binary mixtures of spheres is also commonly represented in terms of the apparent volume (i.e., total volume of the solid phase and porosity) occupied by unit volume of solid [4]. The apparent volume is defined as

$$V_a = \frac{1}{1 - P} \qquad (6.2)$$

where P is the fractional volume of the voids (i.e., the porosity). As shown in Fig. 6.6, the line CRF represents the theoretical curve for the packing of a binary mixture in which the size of the large spheres is very much greater that of the small spheres.

So far, our approach to the packing of binary mixtures has been to fill the interstitial holes between the large spheres with a large number of very small spheres. Another approach is to insert into each hole a single sphere with the largest possible diameter that would fit into the hole. For an aggregate of monosize spheres in dense random packing, computer simulations reveal that, with this approach, the maximum packing density of the binary mixture is 0.76. Although this value is smaller than the maximum packing density (0.87) obtained by filling the interstices with a large number of fine spheres, it may represent a more realistic upper limit to the packing of real powder mixtures.

The packing density of mixtures of spheres can be increased further by going to ternary mixtures, quaternary mixtures, and so on. For example, if each interstitial hole in the binary mixture (packing density 0.87) is filled with a large number of very fine spheres in dense random packing, the maximum packing density becomes 0.95. Using the same approach,

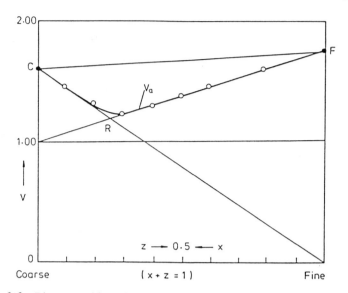

Figure 6.6 Binary packing of spheres plotted in terms of the apparent volume occupied by unit volume of the solid phase. In this representation, straight lines are obtained for the theoretical case CRF when the ratio of large sphere diameter to small sphere diameter approaches infinity. The data of Westman and Hugill [4] for a diameter ratio of 50 are also shown.

the maximum packing density of quaternary mixtures is 0.98. Following this packing scheme, McGeary experimentally achieved packing densities of 0.90 for a ternary mixture and 0.95 for a quaternary mixture of steel spheres that were compacted by vibration [3].

C. Packing of Powders in Practice

Most powders used for the fabrication of ceramics have a continuous distribution of particle sizes between some minimum and maximum sizes. Even in the case of spherical powders prepared by chemical precipitation from solution (see Chapter 2), although the spread in particle sizes may be small, the powder is not monosize. This spread in particle size has a significant influence on the packing density of the consolidated powder. We will therefore examine how a spread in the particle size about a certain mean size influences the packing of powders. The simplest case to consider is that of powders consisting of spherical particles.

As described in Chapter 3, the lognormal distribution function provides a good approximation to the particle size distribution of many pow-

Powder Consolidation

ders. Recall that the spread in the particle size of the distribution is represented by the standard deviation S. The packing density of spherical particles with a lognormal distribution of sizes has been investigated for various values of S. Both theoretical calculations and experiments show that the packing density increases as the standard deviation of the distribution increases [5], that is, as the particle size distribution becomes broader (Fig. 6.7). The packing density in dense random packing reaches fairly high values for a wide distribution of particle sizes. For spherical particles, the data of Fig. 6.7 can be approximated by an equation of the form

$$\text{Packing density} = 0.96 - \frac{0.28}{S} \tag{6.3}$$

For the case of loose random packing, the packing density also increases

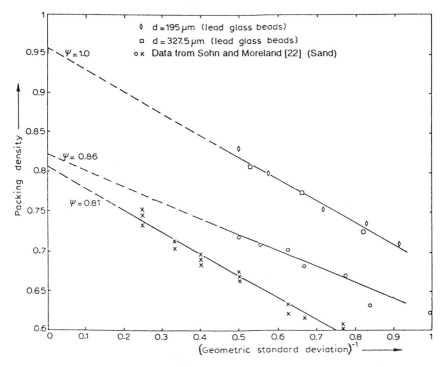

Figure 6.7 Packing density versus the reciprocal of the standard deviation for particles with a lognormal size distribution. The parameter ψ is a measure of the sphericity of the particles and is equal to the reciprocal of the shape factor. (From Ref. 5.)

with the standard deviation of the distribution, but the magnitude of the increase is significantly smaller than in dense random packing.

In view of the high packing densities achieved, it seems that we may be better off, as far as the consolidated powder form is concerned, with powders having a wide distribution of particle sizes (i.e., polydisperse powders). However, the packing density by itself is a misleading parameter for predicting the densification behavior and microstructural evolution of the body during firing. A more important consideration is the packing uniformity or, equivalently, the spatial scale over which density fluctuations occur in the consolidated powder form. Computer simulations [6] show that as the width of the particle size distribution increases, the scale over which density fluctuations occur also increases (Fig. 6.8). It is this increasing scale of density fluctuations that causes many of the problems in the firing stage.

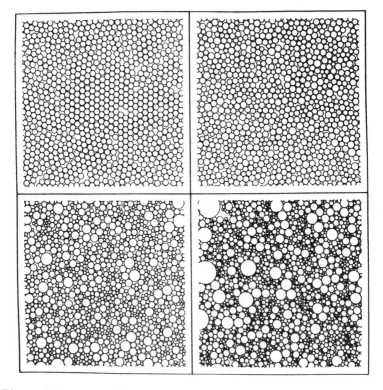

Figure 6.8 Four typical packings of multisized circles. (From Ref. 6, with permission.)

Powder Consolidation

Powders used in the industrial production of ceramics very rarely have spherical particles. Experiments with distributions of spherical and irregularly shaped particles show a reduction in the packing density for irregularly shaped particles (Fig. 6.7). The surfaces of the particles are also rarely smooth. Interparticle friction can lead to a small reduction in the packing density of particles with rough surfaces.

Commercial powders, which have a wide distribution of particle sizes, can be classified to provide fractions with the desired size range. One approach is to use the colloidal techniques described in Chapter 4 to disperse the powder and to remove hard agglomerates and large particles by sedimentation [7]. The supernatant can then be decanted and fractionated into various size fractions. Although in many cases this approach may not be economical for industrial production, the benefits for the achievement of uniform packing in the green body are clear from the micrographs shown in Fig. 6.9. The unclassified powder contains large agglomerates that produce regions with very nonuniform packing. The classified powder shows fairly uniform packing with a high packing density.

The blending of powders with discrete sizes, as we saw earlier, can be used to achieve high packing densities. In practice, little is gained beyond the use of ternary mixtures because the finer particles do not locate into their ideal positions to maximize the packing density. Although less severe in such powder blends than in the case of powders with a wide particle size distribution, the problem of the packing uniformity of the consolidated body still needs to be considered seriously. Two requirements must be satisfied for this approach to be successful. First, uniform mixing of the powder fractions by mechanical or colloidal methods must be achieved. Second, the mixture must be consolidated to produce uniform packing. The objective is a consolidated powder form in which the small pores are fairly uniformly spaced and the large voids are eliminated.

D. Packing of Mixtures of Powders and Short Fibers

Ceramic matrix composites reinforced with short single-crystal fibers (called whiskers) have been investigated extensively in recent years because of the improvement in mechanical properties that they offer compared to monolithic ceramics. Common material systems include matrices of Al_2O_3 or Si_3N_4 and whiskers of SiC. One of the advantages of the use of short fibers is that the composites can be produced by conventional methods used for unreinforced ceramics. However, the nature of the short fibers can present additional problems in the processing of the composites.

Packing uniformity and a high enough density remain our basic requirements for the green composite. More specifically, we require a uni-

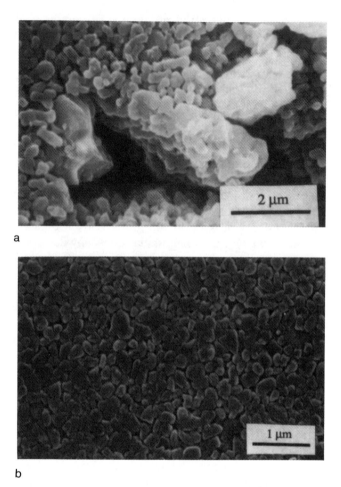

Figure 6.9 Scanning electron micrographs of an unclassified alumina powder (a) and the powder after classification by sedimentation in water (b). (From Ref. 7, with permission.)

form distribution of the whiskers and the elimination of large voids in the composite. Experiments performed with cylindrical rods indicate that short fibers pack to very low densities, with the packing density decreasing with increasing aspect ratio (i.e., ratio of length to diameter). Figure 6.10 shows data for the apparent volume of unit mass of short fibers with various aspect ratios [8]. In addition to the low packing density, whiskers

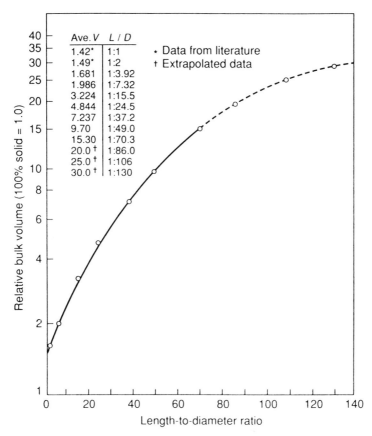

Figure 6.10 Packing curve for fibers with varied length-to-diameter (L/D) ratios. (From Ref. 8, with permission.)

with high aspect ratio (e.g., >50–100) also tend to tangle and to form bundles or loose clumps, leading to poor distribution of the whiskers and large voids spanned by whiskers in the green composite. However, whiskers with aspect ratios less than ≈30 flow fairly easily and show behavior close to that of a powder. Whiskers with high aspect ratios therefore should not be used in the production of ceramic composites. It turns out, as we shall see below, that whisker aspect ratios in the range of 15–20 may be most suitable for the production of short-fiber-reinforced composites for mechanical applications.

The packing of model mixtures consisting of cylindrical rods and spherical particles has been studied experimentally, and some of the results are presented in Table 6.3. The parameter R represents the ratio of the diameter of the particle to the diameter of the rod. The results indicate that efficient packing is promoted by a low volume fraction of whiskers that have a low aspect ratio and are thin compared to the particle size of the matrix powder.

For whisker-reinforced ceramic composites, theoretical models and experimental investigations indicate that very little enhancement in the mechanical properties is achieved for whisker aspect ratios above $\approx$15–20. Most commercial whiskers, however, contain a large fraction with aspect ratios greater than this value. Normally, ball milling is used to reduce the aspect ratios of the whiskers. For closer control of the aspect ratio, the ball-milled whiskers can be fractionated to produce fractions with the desired range of aspect ratios. Good mixing of the whiskers and the matrix powder is achieved by ball milling for several hours or by colloidal dispersion techniques.

Table 6.3 Experimental Solid Contents at 25%, 50%, and 75% Fiber Loading for Fiber–Sphere Packing

Fiber L/D	Percent fibers	R value									
		0	0.11	0.45	0.94	1.95	3.71	6.96	14.30	17.40	∞
3.91	25	68.5	68.5	65.4	61.7	61.0	64.5	70.0	74.6	76.4	82.0
	50	76.4	74.6	67.2	61.7	60.2	64.1	67.5	72.5	74.5	75.7
	75	78.2	69.5	64.5	61.0	59.5	62.5	64.2	66.7	67.2	67.1
7.31	25	68.5	68.5	64.5	61.0	58.5	59.9	64.5	73.5	74.6	80.6
	50	76.4	71.4	67.5	58.8	55.5	56.6	58.8	65.4	67.1	67.1
	75	66.3	61.7	60.0	55.0	52.8	53.5	54.6	57.2	58.2	57.4
15.52	25	68.5	66.7	63.7	59.9	54.6	50.3	50.5	54.1	57.5	65.0
	50	61.7	55.6	51.8	50.7	45.5	42.0	42.4	44.3	44.3	48.1
	75	41.0	40.4	37.9	38.2	37.3	35.7	35.5	36.0	36.8	38.2
24.50	25	68.5	66.5[a]	61.5[a]	55.5[a]	47.5	45.5	40.2	42.7	44.7	50.5
	50	40.0	39.0[a]	38.0[a]	36.0[a]	34.0[a]	32.7	30.3	31.8	31.8	33.5
	75	26.4	26.3[a]	26.2[a]	25.8[a]	25.5[a]	25.2	24.3	25.0	25.6	26.2
37.10	25	50.0	48.0[a]	45.0[a]	42.0[a]	39.4	37.7	33.8	33.1	39.2	41.3
	50	25.7						22.6	22.6	22.6	25.6
	75										

[a] Estimated values (extrapolated data).
Source: Ref. 8, reprinted with permission.

Powder Consolidation

Table 6.4 Feed Materials and Shapes of the Consolidated Powder Form for the Common Ceramic Forming Methods

Forming method	Feed material	Shape of powder form
Dry or semidry pressing		
Die compaction	Powder or free-flowing granules	Small simple shapes
Isostatic pressing	Powder or fragile granules	Larger, more intricate shapes
Casting of a slurry		
Slip casting	Free-flowing slurry with low binder content	Thin intricate shapes
Tape casting	Free-flowing slurry with high binder content	Thin sheets
Deformation of a plastic mass		
Extrusion	Moist mixture of powder and binder solution	Elongated shapes with uniform cross section
Injection molding	Granulated mixture of powder and solid binder	Small intricate shapes

6.3 POWDER CONSOLIDATION METHODS: A PREVIEW

The common methods used to consolidate ceramic powders to form the green body are summarized in Table 6.4. The specific method to be adopted will depend in each case on the shape and size required in the consolidated powder form. For example, relatively simple shapes such as disks or tiles can be made by uniaxial pressing in a die, but for more complex shapes, rods, or sheets, the use of other methods is necessary.

6.4 IMPORTANCE OF ADDITIVES IN THE FORMING PROCESS

In the forming methods described above, the use of certain additives, in some cases in concentrations as low as a fraction of a percent by weight, is vital for controlling the characteristics of the feed material and hence for controlling the packing uniformity of the consolidated body. In methods such as tape casting and injection molding, the selection of suitable additives forms one of the most vital parts of the consolidation process. The additives, in most cases polymeric in structure, serve a variety of functions, which may be divided into four main categories:

Table 6.5 Types of Additives Commonly Used in the Production of the Consolidated Powder Form

Forming method	Binder	Plasticizer	Dispersant	Lubricant
Die compaction	X			X
Isostatic pressing	X			
Slip casting	X		X	
Tape casting	X	X	X	
Extrusion	X			X
Injection molding	X	X		X

1. Binders, when used in small concentrations, serve primarily to provide bridges between the particles. In this way they aid the granulation of a powder (e.g., as the feed material for die compaction) and serve to provide strength in the green body. In fairly high concentrations, they serve to provide plasticity in the feed material during deformation (e.g, in injection molding).
2. Plasticizers soften the binder in the dry of nearly dry state, thereby increasing the flexibility of the green body (e.g., tapes formed by tape casting).
3. Dispersants, also referred to as deflocculants, serve to stabilize a slurry by increasing the repulsion between the particles. As a result, they strongly influence the viscosity of the slurry.
4. Lubricants are used to reduce the friction between the particles of the feed material and between the powder and the die during compaction or deformation. The reduced friction allows easier movement of the particles so that a higher and more uniform packing density is achieved.

Various combinations of these four types of additives are used in each of the forming methods as summarized in Table 6.5.

6.5 SELECTION OF ADDITIVES

Additives, as discussed above, serve a variety of functions. Furthermore, there are a large number of substances that can be used for any given additive function. There is therefore no simple way, especially in the case of the casting methods, of selecting an additive or set of additives for the forming process. Most successful additives have been found by a trial-and-error approach. However, a few practical guidelines can be formulated.

A. Solvents

Except for the case of injection molding, the additives are dissolved in a liquid and incorporated into the powder as a solution. The liquid is therefore important for uniformly dispersing the additive throughout the particles. In the forming methods that employ casting, it also provides fluidity for the slurry. The selection of a solvent is basically a choice between water and an organic liquid. Organic liquids, as the reader will recall, are generally more volatile and less polar than water. In practice, water is generally used for slip casting and for extrusion. Both water and organic liquids are used for tape casting, but organic liquids find much greater use. However, problems with disposal and toxicity of organic solvents may lead to a shift toward greater use of aqueous solvents. Commonly used organic liquids are alcohols (e.g., methanol, ethanol, isopropanol), ketones (e.g., acetone, methyl ethyl ketone), chlorinated hydrocarbons (e.g., trichloroethylene), toluene, xylene, and cyclohexanone or mixtures of two liquids (e.g., a mixture of ethanol and methyl ethyl ketone, a mixture of ethanol and trichloroethylene).

B. Binders

In most of the forming methods, binders are the first additive selected. A large number of organic substances can be used as binder. Some are soluble in water, whereas others are soluble in organic liquids. When the binder is incorporated as a solution, the solubility of the binder in the chosen solvent (water or organic liquid) must be considered.

The use of organic binders in the forming of ceramics has been reviewed by Onoda [9]. Most soluble organic binders are long-chain polymer molecules. The backbone of the molecule consists of covalently bonded atoms such as carbon, oxygen, and nitrogen. Attached to the backbone are side groups located at frequent intervals along the length of the molecule. The chemical nature of the side groups determines, in part, what liquids will dissolve the binder. If the side groups are polar, solubility in water is promoted, whereas if they are nonpolar, solubility in nonpolar solvents is promoted. Binders soluble in polar organic liquids have side groups of intermediate polarity.

The monomer formulas of some important synthetic binders are shown in Fig. 6.11. They include the vinyls, acrylics, and poly(ethylene oxides). The vinyls have a structure consisting of a linear chain in which the side groups are attached to every other carbon atom. The acrylics are similar to the vinyls in that they have the same backbone structure. However, some acrylics have two side groups attached to the carbon atom.

Soluble in Water

—CH—CH₂—
|
OH

Poly(vinyl alcohol)

—CH—CH₂—
|
N
H₂C   C=O
H₂C——CH₂

Polyvinylpyrrolidone

—CH—CH₂—
|
C
O   OH

Poly(acrylic acid)

CH₃
|
—C—CH₂—
|
C
O   OH

Poly(methylacrylic acid)

—CH₂—CH₂—O—

Poly(ethylene glycol) and poly(ethylene oxide)

—CH₂—CH₂—NH—

Polyethylenimine

Soluble in Organic Solvents

Vinyls

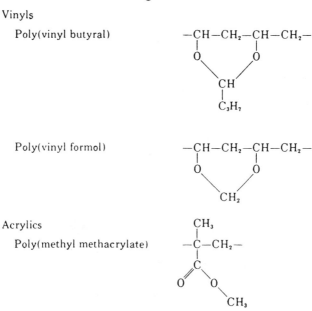

Poly(vinyl butyral)

Poly(vinyl formol)

Acrylics

Poly(methyl methacrylate)

Figure 6.11 Monomer formulas of some synthetic binders. (From Ref. 9, with permission.)

Powder Consolidation

The cellulose derivatives are an important class of naturally occurring binders. The polymer molecule is made up of a ring-type monomer unit having a modified α-glucose structure (Fig. 6.12). The modifications to the polymer occur by changes in the side groups R. The degree of substitution, DS, is the number of sites on which modifications are made in the monomer. Substitutions occur first at the C-5 site, followed by the C-2 site and finally at the C-3 site. The formulas of the R groups in some common cellulose derivatives are shown in Fig. 6.13.

Once the type of liquid (i.e., water or organic liquid) and the class of binder (i.e., water-soluble or soluble in organic liquids) are chosen, the effect of the binder on the rheology of the liquid must be considered. Organic binders increase the viscosity and change the flow characteristics of the liquid. Some can even lead to the development of a gel. The behavior of the suspension produced by adding the powder to the binder solution is influenced significantly by the rheology of the solution. The increase in viscosity of the solution produced by the binder forms one of the primary considerations in the selection of a binder for a specific forming process.

Binders are often arbitrarily classified according to how effectively they increase the viscosity of the solution. The terms low, medium, and high viscosity grades are normally used for specific binders that are available in different molecular weights. From the point of view of comparing different binders, the scheme shown in Fig. 6.14 has been proposed by Onoda [9] to establish definitions of viscosity grades. The classification of several water-soluble binders according to this scheme is shown in Table 6.6. The binder grade is controlled to a certain extent by the chemical structure of the binder. Polymer molecules in solution, as discussed in Chapter 4, take up the conformation of a coil. The size of the coil (or the mean end-to-end distance) depends on the molecular weight of the polymer. Lower molecular weights lead to smaller coils. Higher flexibility

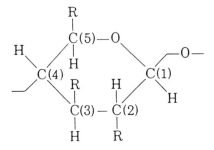

Figure 6.12 The modified α-glucose structure.

Binder	R Group	DS

Soluble in Water

Binder	R Group	DS
Methylcellulose	$-CH_2-O-CH_3$	2
Hydroxypropylmethylcellulose	$-CH_2-O-CH_2-CH(OH)-CH_3$	2
Hydroxyethylcellulose	$-CH_2-O-C_2H_4-O-C_2H_4-OH$ $-CH_2-O-C_2H_4-OH$	0.9–1.0
Sodium carboxymethylcellulose	$-CH_2-O-CH_2-COONa$	
Starches and dextrins	$-CH_2-OH$	
Sodium alginate	$-COONa$	
Ammonium alginate	$-COONH_3$	

Soluble in Organic Solvents

Binder	R Group	DS
Ethyl cellulose	$-CH_2-O-CH_2-CH_3$	

Figure 6.13 Formulas of the side groups in some cellulose derivatives. (From Ref. 9, with permission.)

in the polymer chain backbone also allows the chain to form a smaller coil. Smaller coils exert less viscous drag on the molecules of the liquid and lead to a smaller increase in the viscosity as the concentration increases. The viscosity grades of the vinyl, acrylic, and poly(ethylene oxide) binders are therefore expected to be relatively low and, for approximately the same degree of polymerization, lower than those of the cellu-

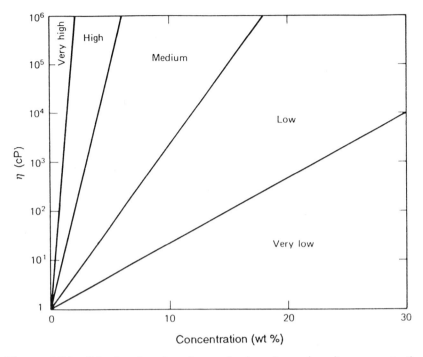

Figure 6.14 Criterion for viscosity grades based on viscosity–concentration relationship. (From Ref. 9, with permission.)

lose binders. Although the molecular weights of the binders shown in Table 6.6 are not known accurately, the viscosity grades appear to be consistent with this line of reasoning.

The amount of binder used in tape casting and injection molding is fairly large. In tape casting, the slurry must have a low enough viscosity for it to be pourable. This in turn means that the binder solution should have a low viscosity. The widespread use of vinyl and acrylic binders (which are low viscosity grade binders) for tape casting with aqueous or organic solvents is therefore not surprising. In the case of injection molding, the binder is not dissolved in a liquid but is mixed directly with the powder. The main function of the binder is to control the plastic flow of the mixture during forming.

The binder as well as the other additives used to aid the forming process must be removed before firing. As we shall discuss later, the binder removal characteristics are also an important consideration in the selection of a binder.

Table 6.6 Viscosity Grades for Some Water-Soluble Binders

	Viscosity grade					Electrochemical type			
	Very low	Low	Medium	High	Very high	Nonionic	Anionic	Cationic	Biodegradable
Gum arabic	•						X		X
Lignosulfonates	•						X		X
Lignin liquor	•						X		X
Molasses	•					X			X
Dextrins	•——					X			X
Polyvinylpyrrolidone	•——•					X			
Poly(vinyl alcohol)	•————•					X			
Poly(ethylene oxide)		•——•				X			
Starch		•————•				X			X
Acrylics		•——•				X			
Polyethylenimine (PEI)		•————•						X	
Methylcellulose			•—————•			X			X
Sodium carboxymethylcellulose			•—————•				X		X
Hydroxypropylmethylcellulose			•—————————•			X			X
Hydroxyethylcellulose			•———————————•			X			X
Sodium alginate				•——•			X		X
Ammonium alginate				•——•			X		X
Polyacrylamide				•————•		X			
Scleroglucan					•	X			X
Irish moss					•		X		X
Xanthan gum					•				X
Cationic galactomanan					•			X	X
Gum tragacanth					•	X			X
Locust bean gum					•	X			X
Gum karaya					•	X			X
Guar gum					•——•	X	X	X	

Source: Ref. 9, reprinted with permission.

C. Plasticizers

The selection of the plasticizer must follow that of the binder because the function of the plasticizer is to modify (i.e., soften) the binder in the dry state. Compared to binders, plasticizers are usually lower molecular weight organic substances. For forming processes in which the binder is introduced as a solution (e.g., tape casting), the plasticizer must be soluble in the same liquid used to dissolve the binder. In the dry state, the binder and plasticizer are homogeneously mixed as a single substance. The plasticizer molecules get between the polymer chains of the binder, thereby disrupting the chain alignment and reducing the van der Waals bonding between adjacent chains. This leads to softening of the binder but also reduces the strength. Some commonly used plasticizers are listed in Table 6.7. Combinations of binder and plasticizer are especially important in the formation of thin sheets by tape casting. Specific combinations of binder and plasticizer will be given later when the tape casting process is described.

D. Dispersants (Deflocculants)

For the forming methods in which a dispersant is also used (e.g., slip casting and tape casting), the selection of a dispersant forms the next stage of the process. While normally used in very small concentrations (e.g., a fraction of a percent by weight), the dispersant has a very important role to play: it stabilizes the slurry against flocculation of the particles so that a slurry with high solids content and low enough viscosity for pouring can be obtained. Stabilization occurs by electrostatic repulsion or by steric repulsion, which were considered in detail in Chapter 4.

Dispersants cover a wide range of available products. They can be either inorganic or organic substances. When water is the solvent, we have a choice between an inorganic dispersant and a water-soluble organic dispersant (Table 6.8). For electrostatic stabilization in water, the stabili-

Table 6.7 Common Plasticizers

Plasticizer	Melting point (°C)	Boiling point (°C)	Molecular weight
Water	0	100	18
Ethylene glycol	−13	197	62
Diethylene glycol	−8	245	106
Triethylene glycol	−7	288	150
Tetraethylene glycol	−5	327	194
Poly(ethylene glycol)	−10	>330	300
Glycerol	18	290	92
Dibutyl phthalate		340	278
Dimethyl phthalate	1	284	194
Octyl phthalate			
Benzyl butyl phthalate			

Table 6.8 Common Dispersants (Deflocculants) Used in Water

Inorganic	Organic
Sodium silicate	Sodium polyacrylate
Sodium carbonate	Ammonium polyacrylate
Sodium borate	Sodium succinate
Tetrasodium pyrophosphate	Sodium polysulfonate
	Sodium citrate
	Ammonium citrate
	Sodium tartrate

zation can be controlled by either changing the pH or adding an electrolyte as a dispersant. The electrolyte produces counterions that modify the strength of the double layer repulsion (Chapter 4). Counterions with higher valences are more effective for causing flocculation (Schulze–Hardy rule); for ions of the same valence, the smaller ions are more effective. For monovalent cations, the effectiveness of flocculation is in the order $Li^+ > Na^+ > K^+ > NH_4^+$, whereas for divalent cations the order is $Mg^{2+} > Ca^{2+} > Sr^{2+} > Ba^{2+}$. This sequence is known as the Hofmeister series. For common anions, the effectiveness of flocculation is in the order $SO_4^{2-} > Cl^- > NO_3^-$. Sodium silicate forms one of the most effective deflocculants for clays; sodium carbonate, calcium carbonate, and magnesium sulfate have also been used for other industrial materials.

In the advanced ceramics sector, the use of ionic polymers (referred to as polyelectrolytes) as dispersants in aqueous solvents is steadily increasing. The principles of stabilizing suspensions with the aid of polyelectrolytes was discussed in Chapter 4. As shown in Fig. 6.15 for sodium polymethacrylate, the sodium salt of poly(methacrylic acid), ionization of the polyelectrolyte produces charged side groups in the polymer (COO^- in this case). Adsorption of the ionized polymer onto surfaces of the parti-

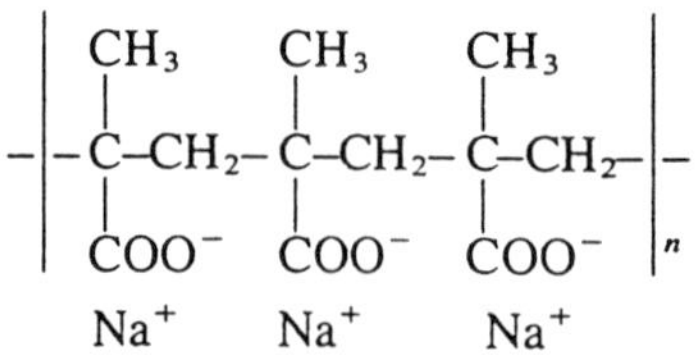

$$-\left[\begin{array}{c} CH_3 \\ | \\ -C-CH_2- \\ | \\ COO^- \\ Na^+ \end{array} \begin{array}{c} CH_3 \\ | \\ C-CH_2- \\ | \\ COO^- \\ Na^+ \end{array} \begin{array}{c} CH_3 \\ | \\ C-CH_2- \\ | \\ COO^- \\ Na^+ \end{array} \right]_n -$$

Figure 6.15 Chemical structure of sodium polyacrylate, the sodium salt of poly(methacrylic acid).

Powder Consolidation

cles may reverse the surface charge (if the surface is positively charged) and eventually lead to an increase in the surface charge. Under carefully controlled conditions (pH of the suspension and degree of adsorption of the ionized polymer), very effective stabilization may be achieved. The chemical compositions of many commercial dispersants remain proprietary. However, two commercial dispersants, Darvan 7 (a sodium polyacrylate) and Darvan C (an ammonium polyacrylate), have been used successfully with many aqueous slurries of oxide powders. For advanced ceramics, a problem with the use of inorganic dispersants is the presence of cation residues after the binder removal process. These cations (e.g., Na^+ and Ca^{2+}), even in very small amounts, can lead to the formation of liquid phases during firing, thereby making microstructural control more difficult.

In organic solvents, dispersants are generally assumed to produce stabilization of the slurry by a mechanism of steric repulsion (Chapter 4). Because of this, all dispersants in organic solvents are often labeled *steric stabilizers*. Organic polymers of high molecular weight that either adsorb or chemically bond onto the surface of the particles provide the most effective steric stabilization. However, many dispersants used in organic solvents are not long-chain polymers and have acidic or basic anchoring groups that can cause the dispersant to have a tight conformation around the particle. The effectiveness of these dispersants for providing steric stabilization is unclear. A reduction in the van der Waals attractive forces combined with a less effective steric repulsion has been suggested as a potential mechanism for stabilization by these smaller molecules. The most commonly used commercial dispersants in organic solvents are OLOA-1200 (a polyisobutene), menhaden fish oil, glyceryl trioleate, and phosphate ester.

Whether water or an organic liquid is used as the solvent, the dispersant must be compatible with the binder. Dispersants can change the viscosity of a binder solution or cause gelation or precipitation of the binder. In aqueous solvents, it is generally found that nonionic binders such as poly(vinyl alcohol) are compatible with most ionic dispersants over a wide pH range. Ionic binders generally have a lower range of compatibility.

E. Lubricants

Lubricants are used in die compaction, extrusion, and injection molding to reduce the friction between the particles themselves or between the particles and the die. They are normally among the last additives to be selected. Under the application of an external pressure, the particles rearrange more easily, thereby leading to a higher and more uniform packing density. When granulated powders are used in die compaction, the lubricant is incorporated with the other additives prior to granulation or mixed with the granules in a separate step. In laboratory practice where only a binder is added to the powder, the internal surfaces of the die are usually coated with a lubricant to reduce die wall friction. Common lubricants are steric acid, stearates, and various waxy substances.

F. Other Potential Additives

In slip casting and tape casting, other additives not considered above are sometimes added. In most cases they are unnecessary. Surfactants, also referred to as wetting agents, adsorb onto the surfaces of the particles and may reduce the surface tension of the liquid, thereby improving the wetting of the particle surfaces. The improved wetting of the particle surfaces leads to better dispersion. In some cases, the use of a dispersant or surfactant will cause a foaming action in the slurry during ball milling. To combat this, an antifoaming agent is used.

6.6 DRY AND SEMIDRY PRESSING METHODS

Die pressing and isostatic pressing are commonly used for the compaction of dry powders, which typically contain <2 wt % water, and semidry powders, which contain 5–20 wt % water. Die compaction is one of the most widely used operations in the ceramics industry. However, the agglomeration of dry powders combined with the nonuniform transmission of the applied pressure during compaction leads to significant variations in the packing density of the green body. To minimize the density variations, die pressing is used for the production of relatively simple shapes (e.g., disks) with a length-to-diameter ratio of <0.5–1.0. Isostatic pressing produces a better uniformity in the packing density and can be used for production of green bodies with relatively complex shapes and with much higher length-to-diameter ratios.

A. Die Compaction

Die compaction involves the simultaneous uniaxial compaction and shaping of a powder or granular material in a rigid die. It allows the formation of relatively simple shapes rapidly and with accurate dimensions. The overall process consists of filling the die, compacting the powder, and ejecting the compacted powder. There are three main modes of compaction, which are defined in terms of the relative motion of the die and the punches. In the single-action mode, the top punch moves but the bottom

Powder Consolidation

punch and the die are fixed, whereas in the double-action mode both punches move but the die is fixed. In the floating die mode the top punch and the die move but the bottom punch is fixed.

Die Filling

For the achievement of reproducible properties and fast pressing rates in industrial practice, good flow properties of the powder during the die-filling stage is essential. Fine powders do not flow very well, and it is often necessary to granulate them to improve their flow properties. The spray-drying technique described in Chapter 2 is often used to produce granules from powder slurries. Since they have a significant influence on the flow and compaction behavior, the characteristics of the granules must be controlled during the granulation process. The important characteristics are the following.

1. *Size, size distribution, and shape.* Nearly spherical granules with sizes greater than $\approx$40–50 μm and with a fairly narrow size distribution normally have good flow properties.
2. *Density of the granules.* A high packing density of the particles within the granule is required for good die filling and compaction.
3. *Friction between the granules and between the granule surfaces and the die walls.* Low friction improves the packing density and packing uniformity in the consolidated powder form.

In order to produce the desirable granule characteristics, a number of additives are incorporated into the powder slurry prior to granulation. These additives consist of a binder, a plasticizer, a dispersant, and in some cases a lubricant, a surfactant, and an antifoaming agent. The chemistry of the additives varies from powder to powder. However, commonly used binders include poly(vinyl alcohol) and Carbowax, while poly(ethylene glycol) is widely used as a plasticizer.

Powder Compaction

The compaction process is a complex many-body problem. However, it can be divided into two stages. The system consists of an assemblage of dense particles or, in the case of a granulated powder, an assemblage of porous granules. Whether we have particles or granules, the initial structure contains large voids on the order of the particle or granule size and voids that are smaller than the particle or granule size. The first stage of the compaction process involves reduction of the large voids by sliding and rearrangement of the particles or granules (Fig. 6.16). In the second stage, the small voids are reduced by fracture of the particles. In the case of granules where a greater amount of binder is present or in the case of

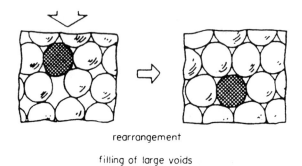

Figure 6.16 Schematic diagram showing the different mechanisms of compaction of particles or granules. (From Ref. 24, with permission.)

relatively soft ceramics (e.g., NaCl), plastic deformation also leads to a reduction of the small voids. The particles also undergo elastic compression during the second stage, which, as we shall see later, has important consequences for the formation of defects when the compact is ejected from the die.

As we can see from the above discussion, the compaction process involves several complex mechanisms. Theoretical analysis of the process is therefore difficult. In view of this difficulty, a number of empirical equations have been developed to account for the experimental data. None have been found to be generally applicable. One equation that has

Powder Consolidation

the advantage of simplicity and is as good as any of the others is

$$p = \alpha + \beta \ln \left(\frac{1}{1 - \rho}\right) \tag{6.4}$$

where p is the applied pressure, ρ is the relative density, and α and β are constants that depend on the initial density and the nature of the material. The empirical expressions have often been criticized as being merely curve fitting. However, they have served to focus attention on the complexity of the variables and mechanisms of the process.

A severe problem in die compaction, as outlined earlier, is that the applied pressure is not transmitted uniformly to the powder because of friction between the powder and the die wall. The applied pressure is transmitted to the die wall, giving rise to a normal (or radial) stress and a shear (or tangential) stress at the die wall. The shear stress opposes the applied stress at the die wall, leading to stress gradients. The ratio of the shear stress to the normal stress is the *friction coefficient* between the powder and the die wall. Strijbos and coworkers [10] have made extensive studies of die wall friction using an apparatus shown schematically in Fig. 6.17. The parameters investigated included the mean powder particle size d_p, the roughness of the wall R_w, the hardness of the powder H_p, the hardness of the wall H_w, and the use of lubricants. For ferric oxide powder (Vickers hardness ≈600) and walls of tungsten carbide and hardened and nonhardened tool steels (Vickers hardness ≈1300, 600, and 200, respec-

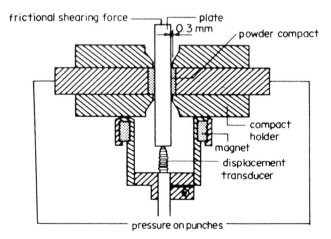

Figure 6.17 Powder–wall friction apparatus. (From Ref. 10.)

tively), some data for the friction coefficient f_{dyn} as a function of d_p/R_w for the three values of H_p/H_w are shown in Fig. 6.18. For particles smaller than the roughness, f_{dyn} is as high as 0.6. A layer of the fine particles sticks to the die wall, so there is no direct contact between the stationary powder compact and the moving wall in the apparatus shown in Fig. 6.17. In this case, R_w and H_w have no effect on f_{dyn}, and the high value of f_{dyn} is a reflection of the friction between the powder particles. Failure occurs within the powder compact and not at the die wall. For particles larger than the wall roughness, f_{dyn} is between 0.2 and 0.4 and is dependent on both the powder parameters and the die wall parameters. The low values of f_{dyn} are a reflection of the friction between the particles and the relatively smooth wall. Failure in this case occurs at the powder/wall interface.

The friction coefficient between the powder and a rough wall is also dependent on the direction of the grooves in the wall. The friction is lower if the grooves run in the direction of relative motion between the powder and the die wall. The effect of die–wall lubricants (e.g., stearic acid) can

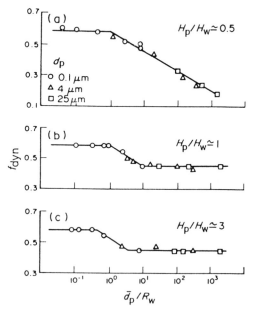

Figure 6.18 Dynamic powder–wall friction coefficient versus the ratio of powder particle size to wall roughness, for three values of the ratio of particle hardness to wall hardness. The powder is ferric oxide, and the walls are (a) tungsten carbide, (b) hardened tool steel, and (c) nonhardened tool steel. (From Ref. 10.)

Powder Consolidation

be fairly complex. For fine particles ($d_p/R_w < 1$), the coefficient of friction decreases gradually as the thickness of the lubricant increases, and the magnitude of the decrease can be fairly significant when the thickness of the lubricating layer becomes larger than the particle size. For coarse particles ($d_p/R_w > 1$), the presence of a lubricant causes only a small reduction in the die–wall friction, and almost no dependence on the thickness of the lubricating layer is observed.

Density variations in sections of the compacted powder have been measured by microscopy or by X-ray radiography. Figure 6.19 shows the density variations in a section of a cylindrical manganese-zinc ferrite powder compact (diameter 14 mm) produced by die compaction in the single action mode of pressing. Relatively large variations along the sides of the compact caused by die–wall friction are very noticeable.

Ejection of the Powder Compact

As discussed earlier, the powder undergoes elastic compression during the compaction process. The stored elastic energy leads to an expansion of the compact when the applied pressure is released. This expansion is referred to as strain recovery, strain relaxation, or springback. Strain recovery is almost instantaneous on release of the pressure. The amount of strain recovery is generally higher for higher amounts of organic additives in the powder and for higher applied pressure. Although a small amount of strain recovery is desirable to cause the compact to separate from the punch, an excessive amount can lead to defects. Ejection of the

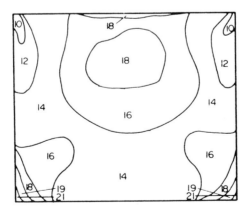

Figure 6.19 Density variation in the single-action die pressing of manganese ferrite powder. The numbers represent regions of approximately the same density. (From Ref. 24, with permission.)

powder compact from the die is resisted by friction between the compact and the die wall. Lubricants added to reduce die–wall friction during the compaction process have an additional benefit in that they reduce the pressure required for ejection.

Control of Density Gradients and Compaction Defects

After the completion of the die compaction process, we require that the green body be free of defects and that gradients in the packing density be as low as possible. Density gradients in the green body, as noted earlier, can lead to heterogeneities in the fired body that cause a degradation in properties. They can also lead to defects in the compact on ejection from the die. A number of factors can be controlled to reduce the extent of density gradients in the powder compact. Good filling of the die with the powder reduces the amount of internal movement of the powder during the compaction process. The use of lubricants to reduce the internal friction between the particles and the die–wall friction can lead to significant improvements. Stress gradients (and hence density gradients) due to die–wall friction are enhanced with increasing length-to-diameter ratio (L/D) of the compact. For the single-action mode of die compaction, L/D should be less than 0.5, while for the double-action mode it should be less than 1.

Two of the most common defects in die-compacted powders are delamination and end capping (Fig. 6.20). These defects are caused by differential strain recovery during the ejection of the compact from the die. The use of a small amount of binder to increase the compact strength, reduction of the applied pressure to reduce the extent of the springback, and the use of a lubricant to reduce die–wall friction can significantly reduce the tendency for delaminations and end caps to form.

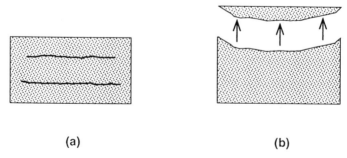

(a)   (b)

Figure 6.20 Two of the more common defects in die compaction of dry or semidry powders: (a) delamination and (b) end capping.

Powder Consolidation

B. Isostatic Compaction

Isostatic pressing is the application of a uniform hydrostatic pressure to the powder contained in a flexible rubber container. Compared to die compaction, the amount of powder movement is less, and the die walls are absent. Isostatic pressing can therefore solve many of the problems of pressing relatively complex shapes to a uniform density. There is still a die-filling requirement, but in this case it is necessary to profile the powder or granules to produce an article of the desired shape.

Isostatic pressing methods are generally divided into two classes: wet-bag pressing and dry-bag pressing (Fig. 6.21). In wet-bag pressing, a flexible rubber mold is filled with the powder and is then submerged in a

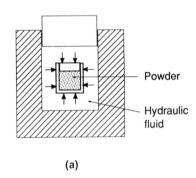

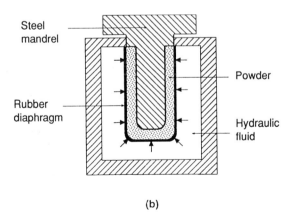

Figure 6.21 Two modes of isostatic pressing: (a) wet-bag pressing and (b) dry-bag pressing.

pressure vessel filled with oil and pressed. After the pressure is released, the mold is removed from the pressure vessel and the green body is retrieved. Wet-bag pressing is used for the formation of fairly complex shapes and for large articles. In dry-bag pressing, the mold is fixed in the pressure vessel and need not be removed. The pressure is applied to the powder situated between a fairly thick rubber mold and a rigid core. After release of the pressure, the compacted powder is removed from the mold. Dry-bag pressing is easier to automate than wet-bag pressing. It has been used for the formation of spark plug insulators by compressing a porcelain powder mixture around a metal core as well as for plates and hollow tubes. Compared to die compaction, the formation of defects in isostatically pressed powder compacts is much less severe. However, delamination and fracture of the compact (caused by springback) can occur if the pressure is released too rapidly after compaction.

6.7 CASTING METHODS

The casting methods, e.g., slip casting along with the associated pressure casting technique and tape casting, involve the consolidation of powders from a concentrated slurry. As we have seen, the interparticle forces in the slurry can be carefully controlled by colloidal techniques. Compared to the other forming methods, the casting methods have the capability for producing a fairly uniform packing density in the green body. However, they are limited to the production of relatively thin articles. Slip casting offers a route for the production of complex shapes and is widely used in the traditional clay-based industry, for example, for the manufacture of pottery and sanitary ware. It has been steadily introduced over the past 50 years or so to the production of technical ceramics (e.g., crucibles and tubes of alumina and zirconia). In tape casting (sometimes referred to as doctor blading), the slip is spread on a flat surface and the liquid is removed by evaporation. Tape casting is widely used for the production of thin sheets or substrates for the electronics packaging industry. Sheets with thicknesses from as low as 10 μm to as great as 1 mm are prepared by tape casting. As mentioned earlier, for both methods we require slips with the highest concentration of solids but with a low enough viscosity for pouring. However, the slip for tape casting differs from that used for slip casting in that it contains a far greater amount of binder (and plasticizer), which must provide the strength and flexibility in the thin sheets after the liquid is removed.

The degree to which the particles are dispersed in the slurry has a profound effect on the microstructural uniformity of the green body. To avoid excessive shrinkage of the cast during removal of the fluid, the

Powder Consolidation

slurry should contain the highest possible fraction of particles. Casting also requires flow. The importance of polymeric additives for achieving good dispersion of the particles and for controlling the flow properties of the slurry has been described in Chapter 4. Here we will consider the more practical issues associated with the use of polymeric additives. Control of the properties of the slurry is also essential for reproducible and reliable fabrication. The high solids content of the slurry may also produce undesirable rheological behavior such as dilatancy (Chapter 4). Proper control and characterization of the rheological properties of the slurry can be very valuable in controlling of the quality of the slurry.

A. Slip Casting

In slip casting, a slurry is poured into a microporous plaster of Paris mold. The porous nature of the mold provides a capillary suction pressure of ≈ 0.1–0.2 MPa, which draws the liquid from the slurry into the mold. The origin of the capillary suction pressure is similar to that of capillary rise described in Chapter 5. A consolidated layer of solids, referred to as a cast or cake, forms on the walls of the mold (Fig. 6.22). After a sufficient thickness of the cast is formed, the surplus slip is poured out and the mold and cast are allowed to dry. Normally, the cast shrinks away from the mold during the drying process and can be easily removed. Following binder burnout, the cast is fired to produce the final article.

Slip Casting Mechanics

As shown in Fig. 6.22, the filtrate (i.e., the liquid) passes through two types of porous media: (i) the consolidated layer and (ii) the mold. In

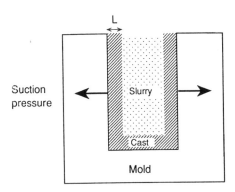

Figure 6.22 Slip casting in a porous mold.

some models for slip casting, the resistance of the mold to the flow of the liquid is neglected. However, such models represent only a first approximation to the process. A more rigorous model in which the resistance of both media to flow is taken into account was analyzed by Tiller and Tsai [11]. Some features of the model are outlined below.

The flow of liquid through a porous medium is described by Darcy's law (Chapter 5) which, in one dimension can be written as

$$J = \frac{K(dp/dx)}{\eta_L} \tag{6.5}$$

where J is the flux of liquid, dp/dx is the pressure gradient in the liquid, η_L is the viscosity of the liquid, and K is the permeability of the porous medium. According to Eq. (6.5), a pressure gradient is essential for flow to occur. The pressure of the liquid in the cast and in the mold must therefore be known. The forces acting across any plane of the cast must satisfy a force balance. For an area A, a simple force balance yields

$$Ap(t) = Ap_L(x, t) + F_S(x, t) \tag{6.6}$$

where p is the total filtration pressure, which, in general, is a function of time. The local pressure in the liquid, p_L, and the force on the particles, F_S, are functions of distance and time. In slip casting, the pressure p is assumed to be constant and equal to the capillary suction pressure of the mold. Sometimes it is convenient to describe a fictitious effective pressure in the cast such that $p_S(x, t) = F_S(x, t)/A$. Equation (6.6) can then be written

$$p = p_L + p_S \tag{6.7}$$

The variation of p_L and p_S with distance from the boundary between the mold and the cast is shown schematically in Fig. 6.23. If the cast is incompressible, then p_L is a straight line. The pressure p can also be written

$$p = \Delta p_m + \Delta p_c \tag{6.8}$$

where Δp_m and Δp_c are the pressure drops in the mold and in the cast, respectively. The flux of liquid must be the same in both the cast and the mold. Consequently,

$$J\eta_L = \frac{K_m}{L_1} \Delta p_m = \frac{K_c}{L} \Delta p_c \tag{6.9}$$

where L is the thickness of the cast, L_1 is the thickness of the mold saturated with liquid, and K_m and K_c are the permeabilities of the mold and cast, respectively.

Powder Consolidation

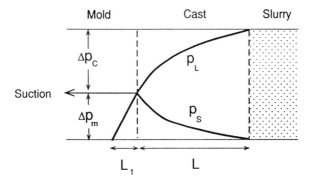

Figure 6.23 Pressure distribution across the cast and the mold in slip casting.

After integrating Eq. (6.5) and applying the appropriate boundary conditions, Tiller and Tsai found that the thickness of the consolidated layer is related to the time of casting by the equation

$$L^2 = 2Hpt/\eta_L \tag{6.10}$$

The function H depends on the properties of both the consolidated layer and the mold and is given by

$$H = \cfrac{1}{\left(\cfrac{V_c}{V_s} - 1\right)\left[\cfrac{1}{K_c} + \cfrac{V_c/V_s - 1}{P_m K_m}\right]} \tag{6.11}$$

where V_c is the volume fraction of solids in the cast, V_s is the volume fraction of solids in the slip, P_m is the porosity of the mold, and the other parameters have been defined earlier.

Effect of Mold and Slurry Parameters

Equations (6.10) and (6.11) provide the most detailed dependency of the casting rate on the mold and slurry parameters. We can see that for a given system of mold and slurry, the function H is constant and Eq. (6.10) predicts that the thickness of the cast, L, is proportional to $t^{1/2}$, a result that is also obtained from the simpler models. The rate of consolidation therefore decreases with time, and this limits the usefulness of the slip casting route to a certain thickness of the cast. Above this value, further increases in the thickness are very time-consuming. The rate of consolidation also depends on the viscosity of the filtrate, η_L. A reduction in η_L achieved by increasing the temperature of the slurry or, in a less practical way, by using a liquid with a lower viscosity leads to an increase in the consolidation rate. For aqueous slips, an increase in the temperature improves the stability of the slip and leads to a cast with a higher packing density. This decreases the consolidation rate because the permeability of the cast, K_c, decreases. However, the reduction in η_L with increasing temperature has a far greater effect, and the overall result is an increase in the casting rate. Equation (6.10) implies that L is proportional to the capillary suction pressure p. If there were no other effects, an increase in p would always lead to a shorter time for a given thickness of the cast. The capillary suction pressure varies inversely as the pore radius of the mold, and we may think that a decrease in the pore radius would lead to an increase in the casting rate. However, the permeability of the mold, K_m, also decreases with a decrease in the radius of the mold. The analysis of Tiller and Tsai outlined earlier predicts an optimum pore size of the mold for the maximum rate of casting.

From the point of view of processing, the microstructural uniformity of the green body is of great importance. The microstructural uniformity of the cast is dependent on the stability of the slip. A flocculated slip leads to a cast with a fairly high porosity. Moreover, the compressible nature of the highly porous cast coupled with the variation in the effective pressure in the cast (p_S in Fig. 6.23) leads to the production of density variations in the cast. Because of the way in which p_S varies, the density decreases with distance from the boundary between the mold and the cast. Nonuniformities in the green body microstructure, as we will recall, hinder microstructural control during the firing process. A well-dispersed slip containing no agglomerates and stabilized by electrostatic or steric repulsion leads to the formation of a fairly dense cast with better microstructural uniformity. In practice, the fairly dense cast formed from a well-dispersed slip has a low permeability. The rate of casting is therefore low. For industrial operations where such low casting rates are economical, the slip is only partially deflocculated.

In addition to the degree of dispersion of the slip, other parameters such as the size, size distribution, and shape of the particles can also influence the compressibility and hence the packing density of the cast. Figure 6.24 schematically summarizes the effects of some of these variables on the packing density of the cast. For fairly coarse particles (greater than $\approx$10–20 μm), colloidal effects are insignificant and the degree of dispersion of the particles has no effect on the packing density. Large spheres of the same size produce casts with a packing density of $\approx$0.60–0.65, close to that for dense random packing. Irregular particles produce beds with a lower packing density. As the particle size decreases below $\approx$10 μm, colloidal effects control the packing density. At one extreme, well-dispersed slips produce casts with high packing density, while

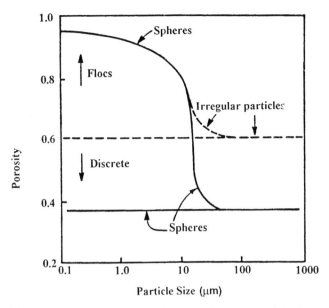

Figure 6.24 Schematic diagram showing the effects of particle size, shape, and degree of flocculation on the porosity of the cast produced in slip casting. (From Ref. 11, with permission.)

at the other extreme, flocculated slips yield a low packing density. Particles with a distribution in sizes may produce casts with a packing density that is higher than that for particles with the same size.

Because of the widespread use of slip casting, many examples of slip compositions can be found in the ceramics literature. For a given powder, a variety of dispersants and binders can be used effectively. However, to provide the reader with an indication of the quantities of the components in the slurry, we will consider the case of Al_2O_3. Slip casting of Al_2O_3 may be performed in either acidic or basic conditions. Polyelectrolytes, e.g., Darvan 7, a sodium salt of poly(methacrylic acid) or Darvan C, an ammonium salt of the same acid, are very effective dispersants. A typical composition (in volume percent) may be as follows: Al_2O_3 (40–50); water (50–60); dispersant (0.25–1); binder (0–0.5).

B. Pressure Casting

Equation (6.10) indicates that for a given slurry, as the filtration pressure p increases, the time taken to produce a given thickness of cast, L, decreases. The casting can therefore be speeded up by the application of an external pressure to the slurry. This is the principle of pressure casting, also referred to as pressure filtration. The plaster of Paris molds used in slip casting are, however, weak and cannot withstand pressures greater than ≈0.5 MPa. Plastic or metal molds must be used in pressure casting. A schematic of the main features of a laboratory-scale pressure casting device is shown in Fig. 6.25. Particles in the slurry form a consolidated layer (the cast) on the filter as the liquid is forced through the system. Compared to the filter, the cast provides a much greater resistance to flow of the liquid. Using Eqs. (6.10) and (6.11), the kinetics of pressure casting can be written

$$L^2 = \frac{2ptK_c}{\eta_L(V_c/V_s - 1)} \tag{6.12}$$

The permeability of the cast can be estimated by one form of the Carman–Kozeny relation

$$K_c = \frac{D^2(1 - V_c)^3}{36\alpha V_c^2} \tag{6.13}$$

where D is the particle size and α is a constant, equal to 5 for many systems, that defines the shape and tortuosity of the flow channels.

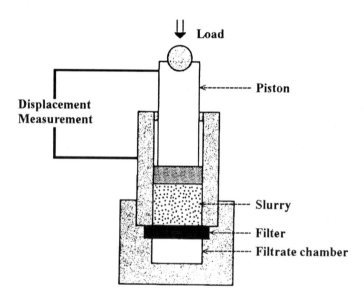

Figure 6.25 A laboratory scale pressure casting apparatus.

Powder Consolidation

The kinetics and mechanics of pressure casting have been investigated by Lange and Miller [12], and we shall use their results to illustrate some of the main features of the process. As observed for slip casting, the structure of the cast is also dependent on the degree of dispersion of the particles in the slurry. Figure 6.26 shows that the highest packing density is obtained with well-dispersed slurries. Furthermore, the packing density obtained with well-dispersed slurries is independent of the applied pressure above ≈0.5 MPa. Dynamic models for particle packing that incorporate rearrangement processes have not been developed so far. However, the high packing densities obtained with dispersed slurries at low applied pressures indicate that the repulsive forces between the particles must facilitate rearrangement. For flocculated slurries, the packing density is dependent on the applied pressure and appears to obey a law (the relative density increasing linearly with the logarithm of the applied pressure) commonly observed in the die compaction of dry powders.

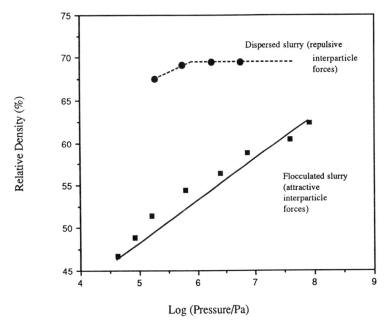

Figure 6.26 Relative density of different bodies produced from the same Al_2O_3 powder by filtration at different applied pressures. Bodies were consolidated from either dispersed (pH 2) or flocculated (pH 8) aqueous slurries containing 20 vol % solids. (From Ref. 2, with permission.)

On removal of the applied pressure, the cast produced from either a dispersed slurry or a flocculated slurry expands, i.e., undergoes strain recovery, due to the stored elastic energy. The nature of the strain recovery is, however, different from that observed in the die compaction of dry powders (Section 6.6C). The strain recovery for compacts produced by pressure casting is time-dependent. This phenomenon of time-dependent strain recovery arises because fluid (liquid or gas) must flow into the compact to allow the particle network to expand and relieve the stored strain. The magnitude of the recovered strain increases with increasing consolidation pressure in a nonlinear manner and can be described by a Hertzian elastic stress–strain relation of the form

$$\sigma = \beta \epsilon^{3/2} \tag{6.14}$$

where σ is the stress, ϵ is the strain, and β is constant for a given particulate system. For Al_2O_3, the recovered strain can be fairly large (2–3%) for moderately low pressures in the range of 50–100 MPa. A consequence of the fairly large strain recovery is the tendency for cracking to occur in the compact. However, this can be relieved to a certain extent by the reduction of the consolidation pressure and by the use of a small amount of binder (<2 wt %).

Compared to slip casting, pressure casting offers the advantages of greater productivity through shorter consolidation times and the need for less floor space for setting up the molds. However, the molds used in pressure casting are more expensive, and the formation of intricate shapes is not as easy. In general, pressure casting is finding increasing use in the formation of small articles with fairly simple shapes.

C. Tape Casting

In tape casting, also called the doctor-blade process, the slurry is spread over a surface covered with a removable sheet of paper or plastic using a carefully controlled blade referred to as a doctor blade. For the production of long tapes, the blade is stationary and the surface moves (Fig. 6.27), whereas for the production of short sheets in the laboratory, the blade is pulled over a stationary surface. After drying, the tape is peeled off from the surface, cut to shape, heated to remove the organic additives (referred to as binder burnout), and finally fired to produce a dense tape. During drying, the tape adheres to the surface so that the shrinkage occurs in the thickness. Almost no shrinkage occurs in the binder burnout stage. During firing, the shrinkage in the lateral direction is much greater than that in the thickness.

Although the drying and firing of the tape are clearly important, the success of the tape casting process depends critically on the formulation

Powder Consolidation

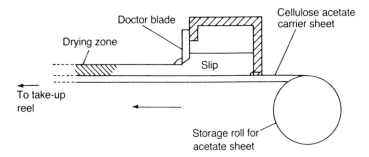

Figure 6.27 Schematic diagram of the tape casting process.

of the slurry and the burnout of the organics. Because it is common to almost all of the common forming methods, we shall provide a general consideration of the binder burnout stage later in this chapter (Section 6.9). To illustrate the important factors in the formulation of the slurry, we shall consider the tape casting of alumina [13]. However, the various steps in the process are very similar for any ceramic. Figure 6.28 shows a schematic flow diagram of the various steps in the tape casting process; the starting materials for the preparation of the slurry are listed in Table 6.9. Except for the powder, the components of the first stage constitute the dispersion system. The solvent must be capable of dissolving all the organic additives in the system. It must have a fairly high vapor pressure

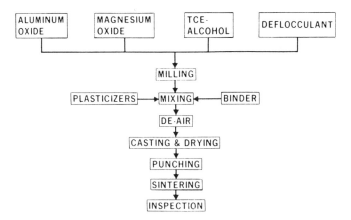

Figure 6.28 Schematic flow diagram for the tape casting of an alumina substrate material. (From Ref. 13, with permission.)

Table 6.9 Materials and Milling Procedure for the Preparation of Alumina Slurries for Tape Casting

Milling step	Material	Function	% (by weight)
Stage 1: Mill for 24 h	Alumina powder	Substrate material	59.6
	Magnesium oxide	Grain growth inhibitor	0.15
	Menhaden fish oil	Dispersant	1.0
	Trichloroethylene	Solvent	23.2
	Ethyl alcohol	Solvent	8.9
Stage 2: Add to above and mill for 24 h	Poly(vinyl butyral)	Binder	2.4
	Poly(ethylene glycol)	Plasticizer	2.6
	Octyl phthalate	Plasticizer	2.1

Source: Ref. 13.

so that it can be removed below $\approx 100°C$, and it must be relatively safe and inexpensive. Although trichloroethylene meets these requirements, the addition of ethanol increases the solubility of the binder and dispersant. The dispersant is menhaden fish oil, a polymeric fatty acid containing anchor groups that bond strongly onto the particle surface. The remainder of the polymer, consisting of flexible tails, dissolves freely in the solvent. The anchor groups must displace other species from the surface. In this case, the competition is from hydroxyl groups on the ethanol, on the binder, and on the ends of the poly(ethylene glycol) chains.

The second stage of the slurry preparation involves incorporation of the binder and plasticizer. Poly(vinyl butyral) with a molecular weight of $\approx 30,000$ acts as the binder. It has a glass transition temperature of $\approx 50°C$ (which means that it is fairly deformable above this temperature), but this is reduced to below room temperature by the use of the plasticizer. The high molecular weight coupled with the low glass transition temperature of the binder provides enough strength and flexibility to the dried tape. The binder is also compatible with the dispersant, which, as noted earlier, is important to prevent these two additives from separating after the solvent is removed. Another important characteristic of the binder is that it can be removed fairly cleanly during the burnout stage.

The tape is normally dried by evaporating the solvent to produce a system in which the particles are bonded by a polymer phase. The dried tape consists of approximately 50 vol % solids, 30 vol % polymer, and 20 vol % porosity. Despite the relatively low density of the tape after removal of the organic matter, the fired tape can reach nearly full density.

Table 6.10 Slurry Formulation for Tape Casting of Barium Titanate Powder

Material	Function	% (by weight)
Barium titanate	Dielectric material	77.2
Phosphate ester	Dispersant	0.3
Methyl ethyl ketone	Solvent	7.4
Ethanol	Solvent	3.2
Acrylic resin	Binder	7.1
Benzyl butyl phthalate	Plasticizer	2.2
Poly(ethylene glycol)	Plasticizer	2.2

Tape casting is also widely used to prepare $BaTiO_3$ tapes for the manufacture of capacitors. A typical tape casting system for $BaTiO_3$ is given in Table 6.10.

6.8 PLASTIC-FORMING METHODS

Plastic deformation of a moldable powder–additive mixture is employed in a number of forming methods for ceramics. Extrusion of a moist clay–water mixture is used extensively in the traditional ceramics sector for forming components with a regular cross section such as solid and hollow cylinders, tiles, and bricks. It is also used to a lesser extent as a forming method for some advanced ceramics such as alumina substrates, ceramic capacitor tubes, and piezoelectric transducers. Injection molding is potentially a useful method for the mass production of ceramic articles with complex shapes. The development of Si_3N_4 and SiC for mechanical applications at high temperatures, coupled with the interest in the development of a ceramic gas turbine engine in the 1960s and 1970s, generated considerable interest in the method. Considerable developmental efforts are continuing, but injection molding has not yet materialized into a significant forming process for ceramics, mainly because of two factors: the high costs of the dies and the difficulties in removing the binder efficiently from the injection molded body. Extrusion and injection molding are the dominant forming methods for plastics, and the application of these two methods to ceramics has benefitted considerably from the design and tooling principles developed in the plastics industry.

A. Extrusion

In extrusion, a powder mixture in the form of a stiff paste is compacted and shaped by forcing it through a nozzle. Industrially, the process is usually carried out in a deairing pug mill, which is a screw-fed extruder with a means for extracting air from the sample. A requirement is that the body should exhibit plastic behavior, that is at lower stresses behave like a rigid solid and deform only when the stress reaches a certain value called the *yield stress*. Furthermore, the extruded body must be sufficiently strong to be transported to a drying rack without significant distortion.

Clay particles develop desirable plastic characteristics when mixed with a controlled amount of water; depending on the nature of the clay, the water content is in the range of 15–30 wt %. The powders of the advanced ceramics, when mixed with water, do not possess the desirable plastic characteristics found in the clay–water system. For this reason, they are thoroughly mixed with a viscous liquid such as water containing a few percent of an organic binder to provide the desired plastic characteristics. Since the extruded body must also have sufficient strength, the binder is generally selected from the medium to high viscosity grades, e.g., methylcellulose, hydroxyethylcellulose, polyacrylimides, or poly(vinyl alcohol).

As the schematic of Fig. 6.29 indicates, considerable attention goes into the design of the extruder barrel and screw to enable them to perform the various functions directly associated with the process [14]. The barrel must be strong enough to withstand the pressures developed. The screw has to mix the powder and other additives into a homogeneous mass and generate sufficient pressure to transport the mixture against the resistance of the die. The yield stress of the plastic mixture provides a measure of the stress that must be generated in the extruder to shape the body. However, for difficult geometries (e.g., a honeycomb with thin walls) the pressure in the extruder can reach values as high as ≈ 8 MPa. The high energy consumed in the process leads to the heat-up of the mixture, which is often undesirable. To reduce the heat-up of the powder mixture, the extruder must be cooled or the output reduced. Alternatively, the heat buildup may be reduced by modification of the screw geometry, e.g., by the use of a screw with a thick shaft and a large enough number of flights (or helices), such that

$$\frac{L_S}{H} \geq \frac{0.45 \, \Delta p}{\sigma_y} \qquad (6.15)$$

where L_S is the shaft length of the screw, H is the distance between two flights (see Fig. 6.29), Δp is the pressure in the extruder, and σ_y is the yield stress of the plastic mixture.

Shaping of the extruded body is achieved with the head of the extruder screw and the die. The main functions of the head of the extruder

Powder Consolidation

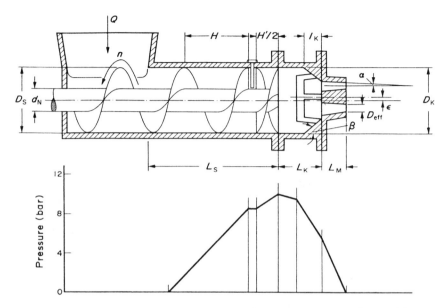

Figure 6.29 Pressure variation along extrusion press. (From Ref. 14, with permission.)

screw are to change the rotational flow of the mixture produced by the screw into an axial flow for extrusion and to produce uniform flow of the mixture into the die. In the release of the body from the extruder, the die serves three main functions. It must generate the desired cross section, ensure a smooth surface free of cracking, and allow uniform flow throughout the entire cross section of the extruded body.

A variety of defects can occur in the extruded body. These include insufficient strength for handling, cracks, delaminations, surface roughness, and surface cracking. The sources of these and other defects are discussed in detail by Reed [15].

B. Injection Molding

The overall fabrication of ceramic articles by injection molding involves the following steps: selecting the powder and the binder; mixing the powder with the binder, producing a homogeneous feed material in the form of granules, injection molding the green body, removing of the binder at lower temperatures, and, finally, heating at higher temperatures to produce the final article. The reader will realize that the individual steps must not be considered in isolation; the critical factors in one step significantly influence the subsequent steps.

The desirable powder characteristics for injection molding are, in general, no different from those described in Chapter 2 for powder processing. We shall consider the binder removal step in the next section because it involves features common to other forming methods (e.g., tape casting). The firing process will be considered in detail in the remaining chapters of this book. Here, we shall consider the key factors in binder selection, preparation of the feed material, and injection molding of the green body.

Binder System

Although the binder plays a transient role in the overall fabrication route, careful selection of a binder is vital to the success of the injection molding operation. The binder must provide the desired rheological properties to the feed material so that the powder can be consolidated into the desired shape and then must be removed completely from the shaped powder form, prior to firing, without the disruption of the particle packing or any chemical reaction with the powder. A good binder must therefore have desirable rheological characteristics, chemical characteristics, and binder burnout behavior. In addition, it must possess many qualities suitable for manufacturing such as environmental safety and low cost. A single binder cannot satisfy all of these requirements, so in practice a combination of several additives is used. It turns out that the use of a binder system consisting of several additives may be beneficial for the binder burnout step in that the additives can be removed progressively as dictated by their volatility or chemical reactivity. The removal of the first binder creates a network of porosity through which the decomposition products of the second binder can be removed more easily.

Binder systems for injection molding have generally been developed on a trial-and-error basis. However, a number of key issues arise in their formulation [16,17]. One of these is the type of binder system. Although either thermoplastic polymers or thermosetting polymers can be used as the binder, we shall focus our discussion on the thermoplastic binder system because it is used in most injection molding operations. Another issue is the binder formulation. The binder system generally consists of three components: a major binder, a minor binder, and small amounts of additives used as processing aids. The major binder controls the rheology of the feed material during injection molding to produce a body free of defects and also controls the strength of the green body and the binder burnout behavior. The minor binder is used to modify the flow properties of the feed material for good filling of the mold. The processing aids may include one or all of the following: a surfactant to improve the wetting between the particle surfaces and the polymer melt, a lubricant to reduce

Powder Consolidation

interparticle and die–wall friction, and a plasticizer to control the viscosity of the feed material. Table 6.11 lists some commonly used additives employed in the formulation of thermoplastic binder systems in the injection molding of ceramics.

Another important issue is the amount of binder in the feed material. Assuming that in the injection molding operation the particles achieve a packing arrangement close to that of dense random packing, then the amount of binder required to fill the interstices between the particles is roughly 35–40 vol %. However, other factors must be taken into account. While a high packing density of the particles is desirable, the mixture must also have a low enough viscosity for mold filling. Too little binder leads to a high viscosity and defects in the molded body, whereas excess binder leads to distortion of the molded body and disruption of the particle packing arrangement in the binder burnout step. Rheological characterization of binder systems and of powder–binder mixtures has an important role to play in striking the best balance between formability and binder content. In practice, depending on the characteristics of the powder and binder, the amount of binder ranges from 15 to 50 vol %.

As an example of the formulation of the feed material, we consider the injection molding of silicon powder. As outlined in Chapter 1, the nitridation of a consolidated powder form of silicon can be used for the production of reaction-bonded silicon nitride. The capability for the formation of complex shapes by injection molding coupled with the very negligible dimensional change accompanying the reaction bonding process can produce significant fabrication benefits. An approximate formulation of the feed material for injection molding is given in Table 6.12.

Mixing of the Feed Material

Homogeneous mixing of the powder and binder is essential. Mixing is usually performed by adding the powder to the molten binder in a shear

Table 6.11 Substances Commonly Employed in the Formulation of Thermoplastic Binder Systems for the Injection Molding of Ceramics

Major binder	Minor binder	Plasticizer	Other additive
Polypropylene	Microcrystalline	Dimethyl phthalate	Stearic acid
Polyethylene	wax	Diethyl phthalate	Oleic acid
Polystyrene	Paraffin wax	Dibutyl phthalate	Fish oil
Poly(vinyl acetate)	Vegetable oil	Dioctyl phthalate	Organosilane
Poly(methyl methacrylate)		Carnauba wax	Organotitanate

Table 6.12 Composition of a Typical Feed Material for the Injection Molding of Silicon Powder

Material	Function	% (by weight)
Silicon powder	Ceramic powder form	82
Polypropylene	Major binder	12
Microcrystalline wax	Minor binder	4
Stearic acid	Lubricant	2

mixer. Initially, a high torque develops as the powder is added to the binder due to high friction between the mixer and the powder agglomerates that have not yet been coated with the binder. As mixing continues, the torque decreases gradually as more powder is wetted by the binder and more powder agglomerates are broken down. Contamination by abrasion is a common problem during the initial period of high torque. Although a high shear rate is normally used in the final stages of mixing, insufficient breakdown of the agglomerates and therefore inhomogeneous mixing are also serious problems.

After cooling, the mixture is passed though a cutting mill, where it is converted into granules with a diameter of a few millimeters. These granules form the feed material for the injection molding stage.

Injection Molding of the Green Body

The injection molding of ceramic green bodies is basically identical to that of plastics. However, the significantly greater hardness of ceramics necessitates the use of hardened tool steels for the construction of the molds in order to reduce abrasive wear. Furthermore, the feed material for ceramic injection molding consists of two very different phases: a hard, rigid powder and a much softer, highly deformable polymeric binder. It may be expected that the considerations involved in filling the mold and the stress distribution during molding would become more complex than in the case of injection molding of plastics. However, little work has been done in the modeling of injection molding of ceramics. Instead a more practical approach is used in which the process variables are varied systematically to produce a useful green body.

Figure 6.30 illustrates the principle of operation of an injection molding machine and some of the parameters that influence its functioning [18]. The feed material in the form of granules is fed into the machine, transported by a screw or plunger to the injection chamber where it is heated to produce a viscous mass, and then injected, under pressure, into

Powder Consolidation

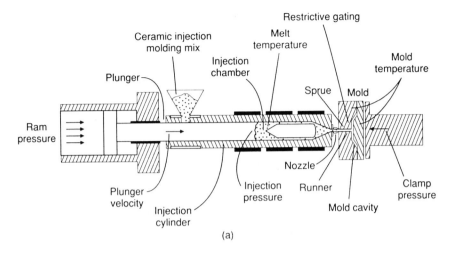

Figure 6.30 (a) Schematic diagram of a plunger injection molding machine identifying the principal machine variables. (b) Relation of fundamental injection molding variables to plunger machine variables. (From Ref. 18, with permission.)

the mold cavity. When the cavity is filled, the mold is cooled and the green body is ejected. It is obvious from Fig. 6.30b that a given machine variable can influence more than one of the fundamental variables. This, combined with the large number of variables, makes the optimization of the process a considerable task.

With regards to competition with the other forming methods considered in this chapter, injection molding may be best compared with die compaction. Because of the fluidity of the mixture, injection molding provides better die filling for complex shapes and a more homogeneous green body. However, the injection molding process has a much longer processing cycle and is much more complicated to optimize. The time necessary to remove the binder from relatively thick (>5–10 cm) green bodies can be very long. The molds used in injection molding, as outlined earlier, are very expensive. In general, therefore, the injection molding process is applied selectively to the forming of small articles with complex shapes.

6.9 DRYING OF CAST OR EXTRUDED ARTICLES

Articles produced by casting or plastic forming must be dried prior to binder burnout and firing. While important differences exist between gels and granular solids produced by casting or extrusion, several features of drying discussed in Chapter 5 with particular reference to polymeric gels are applicable to granular solids. We shall not repeat the discussion. Instead, it will be assumed that the reader is familiar with the discussion of drying presented in Chapter 5 and briefly consider the application of the ideas to the drying of cast or plastically formed articles. It will be recalled that cast or extruded articles contain considerably less water than gels so that the shrinkage is considerably less. Furthermore, because of the much larger pores, the permeability of granular solids is much higher than that of gels. Compared to gels, the significantly reduced shrinkage and much higher permeability alleviate the drying problems in granular solids. However, inhomogeneities in the body produced by imperfect forming operations are an additional source of problems in granular solids.

A. The Physical Process of Drying

The drying curves for moist granular bodies produced by casting or plastic forming show the same general features as those described in Chapter 5 for gels. The evaporation rate (Fig. 5.8) shows a constant rate period (CRP) followed by a falling rate period (FRP). In some cases, two parts can be distinguished in the FRP: the first falling rate period (FRP1) and the second falling rate period (FRP2). Drying curves for clay-based and other granular ceramics are commonly plotted as a function of the moisture content of the body. As defined by Eq. (5.19), the moisture content may be expressed as a percentage of the dry weight of the solid (dry basis) or the initial weight of the moist material, i.e., solid plus water (wet basis). Here the dry basis will be used. The moisture content of articles produced by casting or plastic forming is typically in the range of 20 to 35%.

Powder Consolidation

In the CRP, the rate of evaporation is independent of the moisture content (Fig. 6.31) and can be described by Eq. (5.30):

$$\dot{V}_E = H(p_w - p_A) \tag{5.30}$$

where p_w is the vapor pressure of the liquid at the surface p_A is the ambient vapor pressure and H is a factor that depends on the temperature, the velocity of the drying atmosphere and the geometry of the system. Initially, the particles are surrounded by a layer of liquid. As evaporation starts, a dry surface region is created and the liquid stretches to cover the dry region. A tension develops in the liquid and this is balanced by compressive stresses on the solid phase. The compressive stresses cause the body to contract and the liquid meniscus remains at the surface. As drying proceeds, the particles achieve a denser packing and the body becomes stiffer. The liquid meniscus at the surface deepens and the tension in the liquid increases. Eventually, the particles, surrounded by a thin layer of bound water, touch and shrinkage stops. This point is called the leatherhard moisture content.

When shrinkage stops, further evaporation drives the meniscus into the body and the rate of evaporation falls. The point at which the rate starts to fall is the critical moisture content. Although defined with respect to different drying properties (i.e., evaporation rate and shrinkage), the

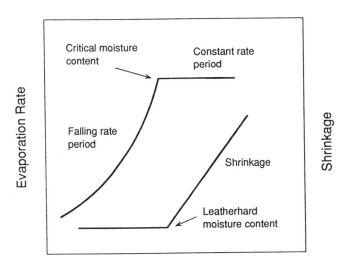

Figure 6.31 Schematic diagram showing (a) the evaporation rate and (b) the shrinkage versus the moisture content during drying of a moist granular material.

critical moisture content and the leatherhard moisture content are roughly equal (Fig. 6.31). In the first falling rate period, the liquid is still evaporating from the surface (Fig. 5.14). Contiguous pathways exist along which liquid can flow to the surface. Eventually, the liquid near the outside of the body becomes isolated into pockets. Flow to the surface stops and the liquid is removed predominantly by diffusion of the vapor. At this stage, drying enters the second falling rate period.

B. Cracking and Warping

Cracking and warping during drying are caused by differential strain due to (i) pressure gradients in the liquid and (ii) inhomogeneities in the body produced by imperfect forming operations. As discussed in Chapter 5, a tension develops in the liquid as it stretches to cover the dry region exposed by evaporation of the liquid. If the tension, p, in the liquid is uniform, there is no stress in the solid phase [Eq. (5.33)]. However, when p varies through the thickness, the body tends to contract more where p is high and the differential strain causes warping or cracking.

If the evaporation rate is very high, the tension in the liquid can reach its maximum value given by Eq. (5.26) and the total stress at the surface of the body is given by Eq. (5.34):

$$\sigma_x = \frac{2\gamma_{lv}\cos\theta}{a} \tag{5.34}$$

where γ_{lv} is the surface tension of the liquid, θ is the contact angle and a is the pore radius. For clay/water mixtures with a particle size in the range 0.5 to 1 μm, assuming $\theta = 0$, $\gamma_{lv} = 0.07$ J/m² and the pore size equal to half the particle size, gives $\sigma_x \approx 0.5$ MPa. Unfired ceramics containing little binder are very weak and this stress is sufficiently high to cause cracking.

As discussed in Chapter 5, during the constant rate period, the boundary condition at the surface of the body is given by Eq. (5.32):

$$\dot{V}_E = \frac{K}{\eta_L} \nabla p \big|_{surface} \tag{5.32}$$

where $\dot{V}_E$ is the evaporation rate, K is the permeability of the body, η_L is the viscosity of the liquid and ∇p is the pressure gradient in the liquid. According to Eq. (5.32), fast evaporation rate leads to high ∇p. To avoid cracking or warping the body must be dried slowly. However, the "safe" drying rates may be so slow that uneconomically long drying times are needed. To increase the safe drying rates, a few procedures can be used. According to Eq. (5.32), for a given $\dot{V}_E$, ∇p decreases with higher K and

Powder Consolidation

lower η_L. As outlined earlier, K increases roughly as the square of the particle size or the pore size. One approach is to use a larger particle size or to mix a coarse filler with the particles. However, this approach is often impractical because it can lead to a reduction in densification in the firing stage. Another approach, based on a decrease in η_L is used practically in "high humidity drying" where the process is carried out at slightly elevated temperatures ($\approx 70°C$) and high ambient humidity in the drying atmosphere. Increase in the temperature leads to a decrease in η_L (by a factor of ≈ 2) and also increases the rate of drying. However, the increase in the drying rate is counteracted by increasing the ambient humidity (see Eq. 5.30). In this way a reasonable drying rate is achieved while keeping ∇p small. The sequence of operation in a high humidity dryer involves increasing the ambient humidity followed by increasing the temperature and, after drying, decreasing the humidity followed by decreasing the temperature.

Even if the moist body were dried uniformly and at a safe rate, inhomogeneities produced by imperfect forming operations can still lead to cracking and warping. As Fig. 6.31 indicates, regions with higher moisture content (above the leatherhard value) undergo higher shrinkage. Differen-

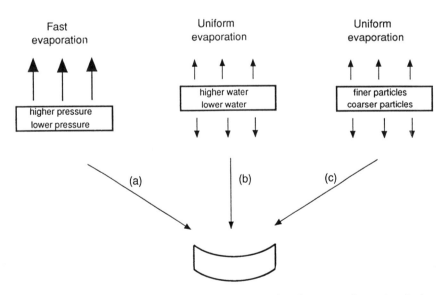

Figure 6.32 Schematic diagram illustrating the development of warping during drying due to (a) pressure gradient in the liquid, (b) moisture gradient in the body, and (c) segregation of particles.

tial shrinkage due to moisture gradients is a common source of cracking and warping. Gradients in the particle size caused, for example, by preferential setting can also lead to differential shrinkage. Figure 6.32 summarizes the effects of differential shrinkage caused by (a) pressure gradients in the liquid, (b) moisture gradients and (c) preferential settling.

6.10 BINDER REMOVAL

As mentioned earlier, the binder and other additives (referred to in this section simply as the binder) must be removed from the green body prior to firing at elevated temperatures to produce the final article. Ideally, we would like to remove the binder completely without disrupting the packing of the particles or producing any new defects in the green body. Residual contaminants such as carbon and inorganic ions and defects such as cracks and large voids affect the microstructural evolution during firing and hence the properties of the fabricated body. Binder removal is a critical step in ceramic processing, especially in the case of forming methods such as tape casting and injection molding where the binder content in the green body is relatively high. The process of binder removal is also referred to as *debinding*.

Binder removal can be accomplished thermally by heating the green body at slightly elevated temperatures or by dissolution of the binder in a solvent. While the solvent extraction approach finds some use in practice, it suffers from a number of limitations such as cost, disposal, and safety. Here, we shall consider the thermal debinding process commonly referred to as binder burnout.

Thermal debinding of ceramics is a complex process that is influenced by both chemical and physical factors. Chemically, the composition of the binder determines the decomposition temperature and the decomposition products. Physically, the removal of the binder is controlled by heat transfer into the body and mass transport of the decomposition products out of the body. In practice, complications arising from the use of binder systems consisting of three or four additives that differ in volatility and chemical decomposition must also be taken into account. Furthermore, the ceramic powder may alter the decomposition of the pure polymer. In view of its complexity, a detailed discussion of the thermal debinding process cannot be provided in the limited space remaining in this chapter. Instead, we consider the basic features of the process for a simplified system consisting of a powder compact with a single binder, e.g., a high molecular weight polymer such as poly(methyl methacrylate) or polyethylene. Later, some of the more important practical issues will be outlined.

A. Stages and Mechanisms of Thermal Debinding

The process of thermal debinding of a thermoplastic binder can be roughly divided into three stages. Stage 1 involves initially heating the binder to a point where it softens, e.g., ≈150–200°C. Chemical decomposition and binder removal are negligible in this stage, but a number of processes occur that can have serious consequences for the control of the shape and structural uniformity of the body. These processes include shrinkage, deformation, and bubble formation. Shrinkage occurs by a rearrangement process as the particles try to achieve a denser packing under the action of the surface tension of the polymer melt. The magnitude of the shrinkage occurring during this stage of thermal debinding increases with decreasing packing density of the particles in the green body. Deformation is promoted by a lower packing density in the green body, higher binder content, and lower melt viscosity. Bubble formation results from the decomposition of the binder as well as from residual solvent, dissolved air, or air bubbles trapped within the green body during forming. The formation of bubbles is a possible source of failure or defect formation during thermal debinding.

In stage 2, typically covering a temperature range of 200–400°C, most of the binder is removed by evaporation and chemical decomposition. Appreciable capillary flow of the molten polymer can accompany the evaporation process. The nature of the decomposition reactions depends on the chemical composition of the polymer and on the atmosphere. In inert atmospheres such as nitrogen or argon, polymers such as polyethylene undergo thermal degradation by chain scission at random points in the main chain to form smaller chain segments (Fig. 6.33a). The formation of smaller chain segments leads to a reduction in the polymer viscosity. With continued thermal degradation, the chain segments become small enough (i.e., their volatility increases) that evaporation is promoted. Other polymers, e.g., poly(methyl methacrylate) undergo depolymerization reactions to produce a high percentage of volatile monomers (Fig. 6.33b). In oxidizing atmospheres, degradation by oxidation occurs in addition to thermal degradation. Oxidative degradation commonly occurs by a free radical mechanism to produce decomposition products that contain a high percentage of volatile, low molecular weight compounds such as water, carbon dioxide, and carbon monoxide. Compared to thermal degradation, oxidation reactions generally lead to decomposition at lower temperatures and to an increase in the rate at which the binder is removed.

Finally, in stage 3, the small amount of binder still remaining in the body is removed by evaporation and decomposition at temperatures above ≈400°C. The highly porous nature of the body facilitates the removal of

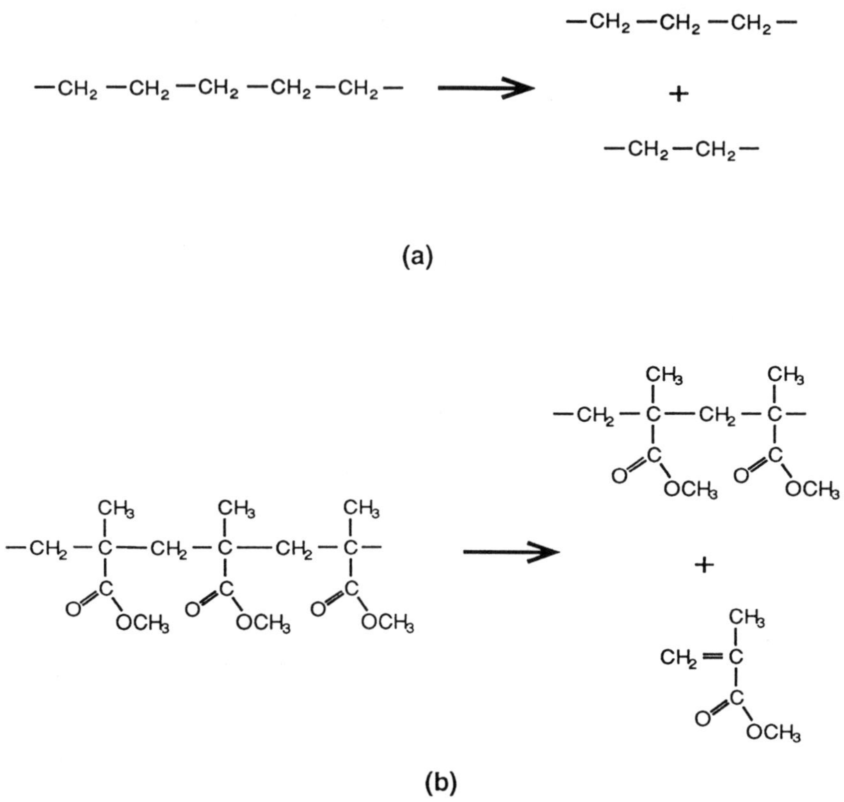

Figure 6.33 Mechanisms of thermal degradation of polymers showing (a) chain scission at random points in the chain, e.g., polyethylene, and (b) depolymerization (unzipping) to produce monomers, e.g., poly(methyl methacrylate).

the binder. However, the atmosphere must be carefully chosen to avoid the retention of an excessive amount of binder residue.

B. Models for Thermal Debinding

From the previous section we can identify two mechanisms for thermal debinding of a simple system containing a single high molecular weight polymeric binder: thermal degradation and oxidative degradation. However, practical binder systems, as outlined earlier, normally contain low molecular weight, volatile components that can undergo evaporation without thermal degradation. In general, three models need to be considered: (i) thermal degradation, (ii) oxidative degradation, and (iii) evaporation.

Powder Consolidation

Thermal degradation leads to the production of volatile, low molecular weight products throughout the binder phase. The rate of evaporation at the surface and the rate of transport of the degradation products through the body determine the concentration profile of the products. Volatile products present, for example, in the center of the body must not be allowed to reach temperatures above their boiling point because this would lead to the formation of bubbles and hence, defects. Thermal degradation therefore involves removal of the binder by evaporation of a liquid.

Binder removal during thermal degradation may be quite similar to the drying of a moist granular material. Let us consider a model in which interconnected pores of two different radii are present (Fig. 6.34a). Even though the pores have different radii (r_L and r_S), initially liquid evaporates from them at the same rate so that the radii of the menisci (r_m) are equal. The capillary tension in the liquid is given by the equation of Young and Laplace

$$p = \frac{2\gamma_{lv}}{r_m} \tag{6.16}$$

where γ_{lv} is the surface tension of the liquid. If the radii of the menisci were different, the capillary tension given by Eq. (6.16) would also be different and liquid would flow from one pore to the other until the menisci became equal again.

As evaporation from the surface proceeds, the radius of the menisci decreases. However, almost no shrinkage of the body occurs because the particles are practically touching. A point is reached where the radius of the menisci is equal to the radius of the large pore, i.e., $r_m = r_L$ (Fig. 6.34b). Further evaporation forces the liquid to retreat into the large pore. However, the radius of the meniscus, r_m, will continue to decrease in the small pore and the capillary tension will suck liquid from the large pore (Fig. 6.34c). In this way the large pore empties first and the small pore remains full of liquid. After the large pore has been emptied, the small pore starts to empty.

In practice, we will have a distribution of pore sizes and pore shapes. However, the same principles discussed above will apply. The largest pores empty first and the smallest empty last. We see that considerable redistribution of the liquid would be expected to occur during thermal degradation. Furthermore, the evaporation front is not expected to move uniformly into the body. Instead, pore channels first develop deep in the body as liquid from the larger pores is drawn into the smaller pores.

In the evaporation model, the volatile components (e.g., a low molecular weight plasticizer) are distributed throughout the binder phase. However, this model differs from the thermal degradation model only in that

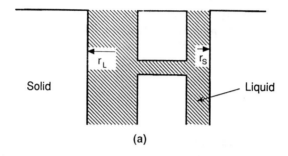

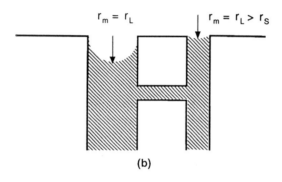

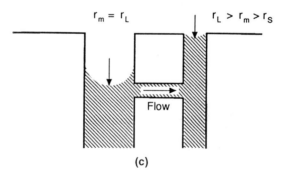

Figure 6.34 Sketch illustrating evaporation and liquid flow during binder removal by thermal degradation. The large pore empties first, while the meniscus is maintained at the surface of the small pore, that the small pore empties. (Adapted from Ref 23.)

Powder Consolidation

the volatile components are already present in the binder at some initial concentration. As the volatile components diffuse to the surface and evaporate, their concentration decreases. In thermal degradation, the volatile products have a concentration that is initially zero but increases with degradation and then decreases as the volatile products diffuse to the surface and evaporate. The model discussed earlier for the thermal degradation case is also expected to describe the main features of the evaporation of volatile, low molecular weight components that do not undergo thermal degradation. Recent observations do indicate that the removal of such components in binder systems leads to the development of porosity and liquid redistribution along the lines outlined for the thermal degradation model [19].

In oxidative degradation, the reaction occurs at the polymer/gas interface, which recedes into the body as degradation proceeds. The reaction products are gaseous and must be removed by diffusion or permeation through the porous outer layer (Fig. 6.35). This model is analogous to the shrinking core model outlined in Chapter 2 for the production of powders by calcination. The mean free path of the gaseous reaction products determines whether diffusion or permeation controls the transport of the gases through the porous outer layer. For the model shown in Fig. 6.35, and assuming that the binder is removed isothermally as a single-component, low molecular weight vapor, the time for removal of the binder depends on several parameters [20]. The dependence is summarized in Table 6.13 for diffusion and permeation control. In both cases, small particle size and low porosity reduce the rate of binder removal. However, small particle size and low porosity (with uniform distribution of the pores) enhance densification during the firing process. A conflict therefore exists between

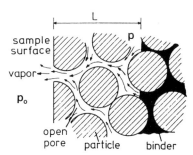

Figure 6.35 Schematic diagram of the model for thermal debinding by oxidative degradation where the binder/vapor interface is at a distance L from the surface of the compact. (From Ref. 20, with permission.)

Table 6.13 Effect of Process Variables on the Time for Thermal Debinding by Oxidative Degradation

Process variable	Diffusion control	Permeation control
Powder particle size	D^{-1}	D^{-2}
Porosity of green body	P^{-2}	$(1-P)^2/P^2$
Thickness of green body	H^2	H^2
Viscosity of gaseous product	—	η
Pressure drop	$(p-p_0)^{-1}$	$p/(p^2-p_0^2)$

rapid removal of the binder and the achievement of high densification. One solution of the conflict might involve the use of a sintering aid that enhances the densification process during firing. Table 6.13 also indicates that a low ambient pressure or a vacuum serves to reduce the time for binder removal. A vacuum, however, does not lead to oxidative degradation. Furthermore, temperature control and transport of heat are poor in a vacuum. The use of an oxidizing gas at reduced ambient pressure may provide adequate degradation as well as good thermal transport.

C. Binder Removal in Practice

The binder content of the green body, as we have observed, depends on the forming method. Green bodies formed by die compaction contain less than ≈5 vol % of binder. The removal of the binder can be accomplished relatively fast and is not a critical step in the overall fabrication process. Tape-cast bodies contain ≈30 vol % binder, whereas the binder content of injection-molded bodies is normally in the range of 30–50 vol %. For these high binder contents, removal of the binder is a critical and time-consuming step. Our foregoing discussion of binder removal is therefore most closely identified with green bodies formed by tape casting and injection molding. However, we considered a relatively simple system containing a single high molecular weight binder. Additional factors need to be addressed.

The decomposition of the polymeric binder in a ceramic green body is more complex than that of the pure binder. In the case of poly(vinyl butyral), Fig. 6.36 shows that oxides can lead to a decrease in the decomposition temperature [21]. CeO_2, for example, reduced the temperature for the greatest weight loss by ≈200°C. Binders that normally burn out completely in the pure state may leave a small amount of residue that cannot be easily removed from the particle surfaces. For a given binder composition, the amount of residue depends on a number of factors such

Powder Consolidation

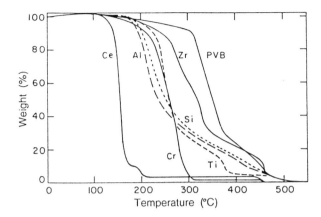

Figure 6.36 Thermogravimetric plots of the decomposition of poly(vinyl butyral) from films containing 2 wt % poly(vinyl butyral) and 98 wt % of various oxides during heating in air at 5°C/min. (From Ref. 21.)

as the powder composition, the gaseous atmosphere, and the structure and chemistry of the powder surface.

Practical binder systems for tape casting and injection molding consist of three or four additives that differ in terms of their volatility and decomposition behavior. The removal of each additive may occur at a different stage of the heating process or may overlap with that for another additive. The situation is therefore more complex than the simple case considered above for a single binder. It is obvious that binder systems in which there is not an appreciable overlap between the temperature ranges for removal of the additives provides better control of the binder removal process. In some injection molding operations, limited attempts have been made to develop such binder systems.

For green bodies formed by tape casting and injection molding, it is essential to achieve some balance between the time for binder removal and the prevention of defects during binder removal. Normally this is achieved by controlling the heating cycle. A very slow heating rate makes the thermal debinding process time-consuming, but fast heating rates lead to defects in the body. For a single binder, the body is heated at a moderate rate (e.g., ≈5°C/min) to the softening temperature of the binder followed by a slow heating rate (e.g., 1°C/min) above this temperature. For practical binder systems containing more than one binder, empirical heating cycles that incorporate several slow heating rate and isothermal stages have evolved to accommodate the more complex requirements.

6.11 MICROSTRUCTURAL CHARACTERIZATION OF THE GREEN BODY

The microstructure of the green body, as noted earlier, has a significant influence on its microstructural evolution during firing. Its characterization is therefore essential. For a given powder system, the important characteristics of the green body are the relative density (or porosity) and the pore size distribution, which is used as a measure of the uniformity of packing. As mentioned earlier, the density alone is a misleading parameter for predicting the sintering characteristics of a powder compact. The density is measured from the mass and external volume. For a body with a regular shape, the external volume can be measured from the dimensions. Porosity and pore size distribution are normally measured by mercury porosimetry or, for very fine pores, by gas adsorption. Scanning electron microscopy of a fracture surface provides a rough visual guide to the uniformity to the powder packing and is also easy to perform. The techniques of mercury porosimetry, gas adsorption, and scanning electron microscopy, which are also commonly used for powder characterization, have been described in Chapter 3.

6.12 CONCLUDING REMARKS

In this chapter we considered the common methods used for the production of the consolidated powder form (green body). Powder consolidation forms the last major processing step before the firing process in which the microstructure of the final article is developed. Since heterogeneities present in the green body cannot be easily removed in the firing stage, careful control of the green body microstructure prior to firing provides considerable benefits for microstructural control in the final article. The forming method used for the production of a ceramic article depends on the size and shape of the article. However, for any forming method, careful manipulation of the powder characteristics and the consolidation parameters provides a useful approach for the optimization of uniformity of the green body microstructure. Polymeric binders play an important role in powder consolidation. Binder selection and binder burnout procedures remain largely empirical, and the literature on the use of binders can be somewhat confusing. It is hoped that current and future work will lead to a reduction in the empiricism in the use of binders.

PROBLEMS

6.1 Find the radius, r, of a spherical particle that would fit precisely in the interstices of a hexagonal close packed arrangement of spheres of radius R. Determine the density of a compact of spherical Al_2O_3

Powder Consolidation

particles if 25 wt % of the particles fits precisely in the interstices of the hexagonal close packed larger fraction. (The theoretical density of Al_2O_3 is 3.96 g/cm^3.)

6.2 Show that the representations of the random packing of binary mixtures of spheres in terms of the packing density (Fig. 6.5) and in terms of the apparent volume (Fig. 6.6) are equivalent.

6.3 An aqueous Al_2O_3 slip has a density of 2.5 g/cm^3. Estimate the concentration of Al_2O_3 in the slip on (i) a weight basis and (ii) a volume basis.

6.4 What would be the effect on slip casting behavior when using:
 (i) a cold mold and a warm slip?
 (ii) a warm mold and a cold slip?
 (iii) a warm mold and a warm slip?

6.5 Compare the advantages and disadvantages of pressure casting with those of slip casting.

6.6 The slip in Problem 6.3 is pressure cast under an applied pressure of 1.5 MPa. Assuming that the density of the cast is equal to that for dense random packing of monosize spheres and the particle size of the powder is 1 μm, use Eqs. (6.12) and (6.13) to estimate the time required for the formation of a 1 cm thick cast.

6.7 A slip for tape casting contains, on a weight basis, 100 parts Al_2O_3, 10 parts of nonvolatile organics and 35 parts toluene. After drying, the tape which adheres to the surface on which it is cast is 60% of the original thickness of the cast. Estimate the porosity of the tape (i) after drying and (ii) after binder burnout. On firing, there is a shrinkage of 7.5% in the thickness of the tape and the density is 95% of the theoretical. Estimate the shrinkage in the plane of the tape.

6.8 (a) Consider an injection molded article consisting of ceramic powder and polymeric binder: develop a relationship between the ceramic solids content of the article and its linear shrinkage when fired to a specific endpoint density, assuming that the binder is removed completely during pyrolysis.
 (b) Repeat the calculation in part (a) assuming that, on pyrolysis, the binder produced an 80 wt% yield of ceramic powder having the same composition as the starting powder.
 (c) Assuming an endpoint density equal to the theoretical value of the ceramic, plot the shrinkage versus solids content for parts (a) and (b).

REFERENCES

1. D. J. Cumberland and R. J. Crawford, *The Packing of Particles*, Elsevier, New York, 1987.

2. F. F. Lange, *J. Am. Ceram. Soc.*, *72*:3 (1989).
3. R. K. McGeary, *J. Am. Ceram. Soc.*, *44*:513 (1961).
4. A. E. R. Westman and H. R. Hugill, *J. Am. Ceram. Soc.*, *13*:767 (1930).
5. R. J. Wakeman, *Powder Technol.*, *11*:297 (1975).
6. R. Burk and P. Apte, *Am. Ceram. Soc. Bull.*, *66*:1390 (1987).
7. A. Roosen and H. K. Bowen, *J. Am. Ceram. Soc.*, *71*:970 (1988).
8. J. V. Milewski, *Ad. Ceram. Mater.*, *1*:1 (1986).
9. G. Y. Onoda, Jr., in *Ceramic Processing Before Firing* (G. Y. Onoda, Jr. and L. L. Hench, Eds.), Wiley, New York, 1978, pp. 235–251.
10. S. Strijbos, A. Broese van Groenou, and P. A. Vermeer, *J. Am. Ceram. Soc.*, *62*(1–2):57, (1979); A. Broese van Groenou, *Powder Metallurgy Int.*, *10*:206 (1978).
11. F. M. Tiller and C.-D. Tsai, *J. Am. Ceram. Soc.*, *69*:882 (1986).
12. F. F. Lange and K. T. Miller, *Am. Ceram. Soc. Bull.*, *66*:1498 (1987).
13. R. E. Mistler, D. J. Shanefield, and R. B. Runk, in *Ceramic Processing Before Firing* (G. Y. Onoda, Jr. and L. L. Hench, Eds.) Wiley, New York, 1978, pp. 411–448.
14. C. O. Pels Leusden, in *Concise Encyclopedia of Advanced Ceramic Materials* (R. J. Brook, Ed.) Pergamon, Oxford, 1991, pp. 131–135.
15. J. S. Reed, *Introduction to the Principles of Ceramic Processing*, Wiley, New York, 1988, pp. 355–373.
16. R. M. German, K. F. Hens, and S.-T. P. Lin, *Am. Ceram. Soc. Bull.*, *70*:1294 (1991).
17. M. J. Edirisinghe and J. R. G. Evans, *Int. J. High Technol. Ceram.*, *2*:3 (1986).
18. J. A. Mangels and W. Trela, in *Advances in Ceramics*, Vol. 9 (J. A. Mangels, Ed.), American Ceramic Society, Columbus, OH, 1984, pp. 220–233.
19. M. J. Cima, J. A. Lewis, and A. D. Defoe, *J. Am. Ceram. Soc.*, *72*:1192 (1989).
20. R. M. German, *Int. J. Powder Metall.*, *23*:237 (1987).
21. S. Masia, P. D. Calvert, W. E. Rhine, and H. K. Bowen, *J. Mater. Sci.*, *24*:1907 (1989).
22. H. Y. Sohn and C. Moreland, *Can. J. Chem. Eng.* *46*:162 (1968).
23. G. W. Scherer, *J. Am. Ceram. Soc.*, *73*:3 (1990).
24. J. T. A. M. Welzen, in *Concise Encyclopedia of Advanced Ceramic Materials* (R. J. Brook, Ed.) Pergamon, Oxford, 1991, pp. 112–120.

FURTHER READING

G. Y. Onoda, Jr. and L. L. Hench (Eds.), *Ceramic Processing Before Firing*, Wiley, New York (1978).

J. A. Mangels and G. L. Messing (Eds.), *Advances in Ceramics, Vol. 9: Forming of Ceramics*, American Ceramic Society, Columbus, OH (1984).

F. F. Y. Wang (Ed.), *Treatise on Materials Science and Technology, Vol. 9: Ceramic Fabrication Processes*, Academic Press, New York, (1976).

＃ 7
Sintering of Ceramics: Fundamentals

7.1 INTRODUCTION

The sintering process, as outlined in Chapter 1, plays a prominent role in the fabrication of ceramics. Almost all ceramic bodies must be fired at elevated temperatures to produce a microstructure with the desired properties. This widespread use of the sintering process has led to a variety of approaches to the subject. Since the remainder of this book will be concerned primarily with the sintering process, it may be useful at the outset to outline our objectives with respect to the study of sintering. In practice, the ceramist, wishing to prepare a material with a particular set of properties, identifies the required microstructure and tries to design processing conditions that will produce this required microstructure. The objective of sintering studies should therefore be to understand how the processing variables influence the microstructural evolution during sintering. In this way, useful information can be provided for the practical effort of designing processing conditions for the production of the required microstructure.

At present, the sintering process in glasses that do not crystallize prior to complete densification is fairly well understood at a quantitative level. However, the situation for polycrystalline ceramics is quite different. Although the sintering process is understood qualitatively, the database and models are inadequate to quantitatively predict the sintering behavior for most systems of interest.

We start our treatment of the firing process by examining, in this chapter, some of the fundamental concepts in sintering. The discussion will be concerned primarily with the sintering of polycrystalline ceramics. For sintering to occur, there must be a decrease in the free energy of the system. The curvature of the free surfaces and, when it is used, the applied pressure provide the main motivation or driving force for sintering to occur. However, to accomplish the process within a reasonable time, we must also consider the kinetics of matter transport. In crystalline solids, matter is transported predominantly by diffusion of atoms, ions, or other charged species. There are several paths by which solid-state diffusion can occur. These paths define the mechanisms of diffusion and therefore the mechanisms of sintering. The rate of diffusion through the lattice depends on the type and concentration of defects in the solid. We must therefore have an understanding of the defect structure and the changes in the concentration of the defects (the defect chemistry). We must also understand how the defects and defect chemistry can be controlled by variables in the firing process [e.g., the temperature, gaseous atmosphere, and dopants (impurities)].

In order to predict how the rate of sintering depends on the primary processing variables, equations for the flux of matter must be formulated and solved subject to the appropriate boundary conditions. Matter transport can be viewed in terms of the flux of atoms (ions) or, equivalently, in terms of the counterflow of vacancies. While the concentration of the diffusing species can be analyzed (e.g., as in Fick's laws of diffusion), the flux equations take a relatively simple form when expressed in terms of the chemical potential (the molar Gibbs free energy). In this view, matter transport occurs from regions of high chemical potential to regions of low chemical potential.

In inorganic solids, the different species diffuse at different rates. However, matter transport must take place in such a way that charge neutrality in the solid is preserved. Some coupling of the diffusion process therefore occurs. The coupled diffusion of charged species is referred to as *ambipolar diffusion*.

7.2 SINTERING STUDIES: SOME GENERAL CONSIDERATIONS

The objective of sintering studies, as we outlined earlier, is to provide a better understanding so that microstructures can be fabricated in a more reproducible and predictable manner. One approach to this understanding involves connecting the behavior or changes in behavior during firing to controllable variables and operations. This can be achieved (i) empirically, by making measurements of the sintering behavior under controlled conditions, and (ii) theoretically, by modeling the process. The theoretical analysis and experiments performed during the past 40 years or so have provided an excellent qualitative understanding of sintering in terms of the driving force for the process, the mechanisms, and the effect of the principal processing variables such as particle size, temperature, and applied

Sintering Fundamentals

pressure. However, the database and models are far less successful at providing a quantitative description of the sintering behavior for most systems of interest. For this shortcoming, the models have received some criticism.

Table 7.1 lists some of the important parameters in firing that may serve to illustrate the scope of the problem [1]. In general, the processing and material parameters provide a useful set of variables for model experimental and theoretical studies. Some parameters, such as the firing temperature, applied pressure, average particle size, and gaseous atmosphere,

Table 7.1 Some Important Parameters in the Firing of Ceramics

Behavior	*Processing and material parameters*
General morphology	Powder preparation: particle size, shape, and size distribution
Pore evolution: size, shape, interpore distance	Distribution of dopants or second phases
Grain evolution: size and shape	Powder consolidation: density and pore size distribution
Density: function of time and temperature	Firing: heating rate and temperature
Grain size: function of time and temperature	Applied pressure
Dopant effects on densification and grain growth	Gaseous atmosphere
	Characterization measurements
Models	Neck growth
Neck growth	Shrinkage, density, and densification rate
Surface area change	Surface area change
Shrinkage	Grain size, pore size, and interpore distance
Densification in the later stages	Dopant distribution
Grain growth: porous and dense systems, solute drag, pore drag, pore breakaway	Strength, conductivity, and other microstructure-dependent properties
Concurrent densification and grain growth	
Database	
Diffusion coefficients: anion and cation, lattice, grain boundary, and surface	
Surface and interfacial energies	
Vapor pressure of components	
Gas solubilities and diffusivities	
Solute diffusivities	
Phase equilibria	

Source: Adapted from Ref. 1.

can be controlled with sufficient accuracy. Others, such as the powder characteristics and particle packing in the consolidated powder form, are more difficult to control but have a significant effect on sintering. As described earlier in this book, techniques exist for the preparation and uniform consolidation of powders with controlled characteristics. However, they are rarely used in industrial practice. The sintering behavior of real powder systems is fairly complex. As shown in Table 7.1, while partial information exists in the other areas of behavior, characterization measurements and the database, much critically needed information is severely lacking. This is especially serious in the database for the fundamental parameters such as the surface and grain boundary energies and the diffusion coefficients. This lack of information makes quantitative predictions of the sintering behavior very difficult even for the simplest systems.

To summarize at this stage, the theoretical models and experiments have produced a qualitative understanding that can be very useful in practical sintering. However, the lack of critical information coupled with the complexity of real ceramic systems presents severe impediments to the quantitative prediction of the sintering behavior of polycrystalline ceramics.

7.3 DRIVING FORCES FOR SINTERING

As with all other irreversible processes, sintering is accompanied by a lowering of the free energy of the system. We first identify the sources that give rise to this lowering of the free energy. These sources are usually referred to as the driving forces for sintering. Three possible driving forces are (i) the curvature of the particle surfaces, (ii) an externally applied pressure, and (iii) a chemical reaction.

A. Surface Curvature

In the absence of an external stress and a chemical reaction, surface curvature provides the driving force for sintering. To see why this is so, let us consider, for example, 1 mole of powder consisting of spherical particles with a radius a. The number of particles is

$$N = \frac{3M}{4\pi a^3 \rho} = \frac{3V_m}{4\pi a^3} \tag{7.1}$$

where ρ is the density of the particles, which are assumed to contain no internal porosity, M is the molecular weight, and V_m is the molar volume.

Sintering Fundamentals

The surface area of the system of particles is

$$S_A = 4\pi a^2 N = 3V_m/a \tag{7.2}$$

If γ_{sv} is the specific surface energy (i.e., the surface energy per unit area) or the surface tension of the particles, then the surface free energy associated with the system of particles is

$$E_s = 3\gamma_{sv} V_m/a \tag{7.3}$$

Taking $\gamma_{sv} = 1$ J/m², $a = 1$ μm, and $V_m = 25 \times 10^{-6}$ m³, then $E_s = 75$ J for the mole of material. This surface free energy provides a motivation for sintering. If a fully dense body is produced from the mole of material, E_s represents the decrease in surface free energy of the system (Fig. 7.1a).

B. Applied Pressure

In the absence of a chemical reaction, an externally applied pressure (Fig. 7.1b) normally provides the major contribution to the driving force when

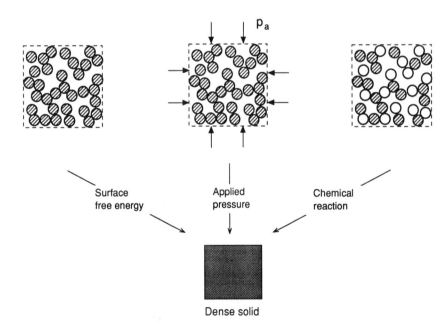

Figure 7.1 Schematic diagram illustrating three main driving forces for sintering: surface free energy, applied pressure and chemical reaction.

the pressure is applied over a significant part of the heating process (e.g., in hot pressing and hot isostatic pressing). Surface curvature also contributes to the driving force, but for most practical situations its contribution is normally much smaller than that provided by the external pressure. The external pressure does work on the system of particles, and for 1 mole of particles the work done can be approximated by

$$W = p_a V_m \tag{7.4}$$

where p_a is the applied pressure and V_m is the molar volume. In Eq. (7.4), W represents the driving force for densification provided by the external pressure. For $p_a = 30$ MPa, which is a typical value of the stress applied in hot pressing, and $V_m = 25 \times 10^{-6}$ m^3, then $W = 750$ J.

C. Chemical Reaction

A chemical reaction may also provide a driving force for sintering (Fig. 7.1c). The decrease in energy accompanying a chemical reaction can be much greater than the driving force provided by an applied stress, which, as we showed above, is significantly greater than that provided by the surface curvature. The change in free energy accompanying a chemical reaction is given by

$$\Delta G^0 = -RT \ln K_{eq} \tag{7.5}$$

where R is the gas constant (8.3 J/mol), T is the absolute temperature, and K_{eq} is the equilibrium constant for the reaction. Taking $T = 1000$ K and $K_{eq} = 10$, then $\Delta G^0 \approx 20,000$ J/mol. In spite of the very high driving force, a chemical reaction is not used deliberately to drive the densification process in advanced ceramics. As we see later in this book, a major problem is that microstructure control becomes extremely difficult when a chemical reaction occurs concurrently with the sintering process (see Chapter 11).

The driving forces provide a motivation for sintering. However, for sintering to actually occur, transport of matter must take place. In crystalline solids, matter transport occurs by a process of diffusion.

7.4 DIFFUSION IN SOLIDS

Solid state diffusion has been discussed in detail in many textbooks (e.g., Shewmon [2]) and review articles (e.g., Howard and Lidiard [3]). We will not repeat this discussion; instead we will outline the important issues that have direct relevance to the understanding of the sintering process.

Sintering Fundamentals

A. Fick's Laws of Diffusion

In an elementary view of diffusion, the movement of the diffusing species (atoms, ions, or molecules) is considered to be driven by gradients in the concentration. The concentration can vary as a function of distance and time. When the concentration is independent of time, the mathematics of the diffusion process is described by Fick's first law, which states that the flux of the diffusing species is proportional to the concentration gradient and occurs in the direction of decreasing concentration. The constant of proportionality, usually denoted by the symbol D, is called the *diffusion coefficient* or the *diffusivity*. Fick's first law may be written

$$\mathbf{J} = -D \nabla C \tag{7.6}$$

where the flux $\mathbf{J}$ is a vector with components (in the Cartesian system) of J_x, J_y, and J_z. The flux $\mathbf{J}$ represents the number of diffusing species crossing unit area, normal to the direction of flux, per second. In one dimension, Eq. (7.6) can be written

$$J_x = -D\frac{\partial C}{\partial x}; \quad J_y = -D\frac{\partial C}{\partial y}; \quad J_z = -D\frac{\partial C}{\partial z} \tag{7.7}$$

A concentration that is independent of time is often experimentally difficult to establish in a solid. It is more often convenient to measure the change in concentration as a function of time t. This is given by Fick's second law, which can be written

$$\frac{\partial C}{\partial t} = \nabla \cdot D \nabla C \approx D \nabla^2 C \tag{7.8}$$

In Cartesian coordinates, Eq. (7.8) becomes

$$\frac{\partial C}{\partial t} = D \left(\frac{\partial^2 C}{\partial x^2} + \frac{\partial^2 C}{\partial y^2} + \frac{\partial^2 C}{\partial z^2} \right) \tag{7.9}$$

Fick's second law can be derived from his first law and an application of the principle of conservation of matter [4]. For the one-dimensional case, consider the region between the two planes $[x_1, (x_1 + dx)]$ shown in Fig. 7.2. The solute concentration C is shown schematically as a function of distance x in Fig. 7.2a. Since $\partial C/\partial x$ at x_1 is greater than $\partial C/\partial x$ at $(x_1 + dx)$, $J(x_1)$ will be greater than $J(x_1 + dx)$, as shown in Fig. 7.2b. If $J(x_1) > J(x_1 + dx)$ and matter is conserved, the solute concentration in the region between x_1 and $x_1 + dx$ must increase. Considering a volume element of unit area normal to the x-axis and dx in thickness, the rate of

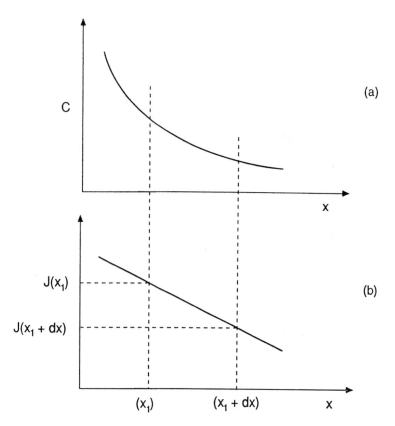

Figure 7.2 (a) Concentration, C, versus distance, x, and (b) the resulting flux, J, versus distance.

change of concentration is

$$\left(\frac{\partial C}{\partial t}\right)_{x_1} dx = J(x_1) - J(x_1 + dx) \tag{7.9a}$$

If dx is small, $J(x_1 + dx)$ can be related to $J(x)$ by the equation

$$J(x_1 + dx) = J(x_1) + \left(\frac{\partial J}{\partial x}\right)_{x_1} dx \tag{7.9b}$$

Substituting Eq. (7.9b) into Eq. (7.9a) and using Eq. (7.7) for J_x gives

Sintering Fundamentals

$$\frac{\partial C}{\partial t} = -\frac{\partial J}{\partial x} = -\frac{\partial}{\partial x}\left(-D\frac{\partial C}{\partial x}\right) = D\frac{\partial^2 C}{\partial x^2} \qquad (7.9c)$$

A similar derivation can be used for the two or three dimensional equation in cartesian, cylindrical or polar coordinates. However, for simplicity, the discussion will be limited to the one dimensional equation. In Eqs. (7.6)–(7.9) we have assumed that D is the same in all directions (i.e., isotropic). For noncubic crystals and anisotropic systems, D will depend on the direction. For dilute concentrations of the diffusing species, D is usually assumed to be independent of concentration. However, it can vary with concentration for systems in which the diffusing species is more concentrated.

Equation (7.8) or (7.9) can be solved for certain boundary conditions that can be approximated experimentally. For example, a common technique to measure D is to deposit a very thin film of a radioactive isotope (e.g., Ag*) on a planar surface of a sample (e.g., Ag) and, after annealing for fixed times, determine the concentration of the diffusing species by measuring the radioactivity as a function of distance. Our experimental system approximates to the case of diffusion of a thin planar source in a semi-infinite solid (Fig. 7.3). The solution (in one dimension) of Eq. (7.8) is

$$C = \frac{C_0}{2(\pi D^* t)^{1/2}} \exp\left(\frac{-x^2}{4D^* t}\right) \qquad (7.10)$$

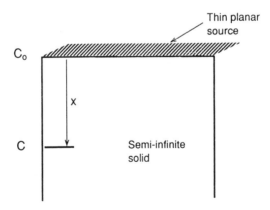

Figure 7.3 Parameters for the diffusion of a thin planar source in a semi-infinite solid. This geometry approximates to the practical situation for the measurement of the tracer diffusion coefficient.

where C is the concentration at a distance x from the surface, C_0 is the original concentration (i.e., at $t = 0$), and t is the annealing time. D^* is the diffusion coefficient of the radioisotope, which, as we see later (Section 7.3C), is referred to as the tracer diffusion coefficient. A plot of the data for ln C versus x^2 yields a straight line with a slope of $-1/(4D^*t)$, from which D^* can be found. Solutions of Eq. (7.9) for a wide variety of geometries and boundary conditions are given by Crank [4].

B. Mechanisms of Diffusion

Crystals are not ideal in structure. At any temperature they contain various imperfections called defects. Defects occur for (i) structural reasons, because the atoms are not arranged ideally in the crystal where all the lattice sites are occupied, and (ii) chemical reasons, because inorganic compounds may deviate from the fixed composition determined by the valence of the atoms. Defects control the rate at which matter is transported through the solid state and therefore determine a number of properties of solids. It is well established that diffusion in solids takes place because of the presence of defects.

The different types of structural defects are normally classified into three groups: (i) point defects, (ii) line defects, and (iii) planar defects. Point defects are associated with one lattice point and its immediate vicinity. They include missing atoms or vacancies, interstitial atoms occupying the interstices between atoms, and substitutional atoms sitting on sites that would normally be occupied by another type of atom. The point defects are illustrated in Fig. 7.4 for an elemental solid (i.e., a solid consisting of one type of atom in the pure state). The point defects that are formed in pure crystals (vacancies and interstitials) are sometimes referred to as native point defects. Line defects, commonly referred to as dislocations, are characterized by displacements in the periodic structure of the lattice in certain directions. Planar defects consist of stacking faults, internal interfaces (e.g., grain boundaries), and free surfaces.

The different types of defects determine the path of matter transport. Diffusion along the major paths gives rise to the major mechanisms of matter transport: lattice diffusion (also referred to as volume or bulk diffusion), surface diffusion, grain boundary diffusion, and dislocation pipe diffusion. In sintering, diffusion through the vapor phase (vapor transport by evaporation and condensation) is usually considered alongside the solid state diffusion mechanisms.

Lattice Diffusion

Lattice diffusion, also referred to as volume diffusion or bulk diffusion, takes place through the movement of point defects. The path of the diffu-

Sintering Fundamentals

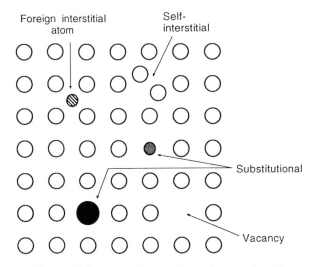

Figure 7.4 Point defects in an elemental solid.

sion is through the bulk of the lattice. The different types of point defects give rise to different mechanisms of lattice diffusion. These mechanisms are considered below for an elemental solid.

(i) **Vacancy Mechanism.** The vacancy mechanism is the most important mechanism in lattice diffusion. An atom on a normal lattice site exchanges its place with a vacant site (Fig. 7.5a). The movement of the atom is opposite that of the vacancy. We can therefore track the movement of the atom (i.e., atom diffusion) or, equivalently, the motion of the vacancy (i.e., vacancy diffusion). If the flux of vacancies is not compensated for by an equal and opposite flux of atoms, then porosity may develop in the crystal due to the accumulation of vacancies. Porosity can develop during the interdiffusion of two atoms that have very different diffusion coefficients.

(ii) **Interstitial Mechanism.** An atom on an interstitial site moves to one of the neighboring interstitial sites (Fig. 7.5b). Since the diffusion of the interstitial atom may involve a considerable distortion of the lattice, this mechanism probably occurs only when the interstitial atom is smaller than the atoms in the regular lattice. The diffusion of interstitial atoms in metals (e.g., carbon atoms in steel) provides the most common examples of the interstitial mechanism.

(iii) **Interstitialcy Mechanism.** If the distortion of the lattice becomes too large for interstitial diffusion to be favorable, then movement of the interstitial atoms may occur by the interstitialcy mechanism. An atom on

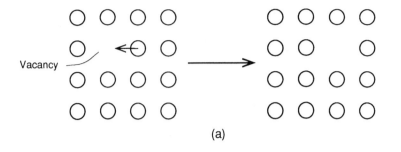

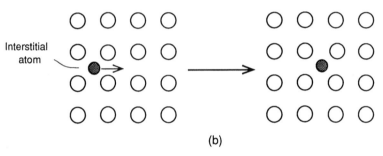

Figure 7.5 Lattice diffusion by (a) the vacancy mechanism, (b) the interstitial mechanism, (c) the interstitialcy mechanism, and (d) the ring mechanism.

the regular lattice site exchanges position with a neighboring interstitial atom (Fig. 7.5c). The two need not be the same type of atom.

(iv) Direct Exchange or Ring Mechanism. Atoms exchange places by rotation in a circle without the participation of a defect (Fig. 7.5d). While it has been proposed that this mechanism may take place in metals, its occurrence in oxides and other inorganic compounds is improbable because of the large energy changes arising from electrostatic repulsion.

Diffusion as a Thermally Activated Process

If we consider the change in energy of an atom as it jumps from one lattice site to another, there is an intermediate stage of higher energy that separates one site from another. For example, for an atom to pass from one interstitial site to another, the lattice must be distorted in the intermediate

Sintering Fundamentals

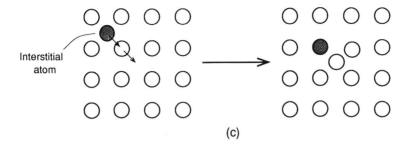

(c)

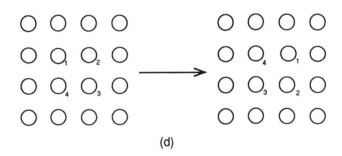

(d)

position in order for the atom to squeeze between the lattice atoms in its way (Fig. 7.6). Only a certain fraction of the atoms will have sufficient energy to surmount this energy barrier. The magnitude of the energy that must be supplied in order to surmount the energy barrier is called the *activation energy* for the diffusion process. Diffusion is one of many processes that are characterized by an energy barrier between the initial and final states. We would therefore expect the diffusion coefficient to depend on temperature in the same way as for other thermally activated processes:

$$D = D_0 \exp\left(\frac{-Q}{RT}\right) \qquad (7.11)$$

where D_0 and Q are constants for a given system. Q is the activation energy for the diffusion process (and is sometimes represented by the

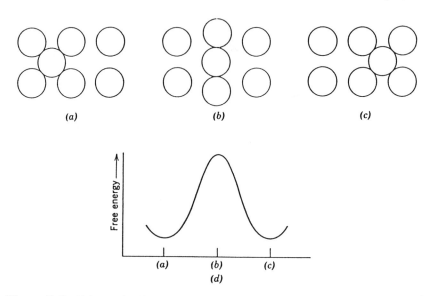

Figure 7.6 Schematic diagram showing the sequence of configurations when an atom jumps from one lattice site to another (a, b, c) and the corresponding change in the free energy of the lattice (d).

symbol ΔH); its value covers a wide range, from ≈ 10 kJ/mol to a few hundred kilojoules per mole.

Surface Diffusion

The free surface of a crystalline solid, as we observed earlier (Fig. 3.15), is not perfectly flat. It contains some vacancies as in the bulk of the crystal, terraces, kinks, edges, and adatoms. The migration of vacancies (as in lattice diffusion) and the movement of adatoms provide the main mechanisms of surface diffusion. The diffusion process is assumed to be confined to a thin surface layer, δ_s, that is one to two atomic diameters thick (≈ 0.3–0.5 nm). Since the atoms on the surface have fewer neighbors than those in the bulk of the lattice and are less tightly bound, we would expect the activation energy for surface diffusion to be less than that for lattice diffusion. This appears to be the case, to judge from the limited data available. Because of the lower activation energy, the relative importance of surface diffusion increases with decreasing temperature. As we shall see in the next chapter, this has important consequences for sintering.

Sintering Fundamentals

Grain Boundary Diffusion

In polycrystalline materials, as we observed earlier, the crystals or grains are separated from one another by regions of lattice mismatch and disorder called *grain boundaries*. Because of the highly defective nature of the grain boundary, we would expect grain boundary diffusion to be more rapid than lattice diffusion in the adjacent grains. The width of the grain boundary region, δ_{gb}, is assumed to be one to two atomic diameters thick (≈ 0.3–0.5 nm), i.e., roughly the same width as for surface diffusion. For a relatively constant grain boundary thickness, the fraction of the solid that is occupied by the grain boundary increases with decreasing grain size. The rate of grain boundary diffusion is therefore dependent on the grain size.

It has often been stated that the diffusion coefficients for lattice diffusion, D_l, grain boundary diffusion, D_{gb}, and surface diffusion, D_s, increase in the order

$$D_l < D_{gb} < D_s$$

and that the corresponding activation energies vary as

$$Q_l > Q_{gb} > Q_s$$

However, these relations may not always be correct.

C. Diffusion Coefficients

Several terms are used in the literature for specifying diffusion coefficients, and it is useful to have a clear understanding of what they mean.

Random diffusion. The reader may have realized that diffusion can occur even in the absence of a chemical composition (or chemical potential) gradient. The atom diffuses by a kind of Brownian motion or random walk. This type of diffusion is sometimes referred to as *random diffusion*.

Self-diffusion coefficient D_{self}. The self-diffusion coefficient refers to the diffusion coefficient of an atom in random diffusion.

Tracer diffusion coefficient D^*. As outlined earlier, the diffusion coefficient of a radioisotope of the solid under investigation (i.e., radioactive silver in silver) can be measured. This measured diffusion coefficient is called the *tracer diffusion coefficient*. Because the amount of radioisotope used is small, changes in the composition of the solid can be ignored. However, the diffusion of the radioactive atom is not completely random; successive jumps are, to a certain extent, correlated and dependent on previous jumps.

Correlation factor f. Because the diffusion of the radioactive atom is not completely random, D^* is not equal to D_{self}. They are related by a correlation factor f such that

$$D_{\text{self}} = D^*/f \tag{7.12}$$

where f is a constant for a given crystal structure. The value of f can be calculated and is in the range of 0.6–1 (see Ref. 2).

Lattice diffusion coefficient D_l. The lattice diffusion coefficient refers to any diffusion process in the lattice of the solid.

Grain boundary diffusion coefficient D_{gb}. Boundary diffusion occurs along a boundary or interface. In sintering, diffusion along the grain boundary between the grains is of great interest.

Surface diffusion coefficient D_s. The surface diffusion coefficient refers to diffusion along a free surface.

Diffusion coefficients for defects. The diffusion coefficients for defects are often specified. The interstitial diffusion coefficient refers to species diffusing by an interstitial mechanism. In sintering, the diffusion coefficient for vacancies, D_v, is normally of greater interest.

Chemical, effective, or interdiffusion coefficients. The chemical, effective, or interdiffusion coefficients refer to diffusion in a chemical composition gradient. These three diffusion coefficients are not identical [4].

Relation Between D_{self} and D_v. When diffusion occurs by a vacancy mechanism, we observed earlier that the diffusion process can be viewed in terms of the diffusion of atoms or, equivalently, in terms of the counterflow of vacancies. The diffusion coefficients for the atoms and vacancies are related but are not equal. We can see this as follows. An atom can jump only if a vacancy is located on an adjacent lattice site. The number of jumps will therefore be proportional to the fraction of lattice sites occupied by vacancies, denoted $[C_v]$. However, the vacancy can jump to any of the occupied nearest neighbor sites. The diffusion coefficient of the atoms is therefore related to that for the vacancies according to

$$D_{\text{self}} = D_v C_v \tag{7.13}$$

Equation (7.13) shows that for diffusion by the vacancy mechanism the vacancy concentration controls the rate of diffusion of the atom. It illustrates an important principle: Diffusion-controlled processes such as solid-state sintering are strongly influenced by the type and concentration of the point defects. We must therefore understand how defects are gener-

Sintering Fundamentals

ated and how they can be manipulated in order to control the rate of diffusion.

7.5 DEFECTS AND DEFECT CHEMISTRY

In our earlier discussion of the mechanisms of diffusion, we considered the main types of point defects present in an elemental solid (such as a metal). The study of point defects in ceramic materials (e.g., oxides) differs from that of point defects in metals in a number of important respects. For example, point defects in ceramics usually possess an effective charge. For an ionic compound with the stoichiometric formula MO consisting of a metal M with a valence of $+2$ and oxygen O (valence -2), the types of point defects that may occur are vacancies and interstitials of both M and O. These point defects may be either charged or neutral. In addition to single defects, it is possible for one or more defects to associate with one another, leading to the formation of defect clusters. There may also be electronic (or valence) defects consisting of quasi-free electrons or holes (missing electrons). If the compound contains a small amount of impurity atoms Mf, substitutional or interstitial defects of Mf will occur, and these may be either charged or neutral.

Another important difference between ceramic and metallic systems is that the composition of the ceramic may become nonstoichiometric by annealing in a suitable gaseous atmosphere (e.g., a controlled oxygen partial pressure). The compound seeks to equilibrate itself with the partial pressure of one of its components in the surrounding atmosphere. This equilibration leads to a change in the composition and a change in the type and concentration of the defects. As an example, annealing of the compound MO in an atmosphere with a low oxygen partial pressure may lead to an oxygen-deficient oxide, MO_{1-x}, in which oxygen vacancies predominate. However, annealing in an atmosphere with a higher oxygen partial pressure may lead to a metal-deficient oxide, $M_{1-y}O$, in which metal vacancies predominate.

Defect chemistry involves the study of changes in the defect concentration as a function of the temperature, composition, and atmosphere. The methods employed are normally applicable to fairly low defect concentrations. In general, a broad distinction is made between (i) intrinsic defects that occur naturally in pure stoichiometric compounds and (ii) extrinsic defects produced by external influences such as impurities and gaseous atmospheres.

In order to succinctly describe the various point defects and to express their formation in terms of equations, it is essential to have an adequate system of notation. The Kroger–Vink notation is most widely used.

A. The Kroger–Vink Notation for Point Defects

In the Kroger–Vink notation, the species in the crystal are denoted with respect to the perfect, idealized lattice [5]. The notation for point defects and lattice positions consists of three symbols: the main symbol, a subscript, and a superscript. For example, in the notation M_L^C, the main symbol M represents the particular type of atom. In the case of a vacancy, the main symbol is V. The subscript L represents the site in the perfect lattice at which the species is located. The superscript C gives the effective charge of the species, i.e., the difference in valence between the species on the L site and the valence of the atom that occupies the L site in the perfect lattice. The effective charge is represented as follows:

Positive effective charge: $C = \cdot$
Negative effective charge: $C = /$
Neutral effective charge: $C = x$

Electronic defects are specified as follows. A quasi-free electron is represented as e', while a missing electron or hole is represented as $h\cdot$. The use of the Kroger–Vink notation is demonstrated in Table 7.2 for some possible defects in Al_2O_3.

B. Defect Reactions

In order to know the concentrations of the defects, we must first understand how to formulate defect reactions. Basically, we can write defect reactions in a way similar to that for chemical reactions once the following conservation rules are observed:

1. *Conservation of mass.* A mass balance must be maintained so that mass is neither created nor destroyed in the defect reaction. Vacancies have zero mass, and electronic defects are considered to have no effect on the mass balance.

Table 7.2 Kroger–Vink Notation for Some Possible Defects in Al_2O_3

$Al_i^{\cdot\cdot\cdot}$	Aluminum ion in the interstitial lattice site
$V_O^{\cdot\cdot}$	Oxygen vacancy
Mg'_{Al}	Magnesium dopant on the normal Al lattice site
Ti_{Al}^x	Titanium dopant on the normal Al lattice site
e'	Quasi-free electron
$h\cdot$	Missing electron or hole

Sintering Fundamentals

2. *Principle of electroneutrality.* The crystal must remain electrically neutral. In writing defect reactions, this means that for the overall reaction the sum of the positive effective charges must be equal to the sum of the negative effective charges.
3. *Site conservation.* The ratio of the number of regular cation sites to the number of regular anion sites in the crystal remains constant. For example, in the compound MO, the ratio of the regular M and O sites must remain in the ratio of 1:1. Sites may be created or destroyed in the defect reaction, but they must occur in such a way that the site ratio in the regular lattice is maintained.

To see how these rules apply, let us consider the addition of MgO dopant to Al_2O_3, a solid solution reaction that we shall refer to on many occasions later in this book. In Al_2O_3, there are two cation sites to every three anion sites. If we incorporate two Mg atoms on cation sites, we must use two Al sites as well as two O sites. Since we have only two O sites, we can tentatively assume that the third O site for site conservation may be vacant. At this stage, on the basis of mass and site balance, we may write

$$2MgO \xrightarrow{Al_2O_3} 2Mg_{Al} + V_O + 2O_O \qquad (7.14)$$

Assuming that the defects are fully ionized, which is believed to be a more realistic solid solution process, application of the principle of electroneutrality gives

$$2MgO \xrightarrow{Al_2O_3} 2Mg'_{Al} + V_O^{\cdot\cdot} + 2O_O^x \qquad (7.15)$$

Another possibility is the formation of Al interstitials (instead of O vacancies), for which we may write

$$3MgO \xrightarrow{Al_2O_3} 3Mg'_{Al} + Al_i^{\cdot\cdot\cdot} + 3O_O^x \qquad (7.16)$$

The reader will verify the conservation of mass, site, and electroneutrality in Eq. (7.16).

C. Defect Concentration

Defect reactions, as we observed above, are considered in a way similar to the way we consider chemical reactions. Let us consider the general reaction in which reactants A and B lead to products C and D:

$$aA + bB \rightleftharpoons cC + dD \qquad (7.17)$$

At equilibrium at a fixed temperature, the law of mass action applies.

Assuming that the activities are equal to the concentrations, application of the law of mass action gives

$$K = \frac{[C]^c [D]^d}{[A]^a [B]^b} \tag{7.18}$$

where the brackets denote the concentrations and K, the equilibrium constant, depends on the absolute temperature T according to

$$K = \exp\left(\frac{-\Delta G}{RT}\right) \tag{7.19}$$

In Eq. (7.19), ΔG is the free energy change for the reaction given by Eq. (7.17) and R is the gas constant. The application of Eqs. (7.18) and (7.19) for the calculation of defect concentrations is illustrated in the following sections.

D. Intrinsic Defects

Two of the more common types of intrinsic defects in ionic crystals are referred to as the Schottky defect and the Frenkel defect.

Schottky Defect

The formation of a Schottky defect involves the transfer of a cation and an anion from their regular lattice sites to an external surface, leaving behind vacancies (Fig. 7.7). At the surface, the cation and anion form extra perfect crystal. For the compound MO and assuming that the defects

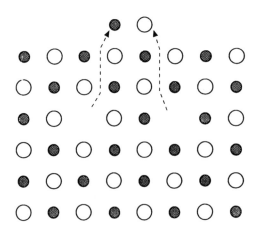

Figure 7.7 Schematic diagram for the formation of a Schottky defect.

Sintering Fundamentals

are fully ionized, the formation of a Schottky defect may be written

$$M_M + O_O \rightleftharpoons V_M'' + V_O^{\cdot\cdot} + M_M + O_O \quad (7.20)$$

In this equation M_M and O_O on both sides of the equation may be canceled, so that the net reaction is usually written

$$0 \rightleftharpoons V_M'' + V_O^{\cdot\cdot} \quad (7.21)$$

where O denotes the perfect crystal. Assuming that the defect reaction has reached equilibrium, we can use Eqs. (7.18) and (7.19) to give

$$K_S = [V_M''][V_O^{\cdot\cdot}] = \exp\left(\frac{-\Delta G_S}{RT}\right) \quad (7.22)$$

where K_S is the equilibrium constant and ΔG_S is the free energy change for the creation of a Schottky defect as defined by Eq. (7.21). For electroneutrality we must have

$$[V_M''] = [V_O^{\cdot\cdot}] \quad (7.23)$$

From Eqs. (7.22) and (7.23), the concentration of the defects is given by $[V_M''] = [V_O^{\cdot\cdot}] = \exp[-\Delta G_S/2RT]$.

Frenkel Defect

For the compound MO and assuming fully ionized defects, the creation of a Frenkel defect may be written

$$M_M \rightleftharpoons M_i^{\cdot\cdot} + V_M'' \quad (7.24)$$

This means that a cation leaves its regular lattice site and occupies an interstitial site, leaving behind a vacancy (Fig. 7.8). If K_F is the equilibrium constant for the reaction defined by Eq. (7.24), then

$$K_F = \frac{[M_i^{\cdot\cdot}][V_M'']}{[M_M]} = \exp\left(\frac{-\Delta G_F}{RT}\right) \quad (7.25)$$

For small defect concentrations, $[M_M]$ is almost constant and is assigned the value of unity. By invoking the electroneutrality condition $[M_i^{\cdot\cdot}] = [V_M'']$, the defect concentrations are found to be

$$[M_i^{\cdot\cdot}] = [V_M''] = \exp\left(\frac{-\Delta G_F}{2RT}\right) \quad (7.26)$$

A defect reaction corresponding to Eq. (7.24) may also be written for the formation of Frenkel defects on the anion sites:

$$O_O \rightleftharpoons O_i'' + V_O^{\cdot\cdot} \quad (7.27)$$

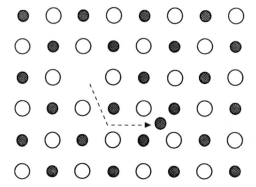

Figure 7.8 Schematic diagram for the formation of a Frenkel defect.

In this case, the defects are often referred to as anti-Frenkel defects. Considerations similar to those for the Frenkel defects apply. Finally, it should be noted that the formation of Frenkel defects and the formation of anti-Frenkel defects are not linked through an electroneutrality condition. The concentration of the cation interstitials therefore need not be equal to the anion interstitial concentration.

E. Extrinsic Defects

Extrinsic defects, as outlined earlier, are caused by external influences such as oxygen partial pressure (which leads to nonstoichiometry) and impurities (or dopants).

Nonstoichiometry

To illustrate the effects of nonstoichiometry, let us consider the response of the compound MO as a function of oxygen partial pressure. At a fixed temperature the compound must be in equilibrium with a specific oxygen partial pressure. The reader may recall that the equilibrium oxygen partial pressure at a temperature T can be calculated by knowing the free energy for the dissociation:

$$MO(s) \leftrightharpoons M(g) + \frac{1}{2} O_2(g)$$

If the oxygen partial pressure in the atmosphere is different from the equilibrium value, then the compound will give up or take up oxygen until equilibrium has been reached. This leads to a change in the cation/anion ratio, i.e., to a change in stoichiometry. When the oxygen partial pressure

Sintering Fundamentals

is greater than the equilibrium value, a situation referred to as *oxygen excess*, the taking up of oxygen is accomplished by the creation of oxygen interstitials or metal vacancies (or a combination of the two). For the oxygen-deficient situation, when the oxygen partial pressure is less than the equilibrium value, the giving up of oxygen is accomplished by the creation of oxygen vacancies or metal interstitials (or a combination of the two).

Kofstadt [5] provides a detailed treatment of the defect chemistry of nonstoichiometric oxides. Here, to illustrate the effects of nonstoichiometry on sintering, we will consider the oxygen-deficient case in which oxygen vacancies are created. For the compound MO, the creation of an oxygen vacancy can be described by the reaction

$$O_O \leftrightharpoons V_O^{\cdot\cdot} + 2e' + \frac{1}{2} O_2 \qquad (7.28)$$

This equation shows that the oxygen vacancies acquire their charge by quasi-free electrons being transported away from the vacant sites. Electroneutrality is preserved because

$$[e'] = 2[V_O^{\cdot\cdot}] \qquad (7.29)$$

Considering $[O_O]$ to be unity (i.e., small defect concentrations), application of the law of mass action gives

$$K = [V_O^{\cdot\cdot}] [e']^2 p_{O_2}^{1/2} \qquad (7.30)$$

where K is the equilibrium constant for the defect reaction and p_{O_2} is the oxygen partial pressure. Substituting from Eq. (7.29) we get

$$[V_O^{\cdot\cdot}]^3 = \frac{K}{4} p_{O_2}^{-1/2} \qquad (7.31)$$

The oxygen vacancy concentration can finally be expressed as

$$[V_O^{\cdot\cdot}] = K_1 p_{O_2}^{-1/6} \qquad (7.32)$$

where K_1 is equal to $(K/4)^{1/3}$.

Equation (7.32) shows that the concentration of oxygen vacancies increases as the oxygen partial pressure decreases (Fig. 7.9). It has important consequences for sintering. If the sintering rate of an oxide were controlled by the diffusion of oxygen vacancies, then lower oxygen partial pressures would be beneficial for sintering. However, as we shall see later, the real situation in sintering is not so straightforward (see Section 7.5F).

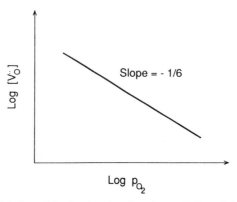

Figure 7.9 Doubly logarithmic plot showing the variation of the oxygen vacancy concentration as a function of the oxygen partial pressure for an oxide MO in the oxygen-deficient condition.

Impurities and Dopants

As we shall describe in detail later in this book, impurities and dopants present in concentrations as low as a fraction of a mole percent have a significant influence on sintering. Here, as a preliminary to the later discussion of their influence, we shall consider the main features of their effects on the defect chemistry. Before doing so, we take the opportunity to define the meaning of a few common terms. The term *impurity* is generally used to describe a substance that is accidentally incorporated into a material as a result of the synthesis or processing operations. *Dopants* refers to cations (or anions) that have been added to the material in controlled quantities to modify the microstructure during the firing process or the properties of the fabricated material. For conciseness, we make no distinction between impurities and dopants in this section and simply use the term dopant. For the oxide MO, dopant cations with valences different from that of M (+2 in this case) are referred to as *aliovalent dopants*, and those with valences equal to that of M are called *isovalent dopants*. Aliovalent dopants in which the cation valency is greater than that of the host crystal are referred to as *donor dopants*; those in which the cation valency is smaller than that of the host are referred to as *acceptor dopants*.

Let us now consider the incorporation of Al_2O_3 dopant into MgO. Assuming that the Al cations substitute on the regular Mg sites, then the incorporation reaction can be written

$$Al_2O_3 \xrightarrow{MgO} 2Al_{Mg}^{\cdot} + V_{Mg}'' + 3O_O \qquad (7.33)$$

Sintering Fundamentals

In this reaction we are assuming that when the Al ions substitute on the regular Mg sites, charge compensation is achieved by the simultaneous creation of vacancies in the regular Mg sites. The reader will also verify the conservation of mass, charge, and electroneutrality in Eq. (7.33). As the incorporation reaction proceeds, the other defect equilibria, such as the creation of Schottky or Frenkel defects, are still present. Suppose that the intrinsic defects in MgO consist of Schottky disorder. Following Eq. (7.22), the concentrations of the intrinsic defects are related by the equation

$$[V_O^{\cdot\cdot}][V_{Mg}''] = K_S \tag{7.34}$$

Applying the electroneutrality condition $[Al_{Mg}^{\cdot}] + 2[V_O^{\cdot\cdot}] = 2[V_{Mg}'']$, but $[Al_{Mg}^{\cdot}] \ll [V_{Mg}'']$ gives

$$[V_O^{\cdot\cdot}] = [V_{Mg}''] = K_S^{1/2} \tag{7.35}$$

According to Eq. (7.35) the concentrations of the intrinsic defects are independent of the concentration of the Al_2O_3 dopant.

As the incorporation reaction described by Eq. (7.33) proceeds, the concentration of the Al in solid solution increases and the extrinsic defects begin to dominate. The neutrality condition now becomes

$$[Al_{Mg}^{\cdot}] = 2[V_{Mg}''] \tag{7.36}$$

Assuming that the Al_2O_3 has been completely incorporated in the reaction, the concentration of Al in solid solution is equal to the total atomic concentration of Al, that is,

$$[Al_{Mg}^{\cdot}] = [Al] \tag{7.37}$$

Since Eq. (7.34) must apply for the cation and anion vacancies, combination of Eq. (7.34) with Eqs. (7.36) and (7.37) yields

$$[V_O^{\cdot\cdot}] = 2K_S/[Al] \tag{7.38}$$

Furthermore, Eqs. (7.36) and (7.37) give

$$[V_{Mg}''] = [Al]/2 \tag{7.39}$$

Equations (7.38) and (7.39) show that the incorporation of Al_2O_3 dopant into MgO leads to changes in the concentration of the defects. With increasing concentration of Al_2O_3, the oxygen vacancy concentration decreases but the magnesium vacancy concentration increases. The consequences for sintering are as follows. If the sintering rate of MgO is controlled by the diffusion of oxygen vacancies, then the addition of Al_2O_3 will act to inhibit the sintering rate. However, if the diffusion of magnesium vacancies is the rate-controlling mechanism, then the addition of

356   Chapter 7

Al_2O_3 will increase the sintering rate. In practice, the influence of dopants is not as clear as this example may indicate. Data for the rate-controlling mechanism are not available for most systems. Furthermore, as we shall discuss later in this book, dopants may have a variety of effects on sintering, which makes an understanding of their role fairly difficult. It is left as an exercise for the reader to determine the effect of an acceptor dopant (e.g., Li_2O) on the defect concentrations of MgO.

Finally, the changes in defect concentration as a function of oxygen partial pressure or dopant concentration are sometimes described semi-quantitatively in terms of a double logarithmic plot known as a Brouwer diagram. The defect concentrations are normally plotted as the ordinate while the abscissa represents the oxygen partial pressure or the dopant concentration. Figure 7.10 shows a Brouwer diagram for the effect of Al_2O_3 dopant on the defect chemistry of MgO that was considered above.

F. Defect Chemistry and Sintering: A Summary

To summarize at this stage, the concentration of the defects and hence the rate of matter transport can be altered by manipulating three accessible variables: the temperature, the oxygen partial pressure (or, in general,

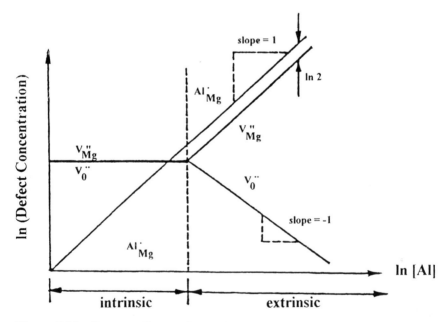

Figure 7.10 Brouwer diagram for MgO doped with Al_2O_3. The concentrations of the defects are shown as functions of the concentration of Al.

Sintering Fundamentals

the gaseous atmosphere), and the concentration of dopants. Writing [X] as the concentration of the defect, the variation of [X] with these three variables can be summarized as

1. *Temperature.* $[X] \sim K \sim \exp\{-\Delta G/(RT)\}$
2. *Atmosphere.* $[X] \sim p_{O_2}^m$, where m is a positive or negative exponent
3. *Dopant Concentration.* $[X] \sim [Mf]^n$, where [Mf] is the dopant concentration and n is a positive or negative exponent.

In practice, the control of the sintering process through manipulation of the defect structure is not so straightforward. A major problem is that in most systems the rate-controlling step in the densification process is not known. Another major problem, as we shall describe in the next chapter, is that several mass transport mechanisms may operate during the sintering process. Some of these mechanisms (e.g., surface diffusion and vapor transport) inhibit the rate of densification. Although changes in temperature, oxygen partial pressure, and dopant may enhance the densification mechanisms, they may also enhance the nondensifying mechanisms. Finally, physical factors such as the homogeneity of particle packing have a considerable influence on the sintering rate and may actually overwhelm the effects produced by changes in the defect structure.

7.6 THE CHEMICAL POTENTIAL

Matter transport during sintering, as discussed earlier, can be analyzed in terms of Fick's first law of diffusion by considering the flux of atoms or vacancies down a concentration gradient. However, many problems can be analyzed in a more consistent manner if sintering is considered as a *chemical diffusion process*. Instead of diffusion down a concentration gradient, the atoms or vacancies are considered to diffuse from regions of higher chemical potential to regions of lower chemical potential, i.e., down a *chemical potential gradient*. We therefore need to understand the meaning of the chemical potential and how it can be determined explicitly for the conditions pertaining to sintering. It is assumed that the reader has an understanding of the elementary aspects of chemical thermodynamics.

A. Definition of the Chemical Potential

Consider first a phase with fixed mass and composition but under conditions of variable temperature T and pressure p. For an infinitesimal reversible process, the change in the Gibbs free energy is

$$dG = \left(\frac{\partial G}{\partial T}\right)_p dT + \left(\frac{\partial G}{\partial p}\right)_T dp \qquad (7.40)$$

The terms in the brackets are defined by the following equations

$$\left(\frac{\partial G}{\partial T}\right)_p = -S; \quad \left(\frac{\partial G}{\partial p}\right) = V \qquad (7.41)$$

where S is the entropy and V the volume of the system. Equation (7.40) can therefore be written

$$dG = -S\,dT + V\,dp \qquad (7.42)$$

Consider now a phase of variable composition consisting of m chemical constituents of which there are n_1 moles of the substance A_1, n_2 moles of A_2, ..., n_m moles of A_m. The resulting change in the Gibbs free energy is now given by

$$dG = \left(\frac{\partial G}{\partial T}\right)_{p,n_1,n_2,\ldots,n_m} dT + \left(\frac{\partial G}{\partial p}\right)_{T,n_1,n_2,\ldots,n_m} dp + \left(\frac{\partial G}{\partial n_1}\right)_{p,T,n_2,n_3,\ldots,n_m} dn_1$$

$$+ \cdots + \left(\frac{\partial G}{\partial n_m}\right)_{p,T,n_1,n_2,\ldots,n_{m-1}} dn_m \qquad (7.43)$$

Since the first two terms on the right-hand side of Eq. (7.43) are at constant mass and composition, we can use Eq. (7.41). Furthermore, consider the effect upon the Gibbs free energy when a small amount of one of the constituents (e.g., the kth constituent) is introduced into the phase, with T, p, and the other n's remaining constant. If dn_k moles of A_k is introduced, the effect on the Gibbs free energy is expressed as

$$\mu_k = \left(\frac{\partial G}{\partial n_k}\right)_{T,p,n_1,n_2,\ldots,n_{k-1},n_{k+1},\ldots,n_m} \qquad (7.44)$$

where μ_k is called the chemical potential of the kth constituent of the phase. We can now write Eq. (7.43) as

$$dG = -S\,dT + V\,dp + \sum_{i=1}^{m} \mu_i\,dn_i \qquad (7.45)$$

Suppose we increase the number of moles of a phase while keeping T, p, and the composition constant. We can write

$$dG_{T,P} = \sum_i \mu_i\,dn_i \qquad (7.46)$$

Since the μ_i depend only on T, p, and composition, they must remain constant. Equation (7.46) can therefore be integrated to give

Sintering Fundamentals

$$G = \sum_i \mu_i n_i \qquad (7.47)$$

We now go on to relate the chemical potential to more familiar variables.

B. Chemical Potential of a Pure Substance

For the case of a pure substance, Eq. (7.47) reduces to

$$G = \mu n \qquad (7.48)$$

so that the chemical potential is the *molar Gibbs free energy* and is a function of T and p only.

C. Chemical Potential of Mixtures of Gases

Consider 1 mol of an ideal gas. At constant temperature T,

$$\left(\frac{\partial G}{\partial p}\right)_T = V = \frac{RT}{p} \qquad (7.49)$$

Integration of this equation yields

$$G(T, p) = G_0(T) + RT \ln p \qquad (7.50)$$

where G_0 is a reference value. We can therefore write

$$\mu(T, p) = \mu_0(T) + RT \ln p \qquad (7.51)$$

For a mixture of ideal gases at constant T and for a constant total pressure, following Eq. (7.51) we can write for each component

$$\mu_i(T, p_i) = \mu_{0i}(T) + RT \ln p_i \qquad (7.52)$$

However, in treating mixtures of gases, it is useful to refer the Gibbs free energy or the chemical potential of a particular component to the total pressure rather than to the partial pressure. This is done by using the molar concentrations of the components, C_i, defined by

$$C_i = \frac{n_i}{\sum n_i} = \frac{p_i}{p} \qquad (7.53)$$

where n_i is the number of moles of each component in the mixture. Equation (7.52) can then be written

$$\mu_i(T, p, C_i) = \mu_{0i}(T, p) + RT \ln C_i \qquad (7.54)$$

Real gases, as the reader will be aware, do not show ideal behavior. However, the deviation from ideal behavior is not too great over a fairly wide range of pressures. Equations (7.52)–(7.54) therefore provide a good approximation to the chemical potential of real gases.

D. Chemical Potential of Liquids and Solids

For liquid and solid solutions, the chemical potential is defined by an expression similar to Eq. (7.52) in which the p_i is replaced by a quantity a_i, called the *activity*; that is,

$$\mu_i = \mu_{0i} + RT \ln a_i \tag{7.55}$$

The activity of pure liquids and solids under some specified standard conditions of temperature and pressure is taken as unity. For other systems, the activity is written

$$a_i = \alpha_i C_i \tag{7.56}$$

where α_i is called the activity coefficient and C_i is the concentration, usually expressed as a mole fraction. Using Eqs. (7.55) and (7.56), the chemical potential can be expressed in terms of the concentration by

$$\mu_i = \mu_{0i} + RT \ln(\alpha_i C_i) \tag{7.57}$$

In the case of ideal solutions, $\alpha_i = 1$.

E. Chemical Potential of Atoms and Vacancies in a Crystal

Factors other than the temperature, pressure, and composition can affect the chemical potential. Even for a pure element, we observed earlier that defects are present in the crystal. The chemical potential depends on the concentration of the defects. The dependence of the chemical potential on the defect concentration is important for sintering because matter transport is driven by gradients in the chemical potential.

To explore the dependence of the chemical potential on the concentration of the defects, let us consider the case of a crystal of a pure element that is perfect except for the presence of vacancies. If there are N_a atoms and n_v vacancies, then the total number of lattice sites in the crystal is $N = N_a + n_v$. The Gibbs free energy of the crystal is [2]

$$G = U + n_v g + pV - TS \tag{7.58}$$

where U is the internal energy and g is the Gibbs free energy for the formation of a vacancy. The configurational entropy of the crystal is given by the Boltzmann relation, which can be written

$$S = k \ln \frac{(N_a + n_v)!}{N_a! \, n_v!} \tag{7.59}$$

Sintering Fundamentals

where k is the Boltzmann constant. The chemical potential of the atoms is defined by

$$\mu_a = \left(\frac{\partial G}{\partial N_a}\right)_{T,p,n_v} \tag{7.60}$$

It is left as an exercise for the reader to show that

$$\mu_a = \mu_{0a} + p\Omega_a + kT \ln\left(\frac{N_a}{N_a + n_v}\right) \tag{7.61}$$

where μ_{0a} is the reference value and Ω_a is the volume of an atom defined by

$$\left(\frac{\partial V}{\partial N_a}\right)_{T,p,n_n} = \Omega_a \tag{7.62}$$

If C_a is the fraction of lattice sites occupied by the atoms in the crystal, then Eq. (7.61) can be written

$$\mu_a = \mu_{0a} + p\Omega_a + kT \ln C_a \tag{7.63}$$

This equation shows that the chemical potential of the atoms depends on the pressure and on the concentration of the atoms in the crystal. However, for small vacancy concentration, the last term on the right-hand side of Eqs. (7.61) and (7.63) is negligible.

The chemical potential of the vacancies is defined by

$$\mu_v = \left(\frac{\partial G}{\partial n_v}\right)_{T,p,N_a} \tag{7.64}$$

Following the same procedure used for μ_a, we obtain

$$\mu_v = \mu_{0v} + p\Omega_v + kT \ln C_v \tag{7.65}$$

where C_v is the vacancy concentration (i.e., the fraction of lattice sites occupied by the vacancies). While the volume of a vacancy Ω_v may be greater or smaller than Ω_a, we will use the rigid lattice approximation and assume that

$$\Omega_a = \Omega_v = \Omega \tag{7.66}$$

F. Chemical Potential Beneath a Curved Surface

The atoms and vacancies beneath a curved surface will have their chemical potentials altered by the curvature of the surface. The difference in chemical potential from one region of the surface to another leads to a diffusional flux of atoms to reduce the free energy of the system. In order to formulate equations for the sintering process, we must understand how the chemical potentials depend on the surface curvature.

Consider a solid consisting of adjoining convex and concave surfaces (Fig. 7.11). We will assume that the solid is a pure element in which vacancies are the only type of point defect present. For the convex surface, the surface can be decreased by reducing the volume of the region beneath it. This can be achieved by reducing the concentration of vacancies. Since a decrease in the surface area leads to a decrease in the surface contribution to the free energy, it is expected that the vacancy concentration will be below normal (e.g., relative to a flat surface). Following a similar type of argument, it is expected that the concentration of vacancies will be above normal beneath the concave surface. The difference in vacancy concentration leads to a diffusional flux of vacancies from the concave to the convex region or, equivalently, a diffusional flux of atoms from the convex to the concave region.

Following Herring [6], we can determine quantitatively the dependence of the chemical potential on the curvature of a surface. Consider a smoothly curved surface as shown in Fig. 7.12. Suppose we build an infinitesimal hump on the surface by taking atoms from beneath the sur-

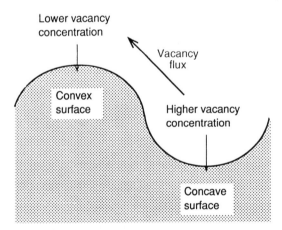

Figure 7.11 Schematic diagram showing the direction of flux for vacancies in a curved surface. The flux of atoms is equal and opposite to that of the vacancies.

Sintering Fundamentals

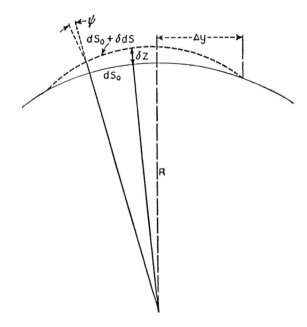

Figure 7.12 Infinitesimal hump formed by building up a curved surface. Full curve: original surface; dashed curve: built-up surface. (From Ref. 5.)

face and adding them to the surface. The change in surface free energy, to a good approximation, is given by

$$\delta\left(\int \gamma_{sv} \, dS\right) = \int \delta(\gamma_{sv}) \, dS_0 + \int \gamma_{sv} \, \delta(dS) \tag{7.67}$$

where δ represents a small change in a quantity, γ_{sv} is the surface tension or specific surface energy of the surface, and dS is the change in surface area. We will assume that the surface is uniform and isotropic so that the first term on the right-hand side of Eq. (7.67) is zero. Referring to the quantities described by Fig. 7.12, $\delta(dS)$ is given by

$$\delta(dS) = (\sec \psi - 1) \, dS_0 + \delta z \left(\frac{1}{R_1} + \frac{1}{R_2}\right) dS_0 \tag{7.68}$$

where R_1 and R_2 are the two principal radii of curvature at dS_0. If Δy is small, then R_1 and R_2 may be taken as constant over the hump. Furthermore, the first term on the right-hand side of Eq. (7.68) can be neglected.

Thus,

$$\delta\left(\int \gamma_{sv} dS\right) = \gamma_{sv}\left(\frac{1}{R_1} + \frac{1}{R_2}\right)\delta v \qquad (7.69)$$

where δv is the volume of the hump.

The change in the volume term of the free energy due to the creation of the small hump is

$$\delta G_v = -p\,\delta v + \mu_v + \mu_v \frac{\delta v}{\Omega} \qquad (7.70)$$

where p is the mean hydrostatic pressure in the crystal just beneath the surface and $\delta v/\Omega$ represents the number of vacancies created by the transfer of atoms to the hump. In equilibrium, the sum of the energy changes defined by Eqs. (7.69) and (7.70) must vanish, so

$$\mu_v = p\Omega + \gamma_{sv}\left(\frac{1}{R_1} + \frac{1}{R_2}\right)\Omega \qquad (7.71)$$

The curvature of the surface, K, is defined as

$$K = \frac{1}{R_1} + \frac{1}{R_2} \qquad (7.72)$$

where $K > 0$ for a convex surface. Substituting for K in Eq. (7.71) yields

$$\mu_v = (p + \gamma_{sv}K)\,\Omega \qquad (7.73)$$

In general, the chemical potential is measured relative to some reference value. Furthermore, as found earlier, μ_v contains a vacancy concentration term. A more general expression for μ_v that incorporates pressure and surface curvature effects is

$$\mu_v = \mu_{v0} + (p + \gamma_{sv}K)\Omega + kT \ln C_v \qquad (7.74)$$

Using the same procedure, the chemical potential of the atoms can be shown to be

$$\mu_a = \mu_{a0} + (p + \gamma_{sv}K)\Omega + kT \ln C_a \qquad (7.75)$$

As outlined earlier, the term containing C_a is normally very small. Equations (7.74) and (7.75) describe two very important relations in sintering. They show that the chemical potential of the atoms or vacancies depends primarily on the hydrostatic pressure in the solid and on the curvature of the surface. Furthermore, since the curvature term $\gamma_{sv}K$ has the units of pressure or stress, it will produce the same effects as an equivalent exter-

Sintering Fundamentals

nally applied pressure. Pressure and curvature effects can therefore be treated by the same formulation. This concept will be used extensively in the next chapter when the sintering models are considered.

7.7 DIFFUSIONAL FLUX EQUATIONS

In order to analyze theoretically the kinetics of sintering in crystalline materials, we must formulate equations for the diffusional flux of matter and solve them subject to the appropriate boundary conditions. There are basically two approaches to the formulation of the flux equations. In one case, the sintering process is viewed in terms of the diffusion of atoms. In the second case, the process is viewed in terms of the diffusion of vacancies.

A. Flux of Atoms

For a pure elemental solid in which the point defects consist of vacancies, diffusion of atoms or vacancies from one region to another does not produce a change in the total number of lattice sites. In a given region, the number of atoms and the number of vacancies change by equal and opposite amounts. The diffusional flux of atoms is determined by gradients in $(\mu_a - \mu_v)$. The flux equation can be written [3]

$$J_a = - \frac{DC_a}{\Omega kT} \nabla(\mu_a - \mu_v) \tag{7.76}$$

where D is the self diffusion coefficient for the atoms, k is the Boltzmann constant, and the other terms have been defined earlier. The diffusion response can therefore be found from Eq. (7.76) subject to the appropriate boundary conditions.

B. Flux of Vacancies

In this approach, the flux of vacancies is considered and the atomic flux is taken as equal and opposite to the vacancy flux, that is,

$$J_a = -J_v \tag{7.77}$$

For the rigid lattice approximation, J_v is given by [3]

$$J_v = -\frac{D_v}{\Omega} \nabla C_v \tag{7.78}$$

where D_v is the vacancy diffusion coefficient and ∇C_v is the gradient in the vacancy concentration. Determination of J_v requires an expression for C_v. Normally C_v is taken as the *equilibrium* vacancy concentration.

At equilibrium,

$$\left(\frac{\partial G}{\partial n_v}\right)_{T,p,N_a} = 0 \tag{7.79}$$

so that $\mu_v = 0$. Assuming that pressure effects are absent so that only the curvature effects are important, Eq. (7.74) can be written

$$\mu_v = \mu_{v0} + \gamma_{sv} K \Omega + kT \ln C_v \tag{7.80}$$

Putting $\mu_v = 0$ gives

$$C_v = C_{v0} \exp\left(-\frac{\gamma_{sv} K \Omega}{kT}\right) \tag{7.81}$$

where K is positive for a convex surface and C_{v0} is a reference value of the vacancy concentration, normally taken as the value under a flat surface. For $\gamma_{sv} K \Omega \ll kT$, Eq. (7.81) becomes

$$C_v = C_{v0}\left(1 - \frac{\gamma_{sv} K \Omega}{kT}\right) \tag{7.82}$$

For the equilibrium situation, we therefore have

$$J_a = \frac{D_v}{\Omega} \nabla C_v \tag{7.83}$$

where C_v is given by Eq. (7.82). It should be remembered that C_v is the fraction of lattice sites occupied by vacancies.

7.8 VAPOR PRESSURE OVER A CURVED SURFACE

In sintering, matter transport by evaporation and condensation is normally treated with the solid state diffusion mechanisms. The rate of transport from a surface is taken as proportional to the equilibrium vapor pressure over the surface. The vapor pressure can be related to the value of $\mu_a - \mu_v$ beneath the surface. Suppose a number dN_a of atoms is removed from the vapor and added to the surface with an accompanying decrease in the number of vacancies beneath the surface. The free energy change for this virtual operation must be zero, so that

$$\mu_{vap} = \mu_a - \mu_v \tag{7.84}$$

where μ_{vap} is the chemical potential of the atoms in the vapor phase. The vapor pressure is proportional to $\exp(\mu_{vap}/kT)$, so that

$$p_{vap} = p_0 \exp\left(\frac{\mu_a - \mu_v - \mu_0}{kT}\right) \tag{7.85}$$

Sintering Fundamentals

where p_0 is a reference value of the vapor pressure corresponding to some standard value of the chemical potential μ_0. Normally p_0 is taken as the value over a flat surface. Using Eqs. (7.74) and (7.75) developed earlier for the chemical potentials of atoms and vacancies beneath a curved surface, we can write

$$\mu_a - \mu_v = \mu_0 + \gamma_{sv} K \Omega \tag{7.86}$$

Substituting into Eq. (7.85) yields

$$p_{\text{vap}} = p_0 \exp\left(\frac{\gamma_{sv} K \Omega}{kT}\right) \tag{7.87}$$

This equation is sometimes referred to as the Kelvin equation. For $\gamma_{sv} K \Omega \ll kT$, it becomes

$$p_{\text{vap}} = p_0 \left(1 + \frac{\gamma_{sv} K \Omega}{kT}\right) \tag{7.88}$$

The vapor pressure over a curved surface relative to that over a flat surface depends on the curvature of the surface.

7.9 DIFFUSION IN IONIC CRYSTALS: AMBIPOLAR DIFFUSION

In most of our discussion so far in this chapter, we have purposely considered the diffusing species to be uncharged atoms (or vacancies) so that any effects of charges on the diffusing species have been avoided. However, the reader will know that in most crystalline inorganic solids, matter transport occurs by the motion of charged species, e.g., ions. We must therefore examine the consequences arising from the charges on the diffusing species.

For compounds containing more that one type of ion, it would be expected that the different ions would have different diffusion rates. However, matter transport must occur in such a way that the charge neutrality of the solid is preserved in the different regions of the solid (Fig. 7.13). Other effects must also be taken into account. If an external electric field is applied to the system, in addition to diffusion down a concentration gradient, the ions will migrate in response to the field which alters $\nabla \mu$. Even in the absence of an external field, the ions themselves can generate an internal field that will influence the motion.

Consider a diffusing species with a charge z_i (i.e., $z_i = +1$ for a sodium ion or $z_i = -2$ for a doubly charged oxygen ion). In a region where the electric potential is ϕ, the chemical potential of an ion is increased by an amount $z_i e \phi$, where e is the magnitude of the charge on an electron.

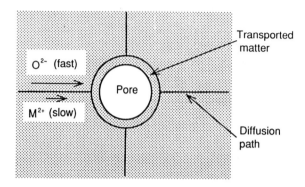

Figure 7.13 Schematic diagram illustrating that the diffusion of ions in an ionic solid must be coupled to preserve the stoichiometry of the matter transported.

From Eq. (7.63) we can write the chemical potential of an ion as

$$\mu_i = \mu_{0i} + kT \ln C_i + z_i e\phi \tag{7.89}$$

where C_i is the fraction of sites occupied by the ion in the crystal. Limiting our discussion to diffusion in one dimension (i.e., the x direction), the flux is given by

$$J_{ix} = -\frac{C_i D'_i}{\Omega_i kT}\frac{d\mu_i}{dx} \tag{7.90}$$

where D' is the diffusion coefficient and the other terms have been defined earlier. At constant temperature,

$$J_{ix} = -\frac{D'_i}{\Omega_i}\frac{dC_i}{dx} - \frac{C_i D'_i z_i e}{\Omega_i kT}\frac{d\phi}{dx} \tag{7.91}$$

Putting

$$D_i = \frac{D'_i}{\Omega_i} \quad \text{and} \quad L_i = \frac{D'_i}{\Omega_i kT} \tag{7.92}$$

Eq. (7.91) becomes

$$J_{ix} = -D_i \frac{dC_i}{dx} - C_i L_i z_i e E \tag{7.93}$$

where $E = d\phi/dx$ is the electric field. The first term on the right-hand side of Eq. (7.93) is the familiar diffusion term given by Fick's first law, and the second term arises from the migration in the electric field.

Sintering Fundamentals

Following the approach of Evans and De Jonghe [7], let us apply Eq. (7.93) to a system consisting of two different types of diffusing ions: one type has a charge z_+ (a positive number) and the other type has a charge z_- (a negative number). This system may correspond to diffusion of metal and oxygen ions in a metal oxide. If no net current passes through the system, the electric current density must be zero, so

$$z_+ J_+ = z_- J_- \tag{7.94}$$

By substituting Eq. (7.93) into Eq. (7.94), we can find an equation for E:

$$-z_+ D_+ \frac{dC_+}{dx} - C_+ L_+ z_+^2 eE = z_- D_- \frac{dC_-}{dx} + C_- L_- z_-^2 eE \tag{7.95}$$

After rearrangement, Eq. (7.95) yields

$$E = -\frac{1}{e(C_+ L_+ z_+^2 + C_- L_- z_-^2)} \left(z_+ D_+ \frac{dC_+}{dx} + z_- D_- \frac{dC_-}{dx} \right) \tag{7.96}$$

There is another condition, i.e., that of electrical neutrality, that must be satisfied. This condition can be expressed as

$$C_+ z_+ = -C_- z_- \tag{7.97}$$

Differentiating this equation and multiplying by D_+ yields

$$z_+ D_+ \frac{dC_+}{dx} = -z_- D_+ \frac{dC_-}{dx} \tag{7.98}$$

Substituting in Eq. (7.96) this equation gives

$$E = -\frac{z_-}{e(-L_+ z_+ C_- z_- + L_- z_- C_- z_-)} (D_- - D_+) \frac{dC_-}{dx} \tag{7.99}$$

Substituting for E in Eq. (7.93) gives for the flux of the negative ions

$$J_- = -\frac{L_- z_- D_+ - L_+ z_+ D_-}{L_- z_- - L_+ z_+} \frac{dC_-}{dx} \tag{7.100}$$

Using Eq. (7.92), we can replace L_i by D_i/kT and Eq. (7.100) becomes

$$J_- = -\frac{D_+ D_- (z_+ - z_-)}{D_+ z_+ - D_- z_-} \frac{dC_-}{dx} \tag{7.101}$$

Finally, applying Eqs. (7.94) and (7.97) to Eq. (7.100) gives

$$J_+ = -\frac{D_+ D_- (z_+ - z_-)}{D_+ z_+ - D_- z_-} \frac{dC_+}{dx} \tag{7.102}$$

By analogy with Fick's first law, we can define an effective diffusion coefficient given by

$$D_{\text{eff}} = \frac{D_+ D_- (z_+ - z_-)}{D_+ z_+ - D_- z_-} \quad (7.103)$$

To illustrate the use of Eq. (7.103), consider the case of Al_2O_3. Assuming that the ions are fully ionized, the diffusing species are Al^{3+} and O^{2-}. The effective diffusion coefficient is

$$D_{\text{eff}} = \frac{5 D_{Al^{3+}} D_{O^{2-}}}{3 D_{Al^{3+}} + 2 D_{O^{2-}}} \quad (7.104)$$

As an illustration of the effect of charge separation, it is instructive to consider two limiting cases. In the first case,

$$\text{If} \quad D_{O^{2-}} \gg D_{Al^{3+}}, \quad \text{then} \quad D_{\text{eff}} = \frac{5}{2} D_{Al^{3+}} \quad (7.105)$$

In the second case,

$$\text{If} \quad D_{Al^{3+}} \gg D_{O^{2-}}, \quad \text{then} \quad D_{\text{eff}} = \frac{5}{3} D_{O^{2-}} \quad (7.106)$$

We see from Eqs. (7.105) and (7.106) that the more slowly diffusing ion determines the rate of matter transport. However, the effect of the faster diffusing species is to accelerate the motion of the slower ion. Physically, the faster diffusing ion, O^{2-}, say, reduces its concentration gradient more rapidly than the slower diffusing ion. However, only a small amount of such diffusion is necessary before a large potential gradient is set up. As illustrated in Fig. 7.14, the potential gradient has the same sign as the

Figure 7.14 Schematic diagram illustrating the mechanism by which the diffusion in an ionic solid (MO) is coupled. It is assumed that the O^{2-} ions diffuse faster than the M^{2+} ions. A small amount of diffusion of the O^{2-} ions sets up a potential gradient that retards the transport of the faster O^{2-} and enhances the transport of the slower M^{2+} ions. The buildup of the potential occurs only to the point where the fluxes are related by Eq. (7.94).

Sintering Fundamentals

concentration gradient. The potential gradient retards the transport of the O^{2-} and enhances the transport of the Al^{3+} by a factor of 2.5, as seen from Eq. (7.105). The buildup of the potential occurs to the point where the fluxes are related by Eq. (7.94).

The coupled diffusion of charged species is referred to as *ambipolar diffusion*. It has important consequences for solid-state sintering as well as for other processes such as the formation of oxide layers on materials. For sintering, the slower (or slowest) diffusing species, as we have observed above, will control the rate of matter transport. However, several mass transport paths may be operating at the sintering temperature (e.g., grain boundary, lattice, and surface diffusion). It is expected that matter transport will occur predominantly along the fastest path. The rate-controlling mechanism is therefore the slowest diffusing species (O^{2-}, say) along the fastest path (the grain boundary, say), i.e., the grain boundary diffusion of O^{2-}, say. An important point is that for the same material at the same temperature the rate-controlling mechanism in sintering may be different from that in other diffusion-controlled processes. A good example is that of ionic conduction, where the fastest diffusing ion normally controls the conductivity.

7.10 CONCLUDING REMARKS

In this chapter we considered a number of issues that are important to an understanding of the sintering process. The concepts introduced will be used extensively in the next chapter dealing with sintering models as well as in subsequent chapters. Matter transport during the sintering of polycrystalline ceramics occurs by diffusion, a thermally activated process. Diffusion can occur along different paths in the solid, giving rise to the different mechanisms of diffusion: lattice, grain boundary, surface, and dislocation pipe diffusion. The rate of diffusion depends on the temperature and on the concentration of defects in the solid. For a given solid, the defect concentration depends on the temperature, the oxygen partial pressure (or atmosphere), and the concentration of dopants or impurities. Matter transport during sintering can be viewed in terms of the flux of atoms (ions) or, equivalently, in terms of the counterflow of vacancies. The flux of the diffusing species is driven by gradients in the chemical potential or by gradients in the vacancy concentration. The chemical potential of atoms or vacancies depends predominantly on the applied pressure and on the curvature of the surface. Pressure and surface curvature effects can be treated within the same formulation. The diffusion of charged species (ions) in inorganic solids is coupled in such a way that the stoichiometry of the compound is maintained. The rate of sintering is controlled by the most slowly diffusing species along the fastest path.

PROBLEMS

7.1 Assuming that a powder has a surface energy of 1 J/m², estimate the maximum amount of energy that is available for densification for particles with a diameter of 1 μm compacted to a green density of 0.60 of the theoretical density. Assume that there is no grain growth during sintering and that the grain boundary energy is 0.5 J/m².

7.2 The measured diffusion coefficient in ZnO is 5.0×10^{-5} cm²/sec at 600°C and 2×10^{-5} cm²/sec at 500°C. What would the diffusion coefficient be at 700°C?

7.3 The self-diffusion coefficient of a metal is 10^{-10} cm²/sec at 800°C. A thin film of a radioactive isotope of the metal is evaporated onto a plane surface of the metal. Assuming that the concentration of the isotope is described by Eq. (7.10)
  (a) Plot the surface concentration as a function of time.
  (b) Show that the average diffusion distance of the isotope after an annealing time t is $(D^*t)^{1/2}$.
  (c) Calculate the average diffusion distance after 10 hours annealing at 800°C.

7.4 The densification of a metal oxide is controlled by the diffusion of oxygen vacancies. The oxide has a native defect structure of oxygen vacancies. Develop the appropriate equations for defect equilibria and show how the densification rate can be increased.

7.5 Nickel oxide, NiO, is metal deficient with cation vacancies predominating. If it is doped with lithium, which goes into solid solution on the regular Ni lattice sites, develop the equations for the defect equilibria and construct the Brouwer diagram.

7.6 A pure oxide MO forms predominantly Schottky defects at the stoichiometric composition. It is oxygen deficient under reducing conditions and metal deficient under oxidizing conditions.
  (a) Develop the equations for the defect equilibria and construct the Brouwer diagram.
  (b) If the oxide is doped with acceptor or donor dopants, develop the new equations for the defect equilibria and construct the Brouwer diagram.

7.7 Estimate the difference in equilibrium vapor pressure for two spherical particles, one with a diameter of 0.1 μm and the other with a diameter of 10 μm, assuming a temperature of 500°C, an atomic radius of 10^{-10} m and a surface energy of 1 J/m². Discuss what would happen if the two particles were placed in the same closed box at 500°C.

Sintering Fundamentals

7.8 Assuming that Schottky defects predominate, derive the expression for the effective diffusion coefficient for ambipolar diffusion in Al_2O_3.

REFERENCES

1. R. L. Coble and R. M. Cannon, in *Processing of Crystalline Ceramics, Mater. Sci. Res.*, Vol. 11, (H. Palmour III, R. F. Davis, and T. M. Hare, Eds.), Plenum, New York, 1978, pp. 151–170.
2. P. G. Shewmon, *Diffusion in Solids*, The Metals Society, Warrendale, PA, 1989.
3. R. E. Howard and A. B. Lidiard, *Rep. Prog. Phys.*, 27: 161 (1964).
4. J. Crank, *The Mathematics of Diffusion*, 2nd ed., Oxford Univ. Press, 1975.
5. P. Kofstad, *Nonstoichiometry, Diffusion and Electrical Conductivity in Binary Metal Oxides*, Krieger, Malabar, FL, 1983.
6. C. Herring, in *The Physics of Powder Metallurgy*, (W. Kingston, Ed.), McGraw-Hill, New York, 1951, pp. 143–179.
7. J. W. Evans and L. C. De Jonghe, *The Production of Inorganic Materials*, Macmillan, New York, 1991, p. 190.

8
Theory of Solid-State and Viscous Sintering

8.1 INTRODUCTION

The reader will recall from Chapter 1 the basic types of sintering processes: solid-state sintering, liquid-phase sintering, viscous sintering, and vitrification. This chapter is devoted to a theoretical analysis of the densification process in solid-state sintering and viscous sintering. The analysis of the densification process, by itself, is not very useful. We must also understand how the microstructure of the powder system evolves during sintering. Microstructural evolution during solid-state sintering forms the subject of Chapter 9. To complete the picture of the basic theory and principles of sintering, liquid-phase sintering and vitrification are considered in Chapter 10.

The analysis of viscous sintering appears relatively simple in principle. Matter transport occurs by a viscous flow mechanism. The path along which matter flows is not specified explicitly. Instead, it is implicitly assumed to be the shortest path. The equations for matter transport are derived on the basis of an energy balance. The models that have been developed to approximate the complex geometry of the real powder system yield satisfactory results.

Compared with viscous sintering, the sintering phenomena in polycrystalline materials are considerably more complex because of the crystalline nature of the grains and the presence of grain boundaries. Matter transport during sintering can occur by at least six different paths that define the mechanisms of sintering. Some mechanisms (referred to as densifying mechanisms) lead to densification of the powder system, whereas others (the nondensifying mechanisms) do not. In practice, more than one mechanism operates during any given regime of sintering. All of the mechanisms lead to growth of the necks between the particles and

Theory of Solid-State and Viscous Sintering

so influence the densification rate. We must therefore understand the separate mechanisms as well as their interaction.

Perhaps the most important consequence of the grain boundaries is the occurrence of grain growth and pore growth during sintering, a process normally referred to as coarsening. The coarsening process provides an alternative route by which the free energy of the powder system can be reduced without densification. It reduces the driving force for densification. The interplay between the two processes is sometimes referred to as a competition between sintering (densification) and coarsening. We shall consider this competition in Chapter 9.

A comprehensive theory of sintering should be capable of describing the entire sintering process as well as the evolution of the microstructure (i.e., grain and pore sizes and their distributions). However, in view of the complexity of the process, it is unlikely that such a theory will be developed. A more realistic approach, which is adopted in this book, is to first develop an understanding of the various phenomena separately and then explore the consequences of their interaction.

A few different approaches have been used to analyze the densification process. The analytical models have received the most attention and provide the basis for the present understanding of sintering. The microstructure of the powder system is approximated by a relatively simple geometrical model, and analytical expressions are derived for the sintering rate as a function of the primary variables such as powder particle size, sintering temperature, and applied pressure. The scaling laws do not assume a specified geometrical model; instead they predict in a general way the dependence of the sintering rate on the change of scale (i.e., particle size) of the powder system. Other approaches are potentially useful but have not achieved the popularity of the analytical models. These approaches include the use of numerical simulations, topological models, and statistical models. Phenomenological equations and sintering maps attempt to represent sintering data in terms of equations or diagrams but provide very little insight into the process.

The theoretical approaches to sintering have been criticized from time to time because the drastic simplifications assumed in the models make them unsuitable for quantitatively predicting the sintering behavior of real powder systems. Normally, the models assume uniform packing of spherical particles of the same size, the occurrence of a single mass transport mechanism, and no grain growth. At best, they provide only a qualitative understanding of sintering. However, in spite of these shortcomings, the role of the theoretical models in the development of our understanding of sintering should not be overlooked.

8.2 SINTERING OF POLYCRYSTALLINE AND AMORPHOUS MATERIALS: A PREVIEW

Prior to our analysis of the sintering of polycrystalline and amorphous materials, it is worth understanding some of the fundamental differences between the sintering characteristics of these two types of materials.

A. Sintering Mechanisms

Polycrystalline materials sinter by diffusional transport of matter whereas amorphous materials sinter by viscous flow. In polycrystalline materials, matter transport takes place along definite paths that define the mechanisms of sintering. We will recall that matter is transported from regions of higher chemical potential (referred to as the *source* of matter) to regions of lower chemical potential (referred to as the *sink*). There are at least six different mechanisms of sintering in polycrystalline materials (Fig. 8.1). All of these lead to growth of necks between the particles. Neck growth produces bonding between the particles, so the strength of the consolidated powder form increases during sintering. Only certain of the mechanisms, however, lead to shrinkage or densification. In these so-called densifying mechanisms, matter is removed from the grain boundaries (mechanisms 4 and 5) or from dislocations within the neck region (mechanism 6). The mechanisms that do not cause densification (mechanisms 1–3) are sometimes referred to as nondensifying mechanisms. The nondensifying mechanisms cannot simply be ignored. When they occur, they reduce the curvature of the neck surface (i.e., the driving force for sintering) and so reduce the rate of the densifying mechanisms.

In amorphous materials, viscous flow leads to neck growth as well as to densification. However, the path by which matter flows is not clearly defined as for polycrystalline materials. The geometrical changes that accompany viscous flow are fairly complex. As we shall see later, severe simplifying assumptions are made in the development of the equations for matter transport. For the sintering of spheres, Fig. 8.2a shows a schematic of two possible flow fields for matter transport [1]. While the form shown on the left-hand side may be expected in real systems, recent simulations (Fig. 8.2b) show nearly vertical translation of most of the sphere (as on the right-hand side of Fig. 8.2a) but most of the energy dissipation occurs near the neck. Table 8.1 summarizes the sintering mechanisms in polycrystalline and amorphous solids.

B. Flux Equations

Since all of the sintering mechanisms produce neck growth, we can use the rate of increase of the neck width as a measure of the sintering rate

Theory of Solid-State and Viscous Sintering

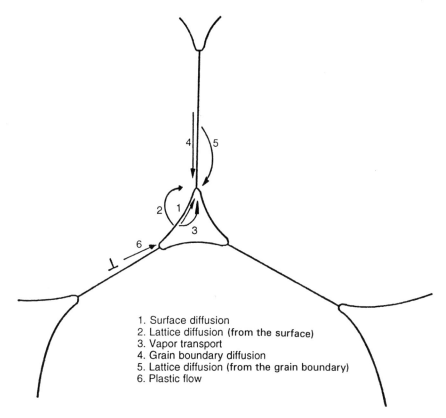

Figure 8.1 Six distinct mechanisms can contribute to the sintering of a consolidated mass of crystalline particles: (1) surface diffusion, (2) lattice diffusion from the surface, (3) vapor transport, (4) grain boundary diffusion, (5) lattice diffusion from the grain boundary, and (6) plastic flow. Only mechanisms 1–3 lead to densification, but all cause the necks to grow and so influence the rate of densification.

and also for comparing the relative rates of the different mechanisms. As we shall see later, the measurement of neck growth in model systems (e.g., two spheres, a sphere on a flat plate, and two wires) was used extensively in the early development of sintering for checking the validity of the theoretical predictions. For the densifying mechanisms we can, in addition, use the rate of shrinkage (or densification) as a measure of the sintering rate.

In Chapter 7 we outlined the basic ideas in the formulation of the flux equations for diffusive transport of matter. The reader will recall that the flux of matter or the counterflow of vacancies can be employed in the

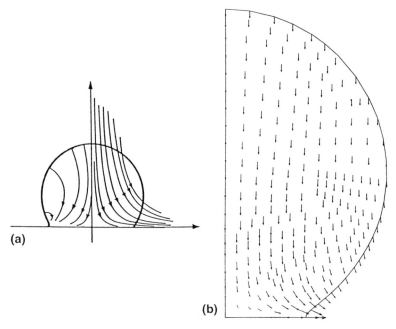

Figure 8.2 Densification of glass powders occurs by viscous flow. The matter transport paths are not clearly defined as for crystalline powders. (a) Two possible flow fields for viscous sintering of spheres are illustrated. (From Ref. 1.) (b) Simulations by finite-element analysis indicate that the flow field on the right-hand side is more realistic. (From Ref. 37, with permission.)

analysis. The flux of matter or the counterflow of vacancies is driven by the gradient in the chemical potential between the source and the sink. A flux equation is formulated for a specified path (i.e., mechanism) and solved subject to the appropriate boundary conditions.

The formulation of the equations for viscous sintering is quite different from that for the diffusion mechanisms. The transport path is assumed implicitly to be the shortest path. A drastic simplification of the structure is used, and the flow of matter is assumed to be governed by an energy balance concept first put forward by Frenkel [2]. When viscous flow occurs, energy is dissipated. Frenkel suggested that the rate of sintering (or more generally the strain rate) could be determined by equating the rate of dissipation of energy by viscous flow to the rate of change of energy due to the reduction in surface area. To summarize, Frenkel's energy

Theory of Solid-State and Viscous Sintering

Table 8.1 Mechanisms of Sintering in Polycrystalline and Amorphous Solids

Type of solid	Mechanism	Source of matter	Sink of matter	Densifying	Nondensifying
Polycrystalline	Surface diffusion	Surface	Neck		X
	Lattice diffusion	Surface	Neck		X
	Vapor transport	Surface	Neck		X
	Grain boundary diffusion	Grain boundary	Neck	X	
	Lattice diffusion	Grain boundary	Neck	X	
	Lattice diffusion	Dislocations	Neck	X	
Amorphous	Viscous flow	Unspecified	Unspecified	X	

balance concept is stated as

$$\text{Rate of energy dissipation by viscous flow} = \text{rate of energy gained by reduction in surface area} \quad (8.1)$$

As we shall see later, the equations based on this energy balance concept have been very successful in describing the sintering kinetics of amorphous materials.

C. Significance of Grain Boundaries

The presence of grain boundaries, as noted earlier, leads to profound effects in the sintering of polycrystalline materials. Compared to the sintering of amorphous materials, the analysis and understanding of the phenomena are considerably more complex.

The grain boundary has a specific energy called the grain boundary energy γ_{gb}. During sintering, part of the energy decrease resulting from a decrease in the free surface area goes into the creation of new grain boundary area (Fig. 8.3). If ΔA_{sv} and ΔA_{gb} are the changes in the surface area and grain boundary area, respectively, then the total change in energy is

$$\Delta E = \gamma_{sv} \Delta A_{sv} + \gamma_{gb} \Delta A_{gb} \quad (8.2)$$

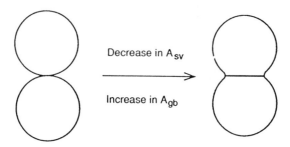

Figure 8.3 In the sintering of crystalline powders, a portion of the energy gained by reduction in surface area is used up in the formation of the grain boundaries.

where γ_{sv} is the specific surface energy of the solid-vapor interface. (In this equation ΔA_{sv} is negative because the free surface area decreases.) The driving force for sintering will therefore be somewhat lower than that calculated in Chapter 7 [see Eq. (7.3)] when the grain boundaries were neglected.

Under certain conditions, ΔE in Eq. (8.2) may actually be positive. This occurs when

$$|\gamma_{gb} \Delta A_{gb}| > |\gamma_{sv} \Delta A_{sv}| \qquad (8.3)$$

For the conditions described by Eq. (8.3), there would be an incentive for the solid-vapor surface area to increase, i.e., for the pores to grow. This means that the density of the powder system will decrease. The theoretical models for densification normally assume a uniformly packed powder system in which the pores decrease in size. We therefore postpone an analysis of pore growth until the next chapter when we consider the interactions of pores with grain boundaries. As we shall describe in detail, the conditions for pore growth normally become favorable when the pore size is significantly greater than the grain size. To summarize at this stage, for amorphous materials it is thermodynamically feasible for all of the pores to shrink. However, for polycrystalline materials, it may not be thermodynamically feasible for the pores to shrink under certain conditions.

The presence of the grain boundaries also dictates the equilibrium shapes of the pores in a polycrystalline material. Consider a hypothetical pore surrounded by three grains in Fig. 8.4. The forces must balance at the junction where the surfaces of the pores meet the grain boundary. They are normally represented by the tension in the interface, i.e., the tension in the solid/vapor interface and the tension in the grain boundary. By analogy with the surface tension of a liquid, a tension arises because

Theory of Solid-State and Viscous Sintering

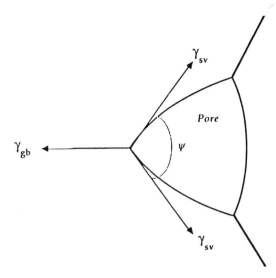

Figure 8.4 The equilibrium shapes of the pores in polycrystalline solids are governed by the balance between the surface and interfacial forces at the point where the grain boundary intersects the pore. γ_{sv} is the surface tension, γ_{gb} is the grain boundary tension, and ψ is the dihedral angle.

an increase in the area of the interface leads to an increase in energy. At the junction, the tension in the solid/vapor interface is tangential to that interface, whereas the tension in the grain boundary is in the plane of the boundary. The balance of forces therefore leads to

$$\gamma_{gb} = 2\gamma_{sv}\cos(\psi/2) \tag{8.4}$$

where ψ is called the dihedral angle.

Perhaps one of the most important consequences of the presence of the grain boundaries is that the growth of the grains provides a process by which the powder system can decrease its energy (by reducing the total grain boundary area). Grain growth is normally accompanied by pore growth, the overall process being described as coarsening. Coarsening therefore occurs concurrently with sintering. The analysis of sintering with concurrent grain growth is fairly complex. A few attempts have been made to develop models with simultaneous densification and grain growth, but they have failed to provide any new insight because the physics of the process is lost in the complex calculations. The usual approach is to analyze sintering and grain growth separately. The understanding gained from the separate analyses is then used to explore the consequences of their interaction.

8.3 THEORETICAL ANALYSIS OF SINTERING

Various approaches have been used in the theoretical analysis of sintering. They are summarized in Table 8.2. The development of the analytical models, starting about 1945–1950, represents the first real attempt at a quantitative modeling of the sintering process. The models suffer from severe deficiencies in the description of the sintering of real powder systems. However, in spite of these deficiencies, they provide the basis for the present understanding of sintering. The analytical models will therefore receive the most attention in our treatment. The scaling laws, formulated at about the same time as the early analytical models, have received relatively little attention in the sintering literature. However, they may provide one of the most reliable guides to the understanding of the sintering mechanisms. The numerical simulation approach has a number of attractive features such as the analysis of complex geometries and the occurrence of concurrent mechanisms. However, the results of many analyses cannot be easily visualized or applied. The topological models make limited predictions of the sintering kinetics and are more appropriate to the understanding of the evolution of the microstructure. They will be

Table 8.2 Main Approaches Used in the Theoretical Analysis of Sintering

Approach	Comments	Ref.
Scaling laws	Not dependent on specific geometry. Effects of change of scale on the rate of single mechanism derived.	5
Analytical models	Greatly oversimplified geometry. Analytical equations for dependence of sintering rate on primary variable derived for single mechanism.	7–11
Numerical simulations	Equations for matter transport solved numerically. Complex geometry and concurrent mechanisms analyzed. Results not easily visualized.	19, 20
Topological models	Analysis of morphological changes. Predictions of kinetics limited. More appropriate to microstructural evolution.	3
Statistical models	Statistical methods applied to the analysis of sintering. Simplified geometry. Semiempirical analysis.	4
Phenomenological equations	Empirical or phenomenological derivation of equations to describe sintering data. No reasonable physical basis.	22, 23

Theory of Solid-State and Viscous Sintering

outlined in the next chapter. The statistical models have received almost no attention since they were originally put forward. In view of the limited space available in this chapter, we shall not consider them any further. Finally, the phenomenological equations are used to fit sintering data but add almost no insight into the process.

8.4 SCALING LAWS

The scaling laws were formulated by Herring [5] in 1950. They consider the effect of change of scale on microstructural phenomena during sintering. For a consolidated powder form, the most fundamental scaling parameter is the particle size. The scaling laws attempt to answer the important question: How does the change in scale (i.e., the particle size) influence the rate of sintering?

The scaling laws do not assume a specific geometrical model. Instead, the main assumptions in the model are that during sintering (i) the particle size of any given powder system remains the same and (ii) the geometrical changes remain similar. Two systems are defined as being geometrically similar if the linear dimension of each of the features (e.g., grains, pores) of one system is equal to a numerical factor times the linear dimension of the corresponding feature in the other system. We can summarize this as follows:

$$(\text{Linear dimension})_1 = \lambda (\text{linear dimension})_2 \tag{8.5}$$

where λ is a numerical factor. Geometrically similar systems therefore involve simply a magnification of one system relative to the other. As an example, two geometrically similar systems consisting of a random arrangement of circles are shown in Fig. 8.5.

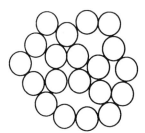

Figure 8.5 An example of two geometrically similar systems consisting of a random arrangement of circles. The systems differ only in scale and involve a simple magnification of one relative to the other.

A. Derivation of the Scaling Laws

To illustrate the derivation of the scaling laws, let us now consider a simple system consisting of two spheres in contact (Fig. 8.6). We are not restricted to this geometry. It is chosen to simplify the illustration. Suppose it takes a time Δt_1 to produce a certain microstructural change (e.g., the growth of a neck to a certain size X_1) in system 1. The question we must attempt to answer is, How long (Δt_2) does it take to produce a geometrically similar change in system 2? For geometrically similar changes, the initial radius of the particle and the neck radius of the two systems are related by

$$R_2 = \lambda R_1; \quad X_2 = \lambda X_1 \tag{8.6}$$

The time taken to produce a certain change by diffusional flow of matter can be expressed as

$$\Delta t = V/JA \tag{8.7}$$

where V is the volume of matter transported, J is the flux, and A is the cross-sectional area over which the matter is transported. We can therefore write

$$\frac{\Delta t_2}{\Delta t_1} = \frac{V_2 J_1 A_1}{V_1 J_2 A_2} \tag{8.8}$$

As an example of the application of Eq. (8.8), consider matter transport by lattice diffusion.

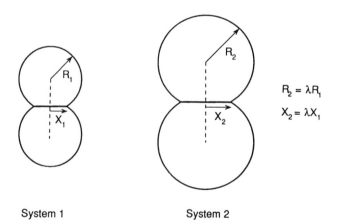

Figure 8.6 Geometrically similar models consisting of two spheres in contact. The linear dimensions of system 2 are a factor λ times those of system 1.

Theory of Solid-State and Viscous Sintering

Scaling Law for Lattice Diffusion

The volume of matter transported is proportional to R^3, where R is the radius of the sphere (see Fig. 8.6). Therefore V_2 is proportional to $(\lambda R)^3$, or $V_2 = \lambda^3 V_1$. For lattice diffusion, the area over which matter diffuses is proportional to R^2. Therefore A_2 is proportional to $(\lambda R)^2$, or $A_2 = \lambda^2 A_1$. The flux J, as discussed in the previous chapter, is proportional to $\nabla \mu$, the gradient in the chemical potential. For a curved surface with a radius of curvature r, μ varies as $1/r$. Therefore J varies as $\nabla(1/r)$ or as $1/r^2$. Now J_2 is proportional to $1/(\lambda r)^2$, so $J_2 = J_1/\lambda^2$. Therefore to summarize the parameters for lattice diffusion:

$$V_2 = \lambda^3 V_1; \quad A_2 = \lambda^2 A_1; \quad J_2 = \frac{J_1}{\lambda^2} \tag{8.9}$$

Using Eq. (8.8), we obtain

$$\frac{\Delta t_2}{\Delta t_1} = \lambda^3 = \left(\frac{R_2}{R_1}\right)^3 \tag{8.10}$$

According to Eq. (8.10), the time taken to produce geometrically similar changes by a lattice diffusion mechanism increases as the cube of the particle size.

Scaling Laws for Other Sintering Mechanisms

The scaling laws for the other mass transport mechanisms can be derived using a procedure similar to that described above for volume diffusion. The derivations are described in detail by Herring [5]. Here, we provide a summary of the scaling laws. The laws can be written in the general form

$$\left(\frac{\Delta t_2}{\Delta t_1}\right) = \left(\frac{R_2}{R_1}\right)^m = \lambda^m \tag{8.11}$$

where m is an exponent that depends on the mechanism of sintering. Table 8.3 gives the values of m for the various sintering mechanisms.

B. Application of the Scaling Laws

An important application of the scaling laws is the determination of how the relative rates of sintering by the different mechanisms depend on the particle size of the system. This type of information is very useful in the fabrication of ceramics with controlled microstructure. As we observed earlier, some mechanisms lead to densification while others do not. The achievement of high density, for example, requires that the rates of the

Table 8.3 Exponents for Herring's Scaling Laws Described by Eq. (8.11).

Sintering mechanism	Exponent m
Surface diffusion	4
Lattice diffusion	3
Vapor transport	2
Grain boundary diffusion	4
Plastic flow	1
Viscous flow	1

densifying mechanisms be enhanced over those for the nondensifying mechanisms. The laws can also be used to develop general sintering equations that are not limited by geometrical details of the model.

Relative Rates of Sintering Mechanisms

To determine the relative rates of the different mechanisms, it is more useful to write Eq. (8.11) in terms of a rate. For a given change, the rate is inversely proportional to the time, so Eq. (8.11) can be written

$$\frac{(\text{Rate})_2}{(\text{Rate})_1} = \lambda^{-m} \qquad (8.12)$$

In a given powder system, let us suppose that grain boundary diffusion and vapor transport (evaporation/condensation) are the dominant mass transport mechanisms. Then the rates of sintering by these two mechanisms vary with the scale of the system according to

$$(\text{Rate})_{gb} \sim \lambda^{-4} \qquad (8.13)$$

and

$$(\text{Rate})_{ec} \sim \lambda^{-2} \qquad (8.14)$$

The variation of the rates of sintering with λ for the two mechanisms is illustrated schematically in Fig. 8.7. We see that for small λ, i.e., as the particle size becomes smaller, the rate of sintering by grain boundary diffusion is enhanced compared to that for vapor transport. Conversely, the rate of sintering by vapor transport dominates for larger λ, i.e., for larger particle sizes. Therefore, the scaling laws predict that smaller particle size is beneficial for densification when grain boundary diffusion and vapor transport are the dominant mechanisms. For the case where surface diffusion and lattice diffusion are the dominant mechanisms, we can use a similar procedure to show that surface diffusion is enhanced as the particle size decreases. It is left as an exercise for the reader to consider

Theory of Solid-State and Viscous Sintering

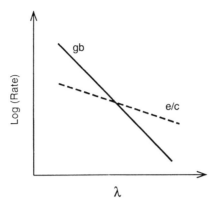

Figure 8.7 Schematic diagram of the relative rates of sintering by grain boundary diffusion and by evaporation/condensation as a function of the scale (i.e., particle size) of the system.

other combinations, for example, (i) lattice diffusion versus grain boundary diffusion and (ii) surface diffusion versus grain boundary diffusion.

C. Sintering Equation Based on the Scaling Laws

Without making any specific assumptions about the geometry of the powder system, we can use the scaling laws and simple physical reasoning to write a sintering equation for the dependence of the sintering rate on particle size and temperature. The effect of particle size is given by Eq. (8.12). For the diffusion mechanisms, the rate of matter transport (i.e., the flux) is also proportional to D/kT, where D is the diffusion coefficient, k is the Boltzmann constant, and T is the absolute temperature. Therefore we can write

$$\text{Sintering rate} = \frac{AD}{kTG^m} \quad (8.15)$$

where A is an unknown constant that contains the appropriate geometrical and physical parameters, G is the particle (or grain) size, and m is the scaling law exponent given in Table 8.3. The diffusion coefficient D has an Arrhenius temperature dependence given by Eq. (7.11).

For viscous sintering, the term D/kT can be replaced by $1/\eta$, where η is the viscosity of the glass, i.e.,

$$\text{Sintering rate} = B/\eta G \quad (8.16)$$

where B is an unknown constant and G is the particle size. The viscosity has a temperature dependence given by an equation of the Vogel–Fulcher form:

$$\eta = \eta_0 \exp\left(\frac{C}{T - T_0}\right) \qquad (8.17)$$

where η_0, C, and T_0 are constants and T is the temperature. Over a fairly narrow range of temperature (50–100°C), η is adequately described by the Arrhenius equation

$$\eta = \eta_0 \exp\left(\frac{Q}{RT}\right) \qquad (8.18)$$

where Q is the activation energy and R is the gas constant.

D. Scaling Laws: Final Comments

Due to the general approach and the simple physical principles employed in the derivation, the scaling law approach has several advantages over the analytical models. Because the geometric details of the powder system did not enter into the derivation, the laws can be applied to particles of any shape and to all stages of the sintering process. However, we must remember the conditions that govern the validity of the laws. The derivation assumed that the particle size of each powder system did not change during sintering. Furthermore, the microstructural changes were assumed to remain geometrically similar in the two systems. The two systems must also be identical in chemical composition so that the mass transport coefficients (e.g., the diffusion coefficients) are the same.

Since the exponent m in Eq. (8.11) or (8.12) depends on the mechanism of sintering, it may seem that the measurement of m would provide information on the mechanism of sintering. In practice, several factors can complicate the task of determining the mechanism. A major factor is the simultaneous occurrence of more than one mechanism. In such a case, the measured exponent may correspond to a mechanism that is entirely different from those operating. For real powder systems, it is normally difficult to maintain the geometrically similar microstructure assumed in the derivation of the scaling laws. Furthermore, the sintering mechanism may change with the size of the particles. While the scaling laws appear to be obeyed reasonably well in simple metallic systems such as nickel wires, copper spheres, and silver spheres, the application of the laws to the sintering of Al_2O_3 produced unexpected results [6]. For powders with a narrow size distribution, high nonintegral values of the exponent m were found. The value of the exponent also varied with the extent of densification, especially in the early stages of sintering.

Theory of Solid-State and Viscous Sintering

To summarize, the use of the scaling laws for the determination of the mechanism of sintering does not appear to be a worthwhile exercise. The laws are, however, extremely useful for understanding the particle size dependence of the sintering mechanisms. They are also extremely useful for determining how the relative rates of the different mechanisms are influenced by the particle size.

8.5 ANALYTICAL MODELS

In the previous section, general sintering equations were developed on the basis of the scaling laws and simple physical reasoning, e.g., Eqs. (8.15) and (8.16). These general equations do not allow us to predict the rate of sintering because they are not exact. They simply predict how the sintering rate varies with particle size and temperature. To obtain more exact equations, sintering models with specified geometrical details must be adopted and the flux equations must be solved under the appropriate boundary conditions. The analytical models assume a relatively simple, idealized geometry, and, for each mechanism, the mass transport equations are solved analytically to provide equations for the sintering kinetics.

Sintering is a continuous process. However, the microstructure of a consolidated powder form changes drastically during sintering. A single model cannot adequately represent the entire process yet still provide the degree of simplicity for the mass transport equations to be solved analytically. It is therefore convenient and useful to divide the process into separate stages. Furthermore, for each stage an idealized geometrical model that has a rough similarity to the microstructure of the powder system is assumed.

A. Stages of Sintering

Sintering is normally thought to occur in three sequential stages referred to as (i) the initial stage, (ii) the intermediate stage, and (iii) the final stage. In some cases an extra stage, stage 0, is included. Stage 0 describes the instantaneous neck formation when the particles are first placed in contact. However, we shall not include this refinement. A stage represents an interval of time or density over which the microstructure is considered to be reasonably well defined. For polycrystalline materials, Fig. 8.8 shows the idealized geometrical structures that are assumed to be representative of the three stages [7]. For amorphous materials, the geometrical structure assumed for the intermediate stage is very different from that of Fig. 8.8c and will be described at the appropriate place later.

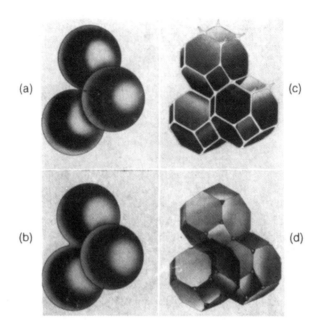

Figure 8.8 (a) Initial stage of sintering; model structure represented by spheres in tangential contact. (b) Near the end of the initial stage. Spheres have begun to coalesce. The neck growth illustrated is for center-to-center shrinkage of 4%. (c) Intermediate stage; dark grains have adopted the shape of a tetrakaidecahedron, enclosing white pore channels at the grain edges. (d) Final stage; pores are tetrahedral inclusions at the corners where four tetrakaidecahedra meet. (From Ref. 7.)

Initial Stage
The initial stage consists of fairly rapid interparticle neck growth by diffusion, vapor transport, plastic flow, or viscous flow. For a powder system consisting of spherical particles, it is represented as the transition between Figs. 8.8a and 8.8b. Shrinkage (or densification) accompanies neck growth for the densifying mechanisms. The large initial differences in surface curvature are removed in this stage. The initial stage is assumed to last until the radius of the neck between the particles has reached a value of ≈0.4–0.5 of the particle radius. For a powder system with an initial density of ≈0.5–0.6 of the theoretical density, this corresponds to a linear shrinkage of 3–5% or an increase in density to ≈0.65 of the theoretical when the densifying mechanisms operate.

Theory of Solid-State and Viscous Sintering

Intermediate Stage

The intermediate stage begins when the pores have reached their equilibrium shapes as dictated by the surface and interfacial tensions (see Section 8.2C). The pore phase is still continuous. An example of the microstructure of a real powder system in the intermediate stage is shown in Fig. 8.9. In the sintering models, the structure is usually idealized in terms of a spaghetti-like array of porosity sitting along the grain edges as illustrated in Fig. 8.8c. Densification is assumed to occur by the pores simply shrinking to reduce their cross section. Eventually the pores become unstable and pinch off, leaving isolated pores; this constitutes the beginning of the final stage. The intermediate stage normally covers the major part of the sintering process; it is taken to end when the density is ≈0.9 of the theoretical.

Final Stage

The microstructure in the final stage can develop in a variety of ways, and we shall consider the subject in detail in Chapter 9. In one of the simplest descriptions, the final stage begins when the pores pinch off and become isolated at the grain corners. The idealized structure is shown in Fig. 8.8d. In this simple description, the pores are assumed to shrink continuously and may disappear altogether. As we outlined in Chapter 1, the removal of almost all of the porosity has been achieved in the sintering of some real powder systems.

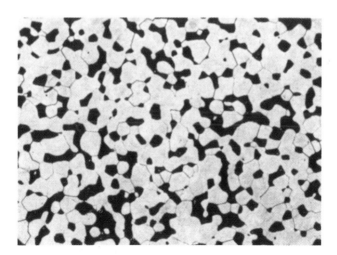

Figure 8.9 Typical intermediate stage structure; alumina gel sintered for 12 h at 1750°C (magnification 250X). (From Ref. 7.)

Some of the main parameters associated with the description of the stages of sintering are summarized in Table 8.4.

B. Modeling the Sintering Process

As outlined earlier, in order to develop simple sintering equations for the various mechanisms, it is necessary to idealize the structure of the powder system in some way to allow it to be analyzed mathematically. In the analytical models, the particles are assumed to be spherical and of the same size. Furthermore, the consolidated powder form is assumed to have a uniform packing arrangement. As we shall see later, these assumptions neglect two important issues in real powder compacts: inhomogeneity of the packing and growth of the grains during sintering.

With these assumptions, a unit of the powder structure can be isolated and analyzed. Furthermore, with the appropriate boundary conditions, the remainder of the powder system can be considered as a continuum having the same macroscopic properties (e.g., shrinkage and densification rate) as the isolated unit. The term *geometrical model* (or simply *model*) is used to refer to the unit. The derivation of the equations for the sintering kinetics follows a simple procedure. For the assumed model, the mass transport equations are formulated and solved under the appropriate boundary conditions.

Table 8.4 Main Parameters Associated with the Stages of Sintering for Polycrystalline Solids

Stage	Typical microstructural feature	Relative density range	Idealized model
Initial	Rapid interparticle neck growth	Up to ≈ 0.65	Two monosize spheres in contact
Intermediate	Equilibrium pore shape with continuous porosity	≈ 0.65–0.90	Tetrakaidecahedron with cylindrical pores of the same radius along the edges
Final	Equilibrium pore shape with isolated porosity	$> \approx 0.90$	Tetrakaidecahedron with spherical monosize pores at the corners

C. Initial Stage Models

Geometrical Parameters

The model for the initial stage consists of two equal-sized spheres in contact, referred to as the two-sphere model. It is normal to consider two slightly different geometries, depending on whether the mechanisms are nondensifying (Fig. 8.10a) or densifying (Fig. 8.10b). The two-sphere

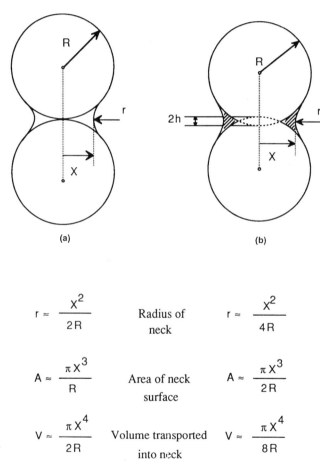

$r \approx \dfrac{X^2}{2R}$	Radius of neck	$r \approx \dfrac{X^2}{4R}$
$A \approx \dfrac{\pi X^3}{R}$	Area of neck surface	$A \approx \dfrac{\pi X^3}{2R}$
$V \approx \dfrac{\pi X^4}{2R}$	Volume transported into neck	$V \approx \dfrac{\pi X^4}{8R}$

Figure 8.10 Geometrical parameters for the two-sphere model used in the derivation of the initial stage sintering equations for crystalline particles. The geometries shown correspond to those for (a) the nondensifying mechanisms and (b) the densifying mechanisms.

model for the densifying mechanisms accounts for interpenetration of the spheres (i.e., shrinkage) as well as neck growth. The main geometrical parameters of the model are the principal radii of curvature of the neck surface, r and X, the area of the neck surface, A, and the volume of material transported into the neck, V. These parameters are summarized in Fig. 8.10, and it is left as an exercise for the reader to derive them. It will be noticed that the parameters for the densifying model (Fig. 8.10b) differ by a small numerical factor compared to those for the nondensifying model (Fig. 8.10a). The neck is assumed to be circular with a radius X and with a principal radius of curvature r. In most of the derivations the grain boundary energy is neglected. This means that r is the same over the section of the neck.

Kinetic Equations

Matter transport by diffusion, as described in Chapter 7, can be analyzed in terms of the flux of atoms or the counterflow of vacancies. In the early development of sintering theory, the approach based on the counterflow of vacancies was used most often. We will start with this approach in which the movement of the vacancies is driven by a vacancy concentration gradient. In Section 8.12 we will outline a more general approach based on the flux of atoms driven by a chemical potential gradient. As an illustration of the derivation of the sintering equations, let us start with the mechanism of grain boundary diffusion.

Grain Boundary Diffusion. According to Eq. (7.83), the flux of atoms into the neck is

$$J_a = \frac{D_v}{\Omega} \nabla C_v \tag{8.19}$$

where D_v is the vacancy diffusion coefficient, Ω is the volume of an atom or vacancy, and ∇C_v is the gradient of the vacancy concentration. (As outlined in Chapter 7, C_v is the fraction of sites occupied by vacancies.) The volume of matter transported into the neck per unit time is

$$\frac{dV}{dt} = J_a A_{gb} \Omega \tag{8.20}$$

where A_{gb} is the cross-sectional area over which diffusion occurs. Grain boundary diffusion is assumed to occur over a constant thickness δ_{gb} so that $A_{gb} = 2\pi X \delta_{gb}$, where X is the radius of the neck. Combining Eqs. (8.19) and (8.20) and substituting for A_{gb} gives

$$\frac{dV}{dt} = D_v 2\pi X \delta_{gb} \nabla C_v \tag{8.21}$$

Theory of Solid-State and Viscous Sintering

Since the neck radius increases radially in a direction orthogonal to a line joining the centers of the spheres, only a one-dimensional solution is necessary. The vacancy concentration gradient between the neck surface and the center of the neck is assumed to be constant so that $\nabla C_v = \Delta C_v / X$, where ΔC_v is the change in the vacancy concentration. Furthermore, the vacancy concentration at the center of the neck is assumed to be equal to that under a flat, stress-free surface. According to Eq. (7.82) we can write

$$\Delta C_v = C_v - C_{v0} = \frac{C_{v0}\gamma_{sv}\Omega}{kT}\left(\frac{1}{r_1} + \frac{1}{r_2}\right) \tag{8.22}$$

where r_1 and r_2 are the two principal radii of curvature of the neck surface. From Fig. 8.10, $r_1 = r$ and $r_2 = -X$, and it is assumed that $X \gg r$. After substituting for ∇C_v, Eq. (8.21) becomes

$$\frac{dV}{dt} = \frac{2\pi D_v C_{v0}\delta_{gb}\gamma_{sv}\Omega}{kTr} \tag{8.23}$$

Using the relations given in Fig. 8.10, $dV/dt = (\pi X^3/2R)dX/dt$ and $r = X^2/4R$. Also the grain boundary diffusion coefficient D_{gb} is defined as equal to $D_v C_{v0}$. Substituting these relations in Eq. (8.23) we obtain

$$\frac{\pi X^3}{2R}\frac{dX}{dt} = \frac{2\pi D_{gb}\delta_{gb}\gamma_{sv}\Omega}{kT}\left(\frac{4R}{X^2}\right) \tag{8.24}$$

Rearranging this equation gives

$$X^5 dX = \frac{16 D_{gb}\delta_{gb}\gamma_{sv}\Omega R^2}{kT} dt \tag{8.25}$$

Integrating and putting in the boundary conditions of $X = 0$ at $t = 0$, Eq. (8.25) becomes

$$X^6 = \frac{96 D_{gb}\delta_{gb}\gamma_{sv}\Omega R^2}{kT} t \tag{8.26}$$

We may also write Eq. (8.26) in the form

$$\frac{X}{R} = \left(\frac{96 D_{gb}\delta_{gb}\gamma_{sv}\Omega}{kTR^4}\right)^{1/6} t^{1/6} \tag{8.27}$$

Equations (8.26) and (8.27) predict that the ratio of the neck radius to the sphere radius increases as $t^{1/6}$.

For this densifying mechanism, the linear shrinkage, defined as the change in length ΔL divided by the original length L_0, can also be found.

As a good approximation (see Fig. 8.10), we can write

$$\frac{\Delta L}{L_0} = -\frac{h}{R} = -\frac{r}{R} = -\frac{X^2}{4R^2} \tag{8.28}$$

where h is half the interpenetration distance between the spheres. Using Eq. (8.27) we obtain

$$\frac{\Delta L}{L_0} = -\left(\frac{3D_{gb}\delta_{gb}\gamma_{sv}\Omega}{2kTR^4}\right)^{1/3} t^{1/3} \tag{8.29}$$

The shrinkage is therefore predicted to increase as $t^{1/3}$.

As another example, let us consider the mechanism of viscous flow. We will recall that for this mechanism an energy balance proposed by Frenkel is used in the derivation of the sintering equations [see Eq. (8.1)].

Viscous Flow. The equation for neck growth between two spheres by viscous flow was derived by Frenkel [2]. The original derivation by Frenkel contained an extra factor π that is omitted in the version given here. For the parameters shown in Fig. 8.11 and assuming that the radius of the sphere remains roughly constant during the viscous flow, the decrease in the surface area of the two spheres is

$$S_0 - S = 8\pi R^2 - 4\pi R^2(1 + \cos\theta) \tag{8.30}$$

Note that this means that the material removed from the plane of contact is uniformly distributed over the surface of the sphere, rather than accumulating at the neck. For small values of θ, i.e., small neck radius, $\cos\theta \approx 1 - \theta^2/2$, so that Eq. (8.30) becomes

$$S_0 - S = 2\pi R^2 \theta^2 \tag{8.31}$$

The rate of change of energy due to the reduction in surface area can be written

$$\dot{E}_s = -\gamma_{sv}\frac{dS}{dt} = 4\pi R^2 \gamma_{sv}\frac{d}{dt}\left(\frac{\theta^2}{2}\right) \tag{8.32}$$

where γ_{sv} is the specific surface energy of the solid/vapor interface. According to Frenkel, the rate of energy dissipation by viscous flow between the two spheres is

$$\dot{E}_v = \frac{16}{3}\pi R^3 \eta \dot{u}^2 \tag{8.33}$$

where η is the viscosity of the glass and $\dot{u}$ is the velocity of motion for

Theory of Solid-State and Viscous Sintering

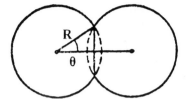

Figure 8.11 Geometrical parameters of the two-sphere model used in the derivation of the initial stage equation for viscous sintering by Frenkel.

the viscous flow given by

$$\dot{u} = \frac{1}{R}\frac{d}{dt}\left(\frac{R\theta^2}{2}\right) = \frac{d}{dt}\left(\frac{\theta^2}{2}\right) \qquad (8.34)$$

This equation is based on the assumption that flow occurs uniformly along the axis joining the centers of the spheres, rather than being concentrated near the neck. Substituting for $\dot{u}$ in Eq. (8.33) and putting $\dot{E}_s = \dot{E}_v$ gives

$$\frac{16}{3}\pi R^3 \eta \dot{u} \frac{d}{dt}\left(\frac{\theta^2}{2}\right) = 4\pi R^2 \gamma_{sv} \frac{d}{dt}\left(\frac{\theta^2}{2}\right) \qquad (8.35)$$

Rearranging, Eq. (8.35) gives

$$\dot{u} = \frac{3}{4}\left(\frac{\gamma_{sv}}{\eta R}\right) \qquad (8.36)$$

Substituting for $\dot{u}$ in Eq. (8.34) and integrating subject to the boundary conditions of $\theta = 0$ at $t = 0$, we obtain

$$\theta^2 = \frac{3}{2}\left(\frac{\gamma_{sv}}{\eta R}\right)t \qquad (8.37)$$

Since $\theta \approx X/R$, where X is the neck radius, Eq.(8.37) yields

$$X^2 = \frac{3}{2}\left(\frac{R\gamma_{sv}}{\eta}\right)t \qquad (8.38)$$

For viscous flow, X or the ratio X/R is predicted to vary as $t^{1/2}$. It is left as an exercise for the reader to determine an equation for the shrinkage produced by viscous flow.

D. Summary of the Initial Stage Sintering Equations

Within the limited space available in this chapter, we cannot give the full derivations of the sintering equations for the other mechanisms. Instead, we provide a summary of the equations and refer the reader to various publications that give the original derivations of the equations or a review of their derivations, e.g., Kuczynski [8], Kingery and Berg [9], Coble [10], and Johnson and Cutler [11].

The equations for neck growth can be expressed in the general form

$$\left(\frac{X}{R}\right)^m = \frac{H}{R^n} t \qquad (8.39)$$

where m and n are numerical exponents that depend on the mechanism of sintering and H is a function that contains the geometrical and material parameters of the powder system. Depending on the assumptions made in the models, a range of values for m, n, and the numerical constant H have been obtained. The values given in Table 8.5 represent the most plausible values for each mechanism.

For the densifying mechanisms, the shrinkage $\Delta L/L_0$ can also be obtained as a function of time. As described earlier for the mechanism of grain boundary diffusion, $\Delta L/L_0$ can be found from the equation for the neck size using the relation

$$\frac{\Delta L}{L_0} = -\frac{X^2}{4R^2}$$

Table 8.5 Plausible Values for the Constants Appearing in Eqs. (8.39) and (8.40) for the Initial Stage of Sintering

Mechanism	m	n	H^b
Surface diffusion[a]	7	4	$56D_s\delta_s\gamma_{sv}\Omega/kT$
Lattice diffusion from the surface[a]	4	3	$20D_l\gamma_{sv}\Omega/kT$
Vapor transport[a]	3	2	$3p_0\gamma_{sv}\Omega/(2\pi mkT)^{1/2}kT$
Grain boundary diffusion	6	4	$96D_{gb}\delta_{gb}\gamma_{sv}\Omega/kT$
Lattice diffusion from the grain boundary	5	3	$80\pi D_l\gamma_{sv}\Omega/kT$
Viscous flow	2	1	$3\gamma_{sv}/2\eta$

[a] Denotes nondensifying mechanism, i.e., $\Delta L/L_0 = 0$.
[b] D_s, D_l, D_{gb}, diffusion coefficients for surface, lattice, and grain boundary diffusion. δ_s, δ_{gb}, thickness for surface and grain boundary diffusion, respectively. γ_{sv}, specific surface energy; p_0, vapor pressure over a flat surface; m, mass of atom; k, Boltzmann constant; T, temperature; η, viscosity.

Theory of Solid-State and Viscous Sintering

Using Eq. (8.39) the shrinkage can be written in the general form

$$\left(\frac{\Delta L}{L_0}\right)^{m/2} = -\frac{H}{2^m R^n} t \tag{8.40}$$

where m, n, and H are given in Table 8.5.

E. Critical Assessment of the Initial Stage Models

It would be of interest to compare the predictions of the analytical models with those of the scaling laws. Taking grain boundary diffusion as an example, according to Eq. (8.25) the rate of neck growth is

$$\frac{1}{X}\frac{dX}{dt} = \left(\frac{16 D_{gb} \delta_{gb} \gamma_{sv} \Omega}{kT}\right)\left(\frac{R^2}{X^6}\right) \tag{8.41}$$

Comparing the rate of neck growth at a fixed ratio X/R, i.e., putting $X/R = C$, where C is a constant, Eq. (8.41) becomes

$$\left(\frac{1}{X}\frac{dX}{dt}\right)_{X/R} = \left(\frac{16}{C^6}\right)\left(\frac{D_{gb} \delta_{gb} \gamma_{sv} \Omega}{kT}\right)\left(\frac{1}{R^4}\right) \tag{8.42}$$

According to this equation, for powder systems that can be described by the two-sphere model, the rate of neck growth (taken at a constant value of X/R) varies as $1/R^4$, i.e., inversely as the fourth power of the particle size. We will recall that this is the same particle (or grain) size dependence predicted by the scaling laws. The temperature dependence of the rate predicted by Eq. (8.42) is also the same as that expected from the scaling laws. To summarize, for grain boundary diffusion, despite the difference in the geometrical details, the predictions of the analytical models for the particle size and temperature dependence are the same as those of the scaling laws. The reader may compare the particle size and temperature dependence as predicted by the analytical models with those of the scaling laws for the other sintering mechanisms.

The form of the neck growth equations given in Table 8.5 indicates that a plot of (i) $(X/R)^m$ versus t or (ii) $\log (X/R)$ versus $\log t$ yields a straight line. By fitting the theoretical predictions to the experimental data, the value of m can be found. A similar procedure can be applied to the analysis of shrinkage if it occurs during sintering. In practice, experimental data for validating the equations are normally obtained by measuring the neck growth in simple systems (e.g., two spheres, a sphere on a plate, or two wires) or the shrinkage in a compacted mass of spherical particles. Since m is dependent on the mechanism of sintering, at first sight it may seem that the measurement of m would provide information on the mechanism of sintering. In line with this view, during the first 20 years or so following the derivation of the early sintering equations by Kuczynski [8] and others, considerable effort was devoted to the experimental verification of the sintering equations and the determination of the sintering mechanism. Although in some cases the choice of the sintering equation seems somewhat arbitrary, almost all of the studies reported excellent agreement between the theoretical predictions and the data. However, the problems in the extraction of meaningful information from the sintering equations began to be appreciated better.

It became clear that the basic assumption in the models of a single dominant mass transport mechanism is not valid in most powder systems. When more than one mechanism operates simultaneously, the measured exponent may correspond to an entirely different mechanism. As an example consider the initial sintering of copper [9]. The neck growth and shrinkage data of copper spheres gave exponents characteristic of lattice diffusion from the grain boundary as the dominant mechanism (Fig. 8.12). However, later analysis showed that surface diffusion was the dominant mechanism, with a significant contribution from lattice diffusion (which gave shrinkage). When more than one mechanism operates simultaneously, the rate of one mechanism influences that of another. The interaction between the mechanisms is therefore very important. An exception is the sintering of glass by viscous flow, where Frenkel's neck growth equation was well verified by Kuczynski [12].

The other important assumptions of the models must also be remembered. For example, the cross section of the neck surface is assumed to be circular. As we shall describe later (see Section 8.6), numerical simulations indicate that this neck geometry is grossly simplified. The models also make simplifying assumptions about the way the matter transported into the neck is distributed over the surface. Matter transported into the neck by grain boundary diffusion must be redistributed over the neck surface to prevent buildup on the grain boundary groove. It is generally assumed that surface diffusion is fast enough to cause redistribution. This assumption has been questioned from time to time in the sintering literature.

The extension of the two-sphere model to real powder systems is valid only if the system consists of spheres of the same size arranged in a uniform pattern. In practice, this system is approached by the uniform consolidation, by colloidal methods, of monodisperse powders (see, e.g., the work of Barringer and Bowen discussed in Chapter 1). Deviations from the uniform packing, the uniform size, and the spherical shape of the particles limit the validity of the model.

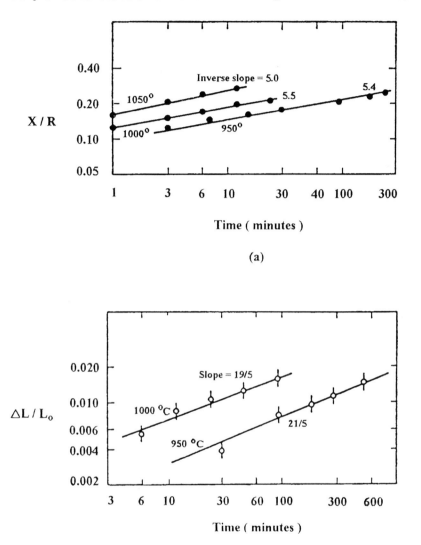

Figure 8.12 Data for (a) neck growth and (b) shrinkage of copper spheres. The exponents are characteristic of those for lattice diffusion. (From Ref. 9.) Later analysis of the data showed that surface diffusion was the dominant mechanism with a major contribution from lattice diffusion that gave shrinkage (see Fig. 8.22.)

In spite of the shortcomings, the role of the initial stage sintering models in the development of the understanding of sintering should not be underestimated. The models form the basis for understanding the mechanisms of sintering. They also provide a qualitative understanding of the initial stage of sintering in terms of the dependence of the sintering rate on the important material and processing parameters such as surface energy, particle size, and temperature. The models also provide a good basis for understanding sintering in the later stages and in nonideal systems.

F. Intermediate Stage Models

The geometrical model for the intermediate stage sintering of polycrystalline systems differs substantially from that for amorphous systems. We will therefore need to consider the models for these two types of systems separately.

Geometrical Model for Polycrystalline Systems

The geometrical model commonly used for the intermediate stage of sintering was proposed by Coble [7]. The powder system is idealized by considering it to consist of a space-filling array of equal-sized tetrakaidecahedra, each of which represents one particle. The pores are cylindrical and occur along the edges of the tetrakaidecahedra, with the axis of the cylinder coinciding with the edge of the tetrakaidecahedra (Fig. 8.8C). The structure is represented by a unit cell consisting of a tetrakaidecahedron with cylindrical pores along its edges.

A tetrakaidecahedron is constructed from an octahedron by trisecting each edge and joining the points to remove the six vertices (Fig. 8.13). The resulting structure has 36 edges, 24 corners, and 14 faces (8 hexagonal and 6 square). The volume of the tetrakaidecahedron is

$$V_t = 8\sqrt{2}\, l_p^3 \qquad (8.43)$$

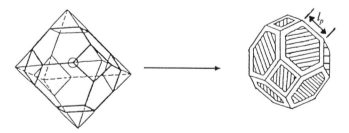

Figure 8.13 Sketch illustrating the formation of a tetrakaidecahedron from a truncated octahedron.

Theory of Solid-State and Viscous Sintering

where l_p is the edge length of the tetrakaidecahedron. If r is the radius of the pore, then neglecting the intersections of the pores, the total volume of the porosity per unit cell is

$$V_p = \frac{1}{3}(36\pi r^2 l_p) \tag{8.44}$$

The volume fraction of pores, i.e., the porosity in a unit cell, is therefore

$$P_c = \frac{3\pi}{2\sqrt{2}}\left(\frac{r^2}{l_p^2}\right) \tag{8.45}$$

Mechanisms

Since the model assumes that the pore geometry is uniform, the nondensifying mechanisms cannot operate. This is because the chemical potential is the same everywhere on the pore surface. We are therefore left with the densifying mechanisms: lattice diffusion and grain boundary diffusion. Plastic flow is not expected to operate because the stresses in the system are not high enough.

Sintering Equations

Using the model based on the tetrakaidecahedron, Coble [7] derived kinetic equations for sintering by lattice diffusion and grain boundary diffusion. We shall outline Coble's analysis starting with the lattice diffusion mechanism.

Lattice Diffusion. Figure 8.14a shows that the cylindrical pores along the edges enclose each face of the tetrakaidecahedron. Because the vacancy flux from the pores terminates on the boundary faces (Fig. 8.14b), Coble assumed radial diffusion from a circular vacancy source. The shape effects on the corner of the tetrakaidecahedron are neglected. For the boundary to remain flat, the vacancy flux per unit area of the boundary must be the same over the whole boundary. According to Coble, the diffusion flux field can be approximated to that of the temperature distribution in a surface-cooled, electrically heated cylindrical conductor.

The flux per unit length of the cylinder is given by

$$J/l = 4\pi D_v \Delta C \tag{8.46}$$

where D_v is the vacancy diffusion coefficient and ΔC is the vacancy concentration difference between the pore and the boundary. Coble also made a number of other assumptions, including the following.

1. The convergence of the flux to the boundary does not qualitatively change the flux equation with respect to its dependence on the pore radius,

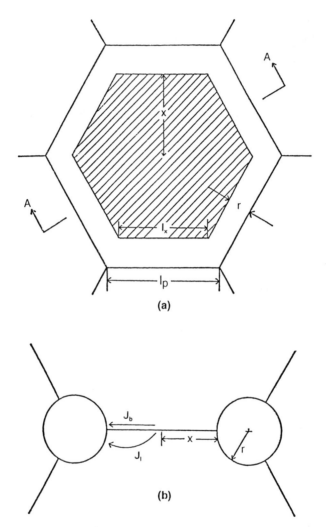

Figure 8.14 The intermediate stage sintering equations for polycrystalline solids are based on a hexagonal neck (a). The section (on A-A) shows a cut through the neck with the atomic flux paths for grain boundary and lattice diffusion (b).

2. The width of the flux field, i.e., the equivalent of l in Eq. (8.46), is equal to the pore diameter.
3. The flux is increased by a factor of 2 due to the freedom of the vacancy diffusion flux to diverge initially, thereby providing additional available area.

Theory of Solid-State and Viscous Sintering

With these assumptions, Eq. (8.46) becomes

$$J = 2(4\pi D_v \Delta C)2r \tag{8.47}$$

Since there are 14 faces in a tetrakaidecahedron and each face is shared by two grains, the volume flux per unit cell is

$$\frac{dV}{dt} = \frac{14}{2} J = 112 \pi D_v \Delta C \tag{8.48}$$

For a cylinder of radius r, the two principal radii of curvature are r and ∞, so the vacancy concentration difference between the pore and the boundary is

$$\Delta C = \frac{C_{v0} \gamma_{sv} \Omega}{kTr} \tag{8.49}$$

Combining Eqs. (8.48) and (8.49) and putting $D_l = D_v C_{v0}$, where D_l is the lattice diffusion coefficient, we obtain

$$dV = \frac{112 \pi D_l \gamma_{sv} \Omega}{kT} dt \tag{8.50}$$

The integral of dV is equal to the total volume of porosity given by Eq. (8.44), that is,

$$\int dV = 12\pi r^2 l_p \Big]_{r_0}^{r} \tag{8.51}$$

Therefore,

$$r^2]_r^0 \approx -10 \frac{D_l \gamma_{sv} \Omega}{l_p kT} t \Big]_t^{t_f} \tag{8.52}$$

where t_f is the time when the pore vanishes. Dividing both sides of this equation by l_p^2 and evaluating the integrand yields

$$P_c \approx \frac{r^2}{l_p^2} \approx \frac{10 D_l \gamma_{sv} \Omega}{l_p^3 kT} (t_f - t) \tag{8.53}$$

In view of the many approximations used by Coble, this equation can be considered only as an order-of-magnitude calculation. The model applies until the pores pinch off and become isolated.

In most sintering studies, the rate of densification is an important parameter. Using the relation between porosity and relative density, i.e., $P = 1 - \rho$, and differentiating Eq. (8.53) with respect to time, we obtain

$$\frac{d}{dt}(P_c) = -\frac{d\rho}{dt} \approx \frac{-10 D_l \gamma_{sv} \Omega}{l_p^3 kT} \tag{8.54}$$

Putting l_p proportional to the grain size, G and writing the densification rate in the form of a volumetric strain rate, Eq. (8.54) becomes

$$\frac{1}{\rho}\frac{d\rho}{dt} \approx \frac{AD_l\gamma_{sv}\Omega}{\rho G^3 kT} \tag{8.55}$$

where A is a constant. We see that the densification rate at a fixed density has the same dependence on the grain size and temperature as that predicted by the scaling law for lattice diffusion.

Grain Boundary Diffusion. Using the same geometrical model as described above for lattice diffusion and modifying the flux equations to account for grain boundary diffusion, Coble derived the equation

$$P_c \approx \frac{r^2}{l_p^2} \approx \left(\frac{2D_{gb}\delta_{gb}\gamma_{sv}\Omega}{l_p^4 kT}\right)^{2/3} t^{2/3} \tag{8.56}$$

Using the same procedure as outlined above for lattice diffusion, Eq. (8.56) can be written

$$\frac{1}{\rho}\frac{d\rho}{dt} \approx \frac{4}{3}\left(\frac{D_{gb}\delta_{gb}\gamma_{sv}\Omega}{G^4 kT\rho(1-\rho)^{1/2}}\right) \tag{8.57}$$

Here again we see that the densification rate at a fixed density has the same dependence on grain size and temperature as that predicted by the scaling law.

Models for the intermediate stage of sintering have also been developed by Johnson [13] and Beeré [14]. These models can be considered refinements of Coble's model. Johnson derived equations for the shrinkage in terms of the average values of the neck radius and the pore radius. His model cannot be used to predict the rate of sintering. Instead, it is meant to help the analysis of sintering data. From the measured values of the average neck radius and pore radius, the boundary and lattice diffusion coefficients and the relative flux of matter due to the two mechanisms can be inferred. Beeré extended the model of Coble by allowing the pores to relax to a minimum free energy configuration. The pores have a fairly complex curvature and meet the grain boundary at a constant dihedral angle satisfying the balance of interfacial tensions (see Section 8.2C). Beeré's model shows the same dependence on grain size and temperature as Coble's model but contains additional terms that involve the dihedral angle and the grain boundary area.

Geometrical Model for Amorphous Systems

The structure shown in Fig. 8.15 was proposed by Scherer [15] as a model for the intermediate stage of sintering of an amorphous material. It consists

Theory of Solid-State and Viscous Sintering

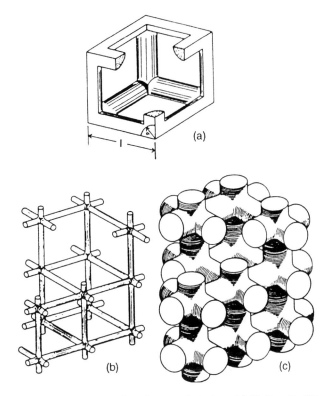

Figure 8.15 Scherer's model for viscous sintering. (a) Unit cell of the structure with edge length l and cylinder radius a. (b) Microstructure with a relative density of ≈0.1. (c) Microstructure with a relative density of ≈0.5. (From Ref. 1.)

of a cubic array formed by intersecting cylinders. Densification is assumed to be brought about by the cylinders getting shorter and thicker. The model can be viewed as an idealized structure in which the cylinders represent strings of spherical particles joined together by necks.

The structure is represented by a unit cell consisting of 12 quarter-cylinders (Fig. 8.15a). The volume of the solid phase in the unit cell is

$$V_s = 3\pi a^2 l - 8\sqrt{2} a^3 \qquad (8.58)$$

where l is the length of the side of the unit cell and a is the radius of the cylinder. Since the total volume of the cell is l^3, the density of the cell, d, is equal to $d_s V_s / l^3$, where d_s is the theoretical density of the solid phase. The relative density ρ, defined as d/d_s, is given by

$$\rho = 3\pi x^2 - 8\sqrt{2} x^3 \qquad (8.59)$$

where $x = a/l$. The inverse of this equation is

$$x = \frac{\pi\sqrt{2}}{8}\left[\frac{1}{2} + \cos\left(\Theta + \frac{4\pi}{3}\right)\right] \tag{8.59a}$$

where $\Theta = (1/3)\cos^{-1}[1 - (4/\pi)^3\rho]$. According to Eq. (8.59), ρ is a function of a/l only. The volume of the solid phase in the unit cell can also be written as ρl^3. This volume does not change and can also be put equal to $\rho_0 l_0^3$, where ρ_0 and l_0 are the initial values of the relative density and length, respectively, of the unit cell. Each cell contains one pore, so the number of pores per unit volume of the solid phase is

$$N = 1/\rho_0 l_0^3 \tag{8.60}$$

The model should be valid until the adjacent cylinders touch, thereby isolating the pores. This occurs when $a/l = 0.5$, i.e., when $\rho \approx 0.94$.

The derivation of the sintering equations for Scherer's model follows a procedure similar to that outlined earlier for Frenkel's initial stage model. The reader is referred to Scherer's original article [15] for a full derivation. The formulation of the equation satisfying Frenkel's energy balance concept gives

$$\int_{t_0}^{t} \frac{\gamma_{sv} N^{1/3}}{\eta} dt = \int_{x_0}^{x} \frac{2\,dx}{(3\pi - 8\sqrt{2}x)^{1/3}\, x^{2/3}} \tag{8.61}$$

where γ_{sv} is the specific surface energy of the solid/vapor interface and η is the viscosity of the solid phase. By making the substitution

$$y = \left(\frac{3\pi}{x} - 8\sqrt{2}\right)^{1/3} \tag{8.62}$$

the integral on the right-hand side of Eq. (8.61) can be evaluated. The result can be written in the form

$$\frac{\gamma_{sv} N^{1/3}}{\eta}(t - t_0) = F_S(y) - F_S(y_0) \tag{8.63}$$

where

$$F_S(y) = -\frac{2}{\alpha}\left[\frac{1}{2}\ln\left(\frac{\alpha^2 - \alpha y + y^2}{(\alpha + y)^2}\right) + \sqrt{3}\arctan\left(\frac{2y - \alpha}{\alpha\sqrt{3}}\right)\right] \tag{8.64}$$

and $\alpha = (8\sqrt{2})^{1/3}$.

Theory of Solid-State and Viscous Sintering

The predictions of the model are normally expressed in terms of a curve of ρ versus $(\gamma_{sv}N^{1/3}/\eta)(t - t_0)$, referred to as the reduced time. The curve is obtained as follows. For a chosen value of ρ, the parameter y is found from Eqs. (8.59a) and (8.62). The function $F_S(y)$ is then found from Eq. (8.64). Finally, the reduced time is obtained from Eq. (8.63). The procedure is repeated for other values of ρ. The predictions of the model are shown by the full curve up to $\rho = 0.94$ in Fig. 8.16. For $0.94 < \rho \leq 1$, the curve is calcualted from the Mackenzie-Shuttleworth model described later. The curve has the characteristic sigmoidal shape observed for the density versus time data in many sintering experiments.

The predictions of Scherer's model have been well validated by the data for many amorphous materials such as colloidal gels, polymeric gels, and consolidated glass particles [1]. In the comparison of the predictions of the model with experimental data, the reader may have noticed that the predictions are plotted versus reduced time whereas the data are obtained as function of measured time (i.e., seconds). To construct such a plot, one must first find the reduced time corresponding to the measured density as described above. A plot of the reduced time versus the measured time should be a straight line with a slope of $\gamma_{sv}N^{1/3}/\eta$. Multiplying

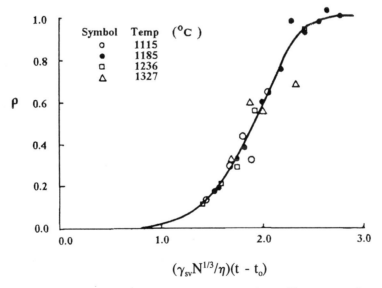

Figure 8.16 Relative density versus reduced time for a silica soot preform sintered in air. (From Ref. 16, with permission.)

that slope by the measured time gives the average reduced time interval for each density. The data points in Fig. 8.16 show a comparison of the sintering results for low-density bodies formed from SiO_2 soot with the predictions of Scherer's model [16].

G. Final Stage Models

For the final stage sintering of polycrystalline materials, the powder system is idealized in terms of an array of equal-sized tetrakaidecahedra with spherical pores of the same size at the corners (Fig. 8.8d). A tetrakaidecahedron has 24 pores (one at each corner). Since each pore is shared by four tetrakaidecahedra, the pore volume in a tetrakaidecahedron is $V_p = (24/4)(4/3)\pi r^3$, where r is the radius of a pore. The porosity in a tetrakaidecahedron is given by

$$P_s = \frac{8\pi r^3}{8\sqrt{2} l_p^3} = \frac{\pi}{\sqrt{2}} \left(\frac{r^3}{l_p^3} \right) \tag{8.65}$$

A more convenient unit cell of the idealized structure can be chosen as a thick-walled spherical shell of solid material centered on a single pore of radius r (Fig. 8.17). The same unit cell, without the grain boundaries, was introduced to model the final stage sintering of amorphous materials. The outer radius of the spherical shell, b, is defined such that the average density of the unit cell is equal to the density of the entire powder system, that is,

$$\rho = 1 - (r/b)^3 \tag{8.66}$$

The volume of the solid phase in a unit cell is $(4/3)\pi(b^3 - r^3)$. Since a unit cell contains a single pore, the number of pores per unit volume of the solid phase is

$$N = \frac{3}{4\pi} \left(\frac{1 - \rho}{\rho r^3} \right) \tag{8.67}$$

Sintering Equations

As in the intermediate stage, the uniform pore geometry assumed in the models precludes the occurrence of nondensifying mechanisms. Final stage sintering models have been developed by Coble [7] and Coleman and Beeré [17] for diffusional mass transport and by Mackenzie and Shuttleworth [18] for viscous flow. We outline only the main results below and refer the reader to the original articles by these authors.

Lattice Diffusion. Coble [7] used a procedure similar to that outlined earlier for the intermediate stage. However, for the final stage the diffu-

Theory of Solid-State and Viscous Sintering

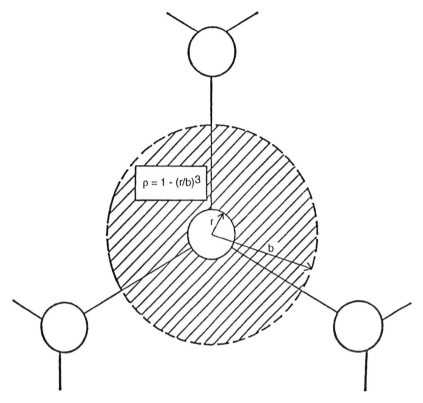

Figure 8.17 A porous solid during the final stage of sintering can be modeled by constructing a spherical shell centered on a single pore. The outer radius b is chosen such that the density of the shell matches that of the porous solid.

sional flux equation was approximated to that corresponding to diffusion between concentric spherical shells. The final result, given in terms of the porosity P_s, versus time is

$$P_s = \frac{6\pi}{\sqrt{2}} \left(\frac{D_l \gamma_{sv} \Omega}{l_p^3 kT} \right) (t_f - t) \tag{8.68}$$

Because of the approximations made in the flux field, this equation is believed to be valid for porosities less than about 2%. For higher porosities (2–5%), a more complex equation was derived by Coble. However, in view of the drastic approximations made, little benefit may be gained from the more complex expression. Comparing Eq. (8.68) with the corresponding equation for the intermediate stage [Eq. (8.53)], the only difference is in the value of the numerical constants. The difference is, however, very small.

A final stage sintering equation for grain boundary diffusion was not derived by Coble. However, as we shall see in Section 8.9, he later developed models for diffusional sintering with an applied pressure from which sintering equations for both intermediate and final stages can be extracted.

Viscous Flow. Mackenzie and Shuttleworth [18] used the concentric sphere model to derive sintering equations for viscous flow of an amorphous solid. The use of the model means that the real system is idealized in terms of a structure consisting of spherical pores of the same size in a solid matrix. The concentric shell maintains its spherical geometry during sintering. Unlike the initial stage, exact equations can be derived in Frenkel's energy balance concept for the rate of reduction of surface area and the rate of dissipation of energy by viscous flow. The result is

$$\int_{t_0}^{t} \frac{\gamma_{sv} N^{1/3}}{\eta} \, dt = \frac{2}{3} \left(\frac{3}{4\pi}\right)^{1/3} \int_{\rho_0}^{\rho} \frac{d\rho}{(1-\rho)^{2/3} \rho^{1/3}} \tag{8.69}$$

This equation can be written

$$\frac{\gamma_{sv} N^{1/3}}{\eta} (t - t_0) = F_{MS}(\rho) - F_{MS}(\rho_0) \tag{8.70}$$

where

$$F_{MS}(\rho) = \frac{2}{3} \left(\frac{3}{4\pi}\right)^{1/3} \left[\frac{1}{2} \ln\left(\frac{1+\rho^3}{(1+\rho)^3}\right) - \sqrt{3} \arctan\left(\frac{2\rho - 1}{\sqrt{3}}\right)\right] \tag{8.71}$$

The form of Eq. (8.70) is similar to that for Scherer's intermediate stage model [Eq. (8.63)], and the predictions for the sintering kinetics can therefore be analyzed in a way similar to that outlined for Scherer's model.

A comparison of the predictions of Mackenzie and Shuttleworth's model with those of Scherer's model is shown in Fig. 8.18. Despite the large difference in the geometry of the models, the agreement is excellent over a wide density range. Significant deviations begin to occur only when ρ falls below ≈ 0.2. Although the Mackenzie and Shuttleworth model is strictly valid for ρ greater than ≈ 0.9, the predictions are applicable over a much wider range. A similar situation exists for Scherer's model. The predictions show excellent agreement not only with Mackenzie and Shuttleworth's model but also with Frenkel's initial stage model. The conclusion we can draw from the predictions of the three models and from the comparison of the model predictions with experimental data is that unlike polycrystalline materials, the sintering behavior of amorphous materials

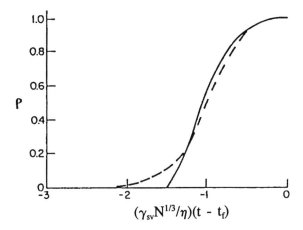

Figure 8.18 Relative density versus reduced time for Mackenzie and Shuttleworth model (solid curve) and Scherer model (dashed curve). For relative densities >0.942, the Mackenzie and Shuttleworth model applies. The curves have been shifted to coincide at the time t_f when sintering is complete. (From Ref. 15, with permission.)

is not sensitive to structural features such as density and pore shape; however, the pore size distribution is important [1].

H. Analytical Models: Final Comments

A striking feature that arises from our consideration of the analytical models is the difference in complexity between the sintering phenomena in polycrystalline materials and those in amorphous materials. The analysis of viscous sintering on the basis of Frenkel's energy balance concept appears relatively simple in principle. The idealization of the structure of amorphous materials leads to analytical solutions that describe the sintering behavior in a very satisfactory manner.

For polycrystalline materials, the sintering phenomena are considerably more dependent on the structural details of the powder system. Because of the many simplifications made in the models, they do not provide an adequate representation of the sintering behavior of real powder systems. The models provide only a qualitative understanding of the variation of the sintering kinetics with parameters such as particle size, temperature, and, as we shall see later, pressure.

The assumptions made in the models must be remembered. The models assume a geometry that is a drastic simplification of a real powder system. They also assume that each mechanism operates separately. Attempts have been made to develop analytical models with more realistic neck geometries (e.g., a catenary) and with simultaneous mechanisms (e.g., surface diffusion and grain boundary diffusion), but the analyses have not succeeded in providing any significant advances over the simple models. The models assume that the powder particles are spherical and of the same size, a regular packing arrangement in the consolidated powder form, and no grain growth. These assumptions are almost never reproduced in real powder systems.

8.6 NUMERICAL SIMULATIONS OF SINTERING

In view of the approximations made in the development of the analytical models, attempts have been made to use numerical simulations to provide a better description of some of the complexities of sintering. Compared to analytical models, most studies employing the numerical simulation approach have attempted to derive neck growth and shrinkage equations for more realistic neck geometries and for simultaneous mechanisms.

The analytical models, as described earlier, generally assume that the cross section of the neck surface is a circle (sometimes referred to as the circle approximation). This circle is also assumed to be tangential to both the grain boundary and the spherical surface of the particle. From a physical point of view, this neck geometry is unsatisfactory because the surface curvature must change discontinuously at the point of tangency between the neck surface and the spherical surface of the particle (Fig. 8.19). A discontinuous change in the surface curvature requires an abrupt change in the chemical potential. The use of a circular neck surface is also at odds with experimental observations. Undercutting of the neck surface occurs, so the surface has the cross section of a bulb rather than that of a circle.

When more than one mechanism operates simultaneously, the rate of one mechanism influences that of another. The interaction between the mechanisms must be taken into account. As an illustration, consider the simultaneous occurrence of surface diffusion and grain boundary diffusion. Matter transport by surface diffusion leads to a reduction in the curvature of the neck surface. The chemical potential of the vacancies (or the vacancy concentration) under the surface is thereby reduced. This causes a reduction in the flux of vacancies between the neck surface and the grain boundary, i.e., a reduction in the rate of grain boundary diffusion. In the same way, matter transport by grain boundary diffusion leads to a reduction in the rate of surface diffusion. The interactions between the mechanisms cannot be easily treated analytically.

Theory of Solid-State and Viscous Sintering

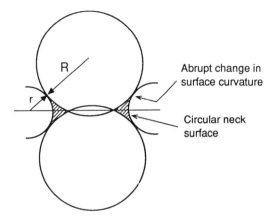

Figure 8.19 The circle approximation used in the analytical models for the geometry of the neck surface. This approximation is physically unrealistic because a discontinuous change must occur at the point of tangency between the neck surface and the spherical surface of the particle.

A. Application of the Method

The numerical simulation of sintering has been described in detail by Nicholls and Mullins [19] for surface diffusion and by Bross and Exner [20] for surface diffusion and grain boundary diffusion operating simultaneously. In general, the three-dimensional situation of real powder systems is reduced to a two-dimensional problem by choosing a geometrical model that can be described by one cross section only, e.g., two cylinders or a row of cylinders. The cylinders need not have the same radius. The differential equations for the flux of matter are then transformed into suitable equations for finite differences, i.e., employing small increments in time and small surface elements. The simulations are finally carried out using a computer to yield results for the neck size or shrinkage.

For a model consisting of a row of cylinders of the same radius, Fig. 8.20 shows the results of Bross and Exner for the neck contours. Two situations were considered: (i) matter transport by surface diffusion only (Fig. 8.20a) and (ii) the simultaneous occurrence of surface diffusion and grain boundary diffusion (Fig. 8.20b). The results for the surface diffusion mechanism are in agreement with those of Nicholls and Mullins. The circle approximation used in the analytical models differs from the contours found by the numerical simulation approach. Undercutting and a continuous change in the curvature of the neck surface are predicted by the nu-

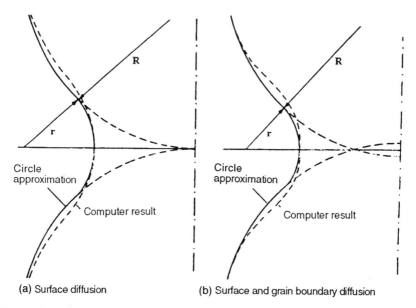

Figure 8.20 Contours of necks between cylinders for sintering by (a) surface diffusion and (b) simultaneous surface and grain boundary diffusion. (From Ref. 20, reprinted with permission from Elsevier Science Ltd.)

merical simulation. Another important finding is that the region of the neck surface influenced by matter transport extends far beyond that given by the circle approximation. The extension is, however, less pronounced when surface diffusion and grain boundary diffusion occur simultaneously.

Numerical simulation of the initial stage of sintering of two spheres for the simultaneous occurrence of grain boundary, lattice, and surface diffusion was performed by Johnson [21]. The simulation gave results for the neck size versus shrinkage that showed good agreement with experimental data for the sintering of iron spheres (12.3 μm radius). The analysis, however, included two approximations that are somewhat unsatisfactory: (i) the circle approximation for the neck surface and (ii) the equality of the chemical potential gradient for grain boundary diffusion and for lattice diffusion. More recent work employing the numerical simulation approach is given in References 37 to 39.

B. Assessment of the Numerical Simulation Approach

The studies performed so far provide good insight into the transport of matter and geometry of the neck surface. The approach has good potential

Theory of Solid-State and Viscous Sintering

for the elucidation of many of the complexities of sintering. However, numerical simulation has two major drawbacks compared with the analytical models: (i) The calculations are fairly complex and (ii) the results cannot be easily visualized or cast into a useful form showing the dependence of the kinetics on the sintering parameters. Because of these drawbacks, the numerical simulation approach has not achieved much popularity.

8.7 PHENOMENOLOGICAL SINTERING EQUATIONS

In the phenomenological approach, empirical equations are developed to fit sintering data, usually in the form of density (or shrinkage) versus time. The equations provide little or no help in understanding the process of sintering. However, they may serve a useful function in some numerical models that incorporate explicit equations for the densification of a powder system.

A simple expression that has been found to be very successful in fitting sintering and hot pressing data is

$$\rho = \rho_0 + K \ln (t/t_0) \tag{8.72}$$

where ρ_0 is the density at an initial time t_0, ρ is the density at time t, and K is a temperature-dependent parameter. Because of the widespread applicability of Eq. (8.72) it is sometimes referred to in the sintering literature as the *semilogarithmic law*. Coble [7] attempted to provide some theoretical justification of this expression in the following way. Using the rate equation for Coble's intermediate or final stage models [see, e.g., Eq. (8.54)] written in the general form

$$\frac{d\rho}{dt} = \frac{AD_l\gamma_{sv}\Omega}{G^3kT} \tag{8.73}$$

where A is a constant that depends on the stage of sintering, and assuming that the grains grow according to a cubic law of the form

$$G^3 = G_0^3 + \alpha t \approx \alpha t \tag{8.74}$$

i.e., it is assumed that $G^3 \gg G_0^3$, Eq. (8.73) becomes

$$\frac{d\rho}{dt} = \frac{K}{t} \tag{8.75}$$

where $K = AD_l \gamma_{sv}/\alpha kT$. On integration of Eq. (8.75) we obtain Eq. (8.72). Because Eq. (8.73) has the same form in the intermediate and final stages, the semilogarithmic law is expected to be valid in both of these stages.

When grain growth is fairly limited, data over a large part of the sintering process can usually be fitted by the equation

$$\frac{\Delta L}{L_0} = Kt^{1/m} \tag{8.76}$$

where K is a temperature-dependent parameter and m is an integer.

Other empirical equations in the sintering literature include one due to Tikkanen and Maekipirtti [22]:

$$\frac{V_0 - V_s}{V_s - V_T} = Kt^n \tag{8.77}$$

where V_0 is the initial volume of the powder compact, V_s is the volume after sintering for time t, V_T is the volume of the fully dense solid, K is a temperature-dependent parameter, and n is an number between 0.5 and 1.0. An equation due to Ivensen [23] is

$$V_s/V_0 = (1 + Kt)^{-n} \tag{8.78}$$

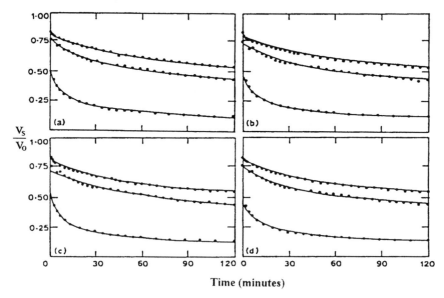

Figure 8.21 Data for UO_2 fitted by (a) Coble's semilogarithmic law, Eq. (8.72); (b) the equation of Tikkanen and Maekipirtti, Eq. (8.77); (c) the equation of Ivensen, Eq. (8.78); and (d) a simple hyperbolic function of the form $V_S/V_0 = K_1/(t + K_2) + K_3$. There is no significant difference in the goodness of the fit. (From Ref. 24.)

Theory of Solid-State and Viscous Sintering

Attempts have sometimes been made to attach physical significance to the empirical equations. However, the difficulties of doing this have been demonstrated by Pejovnik and coworkers [24]. They showed that the sintering data for UO_2 (Fig. 8.21) could be well fitted by any of the four equations (8.72), (8.77), (8.78), and a simple hyperbolic equation of the form $V_s/V_0 = K_1/(t + K_2) + K_3$. Even a statistical test failed to show any significant differences in the goodness of fit. The conclusion from Fig. 8.21 is that more than one empirical equation can provide a good fit to any given set of sintering data. In practice, the choice of any one of the equations appears to be fairly arbitrary. Coble's semiempirical law has the advantage of simplicity and, as discussed earlier, has been very successful in fitting many sintering data.

8.8 SINTERING DIAGRAMS

As discussed earlier, more than one mechanism operates simultaneously during the sintering of polycrystalline systems. The numerical simulation approach attempts to analyze sintering with simultaneous mechanisms, but the results cannot be easily formulated into a practically useful form. A more practical approach to the investigation of sintering with simultaneous mechanisms has been developed by Ashby [25] and Swinkels and Ashby [26]. It involves the construction of sintering diagrams. The earlier diagrams show, for a given temperature and neck size, the dominant mechanism of sintering and the net rate of neck growth or densification. The diagrams are theoretical and are based on the two-particle models described above. Later, a second type of diagram was put forward in which the density rather than the neck size is evaluated. The construction of the density diagrams involves the same principles as those used for the neck size diagrams, and we shall consider only the neck size diagrams in our discussion.

The form that a sintering diagram can take is shown in Fig. 8.22 for the sintering of copper spheres with a radius of 57 µm. The axes are the normalized neck radius X/R, where R is the radius of the sphere, and the homologous temperature T/T_M, where T_M is the melting temperature of the solid. The diagram is divided into various fields. Within each field, a single mechanism is dominant, i.e., it contributes most to neck growth. Figure 8.22, for example, is divided into three fields corresponding to surface diffusion, grain boundary diffusion, and lattice diffusion (from the grain boundary). At the boundary between two fields (shown as solid lines), two mechanisms contribute equally to the sintering rate. Superimposed on the fields are contours of constant sintering time.

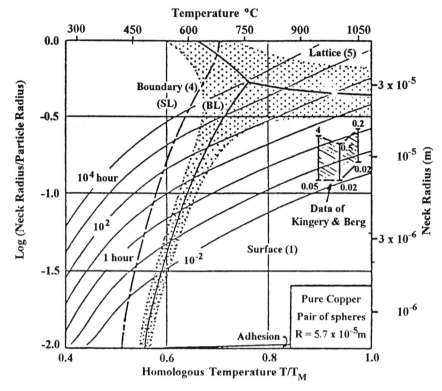

Figure 8.22 Neck size sintering diagram for copper spheres. The data of Kingery and Berg [9] are also shown. (From Ref. 25, reprinted with permission from Elsevier Science Ltd.)

Some diagrams, especially those developed more recently, may contain additional information. For example, broken lines roughly parallel to the temperature axis (not shown in Fig. 8.22) represent the transition between the stages of sintering. On either side of the field boundaries is a shaded band. Outside the shaded band a single mechanism contributes more than 55% of the total neck growth. Within the shaded band, two or more mechanisms contribute in an important way, but none contributes more than 55%. The shaded band provides an idea of the *width* of the field boundaries. A dashed line running roughly parallel to the field boundary between surface diffusion and grain boundary diffusion shows where redistribution of the matter transported by grain boundary diffusion controls the rate of this mechanism. As discussed earlier, matter transported into the neck by grain boundary diffusion requires another mechanism (e.g.,

Theory of Solid-State and Viscous Sintering

surface diffusion) for redistribution over the neck surface. On the side labeled SL, redistribution by surface diffusion controls the rate; on the side labeled BL, boundary diffusion controls the rate.

A. Construction of the Diagrams

The construction of the diagrams is described in detail by Ashby [25]. It requires neck growth equations for specified geometrical models and data for the material constants (e.g., diffusion coefficients, surface energy, and atomic volume) that appear in them. The geometrical models are similar to those outlined earlier when we considered the analytical models. The diagrams are constructed by numerical methods. It is assumed that the total neck growth rate is the sum of all the neck growth rates for the individual mechanisms. The field boundaries at which one mechanism contributes 50% of the neck growth rate are then calculated. The procedure is carried out incrementally for small increases in X/R and T/T_M. The time intervals between steps are calculated and added to give the total time. The total time is used to plot the time contours. Refinements such as the width of the boundaries between the fields are added by modification of the same procedure.

B. Usefulness of the Diagrams

Because the diagrams are based on approximate geometrical models and data for the material parameters that appear in them, their applicability is limited because of two major factors. First, the material parameters are almost never known with sufficient accuracy for ceramic systems. Small changes in the values of the parameters can produce considerable changes in the nature of the diagrams. Second, the geometrical models involve considerable simplifications to the structure of real powder systems. As noted earlier, they provide a poor representation of the densification of real powder systems.

Sintering diagrams have so far been constructed for pure metals and a few very simple ceramics (e.g., NaCl). For practical ceramic systems, small changes in the characteristics of the powders (e.g., purity) can produce significant changes in the material parameters (e.g., diffusion coefficients). Diagrams may therefore have to be constructed for each system, which would involve an enormous amount of work. The diagrams have, however, proved to be fairly useful in visualizing conceptual relationships between the various mechanisms and changes in sintering behavior under different temperature and particle size regimes.

8.9 SINTERING WITH AN EXTERNALLY APPLIED PRESSURE: HOT PRESSING

As discussed in Chapter 1, a difficulty that often arises during sintering is that of inadequate densification. One solution to this difficulty is the application of an external stress or pressure to the powder system during heating. To apply the pressure sintering route successfully, we must understand how the applied stress influences the rate of matter transport.

Models for hot pressing by diffusional transport of matter were formulated by Coble [27,28]. The models incorporate the effects of both surface curvature and applied stress. Two approaches were suggested by Coble. In the first, the analytical models are adapted to include the effects of an applied stress. In the second, the hot pressing process is viewed in a manner analogous to that of creep in solids. The equations developed for the prediction of creep are modified to account for the porosity and surface curvature in powder systems. We illustrate these two approaches separately.

A. Adaptation of the Analytical Sintering Models

Consider the idealized models for the three stages of sintering (Fig. 8.8). The vacancy concentration under the neck surface is not affected by the applied stress, so it is still given by Eq. (8.22); that is,

$$\Delta C_{vp} = \frac{C_0 \gamma_{sv} \Omega}{kT} \left(\frac{1}{r_1} + \frac{1}{r_2} \right) = \frac{C_0 \gamma_{sv} \Omega}{kT} K \tag{8.79}$$

where K is the curvature of the pore surface. For the initial stage of sintering, $K = 1/r = 4R/X^2$, whereas for the intermediate and final stages $K = 1/r$ and $2/r$, respectively, where r is the pore radius. The stress p_a applied to the powder system leads to a stress p_e on the grain boundary. Since the porosity reduces the load-bearing cross-section of the body, p_e is greater than p_a. Let us assume that

$$p_e = \phi p_a \tag{8.80}$$

where ϕ is a factor we shall define in more detail later. The compressive stress on the grain boundary means that the vacancy concentration is less than that of a flat, stress-free boundary; that is,

$$-\Delta C_{vb} = \frac{C_0 p_e \Omega}{kT} = \frac{C_0 \phi p_a \Omega}{kT} \tag{8.81}$$

For the two-sphere model, for the initial stage of sintering (Fig. 8.10) Coble assumed that ϕ is equal to the area of the sphere projected onto

Theory of Solid-State and Viscous Sintering

the punch of the hot pressing die divided by the cross-sectional area of the neck, i.e., $\phi = 4R^2/\pi X^2$. For both the intermediate and final stages, Coble argued that $\phi = 1/\rho$.

Using the parameters K and ϕ, the variation of ΔC_{vp} and ΔC_{vb} is shown schematically in Fig. 8.23. For hot pressing, the difference in the vacancy concentration between the neck and the grain boundary is given

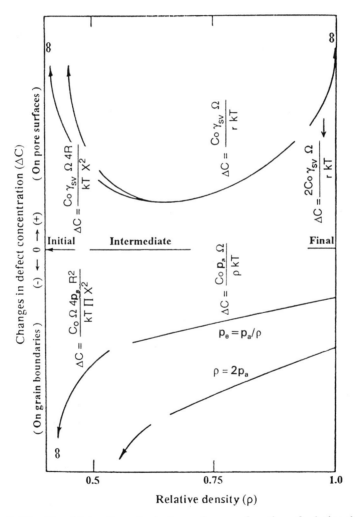

Figure 8.23 The driving force for hot pressing as a function of relative density. (From Ref. 27.)

by

$$\Delta C = \Delta C_{vp} - \Delta C_{vb} \tag{8.82}$$

For the initial stage,

$$\Delta C = \frac{C_0 \Omega 4R}{kTX^2}\left(\gamma_{sv} + \frac{p_a R}{\pi}\right) \tag{8.83}$$

This equation shows that for the initial stage, ΔC for hot pressing is identical to that for sintering except that γ_{sv} is replaced by $\gamma_{sv} + p_a R/\pi$. Since p_a and R are constant, it follows that the hot pressing equations can be obtained from the sintering equations by simply replacing γ_{sv} by $\gamma_{sv} + p_a R/\pi$.

The adaptation of Coble's sintering equations for the intermediate and final stages of hot pressing is not so straightforward. For these two stages, Coble used the approach based on the modification of creep equations. As we shall show later (Section 8.12), if the differential equations for the flux are formulated correctly and the proper boundary conditions are applied, the sintering and hot pressing equations can be derived by the same analysis. However, in view of the importance of Coble's approach to the development of hot pressing theory, we go on to consider his modification of the creep equations.

B. Modification of the Creep Equations

As a prelude to considering their modification by Coble, we outline the form the creep equations can take.

Creep Equations

Consider a single crystal of a pure metal with the cubic structure that has the shape of a rod with a square cross section of edge length $2L$. Let normal stresses p_a act on two sides of the rod as shown in Fig. 8.24a. Nabarro [29] and Herring [30] argued that self-diffusion within the crystal will cause the solid to deform (i.e., creep) in an attempt to relieve the stresses. The creep is caused by atoms diffusing from interfaces subjected to a compressive stress toward those subjected to a tensile stress. Extending this idea of creep to a polycrystalline solid, self-diffusion within the individual grains will cause atoms to diffuse from grain boundaries under compression toward those under tension (Fig. 8.24b). Following our discussion in Chapter 7, we can say that the chemical potential of the atoms in the boundaries where there is a compressive stress is higher than that for the atoms in the boundaries where there is a tensile stress. A diffusive flux occurs driven by the gradient in the chemical potential. By analysis

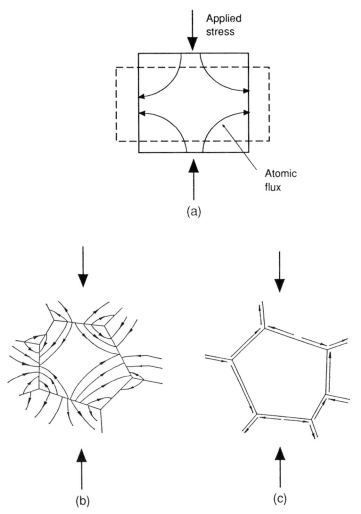

Figure 8.24 (a) A single crystal subjected to a uniaxial stress, showing the direction of the atomic flux. (b, c) A typical grain in a polycrystalline solid showing the expected atomic flux by lattice diffusion (b) and by grain boundary diffusion (c).

of the flux, Herring developed a theory to calculate the creep rate of the solid. The result can be expressed as

$$\dot{\epsilon}_c = \frac{1}{L}\frac{dL}{dt} = \frac{40}{3}\left(\frac{D_l p_a \Omega}{G^2 kT}\right) \tag{8.84}$$

where the creep rate is, by definition, equal to a linear strain rate, p_a is the applied stress, G is the grain size, and the other terms have been defined earlier.

Later, Coble [31] pointed out that creep may also occur by diffusion along the grain boundaries (Fig. 8.24c). For this mechanism, he derived the following equation for the creep rate:

$$\dot{\epsilon}_c = \frac{47.5 D_{gb} \delta_{gb} p_a \Omega}{G^3 k} \tag{8.85}$$

Equations (8.84) and (8.85) have the same linear dependence on the applied stress, but the grain size dependence and the numerical constant are greater for grain boundary diffusion. It should be remembered that the two equations were derived for theoretically dense solids.

The application of a high enough stress may, in some ceramics, activate matter transport by dislocation motion. For this mechanism, the creep rate is given by a general equation of the form

$$\dot{\epsilon}_c = \frac{AD\mu b}{kT}\left(\frac{p_a}{\mu}\right)^n \tag{8.86}$$

where A is a constant, D is a diffusion coefficient, μ is the shear modulus, and b is the Burgers vector. The exponent n depends on the mechanism of the dislocation motion and has values in the range of 3–10.

Hot Pressing Equations

The creep equations outlined above give relations for the linear strain rate and are applicable to fully dense solids. In the hot pressing of powder systems, the data are normally acquired in the form of density versus time from which the densification rate can be determined. Furthermore, considerable porosity is present over a large part of the process. The modification of the creep equations therefore involves two main steps: (i) relating the linear strain rate to the densification rate and (ii) compensating for the porosity.

In hot pressing, the mass of the powder, M, and the cross-sectional area of the die, A, are constants. As the density of the sample, D, increases, the sample thickness L decreases. The variables D and L are

Theory of Solid-State and Viscous Sintering

related by

$$\frac{M}{A} = LD = L_0 D_0 = L_f D_f \qquad (8.87)$$

where the subscripts 0 and f refer to the initial and final values. Differentiating this equation with respect to time gives

$$0 = L\frac{dD}{dt} + D\frac{dL}{dt} \qquad (8.87a)$$

Rearranging this equation and making use of the relation between the relative densification rate and the actual densification rate, i.e., $(1/\rho)d\rho/dt = (1/D)dD/dt$, we obtain

$$\frac{1}{D}\frac{dD}{dt} = \frac{1}{\rho}\frac{d\rho}{dt} = -\frac{1}{L}\frac{dL}{dt} \qquad (8.88)$$

As discussed in Chapter 12, the linear strain rate of the powder system can be obtained by measuring the distance traveled by the punch of the die as a function of time.

In compensating for the porosity of powder systems, we recall from our discussion of Coble's initial stage hot pressing model that the effective stress on the grain boundaries, p_e is related to the externally applied stress p_a by Eq. (8.80). To account for the effects of the applied stress and the surface curvature on the densification rate, Coble argued that the total driving force DF is a linear combination of the two effects; that is,

$$\text{DF} = p_e + \gamma_{sv} K = p_a \phi + \gamma_{sv} K \qquad (8.89)$$

where K is the curvature of the pore. For pores of radius r, $K = 1/r$ for the intermediate stage model and $K = 2/r$ for the final stage model. The driving force given by Eq. (8.90) is substituted for the stress p_a in the creep equations. To summarize at this stage, the hot pressing equations obtained by the modification of the creep equations are given in Table 8.6.

Coble's modification of the creep equations can provide only an approximation to the densification rate during hot pressing. A more rigorous analysis would require additional modifications to the creep models to better represent the situation existing in powder systems. One of these modifications pertains to the atomic flux field. In the creep models, the atomic flux terminates at the boundaries in tension. In the hot pressing of powder systems, the flux terminates at the pore surfaces. Another modification would be to incorporate the difference in the path length for diffusion. In the creep models, the grain boundary area remains constant

Table 8.6 Hot Pressing Equations Obtained by Modification of the Creep Equations

Mechanism	Intermediate stage	Final stage
Lattice diffusion	$\dfrac{1}{\rho}\dfrac{d\rho}{dt} = \dfrac{40}{3}\left(\dfrac{D_l\Omega}{G^2kT}\right)\left(p_a\phi + \dfrac{\gamma_{sv}}{r}\right)$	$\dfrac{1}{\rho}\dfrac{d\rho}{dt} = \dfrac{40}{3}\left(\dfrac{D_l\Omega}{G^2kT}\right)\left(p_a\phi + \dfrac{2\gamma_{sv}}{r}\right)$
Grain boundary diffusion	$\dfrac{1}{\rho}\dfrac{d\rho}{dt} = \dfrac{47.5\, D_{gb}\delta_{gb}\Omega}{G^3kT}\left(p_a\phi + \dfrac{\gamma_{sv}}{r}\right)$	$\dfrac{1}{\rho}\dfrac{d\rho}{dt} = \dfrac{7.5\, D_{gb}\delta_{gb}\Omega}{G^3kT}\left(p_a\phi + \dfrac{2\gamma_{sv}}{r}\right)$
Dislocation motion[a]	$\dfrac{1}{\rho}\dfrac{d\rho}{dt} = \dfrac{AD\mu b}{kT}\left(\dfrac{p_a\phi}{\mu}\right)^n$	$\dfrac{1}{\rho}\dfrac{d\rho}{dt} = \dfrac{BD\mu b}{kT}\left(\dfrac{p_a\phi}{\mu}\right)^n$

[a] A and B are numerical constants; n is an exponent that depends on the mechanism of dislocation motion.

and is related to the grain size. In powder systems, both the grain boundary area and the path length for diffusion change during densification.

C. Hot Pressing Mechanisms

The mechanisms discussed earlier for sintering also operate during hot pressing. However, the non-densifying mechanisms can be neglected because they are not enhanced by the applied pressure whereas the densifying mechanisms are significantly enhanced. New mechanisms can also be activated by the applied pressure. Particle rearrangement contributes to the densification during the initial stage but is difficult to analyze. It is generally recognized that grain boundary sliding is necessary to accommodate the diffusion controlled grain shape changes that occur in the intermediate and final stages. As illustrated in Fig. 8.25 for the two-dimensional situation, a representative element of the powder system (e.g., three hexagonal grains) must follow the overall shape change of the powder compact. In hot pressing, the diameter of the die is fixed so that compaction occurs predominantly in the direction of the applied pressure. The shape of the grains will be flattened in the direction of the applied pressure. The grains must slide over one another to accommodate the shape change. Grain boundary sliding and diffusional transport of matter are not independent mechanisms. They occur sequentially so that the slower mechanism controls the rate of densification. To summarize at this stage, the densifying mechanisms in hot pressing are: (i) lattice diffusion, (ii) grain boundary diffusion, (iii) plastic deformation by dislocation motion, (iv) viscous flow, (v) grain boundary sliding, and (vi) particle rearrangement.

Theory of Solid-State and Viscous Sintering

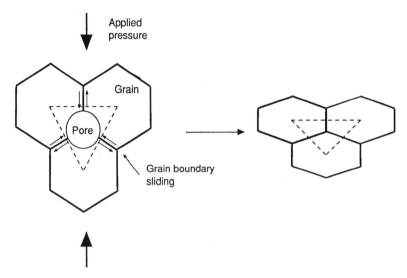

Figure 8.25 Sketch illustrating the change in grain shape that occurs during hot pressing. The grains are flattened in the direction of the applied pressure. When matter transport occurs by diffusion, grain boundary sliding is necessary to accommodate the change in grain shape.

According to Table 8.6, when the applied pressure, p_a, is much greater than the driving force for sintering, the densification rate during hot pressing can be written in the general form

$$\frac{1}{\rho}\frac{d\rho}{dt} = \frac{HD\phi^n}{G^m kT} p_a^n \tag{8.90}$$

where H is a constant, D is the diffusion coefficient of the rate controlling species, ϕ is the stress intensification factor, G is the grain size, k is the Boltzmann constant, T is the absolute temperature and the exponents m and n depend on the mechanism of densification. Table 8.7 shows the values of m and n for the various mechanisms. According to Eq. (8.90) and Table 8.7, hot pressing mechanisms can be identified by determining the exponents m and n from kinetic data for densification and grain growth. However, a difficulty is that in many cases more than one mechanism operates during hot pressing. As discussed in Chapter 12, it is more common to identify the dominant mechanism occurring under a given set of conditions such as applied pressure, grain size and temperature and to display the results in the form of a map (called a hot pressing or deformation map). Kinetic data for many ceramics at the moderate pressures (10–50 MPa) used in hot pressing give a pressure exponent n equal to

Table 8.7 Mechanisms of Densification During Hot Pressing and the Associated Exponents and Diffusion Coefficients

Mechanism	Grain size exponent (m)	Pressure exponent (n)	Appropriate diffusion coefficient*
Lattice diffusion	2	1	D_l
Grain boundary diffusion	3	1	D_{gb}
Plastic deformation	0	≥3	D_l
Viscous flow	0	1	—
Grain boundary sliding Particle rearrangement	1	1 or 2	D_l, D_{gb}

* D_l = lattice diffusion coefficient; D_{gb} = grain boundary diffusion coefficient.

≈ 1, which indicates that matter transport is controlled by diffusion. The dominance of diffusion has been attributed to the use of fine powders (which lead to rapid diffusion rates) and to the low density of dislocations in most ceramics (which make plastic flow difficult).

8.10 THE STRESS INTENSIFICATION FACTOR

In the hot pressing models, we found that the factor ϕ defined by Eq. (8.80) arose every time we needed to relate the mean stress on the grain boundary, p_e, to the externally applied stress, p_a. The factor ϕ is referred to as the stress intensification factor or the stress multiplication factor. It should not be confused with a nearly similar term, the stress intensity factor, which has a different meaning in fracture mechanics. The significance of ϕ is such that while p_a is the stress that is measured, p_e is its counterpart that influences the rate of matter transport. The factor ϕ is geometrical in origin and would be expected to depend on the porosity and the shape of the pores. In view of its significance, it would be useful to determine an expression for ϕ in terms of the parameters of the powder system.

Consider a hydrostatic pressure p_a applied to the external surface of a powder system, a model of which is shown schematically in Fig. 8.26. The applied pressure exerts on the surface of the solid an applied load $F_a = A_T p_a$, where A_T is the total external area of the solid, including such areas as may be occupied by pores. The presence of porosity located at the grain boundaries makes the actual grain boundary area A_e lower than the total external area. Assuming a force balance across any plane of the

Theory of Solid-State and Viscous Sintering

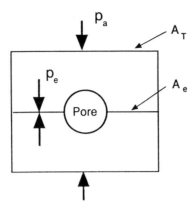

Figure 8.26 For a solid in which part of the grain boundary is occupied by pores, the effective stress on the grain boundary is greater than the externally applied stress. The stress intensification factor ϕ is defined as the ratio of the total external area to the actual grain boundary area.

solid, we obtain

$$p_a A_T = p_e A_e \tag{8.91}$$

Therefore,

$$\phi = \frac{p_e}{p_a} = \frac{A_T}{A_e} \tag{8.92}$$

i.e., the stress intensification factor is equal to the total cross-sectional area divided by the actual grain boundary area.

For spherical pores randomly distributed in a porous solid, ϕ can be easily found. Taking a random plane through the solid, the area fraction of porosity in the plane is equal to the volume fraction of porosity in the solid. Taking A_T equal to unity, then A_e is equal to $1 - P$, where P is the porosity of the solid. Therefore, ϕ is given by

$$\phi = \frac{1}{1 - P} = \frac{1}{\rho} \tag{8.93}$$

where ρ is the relative density of the solid. This expression would be applicable to a glass containing isolated, nearly spherical pores or to a polycrystalline solid in which the equilibrium shapes of the isolated pores are roughly spherical (i.e., for large dihedral angles). When the pores deviate significantly from a spherical shape, this simple expression will no longer hold. In fact, ϕ may be expected to be quite complex. As illustrated in Fig. 8.27 for pores with the same volume but different shapes,

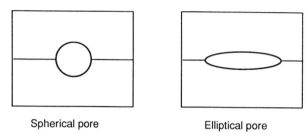

Spherical pore   Elliptical pore

Figure 8.27 The stress intensification factor depends on the porosity as well as the shape of the pores. Pores with lower dihedral angles, e.g., elliptical pores, lead to a stronger dependence on porosity than pores with higher dihedral angles, e.g., spherical pores.

we may expect ϕ to depend not only on the porosity but also on the shape of the pores. As the pore shape deviates from a spherical geometry (i.e., as the dihedral angle decreases), the actual area of the grain boundary decreases. The factor ϕ would therefore be expected to increase.

Computer simulation of the equilibrium shapes of a continuous network of pores has been performed by Beeré [14]. As we discussed earlier in this chapter, the equilibrium shape of a pore is dictated by the balance

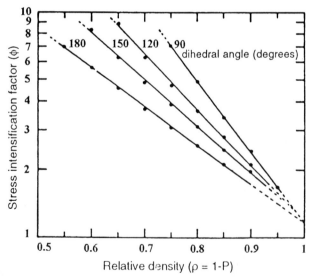

Figure 8.28 Stress intensification factor represented as an exponential function of the porosity. (From Ref. 32, with permission.)

between the interfacial tensions at the grain boundary. From an analysis of Beeré's results, Vieira and Brook [32] showed that the stress intensification factor can be approximated by the expression

$$\phi = \exp(aP) \tag{8.94}$$

where P is the porosity and a is a parameter that depends on the dihedral angle. The parameter a increases with decreasing dihedral angle. A semilogarithmic plot of Beeré's results is shown in Fig. 8.28 for several dihedral angles. For a given dihedral angle, the parameter a is found by taking the slope of the semilogarithmic plot. The expression for ϕ given by Eq. (8.94) has a much stronger dependence on porosity than that given by $\phi = 1/\rho$. It has been verified by recent experimental data [33].

8.11 THE SINTERING STRESS

For the diffusion mechanisms, the equations in Table 8.6 indicate that the densification rate may be written in the general form

$$\frac{1}{\rho}\frac{d\rho}{dt} = \frac{3}{L}\frac{dL}{dt} = \frac{3}{\eta_\rho}\left(p_a\phi + \frac{\alpha\gamma_{sv}}{r}\right) \tag{8.95}$$

where $(1/L)dL/dt$ is the linear strain rate of the sintering solid, η_ρ has the dimensions of viscosity and can be called the densification viscosity, and α is a geometrical constant that depends on the shape of the pore. For cylindrical pores in the intermediate stage sintering models, $\alpha = 1$, whereas for spherical pores in the final stage models, $\alpha = 2$. Equation (8.95) may also be written

$$\frac{1}{\rho}\frac{d\rho}{dt} = \frac{3\phi}{\eta_\rho}(p_a + \Sigma) \tag{8.96}$$

where

$$\Sigma = \alpha\gamma_{sv}/\phi r \tag{8.97}$$

According to Eq. (8.96), the densification rate is proportional to a linear combination of p_a and Σ, where Σ has the dimensions of a stress and depends on the microstructural characteristics of the sintering solid. Because it occurs in a linear combination with p_a, the quantity Σ may be taken to represent an equivalent externally applied hydrostatic stress that would have the same effect on the densification rate as the grain boundaries and pores. It has been referred to by various terms, including *sintering stress*, *sintering pressure*, and *sintering potential*. We shall use the term *sintering stress* in this book.

The expression for the sintering stress given by Eq. (8.97) would be applicable to a glass or to a polycrystalline solid in which the grain boundary energy is small compared with the surface energy (i.e., for large dihedral angles). A further condition is that the pores should have the same radius. For polycrystalline solids in which the grain boundary energy cannot be neglected, the expression for Σ becomes more complex. It contains an additional term that depends on the ratio of the grain boundary energy to the grain size [34].

The definition of the sintering stress as an equivalent externally applied stress has two advantages. First, combinations of the sintering stress with an applied mechanical stress become easier to analyze. We shall outline this approach to the derivation of the sintering equations in Section 8.12. Second, the sintering stress becomes an experimentally accessible parameter. Its variation with material parameters such as the powder particle size and density of the consolidated powder form can be studied to achieve a better understanding of sintering.

A. Measurement of the Sintering Stress

A relatively straightforward technique for the measurement of the sintering stress is similar to the zero-creep technique suggested by Gibbs [35] for the determination of surface energies. The powder system in the shape of a wire or a tape is clamped at one end, and a load is suspended from the other end (Fig. 8.29). The creep of the sample is monitored, and the load required to produce zero creep is determined. The sintering stress is taken as the applied stress at the zero-creep condition because the creep can be assumed to stop when the applied stress just balances the sintering stress. This technique was used by Gregg and Rhines [36] for the measurement of the sintering stress of copper powder compacts in the shape of a wire. They found that at a given density the sintering stress was inversely proportional to the initial particle size of the powder. If the average pore size is assumed to be proportional to the average particle size, then this observation of Gregg and Rhines is consistent with Eq. (8.97). Gregg and Rhines also found that the sintering stress increased with density up to a relative density of 0.95, after which it decreased. The increase can be attributed to the decrease in the pore size as densification proceeds. It is likely that significant coarsening of the microstructure (grain growth and pore growth) caused the observed decrease in the sintering stress above a relative density of 0.95.

For powder compacts in the form of cylindrical pellets, Rahaman et al. [33] used a creep-sintering technique in which a low uniaxial stress is

Theory of Solid-State and Viscous Sintering

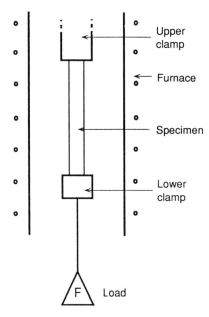

Figure 8.29 Schematic illustration of a method for the measurement of the sintering stress by the zero-creep technique.

applied to the compact during sintering. The application of the uniaxial stress causes the powder compact to undergo creep deformation in addition to densification. In the experiments, the magnitude of the applied stress was low enough (compared with the sintering stress) that it had an insignificant effect on the densification process. Measurement of the axial strain and the radial strain of the compact as a function of time yields data for the creep rate and the densification rate. The creep rate can be described by the equation

$$\dot{\epsilon}_c = \frac{1}{\eta_c} p_z \phi \tag{8.98}$$

where η_c is the creep viscosity and p_z is the applied uniaxial stress. For low p_z, Eq. (8.96) gives for the uniaxial strain rate of the sintering powder compact

$$\dot{\epsilon}_\rho = \frac{1}{3\rho} \frac{d\rho}{dt} = \frac{1}{\eta_\rho} \Sigma \phi \tag{8.99}$$

The ratio of the creep rate to the densification rate is given by

$$\dot{\epsilon}_c/\dot{\epsilon}_\rho = \frac{\eta_\rho}{\eta_c}\frac{p_z}{\Sigma} \tag{8.100}$$

Currently available data suggest that for uniaxial stresses, a close relationship exists between η_c and η_ρ. As a rough approximation, if the two viscosities are assumed to be equal, the sintering stress can be estimated from the data for the ratio of the creep rate to the densification rate and the applied stress. Figure 8.30 shows the sintering stress for CdO determined in this way. The sintering stress is on the order of 1–2 MPa, and for a given initial density ρ_0 it decreases slightly with increasing sintered density. This decrease is attributed to the occurrence of grain growth. For the same grain size, the sintering stress increases with increasing density.

When the applied stress is not insignificant compared with the sintering stress, Eq. (8.96) must be used for the densification rate. It must be remembered that in this equation p_a is a hydrostatic stress. For example, if a uniaxial stress p_z is applied, then the hydrostatic component $p_z/3$ must be used for p_a. A plot of the densification rate at fixed densities versus the applied hydrostatic stress yields the sintering stress by extrapolation to zero densification rate. It should be noted that this procedure neglects the effect of anisotropy (caused by the applied stress) on the densification rate.

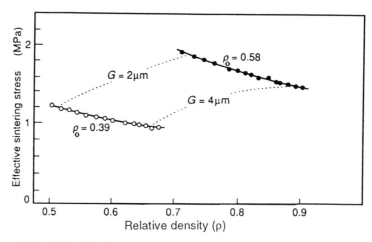

Figure 8.30 The effective sintering stress for CdO determined by a creep-sintering technique. (From Ref. 33.)

Theory of Solid-State and Viscous Sintering

8.12 AN ALTERNATIVE DERIVATION OF THE SINTERING EQUATIONS

In this section we outline how the sintering equations can be derived in an alternative manner to the earlier derivations for the analytical models. The differential equations for the flux of matter are solved subject to the appropriate boundary conditions. In addition, we make use of the concept of sintering stress. Let us consider matter transport by grain boundary diffusion.

A. Grain Boundary Diffusion

To simplify the derivation, a geometrical model consisting of two cylindrical grains is chosen so that the three-dimensional problem can be described by one cross section only. The three-dimensional situation can be visualized by considering the cross section shown in Fig. 8.31a to extend a large distance perpendicular to the plane of the paper. We will assume that the grain boundary remains flat and has a constant thickness δ_{gb}. From Eq. (7.76), the flux of atoms as a function of distance along the neck is

$$J_a = -\frac{D_{gb}}{\Omega kT}\frac{d\mu}{dx} \tag{8.101}$$

where D_{gb} is the grain boundary diffusion coefficient, μ is the chemical potential of the atoms, and the other terms are as defined earlier. The number of atoms transported into the neck per unit time per unit length of the cylinders is

$$J = -\frac{D_{gb}\delta_{gb}}{\Omega kT}\frac{d\mu}{dx} \tag{8.102}$$

The rate of approach of the centers of the cylinders is related to J by

$$J = -\frac{x}{\Omega}\frac{dY}{dt} \tag{8.103}$$

where Y is the distance between the centers of the cylinders. The minus sign is incorporated into the equation because J is a positive quantity and dY/dt is negative (i.e., Y decreases with time). For the boundary to remain flat, dY/dt must be independent of x. Differentiating Eqs. (8.103) and (8.104) with respect to x and equating the right-hand sides gives

$$\frac{D_{gb}\delta_{gb}}{\Omega kT}\frac{d^2\mu}{dx^2} = \frac{1}{\Omega}\frac{dY}{dt} \tag{8.104}$$

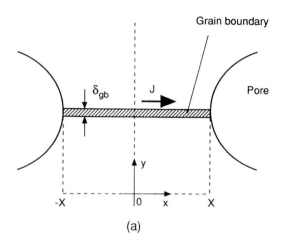

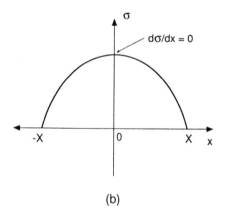

Figure 8.31 (a) Geometrical parameters for matter transport by grain boundary diffusion. (b) Schematic of the stress distribution on the grain boundary during sintering.

Theory of Solid-State and Viscous Sintering

Rearranging, we obtain

$$\frac{d^2\mu}{dx^2} = \frac{kT}{D_{gb}\delta_{gb}} \frac{dY}{dt} \tag{8.105}$$

From Chapter 7, we recall that the chemical potential of an atom under a flat surface that is acted on by a normal stress σ is given by

$$\mu = \sigma\Omega \tag{8.106}$$

Differentiating the expression with respect to x gives

$$\frac{d^2\mu}{dx^2} = \Omega \frac{d^2\sigma}{dx^2} \tag{8.107}$$

Equating the right-hand sides of Eqs. (8.105) and (8.107) and rearranging, we obtain

$$\frac{d^2\sigma}{dx^2} = \frac{kT}{D_{gb}\delta_{gb}\Omega} \frac{dY}{dt} \tag{8.108}$$

Integrating this equation gives

$$\sigma = \left(\frac{kT}{D_{gb}\delta_{gb}\Omega} \frac{dY}{dt}\right) \frac{x^2}{2} + C_1 x + C_2 \tag{8.109}$$

where C_1 and C_2 are constants of the integration that have to be determined from the boundary conditions.

The first boundary condition is that σ must be continuous and symmetric about $x = 0$. This means that

$$\frac{d\sigma}{dx} = 0 \quad \text{at } x = 0$$

so that $C_1 = 0$. The stress distribution described by Eq. (8.109) is therefore parabolic and can be sketched as shown in Fig. 8.31b. The second boundary condition is that the stresses must balance at $x = X$ and at $x = -X$. We will recall that according to the definition of sintering stress, the effects of the pores and grain boundaries were artificially set at zero and replaced by an equivalent external stress equal to Σ. In this representation, the stress acting at the surface of the pore must be equal to zero. Therefore our second boundary condition is

$$\sigma = 0 \quad \text{at } x = X; \quad x = -X$$

Substituting either of these into Eq. (8.109) gives

$$C_2 = -\beta X^2/2 \tag{8.110}$$

where

$$\beta = \frac{kT}{D_{gb}\delta_{gb}\Omega}\frac{dY}{dt} \tag{8.111}$$

The third boundary condition is that the average stress on the grain boundary must be equal to $\phi\Sigma$. This condition can be expressed as

$$\frac{1}{X}\int_0^x \sigma(x)dx = \phi\Sigma \tag{8.112}$$

Carrying out the integration and substituting for C_2 from Eq. (8.110) gives

$$-\beta X^2/3 = \phi\Sigma \tag{8.113}$$

Substituting for β from Eq. (8.111) and rearranging, we obtain

$$\frac{dY}{dt} = \frac{3D_{gb}\delta_{gb}\Omega}{kTX^2}\phi\Sigma \tag{8.114}$$

Putting the center-to-center distance between the cylinders, Y, equal to the grain size G and considering the solid as made up of a planar row of cylinders, the uniaxial strain rate of the sintering solid is given by

$$\dot{\epsilon}_\rho = \frac{1}{Y}\frac{dY}{dt} = \frac{3D_{gb}\delta_{gb}\Omega}{GX^2kT}\phi\Sigma \tag{8.115}$$

A problem with this equation is that the measurement of the neck size or interpore distance X is fairly laborious. However, it can be related to the grain size (which is easier to measure) through the stress intensification factor ϕ. For our model of two cylindrical grains, $\phi = G/X$.

Substituting for X in Eq. (8.115) gives

$$\dot{\epsilon}_\rho = \frac{3D_{gb}\delta_{gb}\Omega}{G^3kT}\phi^3\Sigma \tag{8.115a}$$

For an isotropic solid

$$\phi = G^2/X^2 \tag{8.116}$$

It can be shown without too much difficulty that for a model consisting of spherical grains separated by a grain boundary in the form of a circular disk, the densification rate is given by

$$\dot{\epsilon}_\rho = \frac{8\pi D_{gb}\delta_{gb}\Omega}{G^3kT}\phi^2\Sigma \tag{8.117}$$

B. Lattice Diffusion

The derivation of the sintering equation for lattice diffusion is more difficult to analyze. The usual approach involves a simplification in which Eq. (8.117) is modified by replacing D_{gb} by D_l and δ_{gb} by $2X$. Making these substitutions and remembering that $\phi = G^2/X^2$ gives

$$\dot{\epsilon}_\rho = \frac{6D_l\Omega}{G^2kT} \phi^{3/2}\Sigma \tag{8.118}$$

C. General Isothermal Sintering Equation

For matter transport by diffusion, a general equation for the densification rate can be written

$$\dot{\epsilon}_\rho = \frac{HD\phi^{(m+1)/2}}{G^m kT}(p_a + \Sigma) \tag{8.119}$$

where H is a numerical constant that depends on the model, p_a is an externally applied hydrostatic stress, and

$$D = D_{gb}; \quad m = 3 \quad \text{for grain boundary diffusion}$$
$$D = D_l; \quad m = 2 \quad \text{for lattice diffusion}$$

Compared with Coble's hot pressing equations (Table 8.5) obtained by modifying the creep equations for dense solids, Eq. (8.119) has the same dependence on the material and physical parameters except for the value of the exponent for ϕ. The difference in the exponents arises because Coble assumed that $G = X$ whereas the present derivation related G to X through the stress intensification factor.

D. General Isothermal Creep Equation for Porous Solids

The procedure used in the derivation of Eq. (8.117) can also be used to analyze diffusional creep of porous solids. However, there is one major difference: the mean stress on the grain boundary, Eq. (8.112), is now $p_z\phi$, where p_z is an applied uniaxial stress. Continuing the development as outlined for the case of densification, we can obtain a general creep equation of the form

$$\dot{\epsilon}_c = \frac{H'D\phi^{(m+1)/2}}{G^m kT} p_z \tag{8.120}$$

where H' is a numerical constant and the other parameters are exactly the same as those defined by Eq. (8.119).

8.13 CONCLUDING REMARKS

In this chapter we have considered the main theoretical approaches for the description of solid-state sintering and viscous sintering. The models for the densification of glasses by viscous flow assume Frenkel's energy balance concept in the derivation of the equations for the sintering rate. The predictions of the models give good agreement with the data for real powder compacts of glass particles. The sintering phenomena in polycrystalline ceramics are considerably more complex than those in glasses. Matter transport is assumed to occur along specified paths that define the mechanisms of sintering. At least six different mechanisms may operate in polycrystalline systems. We need to distinguish between densifying mechanisms and nondensifying mechanisms. The analytical models yield explicit expressions for the dependence of the sintering rate on the primary variables such as particle size, temperature, and applied pressure. However, in view of the drastic simplification used in their development, the models cannot provide a quantitative description of sintering. The stress intensification factor and the sintering stress are important parameters in the understanding of the sintering process. These parameters were used in an alternative derivation of the sintering equations that involved the solution of the differential equation for the flux of atoms. Following the same procedure, the development of a general equation for the diffusional creep of porous solids was outlined. The analysis of the sintering rate is, by itself, not very useful. We must also understand how the microstructure of the powder system evolves during sintering. The results of the present chapter must therefore be combined with those of the following chapter in order to develop a more useful understanding of solid-state sintering.

PROBLEMS

8.1 For the idealized solid-state sintering theory, make a table of the three stages of sintering and for each stage give the mechanisms of sintering, stating which mechanisms lead to densification and which do not.

8.2 Derive the equations for the neck length, X, the surface area of the neck, A, and the volume of material transported into the neck, V, given in Fig. 8.10.

8.3 Show that in Eq. (8.11), $m = 4$ for grain boundary diffusion and $m = 2$ for vapor transport.

8.4 Derive the equations given in Table 8.5 for
   (a) lattice diffusion from the grain boundary,
   (b) vapor transport and
   (c) surface diffusion

Theory of Solid-State and Viscous Sintering

8.5 The lattice diffusion coefficient for Al^{3+} ions in Al_2O_3 at 1600°C is $\approx 2 \times 10^{-9}$ cm^2/sec and the activation energy is 580 kJ/mol. Assuming that sintering is controlled by lattice diffusion of Al^{3+} ions, estimate the initial rate of densification for an Al_2O_3 powder compact of 1 μm particles at 1500°C.

8.6 Consider a pore with a constant volume. Assuming that the pore takes on the shape of an ellipsoid of revolution, calculate the surface area of the pore as a function of the ratio of the small axis to the long axis.

8.7 Derive Eq. (8.58) for the volume of the solid phase in a unit cell of Scherer's model.

8.8 For the sintering of a single spherical pore in an infinite viscous matrix, derive an expression for the radius of the pore as a function of the sintering time.

8.9 A pressure of 30 MPa is applied to a powder during hot pressing. Assuming that the stress intensification factor is given by Eq. (8.94) and the pores achieve their equilibrium shape, estimate the stress on the grain boundaries at a relative density of 0.75 when the equilibrium dihedral angle is
  (a) 90° and
  (b) 180°.
Compare the results with the value estimated using Eq. (8.93) for the stress intensification factor.

REFERENCES

1. C. J. Brinker and G. W. Scherer, *Sol-Gel Science*, Academic, New York, 1990, Chap. 11.
2. J. Frenkel, *J. Phys. (Moscow)*, 5:385 (1945).
3. F. N. Rhines and R. T. DeHoff, *Mater. Sci. Res.*, 16:49 (1984).
4. G. C. Kuczynski, *Z. Metallk.*, 67:606 (1976).
5. C. Herring, *J. Appl. Phys.*, 21:301 (1950).
6. H. Song, R. L. Coble, and R. J. Brook, *Mater. Sci. Res.*, 16:63 (1984).
7. R. L. Coble, *J. Appl. Phys.*, 32:787, 793 (1961).
8. G. C. Kuczynski, *Trans. AIME*, 185:169 (1949).
9. W. D. Kingery and M. Berg, *J. Appl. Phys.*, 26:1206 (1955).
10. R. L. Coble, *J. Am. Ceram. Soc.*, 41:55 (1958).
11. D. L. Johnson and I. B. Cutler, *J. Am. Ceram. Soc.*, 46:541 (1963).
12. G. C. Kuczynski, *J. Appl. Phys.*, 20:1160 (1949).
13. D. L. Johnson, *J. Am. Ceram. Soc.*, 53:574 (1970).
14. W. Beeré, *Acta Metall.*, 23:131, 139 (1979).
15. G. W. Scherer, *J. Am. Ceram. Soc.*, 60:236 (1977).
16. G. W. Scherer and D. L. Bachman, *J. Am. Ceram. Soc.*, 60:239 (1977).

17. S. Coleman and W. Beeré, *Phil. Mag.*, *31*:1403 (1975).
18. J. K. Mackenzie and R. Shuttleworth, *Proc. Phys. Soc. (Lond.)*, *62*:833 (1949).
19. F. A. Nicholls and W. W. Mullins, *J. Appl. Phys.*, *36*:1826 (1965).
20. P. Bross and H. E. Exner, *Acta Metall.*, *27*:1013 (1979).
21. D. L. Johnson, *J. Appl. Phys.*, *40*:192 (1969).
22. M. H. Tikkanen and S. A. Makipirtti, *Int. J. Powder Metall.*, *1*:15 (1965).
23. V. A. Ivensen, *Densification of Metal Powders During Sintering*, Consultants Bureau, New York, 1973.
24. S. Pejovnik, V. Smolej, D. Susnik, and D. Kolar, *Powder Metall. Int.*, *11*: 22 (1979).
25. M. F. Ashby, *Acta Metall.*, *22*:275 (1974).
26. F. B. Swinkels and M. F. Ashby, *Acta Metall.*, *29*:259 (1981).
27. R. L. Coble, in *Sintering and Related Phenomena* (G. C. Kuczynski, N. A. Hooton, and C. F. Gibbon, Eds.), Gordon and Breach, New York, 1967, p. 329.
28. R. L. Coble, *J. Appl. Phys.*, *41*:4798 (1970).
29. F. R. N. Nabarro, in *Report of a Conference on the Strength of Solids*, Physical Society, London, 1948, p. 75.
30. C. Herring, *J. Appl. Phys.*, *21*:437 (1950).
31. R. L. Coble, *J. Appl. Phys.*, *34*:1679 (1963).
32. J. M. Vieira and R. J. Brook, *J. Am. Ceram. Soc.*, *67*:245 (1984).
33. M. N. Rahaman, L. C. De Jonghe, and R. J. Brook, *J. Am. Ceram. Soc.*, *69*:53 (1986).
34. L. C. De Jonghe and M. N. Rahaman, *Acta Metall.*, *36*:223 (1988).
35. J. W. Gibbs, in *The Scientific Papers of J. Willard Gibbs*, Dover, New York, 1961.
36. R. A. Gregg and F. N. Rhines, *Metall. Trans.*, *4*:1365 (1973).
37. A. Jagota and P. R. Dawson, *J. Am. Ceram. Soc.*, *73*: [1] 173 (1990).
38. A. Jagota, *J. Am. Ceram. Soc.*, *77*:[8]2237 (1994).
39. J. Svoboda and H. Riedel, *Acta Metall. Mater.*, *43*:[1]1 (1995).

FURTHER READING

S. Somiya and Y. Moriyoshi (Eds.), *Sintering Key Papers*, Elsevier Applied Science, New York (1990).

H. E. Exner, *Reviews on Powder Metallurgy and Physical Ceramics*, *1*:(1–4) (1979), pp. 1–251.

G. C. Kuczynski, N. A. Hooton and C. F. Gibbon, *Sintering and Related Phenomena*, Gordon and Breach, New York (1967).

C. A. Handwerker, J. E. Blendell and W. A. Kaysser, *Sintering of Advanced Ceramics*, Ceramic Transactions, Vol. 7, American Ceramic Society, Westerville, OH (1990).

Several volumes in the series: *Materials Science Research* (Plenum, New York) contain useful papers on sintering.

9
Grain Growth and Microstructural Control

9.1 INTRODUCTION

The engineering properties of ceramics, as outlined in Chapter 1, are strongly dependent on the microstructure. The important microstructural features are the size and shape of the grains, the amount of porosity, the pore size, the distribution of the pores in the structure, and the nature and distribution of any second phases. We must control these microstructural features in order to achieve the desired properties in the fabricated article. For most applications, microstructural control usually means the achievement of as high a density, as small a grain size, and as homogeneous a microstructure as possible. We recall that the microstructure of the fabricated article is influenced to a significant extent by the structure of the consolidated powder form, which in turn depends strongly on the characteristics of the powder and the powder consolidation method. However, even if proper procedures are employed in the production of the consolidated powder form (which is rarely the case), further manipulation of the microstructure is necessary during sintering.

The densification of a polycrystalline powder compact is normally accompanied by a coarsening of the microstructure: the average size of the grains and the average size of the pores become larger. In Chapter 8 we considered the densification process. We must now address the coarsening process and the interplay between the densification and coarsening processes.

To achieve a basic understanding of how grains grow and how such growth can be controlled, it is convenient to first consider a fully dense, single-phase solid. In ceramics, grain growth is divided into two main types: normal and abnormal. In normal grain growth, grain sizes and shapes fall within a fairly narrow range and, except for a magnification factor, the grain size distribution at a later time is similar to that at an earlier time. Abnormal grain growth is characterized by the rapid growth of a few larger grains at the expense of the smaller ones. Simple models have been developed to predict the kinetics of normal grain growth. Most of the models analyze an isolated grain boundary or a single grain and neglect the topological requirements of space filling. In recent years, computer simulation has begun to play an important role in exploring the interplay between the kinetics of grain growth and the topological requirements of space filling.

In the sintering of powder compacts, we must understand the interaction between the pores and the moving grain boundaries. A common approach is to analyze the kinetics of grain growth and of pore growth. However, an understanding of the evolution of the network of porosity in the structure is also important for a broader understanding of microstructural development. The interaction between spherical isolated pores and the grain boundaries can be analyzed in terms of simple models. Since grain growth is more pronounced in the final stage of sintering, this interaction plays a critical role in determining the limits of densification. A key result is that the separation of the boundaries from the pores (a condition that is symptomatic of abnormal grain growth in porous solids) must be prevented if high density is to be achieved. Coarsening also occurs in the earlier stages of sintering but is less pronounced than in the final stage. Since the microstructural evolution in the earlier stages of sintering influences that in the later stages, an understanding of the coarsening of very porous compacts is also important. However, the complexity of the microstructure makes a quantitative analysis of the process very difficult.

The densification process and the coarsening process are by themselves very complex. A detailed theoretical analysis of the interplay between these two processes does not appear to be a fruitful approach. However, simple analyses indicate that the achievement of high density and controlled grain size in the fabricated article is dependent on reducing the grain growth rate, increasing the densification rate, or some combination of the two. Fabrication approaches that satisfy one or both of these conditions include the use of dopants that segregate to the grain boundaries, fine second-phase particles that do not react with the grains, homogeneous packing of powder with a narrow size distribution, sintering with an externally applied pressure (e.g., hot pressing), fast heating rates to the firing temperature (fast firing), and the use of second phases that form a liquid at the firing temperature.

Grain Growth and Microstructural Control

9.2 GRAIN GROWTH: PRELIMINARY CONSIDERATIONS

Before we consider the details of grain growth in dense and porous ceramics, we outline some common terms and general features associated with the process.

A. Grain Growth and Coarsening

Grain growth is the term used to describe the increase in the grain size of a single-phase solid or in the matrix grain size of a solid containing second-phase particles. Grain growth occurs in both dense and porous polycrystalline solids at sufficiently high temperatures. For the conservation of matter, the sum of the individual grain sizes must remain constant; the increase in the average size of the grains is therefore accompanied by the disappearance of some grains, usually the smaller ones.

In porous solids, both the grains and the pores normally increase in size while decreasing in number. There is considerable interaction between the grains and the pores, and the microstructural evolution is considerably more complex than for dense solids. Frequently the term *coarsening* is used to describe the process by which grains and pores grow.

Figure 9.1 Classical picture of a grain boundary. The boundary migrates downward as atoms move less than an interatomic spacing to new positions.

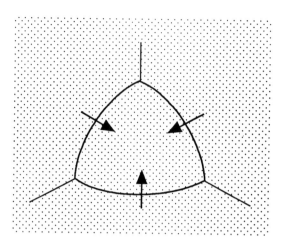

Figure 9.2 Movement of the grain boundary toward its center of curvature.

B. Occurrence of Grain Growth

In the widely accepted picture, the grain boundary is considered to be a region of disorder between two crystalline regions (the grains), as sketched in Fig. 9.1. High-resolution electron microscopy indicates that the thickness of the grain boundary region is ≈1 nm. Grain growth occurs as atoms (or ions) diffuse less than an interatomic distance to new positions. Thus one grain grows at the expense of another. The atoms move from the "convex" surface on one side of the grain boundary to the "concave" surface on the other side more readily than in the reverse direction. The reason for the net flow from the convex to the concave side is that the chemical potential of the atoms under the convex surface is higher than that of the atoms under the concave surface. Atomic flux, as we will recall, occurs down the chemical potential gradient. The result of the net flux is that the boundary moves toward its center of curvature (Fig. 9.2).

C. Driving Force for Grain Growth

As discussed earlier in this book, the atoms in the grain boundary have a higher energy than those in the bulk of the crystalline grain. The grain boundary is characterized by a specific energy, denoted γ_{gb} in this book, typically on the order of 0.2–1.0 J/m². The driving force for grain growth is the decrease in grain boundary energy that results from a decrease in the grain boundary area.

D. Normal and Abnormal Grain Growth

Grain growth in ceramics is generally divided into two types: (i) normal grain growth and (ii) abnormal grain growth, which is sometimes referred to as exaggerated grain growth, discontinuous grain growth, or, in the case of metals, secondary recrystallization. In addition, for some relatively soft ceramics (e.g., NaCl) and metals that have been highly deformed, another class of grain growth, called primary recrystallization, may occur. In primary recrystallization, a new generation of strain-free grains nucleate and grow at the expense of the highly deformed grains. The driving force for primary recrystallization is the decrease in strain energy of the solid. In this book we shall be concerned only with normal and abnormal grain growth, the driving force for which, as noted above, is the decrease in grain boundary energy.

Normal grain growth is generally defined by two main characteristics: (i) The grain sizes and shapes occur within a fairly narrow range and (ii) the distribution in grain sizes at a later time is fairly similar to that at an early time except for a magnification factor. The form of the grain size distribution is therefore time-invariant. This time-invariant characteristic is sometimes expressed as the grain size distribution having the property of scaling or self-similarity.

In abnormal grain growth, a few large grains develop and grow fairly rapidly at the expense of the smaller ones. The grain size distribution may change significantly, giving rise to a bimodal distribution. In such a case, the property of time invariance of the distribution is lost. Eventually, the large grains impinge on each other and may revert to a normal distribution of sizes.

Grain growth in porous ceramics is also described as normal and abnormal. However, the interactions of the pores with the grains must also be taken into account. Thus, normal grain growth in porous ceramics is, in addition, characterized by the pores remaining in the grain boundaries. When the boundaries break away from the pores, leaving them inside the grains, the situation is usually indicative of abnormal grain growth.

Figure 9.3 shows examples of normal and abnormal grain growth in dense and porous ceramics [1,2]. A dense Al_2O_3 structure with normal grains is shown in Fig. 9.3a. If the structure is heated further, a few abnormal grains may start to develop (Fig. 9.3b); they grow rapidly and eventually consume the smaller grains. Figure 9.3c shows a combination of normal and abnormal grain growth in a nickel-zinc ferrite. The pores deep inside the large, abnormal grain to the left should be noticed; for these pores, the diffusion distances are fairly large. As we shall show below, they are difficult to remove during sintering, thereby limiting the final

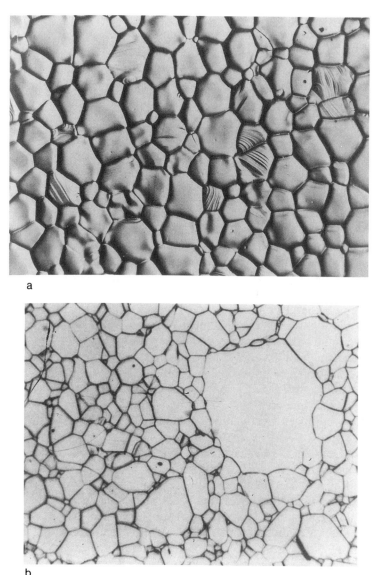

Figure 9.3 (a) A normal grain size distribution in an alumina ceramic. (b) Initiation of abnormal grain growth in an alumina ceramic. (c) Normal and abnormal grain growth in a porous nickel-zinc ferrite. In the fine-grained array to the right, the pores lie almost exclusively on the grain boundaries and can be removed relatively easily. The pores inside the large abnormal grain to the left are difficult to remove. (d) Pores in an alumina ceramic that has undergone abnormal grain growth. [(a, b) from Ref. 1; (c, d) from Ref. 2.]

Grain Growth and Microstructural Control

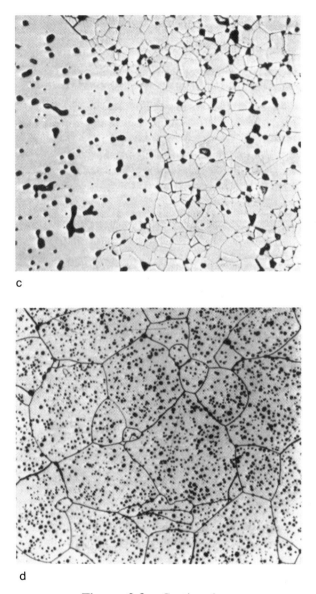

c

d

Figure 9.3 Continued.

density that can be achieved during firing. In the fine-grained array to the right, the pores lie almost exclusively on the grain boundaries. Figure 9.3d shows a sample of Al_2O_3 that has undergone abnormal grain growth.

E. Importance of Controlling Grain Growth

The control of grain growth during firing forms one of the most important considerations in the fabrication of ceramics. Its importance has two main causes. First, the grain size of the fabricated article is one of the major factors that control its engineering properties. Control of grain growth is therefore directly related to the achievement of the desired properties. Second, many engineering properties are improved with higher density, typically close to the theoretical density. Control of grain growth forms an important fabrication approach for the attainment of high density.

Effect of Grain Size on Properties

Few properties are completely independent of grain size. Within the limited space available in this chapter, we can provide only a few well-known relationships for the effect of grain size on properties. The subject is covered in many excellent texts and review articles, including that of Kingery et al. [3]. The fracture strength of many ceramics is found to vary as $1/G^{1/2}$, where G is the grain size. In the diffusional creep of polycrystalline solids, a very strong dependence of the creep rate on the grain size is found [see Eqs. (8.84) and (8.85)]. A wide range of electrical and magnetic phenomena are affected by the grain size, and it is in this area that the manipulation of the grain size has been used most significantly to produce materials with properties suitable for a variety of applications. For example, the electrical breakdown strength of ZnO varistors used in electrical surge suppressors varies as $1/G$. The dielectric constant of $BaTiO_3$ capacitors is found to increase with decreasing grain size (down to ≈ 1 μm).

Attainment of High Density

As we discussed in Chapter 8, densification is caused by the flux of matter from the grain boundaries (the source) to the pores (the sink). For sintering by diffusion mechanisms, the dependence of the densification rate on the grain size G can be written

$$\frac{1}{\rho}\frac{d\rho}{dt} = \frac{K}{G^m} \qquad (9.1)$$

where K is a constant at a given temperature and the exponent m is equal to 3 for lattice diffusion and 4 for grain boundary diffusion. For rapid densification, the distance between the source and the sink must be kept small, i.e., G must remain small (Fig. 9.4a). Rapid grain growth causes a drastic reduction in the densification rate. Prolonged firing times are there-

Grain Growth and Microstructural Control

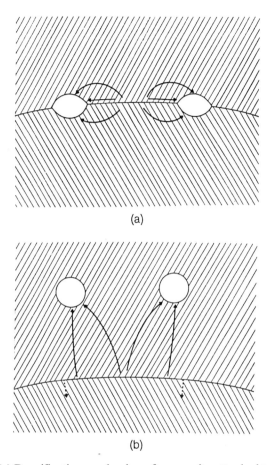

Figure 9.4 (a) Densification mechanisms for porosity attached to a grain boundary. The arrows indicate paths for atomic diffusion. (b) Densification mechanisms for porosity separated from a grain boundary. The solid arrows indicate paths for atomic diffusion, and the dashed arrows indicate the direction of boundary migration. (From Ref. 1.)

fore needed to achieve the required density, and this increases the possibility for abnormal grain growth to occur. When abnormal grain growth occurs, the pores become trapped inside the grains and become difficult or almost impossible to remove. The transport paths become large in the case of lattice diffusion or are eliminated if the mass transport mechanism is grain boundary diffusion (Fig. 9.4b). To summarize at this stage, the

Chapter 9

attainment of high density requires the control of normal grain growth and, more important, the avoidance of abnormal grain growth.

9.3 OSTWALD RIPENING: THE LSW THEORY

Ostwald ripening refers to the coarsening of precipitates (particles) in a solid or liquid medium. However, many features of grain growth and pore growth during firing are shared by the Ostwald ripening process. We will therefore describe the main features of the process.

Let us consider a system consisting of a dispersion of spherical particles with different radii in a medium in which the particles have some solubility (Fig. 9.5). We will recall from Chapter 7 that the chemical potential of the atoms under the surface of a sphere of radius R is given by

$$\mu = \mu_0 + \frac{2\gamma\Omega}{R} \qquad (9.2)$$

where μ_0 is the chemical potential of the atoms under a flat surface, γ is the specific energy of the interface between the sphere and the medium and Ω is the atomic volume. Due to their higher chemical potential, the atoms under the surface of the sphere have a higher solubility than those under a flat surface. Making use of the relation between chemical potential and concentration [Eq. (7.57)] and assuming ideal solutions, we obtain

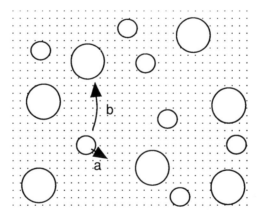

Figure 9.5 Coarsening of particles in a medium by matter transport from the smaller to the larger particles. Two separate mechanisms can control the rate of coarsening: (a) reaction at the interface between the particles and the medium and (b) diffusion through the medium.

Grain Growth and Microstructural Control

$$kT \ln\left(\frac{C}{C_0}\right) = \mu - \mu_0 = \frac{2\gamma\Omega}{R} \tag{9.3}$$

where C is the concentration of solute around a particle of radius R, C_0 is the concentration over a flat surface, k is the Boltzmann constant, and T is the temperature. If $\Delta C = C - C_0$ is small, then $\ln(C/C_0) = \Delta C/C_0$ and Eq. (9.3) becomes

$$\frac{\Delta C}{C_0} = \frac{2\gamma\Omega}{kTR} \tag{9.4}$$

The higher concentration around a particle of smaller radius gives rise to a net flux of matter from the smaller particles to the larger ones. The dispersion of particles coarsens by a process in which the smaller particles dissolve and the larger ones grow. This coarsening process is referred to as Ostwald ripening, in honor of the work of Ostwald in this area around 1900. The driving force for the process is the reduction in the interfacial area between the particles and the medium.

A. The LSW Theory

The basic theory of Ostwald ripening was developed independently by Greenwood [4], Wagner [5], and Lifshitz and Slyozov [6]. It is often referred to as the LSW theory. It applies strictly to a dilute dispersion of particles in a medium.

Returning to Fig. 9.5, we can identify two different steps that may control the rate of coarsening. In one case, the solubility of the particles in the medium or the deposition of the solute onto the particle surfaces may be the slowest step. Under these conditions, the Ostwald ripening process is said to be controlled by the interface reaction. In the other case, the diffusion of atoms through the medium may be the slowest step, and the process is said to be diffusion-controlled.

B. Ostwald Ripening Controlled by the Interface Reaction

In his analysis, Wagner assumed that the rate of transfer of atoms is proportional to the difference between the solute concentration around a particle of radius R given by Eq. (9.4) and an average concentration of the solute, C^*, defined as the concentration that is in equilibrium with particles of radius R^* that neither grow nor shrink. It is also assumed that the change in C^* with increasing R^* can be neglected. The rate of change of the particle radius can be written

$$\frac{dR}{dt} = -\alpha_T \Omega (C_R - C^*) \tag{9.5}$$

where α_T is a transfer constant and the negative sign is included to indicate that for smaller particles, $C_R - C^*$ is positive, but the radius R decreases with time. If the total volume of the particles is constant, then

$$4\pi \sum_i R_i^2 \frac{dR_i}{dt} = 0 \tag{9.6}$$

where the summation is taken over all the particles in the system. Writing Eq. (9.5) as

$$\frac{dR}{dt} = \alpha_T \Omega[(C^* - C_0) - (C_R - C_0)] \tag{9.7}$$

and then substituting this equation into Eq. (9.6) and rearranging gives

$$\sum_i R_i^2 (C^* - C_0) = \sum_i R_i^2 (C_{R_i} - C_0) \tag{9.8}$$

Putting $\Delta C_{Ri} = C_{Ri} - C_0$ and using Eq. (9.4), Eq. (9.8) becomes

$$C^* - C_0 = \frac{2\gamma \Omega C_0}{kT} \left(\frac{\sum R_i}{\sum R_i^2} \right) \tag{9.9}$$

where the summation is taken over all the particles in the system. Substituting Eq. (9.9) into Eq. (9.7) gives

$$\frac{dR}{dt} = \frac{2\alpha_T \gamma \Omega^2 C_0}{kT} \left[\frac{\sum R_i}{\sum R_i^2} - \frac{1}{R} \right] \tag{9.10}$$

Putting $R^* = \sum R_i^2 / \sum R_i$, Eq. (9.10) can be written

$$\frac{dR}{dt} = \frac{2\alpha_T \gamma \Omega^2 C_0}{kT} \left(\frac{1}{R^*} - \frac{1}{R} \right) \tag{9.11}$$

According to this equation, the rate of change of the particle radius is proportional to the difference between the critical curvature and the curvature of the particle.

In Wagner's analysis, the evolution of a system of particles is described by a distribution function $f(R, t)$ such that $f(R, t)dR$ represents the fractional number of particles between R and $R + dR$. The distribution function must satisfy the differential equation

$$\frac{df}{dt} + \frac{\partial}{\partial R}\left[\left(\frac{dR}{dt}\right)f\right] = 0 \tag{9.12}$$

The solution of the coupled differential equations (9.11) and (9.12) is not trivial, and the reader is referred to the original analysis by Wagner. It is

Grain Growth and Microstructural Control

found that the critical radius follows a simple parabolic time law,

$$(R^*)^2 - (R_0^*)^2 = \left(\frac{\alpha_T C_0 \gamma \Omega^2}{kT}\right) t \quad (9.13)$$

where R_0^* is the initial critical radius. When the particle radius is expressed in terms of a reduced size $s = R/R^*$, the distribution function takes the form

$$f(s, t) = \begin{cases} \dfrac{s}{(2-s)^5} \exp\left(\dfrac{-3s}{2-s}\right) & \text{for } 0 < s < 2 \\ 0 & \text{for } s > 2 \end{cases} \quad (9.14)$$

In this distribution, the average radius, $\overline{R}$, taken as the arithmetic mean radius, is given by $\overline{R} = (8/9)R^*$. The maximum particle radius $R_{\max}$ is equal to $2R^*$. Equation (9.14) shows that the distribution function f exhibits the property of self-similarity, since it depends only on R/R^*. Furthermore, the function is independent of the initial size distribution. Regardless of the initial size distribution, the particle size distribution reaches a self-similar distribution given by Eq. (9.14). The distribution function f is sketched in Fig. 9.6.

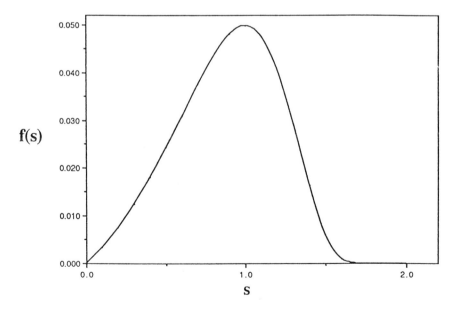

Figure 9.6 The particle size distribution function plotted versus the reduced size s, equal to R/R^*, for Ostwald ripening controlled by the interface reaction.

C. Ostwald Ripening Controlled by Diffusion

The solution of the coarsening problem when the diffusion step is rate-controlling follows along much the same lines as that for reaction control. The rate of change of the particle radius is

$$\frac{dR}{dt} = -D\Omega \frac{dC}{dR} \tag{9.15}$$

where D is the diffusion coefficient for the solute atoms in the medium. For a dilute dispersion of particles in the medium, Eq. (9.15) can be written

$$\frac{dR}{dt} = D\Omega \frac{C^* - C_R}{R} = \frac{DC_0 \gamma \Omega^2}{kTR} \left(\frac{1}{R^*} - \frac{1}{R} \right) \tag{9.16}$$

Once again the coupled differential equations (9.12) and (9.16) have to be solved. The kinetic equation derived by Lifshitz and Slyozov and by Wagner is

$$(R^*)^3 - (R_0^*)^3 = \left(\frac{8DC_0 \gamma \Omega^2}{9kT} \right) t \tag{9.17}$$

For diffusion control, we notice that a cubic growth law is predicted, as opposed to the parabolic law predicted for the case of reaction control. The distribution function for the radius of the particles takes the form

$$f(s,t) = \begin{cases} s^2 \left(\frac{3}{3+s} \right)^{7/3} \left(\frac{3/2}{3/2 - s} \right)^{11/3} \exp\left(-\frac{s}{3/2 - s} \right) & \text{for } 0 < s < 3/2 \\ 0 & \text{for } s > 3/2 \end{cases} \tag{9.18}$$

In this case, the average radius $\bar{R}$ is equal to R^* and the maximum particle radius is equal to $3R^*/2$. As for the case of reaction control, the distribution function given by Eq. (9.18) exhibits the property of self-similarity and is independent of the initial size distribution.

9.4 TOPOLOGICAL AND INTERFACIAL TENSION REQUIREMENTS

A dense polycrystalline solid consists of a space-filling array of grains. As we shall see later, many theories of grain growth consider only the surface tension (or surface curvature) requirements of an isolated grain boundary or an isolated grain. However, in a classic paper in 1952, Smith [7] recognized that for normal grain growth to occur, certain topological

Grain Growth and Microstructural Control

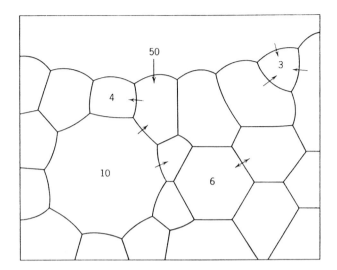

Figure 9.7 Sketch of a section through a dense polycrystalline solid. The sign of the curvature changes as the number of sides increases from less than six to more than six. The arrows indicate the direction in which the boundaries migrate. (Courtesy of J. E. Burke.)

requirements of space filling as well as the balance of interfacial tensions must be satisfied. To keep the analysis simple, we shall consider a two-dimensional section through a dense polycrystalline solid (Fig. 9.7). The structure consists of vertices joined by edges (also called sides) that surround faces. Provided that the face at infinity is not counted, the numbers of faces F, edges E, and vertices V, obey Euler's equation

$$F - E + V = 1 \tag{9.19}$$

For stable topological structures, i.e., those for which the topological features are unchanged by small deformations, the number of edges that intersect at a vertex is equal to 3. For isotropic grain boundary energies, i.e., γ_{gb} the same for all grain boundaries, the edges must meet at an angle of 120°.

Let us consider polygons in which the sides intersect at 120°. Taking N as the number of sides, under such conditions a hexagon ($N = 6$) has plane sides whereas a polygon with $N > 6$ has concave sides and one with $N < 6$ has convex sides (Fig. 9.8). Returning to the two-dimensional sketch shown in Fig. 9.7, we see that grains with straight sides can occur only when $N = 6$. Grains with $N < 6$ have convex boundaries, and those with $N > 6$ have concave boundaries. Since the boundaries migrate toward

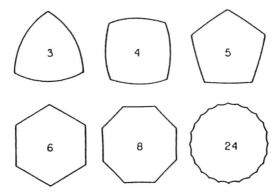

Figure 9.8 Polygons with curved sides meeting at 120°.

their center of curvature, grains with $N > 6$ tend to grow while those with $N < 6$ tend to shrink.

A factor ignored in many of the theoretical models is that the topological properties of the shrinking grain (e.g., faces, edges, and vertices) must be shared out among the remaining grains of the structure. This factor and the topological requirements in three dimensions are discussed in detail by Smith.

9.5 NORMAL GRAIN GROWTH IN DENSE SOLIDS

Although grain growth in porous ceramics is more relevant to fabrication, it is useful to start with the less complex situation present in dense ceramics, where the interactions between the pores and the grains do not arise. Since the early 1950s, the theoretical analysis of normal grain growth in dense polycrystalline solids has received considerable attention. The subject was reviewed in 1988 by Atkinson [8]. Here we provide a discussion of the various approaches that have had a significant impact on advancing our understanding of the grain growth process.

A. The Model of Burke and Turnbull

In one of the earliest analyses, Burke and Turnbull [9] modeled the grain growth process as occurring by transport of atoms across the grain boundary under the driving force of the pressure gradient across the boundary. More correctly, the chemical potential gradient across the boundary should be used. Burke and Turnbull considered an isolated part of the grain boundary and assumed that the grain growth equation derived in

Grain Growth and Microstructural Control

the analysis represents the average behavior of the whole system. Assuming that the instantaneous rate of grain growth is proportional to the average rate of grain boundary migration, v_b (sometimes called the grain boundary velocity), we can write

$$\frac{dG}{dt} \sim v_b \qquad (9.20)$$

where G is the average grain size. The term v_b is assumed to be represented in terms of the product of the driving force and the mobility,

$$v_b = M_b F_b \qquad (9.21)$$

where F_b is the driving force and M_b is the mobility of the boundary, which includes the factors arising from the mechanism of migration. The pressure difference across the boundary is given by the equation of Young and Laplace

$$\Delta p = \gamma_{gb} \left(\frac{1}{r_1} + \frac{1}{r_2} \right) \qquad (9.22)$$

where γ_{gb} is the specific grain boundary energy (i.e., energy per unit area) and r_1 and r_2 are the principal radii of curvature of the boundary. Assuming that the radius of the boundary is proportional to G, then

$$\left(\frac{1}{r_1} + \frac{1}{r_2} \right) = \frac{\alpha}{G} \qquad (9.23)$$

where α is a geometrical constant that depends on the shape of the boundary. The driving force for atomic diffusion across the boundary can be taken as the gradient in the chemical potential:

$$F_b = \frac{d\mu}{dx} = \frac{d}{dx}(\Omega \Delta p) = \frac{1}{\delta_{gb}} \left(\Omega \gamma_{gb} \frac{\alpha}{G} \right) \qquad (9.24)$$

where Ω is the atomic volume and $dx = \delta_{gb}$ is the width of the grain boundary. From Eq. (7.76), the flux of atoms across the boundary is

$$J = \frac{D_a}{\Omega kT} \frac{d\mu}{dx} = \frac{D_a}{\Omega kT} \left(\frac{\Omega \gamma_{gb} \alpha}{\delta_{gb} G} \right) \qquad (9.25)$$

where D_a is the diffusion coefficient for atomic motion across the grain boundary. The boundary velocity becomes

$$v_b \sim \frac{dG}{dt} = \Omega J = \frac{D_a}{kT} \left(\frac{\Omega}{\delta_{gb}} \right) \left(\gamma_{gb} \frac{\alpha}{G} \right) \qquad (9.26)$$

In many analyses the driving force is taken as the pressure difference across the boundary, $\alpha \gamma_{gb}/G$, and v_b is given by

$$v_b \sim \frac{dG}{dt} = M_b \frac{\alpha \gamma_{gb}}{G} \qquad (9.27)$$

Comparing Eqs. (9.26) and (9.27), we see that

$$M_b = \frac{D_a}{kT} \left(\frac{\Omega}{\delta_{gb}} \right) \qquad (9.28)$$

Integrating Eq. (9.27) gives

$$G^2 - G_0^2 = 2\alpha M_b \gamma_{gb} t \qquad (9.29)$$

where G_0 is the initial grain size, i.e., $G = G_0$ at $t = 0$. This is the parabolic grain growth law. It has the same form as Wagner's equation for Ostwald ripening controlled by the interface reaction [Eq. (9.13)].

Equation (9.29) is valid for a fixed temperature. For grain growth at different temperatures, we must include the temperature dependence of M_b as described by Eq. (9.28). M_b depends on the atomic diffusion coefficient, which has an Arrhenius dependence on temperature, i.e., $D_a = D_0 \exp[-Q/kT]$, where Q is the activation energy for the rate-controlling diffusion mechanism. For $G \gg G_0$, we can write

$$G = Kt^{1/2} \qquad (9.30)$$

where $K = (2\alpha M_b \gamma_{gb})^{1/2}$.

In addition to those outlined, several other assumptions are implicit in the derivation of Eq. (9.29). The grain boundary energy γ_{gb} is assumed to be the same for all boundaries and independent of grain size and time. Its dependence on the crystallographic direction is not considered. The grain boundary width δ_{gb} is also assumed to be constant. It would be expected that δ_{gb} would depend on the orientation of the boundary. In ionic compounds, D_a and Ω refer to the rate controlling species (i.e., the slower diffusing species). Finally, the topological considerations of connecting the grains into a space-filling network are ignored.

B. Mean Field Theories

The next major advance in the theoretical analysis of grain growth was made by Hillert [10]. His approach was to consider the change in size of an isolated grain embedded in an environment that represented the average effect of the whole array of grains. Theories based on this type of approach are often referred to in the grain growth literature as *mean field theories*. Hillert's analysis formed the first significant application of the

Grain Growth and Microstructural Control

LSW theory of Ostwald ripening to grain growth. He deduced an expression for the growth rate of grains of radius R of the form

$$v_b \sim \frac{dR}{dt} = \alpha' M_b \gamma_{gb} \left(\frac{1}{R_c} - \frac{1}{R} \right) \qquad (9.31)$$

where α' is a geometrical factor equal to 1/2 in two dimensions (2-D) and 1 in three dimensions (3-D), R is the radius of the circle (2-D) or sphere (3-D) having the same area or volume as the grain, and R_c is a critical grain size such that if $R > R_c$ the grain will grow and if $R < R_c$ the grain will shrink. R_c is expected to be related to the average grain size of the structure, but the exact relation was not derived. Equation (9.31) has the same form as Wagner's expression for the rate of change of the particle radius in Ostwald ripening controlled by the interface reaction [Eq. (9.11)].

Following the procedure of Lifshitz and Slyozov [6] for Ostwald ripening, Hillert derived an equation for the rate of change of the average grain radius given by

$$\frac{dR_c^2}{dt} = \frac{1}{2} \alpha' M_b \gamma_{gb} \qquad (9.32)$$

If the terms on the right-hand side of this equation are taken to be constant, then integration leads to parabolic grain growth kinetics for the average grain size similar to that of Eq. (9.29).

Hillert also derived an equation for the distribution of grain sizes. When the particle radius is expressed in terms of a reduced size, $s = R/R_c$, the distribution function takes the form

$$f(s, t) = (2e)^\beta \frac{\beta s}{(2 - s)^{2+\beta}} \exp\left(-\frac{2\beta}{2 - s} \right) \qquad (9.33)$$

where e is the base of the natural logarithm (equal to 2.718) and $\beta = 2$ in two dimensions or 3 in three dimensions. This distribution function has the property of time invariance. It is fairly sharply peaked in comparison with a lognormal distribution (Fig. 9.9). As described in Chapter 3, the lognormal distribution provides only an approximate fit to particle (grain) size data.

In the grain growth literature, other major theories that use the mean field approach include those of Feltham [11] and Louat [12]. Feltham assumed that the grain size distribution is actually lognormal and time-invariant if plotted as a function of the reduced size, taken as $R/\bar{R}$, where $\bar{R}$ is the average grain size. Using an approach similar to that of Hillert, he obtained parabolic growth kinetics. Louat assumed that the grain boundary motion can be modeled as a random walk process in which the

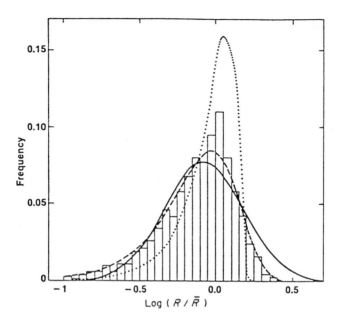

Figure 9.9 The grain size distribution function for three theoretical distributions and that obtained from a computer simulation employing the Monte Carlo procedure: lognormal distribution (solid curve), Hillert's model (dotted curve), Louat's model (dashed curve), and computer simulation (histogram). (From Ref. 15, reprinted with permission from Elsevier Science Ltd.)

fluctuation of the grain size is the cause of the drift to larger average sizes. He deduced that the grain growth kinetics are parabolic. Louat also derived a time-invariant distribution function for the grain sizes given by

$$f(s, t) = As \exp\left(-\frac{s^2}{2}\right) \tag{9.34}$$

where A is a constant and s is equal to $R/\bar{R}$, i.e., the ratio of the grain radius normalized to the average radius. A comparison of this distribution function with that derived by Hillert is shown in Fig. 9.9.

To summarize at this stage, the theories considered so far all predict parabolic grain growth kinetics. It is not clear why, in spite of the different assumptions, they should all predict the same type of growth law. This law has the same form as that predicted by the LSW theory for Ostwald ripening controlled by the interface reaction. The grain size distribution

Grain Growth and Microstructural Control

functions predicted by the theories also show the property of time invariance found in the LSW theory. Because of these major similarities, normal grain growth in dense polycrystalline solids is sometimes considered a special case of interface-controlled Ostwald ripening.

In practice, the grain growth data of dense polycrystalline solids do not follow the predicted parabolic law very well. Writing the equation for grain growth at a fixed temperature as

$$G^m - G_0^m = Kt \tag{9.35}$$

where G is the average grain size, G_0 is the average size at $t = 0$, and K is a constant. The exponent m is often found to be between 2 and 4 and may also depend on the temperature. In ceramics, the value $m = 3$ has been reported most often. Deviations from $m = 2$ are often explained away in terms of additional factors not considered in the models. One common explanation is the segregation of impurities to the grain boundaries. However, the grain growth kinetics of very pure zone-refined metals also show deviations from $m = 2$. Other explanations include the presence of second-phase particles and anisotropy in the grain boundary energy.

A major problem with the grain growth theories considered so far is that they neglect the topological requirements of space filling. In the next section we consider one attempt to deal with the topological requirements of grain growth.

C. Topological Analysis of Grain Growth

The most detailed analysis of the topological features of grain growth was carried out by Rhines and Craig [13]. They realized that in three dimensions the volume of the shrinking grain must be shared with the grains throughout the whole structure. Furthermore, changes in the topological parameters, e.g., the number of faces, edges, and vertices, must also be shared with the other grains. Rhines and Craig introduced two parameters, the sweep constant θ and the structure gradient σ, which have been the subject of some debate. Grain boundaries, as we know, must migrate for the grain growth process to occur. The sweep constant is defined as the number of grains lost when grain boundaries throughout the whole grain structure sweep through the equivalent of unit volume of the solid. There is some doubt about whether the sweep constant is indeed constant. The structure gradient is defined as

$$\sigma = \frac{M_V S_V}{N_V} \tag{9.36}$$

where M_V, S_V, and N_V are the curvature, surface area, and number of grains, respectively, per unit volume of the solid. M_V is defined by the equation

$$M_V = \frac{1}{2} \int_{S_V} \left(\frac{1}{r_1} + \frac{1}{r_2}\right) dS_V \qquad (9.37)$$

where r_1 and r_2 are the two principal radii of curvature. From a limited amount of experimental data, Rhines and Craig found that σ was approximately constant.

In the analysis of the grain growth kinetics, the mean stress on the grain boundary was taken as

$$p = \gamma_{gb}(M_V/S_V) \qquad (9.38)$$

and the mean boundary velocity was taken as the product of the boundary mobility and the force on the boundary, that is,

$$v_b = M_b \gamma_{gb} M_V \qquad (9.39)$$

Considering unit volume of the solid, the volume swept out per unit time is $v_b S_V$. From the definition of the sweep constant, the number of grains lost per unit time is $\theta v_b S_V$. The total volume transferred from the disappearing grains to the remaining grains per unit time is $\theta v_b S_V V_G$, where V_G is the average volume of a grain. The rate of increase in the average volume per grain can be written

$$\frac{dV_G}{dt} = \frac{\theta v_b S_V V_G}{N_V} \qquad (9.40)$$

where N_V, the number of grains per unit volume, is equal to $1/V_G$. Substituting for S_V and v_b from Eqs. (9.36) and (9.39), respectively, gives

$$\frac{dV_G}{dt} = \frac{M_b \gamma_{gb} \theta \sigma}{N_V} \qquad (9.41)$$

Assuming that the terms on the right-hand side of this equation are constant, then integration leads to

$$V_G - V_{G0} = \left(\frac{M_b \gamma_{gb} \theta \sigma}{N_{V0}}\right) t \qquad (9.42)$$

where V_{G0} and N_{V0} are the average grain volume and the average number of grains at $t = 0$. For the grain growth of aluminum at 635°C, the data in Fig. 9.10 show a linear dependence of V_G (measured as $1/N_V$) on time, in agreement with the predictions of Eq. (9.42). Since V_G is proportional

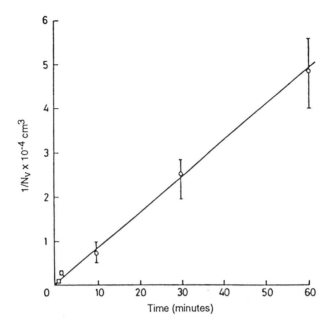

Figure 9.10 Plot of the grain volume ($1/N_V$) versus time of annealing for aluminum showing a linear relationship. (From Ref. 13.)

to G^3, where G is the average grain size, and assuming self-similarity of the grain size distribution, Eq. (9.42) can be written

$$G^3 - G_0^3 = Kt \qquad (9.43)$$

where K is a constant. In comparison with the theories considered earlier that predict an exponent m equal to 2, Eq. (9.43) predicts $m = 3$. We will recall that most experimental data give m values between 2 and 4.

A difficulty with the analysis of Rhines and Craig is the controversial nature of the two parameters introduced, i.e., the sweep constant and the structure gradient. A further difficulty is that the three-dimensional topological features of the grain structure are not very easy to visualize. In recent years, computer simulations have started to play a key role in exploring some of the issues in grain growth, and we outline the approach in the next section.

D. Computer Simulation of Normal Grain Growth

Several computer simulation models for normal grain growth have been developed, and these are reviewed briefly by Atkinson [8]. The simulations of Anderson, Srolovitz, and coworkers [14,15], which employ a Monte Carlo procedure, produce realistic pictures of the growing grains, and the analysis of the patterns shows fairly good agreement with experimental data. The Monte Carlo simulations, while powerful, are fairly complex. At the level of this book, we can provide a description of only the basic features of the method. We shall also limit our discussion to simulations in two dimensions.

An important advantage of the method is that both the topological requirements of space filling and the effects of grain boundary curvature can be incorporated into the simulations. To analyze the complexity of the grain boundary topology, the microstructure is mapped out onto a discrete lattice (Fig. 9.11). Each lattice site is assigned a number between 1 and Q corresponding to the orientation of the grain in which it is embedded. A sufficiently large value of Q (>30) is chosen to limit the impingement of grains of the same orientation. The grain boundary segment is defined to lie between two sites of unlike orientation. The grain boundary energy is specified by defining the interaction between nearest neighbor lattice sites in terms of a Hamiltonian operator. In the two-dimensional analysis, the average area per grain, $\bar{A}$, was monitored, and this can be related to the average grain size G.

An initial simulation of a shrinking circular grain embedded in an infinite matrix ($Q = 2$), i.e., a structure equivalent to the mean field ap-

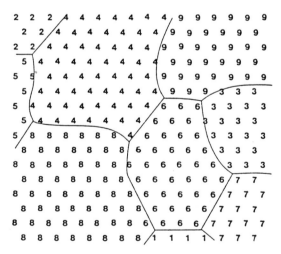

Figure 9.11 Sample microstructure on a triangular lattice where the integers denote orientation and the lines represent grain boundaries. (From Ref. 14, reprinted with permission from Elsevier Science Ltd.)

Grain Growth and Microstructural Control

proximation, showed that the size of the circular grain decreased uniformly according to the equation

$$\overline{A} = A_0 - Kt \tag{9.44}$$

where A_0 is the initial area and K is a constant. The grain growth kinetics of the isolated grain are therefore parabolic, in agreement with the mean field theories and the theory of Burke and Turnbull considered earlier.

For an interconnected network of polycrystalline grains, the kinetics are not, however, parabolic. Large Q (e.g., equal to 64) and long times give grain structures (Fig. 9.12) that resemble those of real systems (e.g., Figs. 9.3a and b). After an initial transient, the grain growth exponent m obtained from the simulations is equal to 2.44. From a survey of the m values determined from the data for six very pure metals, Anderson et al. [14] show that the value obtained from the simulation is in excellent agreement with the average experimental value for the six metals.

The deviation of the value of the grain growth exponent obtained in the simulations ($m = 2.44$) from that predicted by the mean field theories ($m = 2$) was discussed by Anderson et al. [14]. In the mean field approach, the driving force for grain growth, as outlined earlier, is the reduction of the curvature (or area) of the boundary. In the lattice model used in the simulations, the curvature is discretely allocated as kinks on the boundary. Such kinks can be eliminated by two mechanisms: (i) the meeting and annihilation of two kinks of identical orientation (as defined by the lattice) but opposite sign and (ii) adsorption of a kink at a vertex where more than two grains meet. The first mechanism corresponds to grain growth driven purely by curvature, as in the case of the simulation of a circular grain embedded in the infinite matrix The second mechanism can operate only when vertices are present. By absorbing kinks, vertices are capable of decreasing the curvature without causing grain growth. Effectively, the

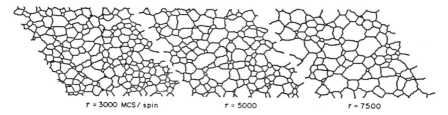

Figure 9.12 The evolution of the microstructure for a $Q = 64$ model on a triangular lattice. (From Ref 14, reprinted with permission from Elsevier Science Ltd.)

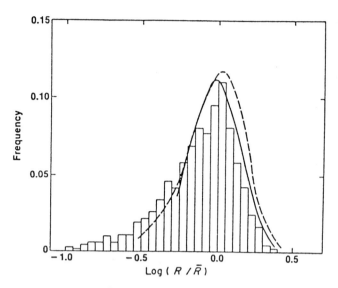

Figure 9.13 The grain size distribution function obtained from a computer simulation (histogram) compared with a lognormal fit to grain growth data for aluminum (solid curve) and with grain growth data for MgO (dashed curve). (From Ref. 15, reprinted with permission from Elsevier Science Ltd.)

growth is slowed, i.e., relative to the case of the circular grain in the infinite matrix.

In a subsequent analysis, Srolovitz et al. [15] showed that when the grain size is normalized to the average grain size, then the grain size distribution function becomes time invariant, as predicted by the mean field theories described earlier. A comparison of the distribution function obtained from the computer simulations with the lognormal distribution, the distribution functions derived by Hillert, and that derived by Louat is shown in Fig. 9.9. The distribution function obtained from the computer simulations agrees most closely with that derived by Louat. As shown in Fig 9.13, it can provide an excellent fit to experimental data.

To summarize, the computer simulations employing the Monte Carlo procedure have shown a remarkable ability to provide fairly realistic pictures of the grain growth process and to provide a good fit to experimental data. The use of these simulations is expected to increase. However, the physical significance of some of the results still needs to be considered more fully. Nevertheless, Srolovitz et al. [15] suggest that the true nature of the grain growth process may lie somewhere between growth driven by the grain boundary curvature and a random walk process.

9.6 ABNORMAL GRAIN GROWTH IN DENSE SOLIDS

Microstructures of polycrystalline ceramics that have been heated for some time often show very large grains in a fine-grained matrix (Fig. 9.3b). The large grains, as we outlined earlier, are referred to as abnormal grains. They are said to have developed as a result of abnormal or runaway grain growth in which the large grains have a much faster growth rate than the fine-grained matrix. The faster growth rate is normally explained in terms of the greater driving force for growth of a large grain in a fine-grained matrix relative to that for a small grain in the same matrix. According to this explanation, as the difference in the number of sides between neighboring grains increases, the curvature of the boundary becomes greater (see Fig. 9.7), thereby providing a larger driving force for growth of the large grain.

Surprisingly, when the evolution of large grains in a fine-grained matrix is analyzed theoretically, it is found that large grains do not undergo fast or runaway grain growth. Using the Monte Carlo procedure outlined in the previous section, Srolovitz et al. [16] simulated the growth of large grains in a matrix of fine (normal) grains. They found that although the large grains did grow, they did not outstrip the normal grains. The normal grains grew at a faster relative rate so the large (abnormal) grains eventually returned to the normal size distribution.

We can also examine the growth of the large grains in the matrix of normal grains analytically [17]. Grain growth in dense polycrystalline solids, as discussed earlier, can be considered a special case of interface-controlled Ostwald ripening. Consider a large grain of radius R in a matrix of fine normal grains. By analogy with Eq. (9.11), we can write

$$\frac{dR}{dt} = 2M_b\gamma_{gb}\left(\frac{1}{R^*} - \frac{1}{R}\right) \tag{9.45}$$

where R^* is the critical radius of the grain structure (i.e., the radius of the grain, which neither grows nor shrinks) and the other terms are as defined earlier. The *relative* growth rate of the large grain can be defined as

$$\frac{d}{dt}\left(\frac{R}{R^*}\right) = \frac{1}{(R^*)^2}\left(R^*\frac{dR}{dt} - R\frac{dR^*}{dt}\right) \tag{9.46}$$

Since the number of normal grains is very large compared to that (one) of the abnormal grains, the time dependence of R^* is still described by Eq. (9.13), which in differential form can be written as

$$\frac{dR^*}{dt} = \frac{M_b\gamma_{gb}}{2R^*} \tag{9.47}$$

Substituting Eqs. (9.45) and (9.47) into Eq. (9.46) and rearranging gives

$$\frac{d}{dt}\left(\frac{R}{R^*}\right) = -\frac{M_b \gamma_{gb}}{2RR^*}\left(\frac{R}{R^*} - 2\right)^2 \quad (9.48)$$

This equation shows that the relative growth rate of the large grains is always negative, except for $R = R^*$, in which case it is zero. Abnormal grains (i.e., with $R > 2R^*$) therefore do not outstrip the normal grain population but rejoin it at the upper limit of $2R^*$. Furthermore, due to irregularities in their shape and to fluctuations, they cannot remain at exactly $2R^*$ after they rejoin the population. They continue to decrease in relative size and are eventually incorporated into the normal distribution.

As we outlined earlier, observations of ceramic microstructures frequently do show examples of large grains growing abnormally. Indeed, a well-known microstructure in the ceramic literature is that of a relatively large single-crystal Al_2O_3 grain in a fine-grained Al_2O_3 matrix (Fig. 9.14). The microstructure, produced by Coble [18], appears to show the single-crystal grain growing much faster than the matrix grains. Furthermore, the seeding of fine-grained ferrites with a large single crystal is often used commercially to fabricate some single-crystal ferrites.

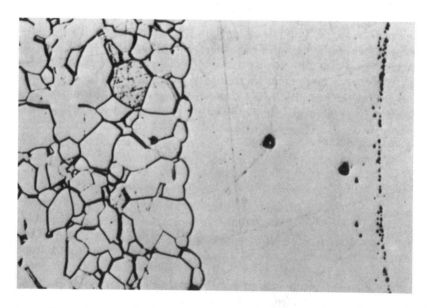

Figure 9.14 Growth of a large Al_2O_3 crystal into a matrix of uniformly sized grains. (Magnification 495×.) (From Ref. 3, with permission.)

Grain Growth and Microstructural Control

In reconciling the experimental observations with the theoretical analysis, it is clear that the reduction of a uniform grain boundary energy alone cannot account for the occurrence of abnormal grain growth. Other factors must be taken into account. Two of these that have been commonly invoked are the effects of impurities in the grain boundaries and an anisotropic grain boundary energy. It is possible that other, more complex factors also have a role to play; however, a consideration of these other factors and their significance must await a better understanding of abnormal grain growth.

9.7 EFFECT OF INCLUSIONS AND DOPANTS ON BOUNDARY MOBILITY

So far we have limited our analysis of grain growth to the case of single-phase solids. The mobility of the boundary, M_b, is assumed to be controlled by the diffusion of the atoms (ions) of the grain across the grain boundary. While impurities (dopants) or second phase particles (inclusions) have been alluded to from a practical point of view in the analysis, their presence has been assumed to have no effect on the basic rate of atomic diffusion across the boundary. Under these conditions, the dopant or inclusions have no effect on the grain boundary mobility, and M_b is referred to as the *intrinsic* grain boundary mobility. In this book it will be assumed to be defined by Eq. (9.28) derived in the analysis of the Burke and Turnbull model.

As we shall see, dopants and inclusions present in sufficient quantities can have a dramatic effect on M_b. In practice, the use of inclusions and, to a much greater extent, the use of dopants form effective approaches for the fabrication of ceramics with high density and controlled grain size. As a prelude to considering their practical use, we examine the mechanism by which they influence M_b.

A. Inclusions: The Zener Relationship

Let us consider a system of inclusions dispersed randomly in a polycrystalline solid in which they are insoluble and immobile. If a grain boundary moving under the driving force of its curvature encounters an inclusion, it will be held up by the particle until the motion elsewhere has proceeded sufficiently far for it to break away. If there are a sufficient number of particles, the boundary will be pinned when it encounters the particles and boundary migration will cease. This situation was considered by Zener in a communication to Smith [19].

For a grain boundary with principal radii of curvature R_1 and R_2, the driving force (per unit area) for boundary motion is

$$F_b = \gamma_{gb}\left(\frac{1}{R_1} + \frac{1}{R_2}\right) \quad (9.49)$$

Assuming that R_1 and R_2 are proportional to the grain size G, then

$$F_b = \alpha\gamma_{gb}/G \quad (9.50)$$

where α is a geometrical shape factor (e.g., $\alpha = 2$ for a spherical grain). When the grain intersects an inclusion, its further movement is hindered (Figs. 9.15a and b). A dimple is formed in the grain boundary, and, compared to the inclusion-free boundary, extra work must be performed for the equivalent motion of the boundary. This extra work manifests itself as a retarding force on the boundary. If r is the radius of the inclusion (Fig. 9.15c), then the retarding force exerted by the inclusion on the boundary is

$$F_d = (\gamma_{gb} \cos \theta)(2\pi r \sin \theta) \quad (9.51)$$

i.e., the retarding force is the grain boundary tension resolved in the direction opposite that of the grain boundary motion times the perimeter of

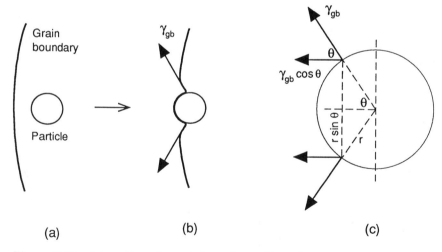

Figure 9.15 Interaction of a grain boundary with an immobile particle. (a) Approach of the boundary toward the particle. (b) Interaction between the grain boundary and the particle leading to a retarding force on the boundary. (c) Detailed geometry of the particle–grain boundary interaction.

Grain Growth and Microstructural Control

contact. The retarding force is a maximum when $\theta = 45°$ and $\sin\theta \cos\theta = 1/2$, i.e.,

$$F_d^{max} = \pi r \gamma_{gb} \tag{9.52}$$

If there are N_A inclusions per unit area of the grain boundary, then the maximum retarding force RF (i.e., per unit area of the boundary) is

$$RF_d^{max} = N_A \pi r \gamma_{gb} \tag{9.53}$$

N_A is difficult to determine; a more accessible parameter is N_V, the number of inclusions per unit volume. If the volume fraction of the inclusions in the solid is f, then

$$N_V = f/(4/3)\pi r^3 \tag{9.54}$$

N_V is related to N_A by the relation

$$N_A = 2rN_V \tag{9.55}$$

Using Eqs. (9.53)–(9.55), we obtain

$$RF_d^{max} = \frac{3f\gamma_{gb}}{2r} \tag{9.56}$$

The net driving force per unit area of the boundary is

$$F_{net} = F_b - RF_d^{max} = \gamma_{gb}\left(\frac{\alpha}{G} - \frac{3f}{2r}\right) \tag{9.57}$$

When $F_{net} = 0$, boundary migration will cease; this occurs when

$$r = (3/2\alpha)G_{max} f \tag{9.58}$$

where G_{max} is the maximum grain size. Equation (9.58) is sometimes known as the Zener relationship.

With all the assumptions used in its derivation, Eq. (9.58) can only be considered an approximation. Nevertheless, it does indicate that a limiting grain size will be reached, the magnitude of which is determined by the size of the inclusion and its volume fraction. For example, if a dense ceramic (e.g., Al_2O_3) contains 10 vol% of inert inclusions (e.g., ZrO_2) with a size of 0.2 μm, the grain size cannot grow to greater than ≈2 μm. It is not clear whether such a limiting grain size is reached in practice. However, Zener's concept provides guidelines for the use of inclusions in the control of grain growth. First, the inclusions must be smaller than the grain. Second, for the same inclusion volume fraction, the inclusions become more effective with decreasing size [Eq. (9.57)]. Third, the impediment to grain growth increases with increasing volume fraction of inclusions. It must be remembered, however, that the use of a large volume fraction of inclusions may significantly alter the properties of the matrix.

If a limiting grain size is reached as predicted by Eq. (9.58), then the further growth of the grains can occur only if (i) the inclusions coarsen by Ostwald ripening, (ii) the inclusions go into solid solution in the matrix, or (iii) abnormal grain growth occurs. For the case of inclusion coarsening when the diffusion step is rate-controlling, r^3 is predicted to be proportional to time t [see Eq. (9.17)]. Since f remains constant, Eq. (9.57) gives

$$G_{max}^3 \sim t \tag{9.59}$$

i.e., G_{max} is predicted to obey cubic growth kinetics.

B. Dopants: The Solute Drag Effect

Let us consider a system in which a small amount of a dopant, e.g., Mg (or MgO), is dissolved in solid solution in a polycrystalline solid, e.g., Al_2O_3. The dopant is sometimes referred to as the *solute*, and the polycrystalline solid is sometimes called the *host*. Suppose an interaction potential exists between the solute ions and the grain boundary, resulting, say, in a small attraction of the solute ions to the grain boundary. (We shall discuss this interaction in greater detail in Chapter 11.) As sketched in Fig. 9.16, the solute ions will tend to have a nonuniform distribution in the region of the grain boundary.

For a hypothetical stationary boundary, the concentration profile of the solute ions will be symmetrical (Fig. 9.17a). The force of interaction due to the solute ions to the right of the boundary balances that due to the ions to the left of the boundary, so that the net force of interaction is zero. If the boundary now starts to move, the dopant concentration profile becomes asymmetric, since the diffusivity of the solute ions across the boundary is expected to be different from that of the host (Fig. 9.17b). This asymmetry results in a retarding force or drag on the boundary that reduces the driving force for migration. Eventually, the boundary may move past the high concentration of solute, sometimes called a *solute cloud* (Fig. 9.17c), and when this occurs its mobility will approach the intrinsic value M_b.

A few models have been developed to analyze grain boundary migration controlled by the drag of solute ions. We shall outline the model of Cahn [20], which can be more directly related to the physical parameters of the process. Cahn's analysis is fairly detailed, and with the limited space available in this chapter we can provide only an outline of the main steps.

Grain Growth and Microstructural Control

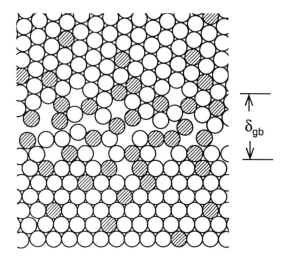

Figure 9.16 Sketch illustrating the nonuniform distribution of dopant atoms (shaded circles) that results from the segregation of the dopant to the grain boundary.

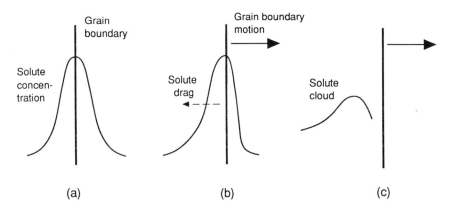

Figure 9.17 Sketch of the solute drag effect produced by the segregation of dopants to the grain boundaries. (a) Symmetrical distribution of the dopant in the region of a stationary grain boundary. (b) For a moving boundary, the dopant distribution becomes asymmetric if the diffusion coefficient of the dopant atoms across the boundary is different from that of the host atoms. The asymmetric distribution produces a drag on the boundary. (c) Breakaway of the boundary from the dopant, leaving a solute cloud behind.

Cahn's model analyzes the problem in one dimension and makes the following assumptions:

1. The dopant concentration C expressed as a fraction of the host atoms is fairly dilute.
2. An interaction potential energy U exists between the solute atoms and the grain boundary; U is independent of the boundary velocity and the dopant concentration but is a function of the distance x from the grain boundary.
3. The boundary velocity v_b is constant; this means that only the steady-state situation is considered.

The chemical potential of the solute atoms is assumed to be given by

$$\mu = kT \ln C(x) + U(x) + \text{constant} \tag{9.60}$$

where $C(x)$ and $U(x)$ are functions of x and the constant is chosen so that $U(\infty) = 0$. For a constant value of v_b, the composition profile of the dopant is expected to reach a steady-state value given by

$$\frac{\partial C}{\partial t} = -v_b \frac{\partial C}{\partial x} \tag{9.61}$$

and the flux of the solute atoms is

$$J = -\frac{D_b C}{\Omega kT} \frac{\partial \mu}{\partial x} \tag{9.62}$$

where D_b is the diffusion coefficient for the solute atoms across the boundary and is dependent on the distance from the boundary. Using Eq. (9.60), we can also write Eq. (9.62) as

$$J = -\frac{D_b}{\Omega} \frac{dC}{dx} + \frac{D_b C}{\Omega kT} \frac{dU}{dx} \tag{9.63}$$

The concentration profile of the dopant atoms, $C(x)$, is calculated from the continuity equation

$$\frac{\partial J}{\partial x} + \frac{1}{\Omega} \frac{\partial C}{\partial t} = 0 \tag{9.64}$$

with the boundary conditions $dC/dx = 0$, $dU/dx = 0$, and $C(x) = C_\infty$ at $x = \infty$. The concentration C_∞ can be taken as that in the bulk of the grain. With the boundary conditions, $C(x)$ must satisfy the equation

$$D_b \frac{dC}{dx} + \frac{D_b C}{kT} \frac{dU}{dx} + v_b(C - C_\infty) = 0 \tag{9.65}$$

Grain Growth and Microstructural Control 479

The solute atom exerts a force $-dU/dx$ on the boundary. The net force exerted by all the solute atoms is

$$F_s = -N_V \int_{-\infty}^{\infty} [C(x) - C_\infty] \frac{dU}{dx} dx \qquad (9.66)$$

where N_V is the number of host atoms per unit volume. In the analysis, the $C(x)$ that satisfies Eq. (9.65) is used to calculate F_s from Eq. (9.66). An approximate solution that satisfies both low and high velocities of the grain boundary is

$$F_s = \frac{\lambda C_\infty v_b}{1 + \beta^2 v_b^2} \qquad (9.67)$$

where the parameters λ and β are defined by

$$\lambda = 4N_V kT \int_{-\infty}^{\infty} \frac{\sinh^2[U(x)/2kT] \, dx}{D_b(x)} \qquad (9.68)$$

and

$$\frac{\lambda}{\beta^2} = \frac{N_V}{kT} \int_{-\infty}^{\infty} \left(\frac{dU}{dx}\right)^2 D_b(x) \, dx \qquad (9.69)$$

The total drag force on the boundary is the sum of the intrinsic drag F_b and the drag due to the dopant atoms, F_s; that is,

$$F = F_b + F_s = \frac{v_b}{M_b} + \frac{\lambda C_\infty v_b}{1 + \beta^2 v_b^2} \qquad (9.70)$$

where M_b is the intrinsic boundary mobility defined by Eq. (9.28). Figure 9.18 shows a sketch of the forces F, F_b, and F_s as a function of v_b [21].

The solute drag effect is normally analyzed by making a distinction between high and low boundary velocity limits. For the low boundary velocity limit that is more likely for grain growth in ceramics, we can neglect the term $\beta^2 v_b^2$ in Eq. (9.70), so that

$$v_b = \frac{F}{1/M_b + \lambda C_\infty} \qquad (9.71)$$

For situations where the dopant segregates to the grain boundary (see Fig. 9.16) and the center of the boundary contributes most heavily to the drag effect, λ can be approximated by

$$\lambda = \frac{4N_V kT \delta_{gb} Q}{D_b} \qquad (9.72)$$

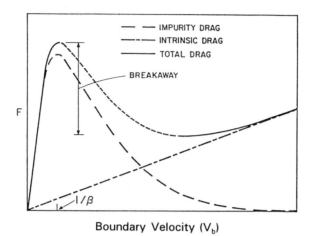

Figure 9.18 Intrinsic and impurity drag forces on a grain boundary moving with a velocity v_b. (From Ref. 21.)

where δ_{gb} is the width of the grain boundary and Q is a partition coefficient (>1) for the dopant distribution between the boundary region and the bulk of the grain, i.e., the atomic concentration of the dopant in the boundary region is QC_∞.

According to Eq. (9.71), dopants are most effective for reducing the grain boundary migration when $\lambda C_\infty \gg 1/M_b$. Taking into account Eq. (9.72), this occurs when the bulk concentration of the dopant, C_∞, is high, the degree of segregation (the factor Q) is high, or the diffusion coefficient for the dopant atoms across the boundary, D_b, is low. As outlined earlier, the use of dopants forms an effective approach for the control of grain growth. However, an understanding of the role of the dopants remains a problem. This lack of understanding limits the effective use of the approach. We shall consider the dopant role in detail in Chapter 11.

If it is assumed that $D_b = D_a$, where D_a is the diffusion coefficient for the host atoms across the boundary, then we can derive an approximate relation for the effective mobility of the boundary. Using Eq. (9.28) for the intrinsic boundary mobility M_b, Eq. (9.72) gives $\lambda = 4N_V \Omega Q/M_b$. Substituting for λ in Eq. (9.71), we obtain

$$v_b = \frac{M_b F}{1 + 4N_V \Omega Q C_\infty} \tag{9.73}$$

Grain Growth and Microstructural Control

The effective mobility of the boundary is therefore

$$M' = \frac{M_b}{1 + 4N_V\Omega QC_\infty} \tag{9.74}$$

With this equation and making several approximations, an equation for the grain growth kinetics limited by solute drag can be deduced. Let C_{avg} be the average concentration of the dopant in the grains. The total number of dopant atoms in a grain is equal to the number of those in the boundary region plus those in the bulk of the grain, i.e.,

$$C_{avg}\alpha G^3 = C_\infty(\alpha G^3 - \alpha'\delta_{gb}G^2) + QC_\infty\alpha'\delta_{gb}G^2 \tag{9.75}$$

where α and α' are geometrical factors that depend on the shape of the grain. Assuming that the dopant is well segregated to the grain boundary such that the first term on the right-hand side of Eq. (9.75) is much less than the second term, we obtain

$$C_{avg} = \left(\frac{\alpha'}{\alpha}\right)\left(\frac{QC_\infty\delta_{gb}}{G}\right) \tag{9.76}$$

Since C_{avg} must be independent of the size of the grain, this equation reveals that C_∞ varies as G. Assuming that $4N_V\Omega QC_\infty \gg 1$ in Eq. (9.74), we find that M' varies as $1/C_\infty$, i.e., as $1/G$. Therefore,

$$v_b \sim \frac{dG}{dt} \sim M'F_b \sim \frac{1}{G^2} \tag{9.77}$$

where F_b varies as $1/G$. On integration, this equation gives $G^3 = G_0^3 + Kt$, where K is a constant; i.e., grain growth limited by solute drag is predicted to follow a cubic law.

To summarize at this stage, the basic features of normal and abnormal grain growth in dense polycrystalline solids have been described. Abnormal grain growth during sintering must be avoided if high densities are to be achieved. The use of inclusions and dopants can be very effective for controlling grain growth. However, before we can apply these two approaches effectively in the fabrication of ceramics, we must understand how the presence of pores influences the grain growth process.

9.8 GRAIN GROWTH IN POROUS SOLIDS

The structure of a porous ceramic, as described in Chapter 8, changes considerably during sintering. As a start in our analysis, we shall consider a rather idealized model consisting of small, isolated spherical pores situated in the grain boundaries. This microstructure may be taken as a rough approximation to the final stage of sintering. Later we shall consider the situation of solids with fairly high porosity typical of the initial and intermediate stages of sintering.

In the analysis, the approach is to consider an isolated region of the grain boundary containing a single pore and assume that the kinetic equations derived for the model represent the average behavior of the whole system. By adopting this approach, we are assuming that the microstructure is homogeneous. Furthermore, the approach may be taken as equivalent to that of Burke and Turnbull for dense solids but with the added complication of the interaction between a pore and the grain boundary. Two cases can be distinguished: (i) the pores are immobile and (ii) the pores can move (mobile pores).

A. Immobile Pores

The case of immobile pores on the grain boundary can be analyzed along the same lines as described earlier for the interaction of immobile inclusions with the grain boundary. In principle, if the drag exerted by the pores is sufficiently high, the boundary will be pinned and boundary migration will stop. A limiting grain size G_{max} will be reached, given by an equation similar to the Zener relationship derived earlier:

$$r = (3/2\alpha)G_{max}P \qquad (9.78)$$

where r is now the radius of the pore, P is the porosity, and α is a geometrical factor that depends on the shape of the grain. If the drag exerted by the pores is insufficient, the boundary will break away, leaving the pores trapped in the grains.

Observations of ceramic microstructures do not provide convincing evidence for Eq. (9.78). Instead, much of the evidence is in favor of pores being able to move.

B. Mobile Pores: Mechanism of Pore Motion

It may initially appear strange that pores can in fact move in a porous solid. However, small isolated pores can be dragged along by the grain boundaries. The reason is that the grain boundary moving under the influence of its curvature applies a force on the pore trying to drag it along. This force causes the pore to change its shape, as illustrated in Fig. 9.19. The leading surface of the pore becomes less strongly curved than the trailing surface. The difference in curvature leads to a chemical potential difference, which causes a flux of matter from the leading surface to the trailing surface. The result is that the pore moves forward in the direction

Grain Growth and Microstructural Control

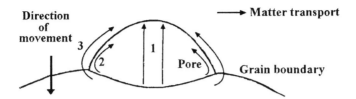

Figure 9.19 Possible transport paths for a pore moving with a grain boundary: (1) vapor transport (evaporation and condensation); (2) surface diffusion; (3) lattice diffusion.

of the boundary motion. Matter transport from the leading surface to the trailing surface can occur by three separate mechanisms: evaporation/condensation (vapor transport), surface diffusion, and lattice diffusion.

C. The Pore Mobility

The flux of matter from the leading surface to the trailing surface of the pore can be analyzed to derive an equation for the pore mobility M_p. By analogy with the case of a moving boundary, the pore velocity v_p is defined by a force–mobility relationship,

$$v_p = M_p F_p \tag{9.79}$$

where F_p is the driving force acting on the pore.

Let us consider a pore of average radius r in which matter transport occurs by surface diffusion from the leading surface to the trailing surface [22]. The net atomic flux is

$$J_s A_s = \left(\frac{D_s}{\Omega kT} F_a\right) 2\pi r \delta_s \tag{9.80}$$

where J_s is the flux of atoms, A_s is the area over which surface diffusion occurs, D_s is the surface diffusion coefficient, Ω is the atomic volume, F_a is the driving force on an atom, δ_s is the thickness for surface diffusion, and the other terms are as defined earlier. If the pore moves forward a distance dx in a time dt, the volume of matter that must be moved per unit time is $\pi r^2 (dx/dt)$. Equating the number of atoms moved to the net flux gives

$$\frac{\pi r^2}{\Omega} \frac{dx}{dt} = -\frac{D_s 2\pi r \delta_s}{\Omega kT} F_a \tag{9.81}$$

The negative sign is inserted in this equation because the flux is opposite to the direction of motion. The work done in moving the pore a distance dx is equal to that required to move $\pi r^2 dx/\Omega$ atoms a distance $2r$, so

$$F_p dx = -F_a \frac{\pi r^2 dx}{\Omega} 2r \tag{9.82}$$

Substituting for F_a from Eq. (9.81) and rearranging, Eq.(9.82) gives

$$F_p = \frac{\pi r^4 kT}{D_s \delta_s \Omega} \frac{dx}{dt} \tag{9.83}$$

Putting the velocity of the pore, v_p, equal to dx/dt, the pore mobility is given by

$$M_p = D_s \delta_s \Omega / \pi kT r^4 \tag{9.84}$$

The pore mobilities for matter transport by vapor transport and by lattice diffusion can be derived by a similar procedure. The formulas are summarized in Table 9.1. For all three mechanisms, M_p is found to have a strong

Table 9.1 Pore and Boundary Parameters

A. Mobilities[a]			
M_p	Mobility of spherical pore; migration by the faster of		
	(a) Surface diffusion		$\dfrac{D_s \delta_s \Omega}{kT\pi r^4}$
	(b) Lattice diffusion		$\dfrac{D_l \Omega}{fkT\pi r^3}$
	(c) Vapor transport		$\dfrac{D_g d_g \Omega}{2kTd_s \pi r^3}$
M_b	Mobility of boundary		
	Pure system		$\dfrac{D_a \Omega}{kT\delta_{gb}}$
	Solute drag		$\dfrac{\Omega}{kT\delta_{gb}} \left(\dfrac{1}{D_a} + \dfrac{4N_V \Omega QC_\infty}{D_b} \right)^{-1}$
B. Forces[b]			
F_p	Maximum drag force of pore		$\pi r \gamma_{gb}$
F_b	Force per unit area of pore-free boundary		$\dfrac{\alpha \gamma_{gb}}{G}$

[a] f = correlation factor; d_g = density in the gas phase of the rate-controlling species; d_s = density in the solid phase of the rate-controlling species.
[b] α = geometrical constant depending on the grain shape.
Source: Adapted from Ref. 1.

Grain Growth and Microstructural Control

dependence on the pore size; it decreases rapidly with increasing pore size.

D. Kinetic Aspects of Pore–Boundary Interactions

The definition of the pore mobility in the previous section allows us to go on to analyze how the interactions between the pores and the grain boundaries influence the kinetics of grain growth. There are two cases that can be considered: (i) the pore becomes separated from the boundary and (ii) the pore remains attached to the boundary.

Case I: Separation of Pores from the Boundary
Pore separation will occur when

$$v_p < v_b \tag{9.85}$$

This condition can also be written as

$$F_p M_p < F M_b \tag{9.86}$$

where F is the effective driving force on the boundary. If F_d is the drag force exerted by a pore, then a balance of forces requires that F_d be equal and opposite to F_p. Considering unit area of the boundary in which there are N_A pores, Eq. (9.86) can be written

$$F_p M_p < (F_b - N_A F_p) M_b \tag{9.87}$$

where F_b is the driving force on the pore-free boundary due to its curvature. Rearranging Eq. (9.87), the condition for pore separation can be expressed as

$$F_b > N_A F_p + \frac{M_p F_p}{M_b} \tag{9.88}$$

Case II: Pores Remain Attached to the Boundary
The condition for pore attachment to the boundary is

$$v_p = v_b \tag{9.89}$$

This equation can also be written as

$$F_p M_p = (F_b - N_A F_p) M_b \tag{9.90}$$

Putting $v_p = F_p M_p = v_b$ in Eq. (9.90) and rearranging gives

$$v_b = F_b \frac{M_p M_b}{N_A M_b + M_p} \tag{9.91}$$

Two limiting conditions can be defined. When $N_A M_b \gg M_p$, then

$$v_b = F_b \frac{M_p}{N_A} \qquad (9.92)$$

The effective driving force on the boundary is $F' = F_b - N_A F_p$, and using Eq. (9.90), $F' = v_p/M_b$. Putting $v_p = v_b = F_b M_p/N_A$ gives $F' = F_b M_p/N_A M_b \ll F_b$. The driving force on the boundary is nearly balanced by the drag of the pores, and the boundary motion is limited by the pore mobility. This condition is referred to as *pore control*.

The other limiting condition is when $N_A M_b \ll M_p$, in which case

$$v_b = F_b M_b \qquad (9.93)$$

The drag exerted by the pores is $N_A F_p = N_A (F_b M_b/M_p) \ll F_b$. The presence of the pores has almost no effect on the boundary velocity, a condition referred to as *boundary control*.

E. Microstructural Diagrams: Grain Size Versus Pore Size

Brook [23] developed a useful method to visualize the effects of the pore–boundary interactions on the coarsening process. The microstructural conditions under which each of the interactions becomes important are determined as functions of the grain size and the pore size. The results are represented on a diagram of the grain size G versus pore size $2r$, where r is the radius of the pore. As an example, we outline the steps in the construction of the diagram when pore migration occurs by surface diffusion.

Equal Mobility Curve

As outlined above, the conditions for attachment of the pore to the boundary fall into two limiting cases: pore control and boundary control. Microstructurally, the conditions that separate these two cases on the G versus $2r$ diagram are roughly represented by a curve, called the *equal mobility curve*, that is defined by the condition that the pore mobility is equal to the boundary mobility. Thus the equal mobility curve is defined by

$$N_A M_b = M_p \qquad (9.94)$$

As an approximation, we can put $N_A \sim 1/X^2$, where X is the interpore distance. Assuming that $X \approx G$ and using the relations for M_b and M_p given in Table 9.1, Eq. (9.94) gives

$$G_{em} = \left(\frac{D_a \pi}{D_s \delta_s \delta_{gb}} \right)^{1/2} r^2 \qquad (9.95)$$

Grain Growth and Microstructural Control

where G_{em} is the grain size defined by the equal mobility condition. Using logarithmic axes for the G versus $2r$ diagram (Fig. 9.20), the equal mobility condition is represented by a straight line with a slope of 2.

Separation Curve

The maximum force that the grain boundary can apply to a pore to drag it along is given by Eq. (9.52), that is,

$$F_p^{max} = \pi r \gamma_{gb} \tag{9.96}$$

The maximum velocity that the pore can attain is therefore

$$v_p^{max} = M_p F_p^{max} \tag{9.97}$$

If the velocity of the boundary with the attached pore were to exceed v_p^{max}, then separation would occur. The limiting condition for separation can therefore be written as

$$v_b = v_p^{max} \tag{9.98}$$

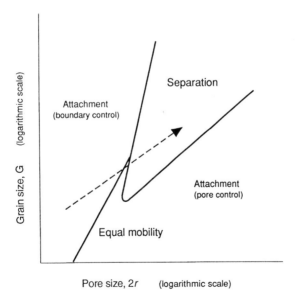

Figure 9.20 Microstructural evolution diagram of the grain size versus pore size, showing the types of pore–boundary interaction when the migration of the pore occurs by surface diffusion. The dashed arrow shows a possible sintering path for a powder compact.

Substituting for v_b from Eq. (9.91), we obtain

$$\frac{M_b M_p}{N_A M_b + M_p} F_b = M_p F_p^{\max} \tag{9.99}$$

Putting $N_A \sim 1/X^2$ and substituting for the other parameters from Table 9.1, after some rearranging Eq. (9.99) gives

$$G_{\text{sep}} = \left(\frac{\pi r}{X^2} + \frac{D_s \delta_s \delta_{gb}}{D_a r^3} \right)^{-1} \tag{9.100}$$

where G_{sep} is the grain size when the boundary separates from the pore. If we make the approximation that $X = G$, Eq. (9.100) can be written

$$\left(\frac{D_s \delta_s \delta_{gb}}{D_a r^3} \right) G_{\text{sep}}^2 - G_{\text{sep}} + \pi r = 0 \tag{9.101}$$

The solution to this quadratic equation determines the separation curve that is sketched in Fig. 9.20.

Complete G versus 2r Diagram

The equal mobility curve and the separation curve can be combined to give the full diagram as sketched in Fig. 9.20. Qualitatively, three different regions are found. For larger, less mobile pores that are closely separated, the pores remain attached to the boundary and control the motion. The microstructure in this region of the diagram is a rough approximation to the earlier stages of sintering when the pores are relatively large compared with the grain size and the continuous porosity is fairly closely spaced. The diagram may therefore provide an explanation for why low boundary mobility and pore attachment are generally observed in the earlier stages of sintering. However, we must also remember that the model used to produce the diagram assumed a structure that is more appropriate to the final stage. Caution must be exercised when applying the model to the earlier stages of sintering. In another region of the diagram, smaller, more mobile pores remain attached to the boundary, and they do not exert a significant drag on the boundary; the boundary migration is therefore controlled by the boundary mobility. Finally, separation occurs for larger pores that are widely separated. For the achievement of high density with controlled grain size, this separation region must be avoided.

In porous ceramics, as discussed earlier, coarsening of the microstructure is widely observed; i.e., the average grain size and the average pore size increase during sintering. For this situation, a possible trajectory for the microstructural evolution runs diagonally upward from left to right (dashed line in Fig. 9.20). According to this representation, it is likely

Grain Growth and Microstructural Control

that the interaction changes from boundary control (with possibly a small region of pore control) to separation.

To allow for the use of quantitative axes for the G versus $2r$ diagram, actual values for the parameters in Eqs. (9.95) and (9.101) must be used. Assuming realistic values for these parameters, Brook [23] obtained the diagram shown in Fig. 9.21 for the case when the pore migration occurred by surface diffusion. Dopants, as discussed earlier, have the effect of decreasing M_b. In the G versus $2r$ diagram, this has the effect of extending the region of pore attachment. In terms of this diagram, a possible explanation of the effectiveness of dopants is that they delay the onset of abnormal grain growth (pore separation) beyond the grain size at which final densification is achieved. It is left as an exercise for the reader to verify the plots in Fig. 9.21 using the values assumed by Brook for the parameters.

A small modification of Brook's analysis was made by Carpay [24], who observed that when separation occurred the assumption $X = G$ was no longer valid. Carpay solved Eq. (9.101) for fixed values of X (in the

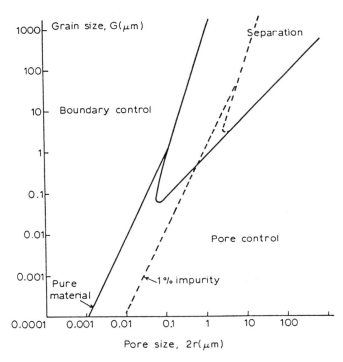

Figure 9.21 The dependence of the type of pore–boundary interaction on microstructural parameters when pores migrate by surface diffusion. The interpore spacing is assumed to be equal to the grain size. (From Ref. 23, with permission.)

range of 1–4 μm). The most significant result of the modification is that after a region of separation, the boundary may again become attached to the pores.

To summarize at this stage, the analysis of the pore–boundary interactions and its representation in terms of the G versus $2r$ diagram required considerable simplifications and assumptions. They provide a useful basis for the qualitative understanding of microstructural control in the final stage of sintering. However, they cannot be used to predict quantitatively the microstructural evolution of real systems.

F. Grain Growth and Pore Coalescence

A consequence of the pores moving along with the boundaries is that the pores may meet and coalesce to produce a larger pore. Figure 9.22 is a sketch of the process [25]. It is believed that this process contributes to the coarsening in the later stages of sintering. As an example, Fig. 9.23 shows the coarsening by such a process in UO_2.

G. Grain Growth Kinetics

For the simplified model assumed earlier, i.e., spherical isolated pores on the grain boundary, we can derive equations for the grain growth kinetics. Consider the situation of grain growth controlled by the pore mobility (pore control). If the pore migration occurs by surface diffusion, Eqs. (9.84) and (9.92) give

$$v_b \sim \frac{dG}{dt} = \frac{F_b}{N_A} \left(\frac{D_s \delta_s \Omega}{\pi k T r^4} \right) \tag{9.102}$$

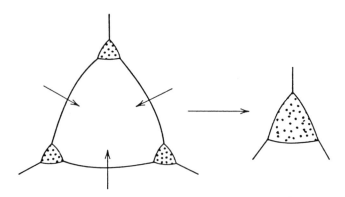

Figure 9.22 Grain growth and pore coalescence.

Grain Growth and Microstructural Control

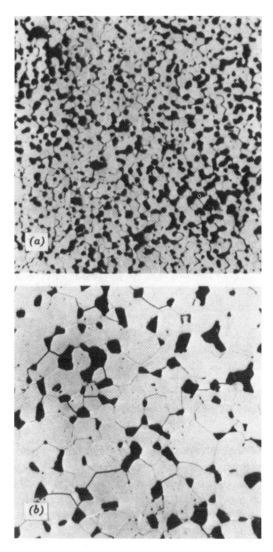

Figure 9.23 Grain growth and pore coalescence in a sample of UO_2 after (a) 2 min, 91.5% dense and (b) 5h, 91.9% dense, at 1600 °C. (Magnification 400×.) (From Ref. 25, with permission.)

Taking $F_b \sim 1/G$, $N_A \sim 1/X^2 \sim 1/G^2$ and assuming coarsening by grain growth and pore coalescence, i.e., $r \sim G$, after rearranging, Eq. (9.102) becomes (at a constant temperature)

$$\frac{dG}{dt} = \frac{K'}{G^3} \tag{9.103}$$

where K' is a constant. Integrating, this equation gives

$$G^4 = G_0^4 + K''t \tag{9.104}$$

where G_0 is the grain size at $t = 0$ and K'' is a constant. Equations for the grain growth kinetics can also be derived by the same procedure for other mechanisms. Writing the equations in the general form of Eq. (9.35), i.e., $G^m = G_0^m + Kt$, the exponent m for the various mechanisms are summarized in Table 9.2. Except for the mechanism of solution of second phase particles ($m = 1$), the m values lie in the range 2–4. In many ceramics, the value $m = 3$ has been commonly observed. This value can correspond to at least five different mechanisms. A conclusion that can be drawn is that the attachment of physical significance to the m values

Table 9.2 Grain Growth Exponent m in the Equation $G^m = G_0^m + Kt$ for Various Mechanisms

Mechanism	Exponent m
Pore control	
Surface diffusion	4
Lattice diffusion	3
Vapor transport (vapor pressure, p = constant)	3
Vapor transport ($p = 2\gamma_{sv}/r$)	2
Boundary control	
Pure system	2
System containing second-phase particles	
Coalescence of second phase by lattice diffusion	3
Coalescence of second phase by grain boundary diffusion	4
Solution of second phase	1
Diffusion through continuous second phase	3
Doped system	
Solute drag (low solubility)	3
Solute drag (high solubility)	2

Note: The pore control kinetics are given for the situation where the pore separation is related to the grain size, i.e., the number of pores per unit area of the boundary $N_A \sim 1/G^2$. Changes in distribution during growth would change the kinetics.
Source: Ref. 1.

Grain Growth and Microstructural Control

determined from grain growth data can be very dubious. Furthermore, the fitting of experimental data to produce m values that are exact whole numbers may not be realistic because the occurrence of simultaneous mechanisms is expected to give m values that are not integers. Finally, we must remember the many approximations used in the derivation of the equations. These approximations include a structurally homogeneous compact, isotropic grain boundary energy, and isolated spherical pores at the grain boundary interfaces. These conditions are rarely achieved in real powder compacts.

9.9 GRAIN GROWTH IN VERY POROUS SOLIDS

Our analysis of coarsening in porous ceramics has so far concentrated on structures that can be considered to be a rough approximation to the final stage of sintering. While generally slower, coarsening is also observed in compacts with far greater porosity. The microstructural changes that occur in the earlier stages of sintering have a significant influence on the evolution in the later stages. The coarsening process in very porous compacts must therefore be addressed. Because of the greater complexity of the microstructure, a detailed analysis of the coarsening of very porous structures has not been performed. Instead, our understanding of the process is at a fairly qualitative level.

As we outlined earlier, grain growth is driven by a decrease in the grain boundary area. For the grain boundary between two spheres (i.e., a rough approximation to the initial stage of sintering), the movement of the boundary will be difficult. As sketched in Fig. 9.24, the balance of

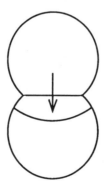

Figure 9.24 Grain growth increases the total grain boundary area between two spheres. The dihedral angle constraint at the surface of the sphere creates a boundary curvature that opposes grain growth.

interfacial tensions requires that the equilibrium dihedral angle of the grain boundary groove at the surface of the sphere be maintained. Movement of the boundary would actually involve a significant *increase* in the grain boundary area. Grain growth will not occur unless other processes that significantly reduce this energy barrier come into play.

One way in which grain growth can occur was suggested by Greskovich and Lay [26] on the basis of observations of the coarsening of Al_2O_3 powder compacts. Figure 9.25 illustrates the development of the porous Al_2O_3 microstructure during sintering at 1700°C. The structure consists of individual grains with a large amount of open porosity. In going from Fig. 9.25a to 9.25b, the relative density increased from 0.31 to 0.40, but

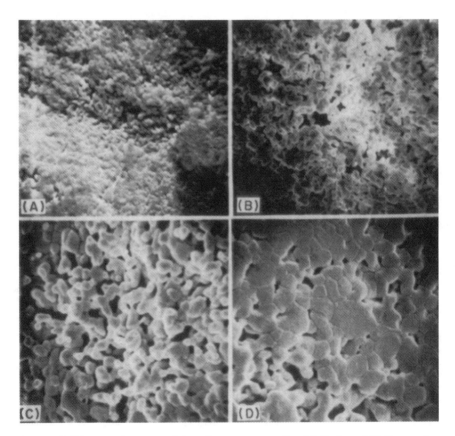

Figure 9.25 Development of the microstructure of an MgO-doped Al_2O_3 powder compact during sintering at 1700°C. (A) Green compact, (B) 1 min, (C) 2.5 min, and (D) 6 min. (Magnification 5000×.) (From Ref. 26, with permission.)

even in such very porous structures the average grain size has nearly doubled.

The model put forward by Greskovich and Lay is shown in Fig. 9.26 for two particles and for a cluster of particles. For two particles that differ in size, surface diffusion is assumed to assist in the rounding of the particles and the growth of necks between the particles. Whether or not

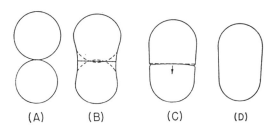

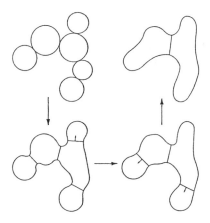

Figure 9.26 Qualitative mechanism for grain growth in porous powder compacts. *Top*: Two particles. (A) Particles of slightly different size in contact; (B) neck growth by surface diffusion between the particles; (C) grain boundary migrating from the contact plane; (D) grain growth. *Bottom*: A cluster of particles. Arrows on the grain boundaries indicate the direction of boundary movement. (From Ref. 26, with permission.)

the boundary can migrate depends on whether or not the structure permits it; that is, there must be a decrease in the total energy of the system for an incremental movement of the boundary. As sketched for the two particles, it is assumed that surface diffusion produces the structural changes for the movement of the boundary to be favorable. The rate of migration of the boundary depends on the difference in initial size between the particles: the greater the difference in size, the greater the curvature of the boundary and the greater the driving force for the boundary to sweep through the smaller grain. Neck growth between the grains is likely to be much slower than the migration of the boundary, so it controls the rate of the overall coarsening process.

Although Greskovich and Lay considered surface diffusion to be the dominant neck growth mechanism, vapor transport may also be important in systems with fairly high vapor pressures. Furthermore, coarsening controlled by neck growth is not expected to be the only process operating in fairly porous systems. A heterogeneously packed powder compact will, after some sintering, contain fairly porous regions as well as fairly dense regions (Fig. 9.25d). For this type of structure, coarsening is likely to be controlled by two separate mechanisms: (i) neck growth in the fairly porous regions and (ii) curvature-driven boundary migration in the fairly dense regions.

Another way in which grain growth can occur in porous compacts was described by Edelson and Glaeser [27]. As sketched in Fig. 9.27a, if large grain size differences exist at the necks between the particles, elimination of grain boundaries by the advancing boundary releases enough energy to make the overall process favorable. In this way, the finer neighboring grains can be consumed in an incremental growth process. Figure 9.27b shows that the growing grain can also entrap porosity, a process that can limit the density of the final article. However, this process is less likely to occur in the earlier stages of sintering; as outlined earlier, the fairly large, continuous pores that are closely spaced provide a significant drag on the boundary and limit its mobility.

9.10 PORE EVOLUTION DURING SINTERING

As our discussion so far in this chapter has shown, a common approach to the treatment of the coarsening process is to consider the kinetics of grain growth. However, a more realistic view of the sintering compact may be that of a network of contacting grains interpenetrated by a network of pores. An understanding of the evolution of the porous network is important for a broader understanding of microstructural evolution in porous systems.

Grain Growth and Microstructural Control

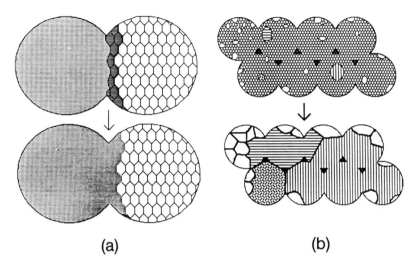

Figure 9.27 Growth of intra-agglomerate grains through polycrystalline necks. (a) Growth of a large grain proceeds across a neck by incremental consumption of smaller grains. (b) Exaggerated intra-agglomerate crystalline growth results in the development of large grains in the early stage of sintering. The resulting grain structure may contain some large grains that grow abnormally during interagglomerate sintering. (From Ref. 27, with permission.)

Some basic aspects of pore evolution have been discussed by Evans and De Jonghe [28]. Figure 9.28 shows an example of a pore network in a partially sintered compact of ZnO with a relative density of 0.73. The pictures are a stereo pair of an epoxy resin replica of the network. The replica was obtained by forcing the epoxy resin, under high pressure, into the partially sintered compact and, after curing of the resin, dissolving the ZnO away in an acid solution. While the closed porosity cannot be viewed by this method, the complexity of the pore network in a real powder compact is readily apparent.

We may expect that for a compact formed from fully dense particles, the porosity is initially all connected. As sintering proceeds, more and more of the open porosity is converted to closed porosity. The conversion is, however, dependent on the packing uniformity of the structure. For a heterogeneously packed structure, the variation of the open and closed porosity may show trends similar to those of Fig. 9.29 for a UO_2 powder compacted in a die [29]. Initially, the total porosity of the UO_2 compact was 0.37, made up of an open porosity of 0.36 and closed porosity of 0.01. During sintering at 1400°C, the open porosity decreased continuously. The

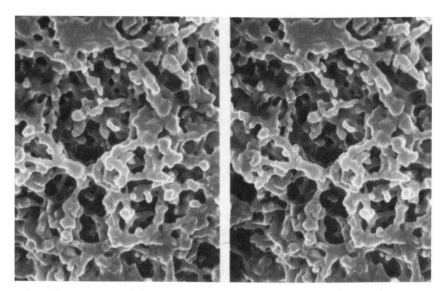

Figure 9.28 Stereographic pair of a replica of the pore space in a partially densified ZnO powder compact (density 73% of the theoretical). (Courtesy of M.-Y. Chu.)

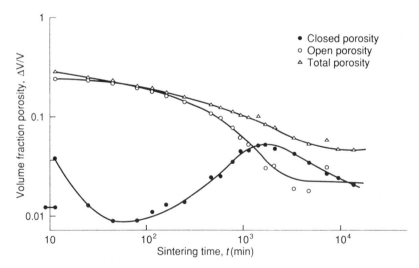

Figure 9.29 Change in porosity during the sintering of a UO_2 powder compact at 1400°C. (From Ref. 29.)

Grain Growth and Microstructural Control

closed porosity increased initially to 0.04 but returned to the value of 0.01. This initial variation in the closed porosity may be due to the sintering of fairly dense agglomerates. When the open porosity had decreased to 0.15, the closed porosity began to increase again and reached a maximum value of 0.05, after which it decreased with further sintering. For a homogeneously packed powder, we may expect the conversion to closed porosity to occur later and more suddenly. Compared to the UO_2 results in Fig. 9.29, a greater fraction of the total porosity may remain as an open network even after the compact has reached relative densities as high as 0.90.

The evolution of the pore network is difficult to analyze quantitatively. Experimentally, very few attempts have been made to produce a detailed stereological characterization of the evolution of the network. Such characterization is very time-consuming, requiring detailed quantitative microscopy of a series of partially sintered samples. On the basis of such stereological observations, Rhines and DeHoff [30] deduced that the evolution of the pore network contained features that were comparable to those of other topological decay processes such as grain growth in fully dense polycrystalline solids (see Section 9.5C). As sketched in Fig. 9.30, the pore network, consisting of channels and junctions, changes by the collapse of pore channels and the re-forming of a new network of lower connectivity. While providing a more realistic picture of the evolution of the pore network, the topological model does not provide information on the kinetics of the coarsening process or on how such coarsening can be controlled.

9.11 THERMODYNAMIC ASPECTS OF PORE–BOUNDARY INTERACTIONS

Earlier we considered the kinetics of the interaction between the pores and the grain boundaries. The thermodynamics of the interaction must also be considered because they determine whether a pore will shrink or grow during sintering.

The equilibrium pore shape, as we discussed in Chapter 8, is dictated by the dihedral angle ψ defined by

$$\cos\left(\frac{\psi}{2}\right) = \frac{\gamma_{gb}}{2\gamma_{sv}} \qquad (9.105)$$

where γ_{sv} and γ_{gb} are the interfacial tensions at the pore surface and in the grain boundary interface, respectively. Let us consider, in two dimensions, a pore with a dihedral angle $\psi = 120°$ that is surrounded by N grains. The number N is called the pore coordination number. Following

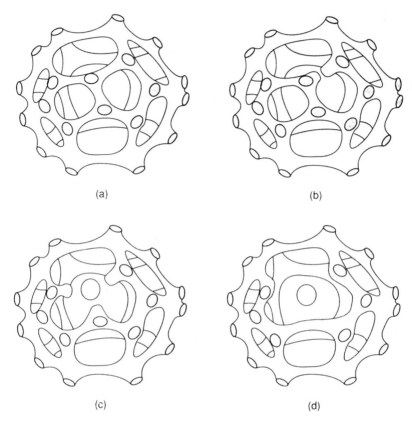

Figure 9.30 Sequence of pore channel collapse in a porous sintering solid. The sequential collapse can be envisioned as the coarsening process leaving a larger pore network as well as isolated pores. (From. Ref. 28, with permission.)

our earlier discussion for the case of a grain surrounded by other grains, the pore has straight sides if $N = 6$, convex sides for $N < 6$, and concave sides for $N > 6$ (Fig. 9.31). Since the surface of the pore will move toward its center of curvature, the pore with $N < 6$ will shrink whereas the one with $N > 6$ will grow. The pore is metastable for $N = 6$, and this number is called the *critical pore coordination number* N_c. We can go on to consider other dihedral angles; for example, $N_c = 3$ for $\psi = 60°$. The general result is that N_c depends on the dihedral angle, i.e., it decreases with a decrease in the dihedral angle.

The geometrical considerations can be extended to three dimensions,

Grain Growth and Microstructural Control

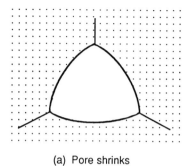

(a) Pore shrinks

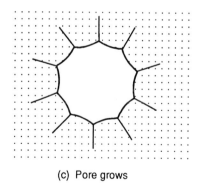

(b) Metastable pore

(c) Pore grows

Figure 9.31 Pore stability in two dimensions for a dihedral angle of 120°.

in which case the pore is a polyhedron. The analysis has been carried out by Kingery and Francois [31]. Taking r_s as the radius of curvature of the circumscribed sphere around a polyhedral pore surrounded by grains, the ratio of the radius curvature of the pore, r, to r_s depends on both the dihedral angle and the pore coordination number, as shown in Fig. 9.32. When the surfaces of the pore become flat (i.e., $r = \infty$), the pore is metastable and there is no tendency for the pore to grow or shrink. The ratio r_s/r is zero, and this condition defines N_c. The value of N_c as a function of the dihedral angle is plotted in Fig. 9.33. As an example of the usefulness of Fig. 9.33, let us consider a pore with a dihedral angle $\psi = 120°$. From Fig. 9.33 we find that $N_c = 12$. For this value of the dihedral angle, the pore surfaces are convex if $N < 12$ and the pore will shrink. However, for $N > 12$, the pore surfaces are concave and the pore will grow. Thus,

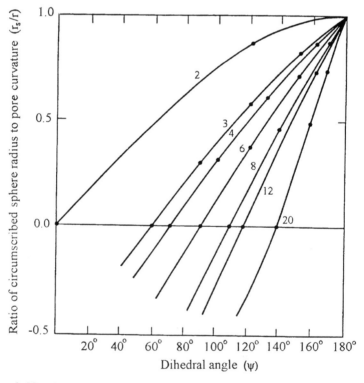

Figure 9.32 Change in the ratio r_s/r with dihedral angle for pores surrounded by different numbers of grains as indicated on the individual curves. (From Ref. 31.)

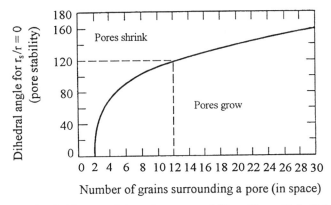

Figure 9.33 Conditions for pore stability. (From Ref. 31.)

knowing the dihedral angle and the pore coordination number, we can determine the curvature of the pore surfaces and the tendency for the pore to shrink or grow.

From this discussion, we see that a poorly compacted powder such as that shown in Fig. 9.34, containing pores that are large compared to

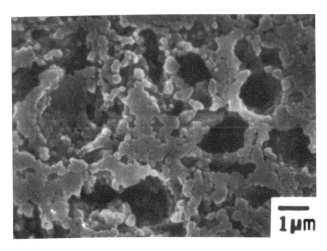

Figure 9.34 Heterogeneously packed powder containing pores that are large compared with the grain size. Under normal sintering conditions such pores will limit the final density of the sample.

the grain size, would be difficult to densify, especially if the dihedral angle is small. One practical solution to the difficulty of densification is to prepare compacts with reasonably high density and a fairly uniform microstructure by cold isostatic pressing or by the colloidal methods discussed in Chapter 6. This problem of the densification of heterogeneous powder compacts will be considered further in Chapter 11.

9.12 INTERACTION BETWEEN DENSIFICATION AND COARSENING

In our consideration of solid-state sintering, we have so far treated the processes of densification and coarsening separately. As sketched in Fig. 9.35 for a final stage microstructure, the two processes occur concurrently. In addressing their interaction, we should remember that each process by itself is fairly complex and that the models provide only a

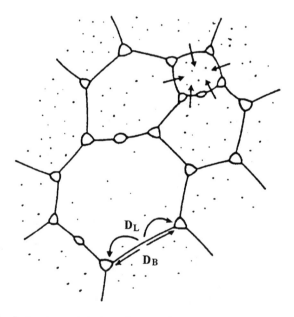

Figure 9.35 Late stage sintering of a powder compact. The two contributions to free energy reduction are (i) densification (lower part of figure, where the arrows show the direction of atom flow) and (ii) coarsening or grain growth (top right, where the arrows show the direction of boundary movement). (From Ref. 34.)

Grain Growth and Microstructural Control

qualitative understanding of each process. In this situation, a detailed theoretical analysis of the coupling of densification and coarsening would not be expected to provide any significant advance in our understanding of sintering. A simpler approach that seeks to provide a qualitative understanding of the interaction may be more useful in practice.

Experimentally, the densification and grain growth data for many ceramic systems show trends similar to those shown in Fig. 9.36 for the sintering of TiO_2. A region of significant densification with limited grain growth is followed by one of reduced densification but with significant grain growth. Generally, coarsening becomes prominent in the late intermediate and final stages of sintering. Gupta [32] found that the average grain size increases slowly and roughly linearly with density up to relative densities of 0.85–0.90, but then it increases much more rapidly above this density value (Fig. 9.37). From the point of view of achieving the desired microstructure in the fabricated article, an understanding of the grain growth and coarsening interaction in the final stage is important.

An approach that builds on the model sintering and grain growth equations was developed by Yan [33] and later by Brook [34]. In this

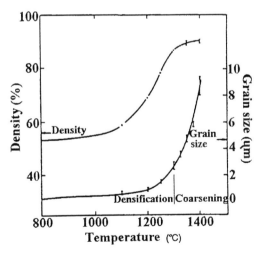

Figure 9.36 The density and grain size of a TiO_2 powder compact as a function of the sintering temperature. Rapid densification with limited grain growth occurs at lower temperatures followed by rapid grain growth with little densification at higher temperatures. (From Ref. 33.)

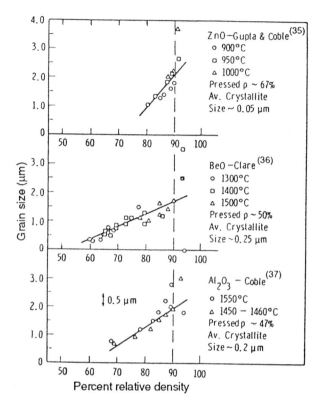

Figure 9.37 Simultaneous plots of density versus grain size for Al_2O_3, BeO, and ZnO. (From Ref. 32, with permission.)

method, an equation for densification by a particular mechanism (e.g., lattice diffusion) is assumed. Taking Coble's final stage equation given in Chapter 8, i.e., Eq. (8.68), we can write

$$\rho - \rho_0 = \frac{6\pi}{\sqrt{2}} \left(\frac{D_l \gamma_{sv} \Omega}{G^3 kT} \right) t \qquad (9.106)$$

where the parameters have the same meaning as defined earlier for Eq. (8.68). Next, a particular mechanism for late stage coarsening is assumed. For grain growth by surface diffusion controlled pore drag, the boundary velocity is given by Eq. (9.102). Also, the product of the number of pores

Grain Growth and Microstructural Control

per unit volume, N_V, and the pore volume must equal the porosity, $P = 1 - \rho$, so that

$$\frac{4}{3}\pi r^3 N_V = 1 - \rho \qquad (9.107)$$

Taking $N_A \sim 1/G^2$, $N_V \sim 1/G^3$, and $F_b \sim 2\gamma_{gb}/G$, Eqs. (9.102) and (9.107) give, after some rearranging followed by integration,

$$G^4 - G_0^4 = \frac{15 D_s \delta_s \gamma_{gb} \Omega}{kT(1-\rho)^{4/3}} t \qquad (9.108)$$

where G_0 is the initial grain size. Selection of the values for ρ_0 and G_0 followed by the use of Eqs. (9.106) and (9.108) and iteration allows the grain size versus density trajectory to be plotted as in Fig. 9.38. According to the results of Fig. 9.38, for a chosen set of initial conditions, the desired endpoint density and grain size in the fabricated article can be reached by a suitable choice of the processing conditions to control $\dot{G}/\dot{\rho}$ (the ratio of the grain growth rate to the densification rate). In general, for $\dot{G} \gg \dot{\rho}$, low density and large grain size are achieved while high density and small grain size are achieved for $\dot{G} \ll \dot{\rho}$.

Although the results of Fig. 9.38 have considerable merit for the fabrication of ceramics with controlled microstructure, it must be remembered

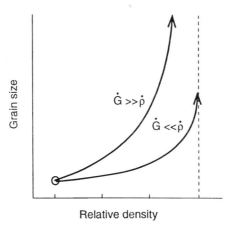

Figure 9.38 The development of grain size and density for two sets of conditions. Starting from similar initial conditions of grain size and density (circle), the lines show how the microstructure develops for samples where (i) the grain growth rate $\dot{G}$ is much greater than the densification rate $\dot{\rho}$, and (ii) $\dot{G}$ is much less than $\dot{\rho}$ (From Ref. 34.)

that the equations used in the method assume a homogeneous microstructure. Heterogeneities such as pores that are large compared to the grain size, density variations, and large grain size distribution will significantly reduce the ability to control the final microstructure.

9.13 FABRICATION ROUTES FOR THE PRODUCTION OF CERAMICS WITH HIGH DENSITY AND CONTROLLED GRAIN SIZE

Most applications of ceramics require products with high density and controlled (small) grain size. The discussion of the previous section indicated that when suitable processing procedures are employed, such an endpoint is achievable through fabrication routes that have the effect of decreasing the ratio $\dot{G}/\dot{\rho}$ or, equivalently, increasing $\dot{\rho}/\dot{G}$. We now examine the principles of these fabrication routes. The technology will be considered in Chapter 12.

A. Sintering with an External Pressure (Hot Pressing)

Compared to sintering, hot pressing produces a greater driving force for densification. The dependence of the densification rate on the driving force can be written

$$\dot{\rho}_{HP} \sim (\Sigma + p_a); \qquad \dot{\rho}_S \sim \Sigma \qquad (9.109)$$

where the subscripts HP and S refer to hot pressing and sintering, respectively, Σ is the sintering stress, and p_a is the applied pressure. Since the grain boundary mobility is unaltered,

$$\dot{G}_{HP} \approx \dot{G}_S \qquad (9.110)$$

For $p_a \gg \Sigma$, we find that $(\dot{\rho}/\dot{G})_{HP} \gg (\dot{\rho}/\dot{G})_S$. According to this discussion, the effectiveness of hot pressing for the production of high densities coupled with small grain size arises from the ability to increase $\dot{\rho}/\dot{G}$. For Al_2O_3 doped with MgO, the measured trajectories for the grain size versus density are shown in Fig. 9.39 for hot pressing and for sintering. The flatter trajectory for the case of hot pressing is consistent with the present discussion.

B. Dopants

The role of dopants is fairly complex, and we shall consider this topic in further detail in Chapter 11. However, we found that dopants that segregate to the grain boundary can reduce the boundary mobility by the solute drag effect. In this case $\dot{G}_{doped} < \dot{G}_{undoped}$. Dopants may also influence

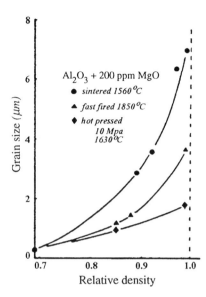

Figure 9.39 Experimental results for microstructural development in Al_2O_3 + 200 ppm MgO showing the grain size versus density trajectories for fabrication by hot pressing, sintering, and fast firing. (♦) Hot pressed, 10 MPa, 1630°C; (●) sintered at 1560°C; (▲) fast fired at 1850°C. (From Ref. 34.)

the densification process, but it appears that this effect is small compared to the effect on the grain growth rate. It appears that the effectiveness of dopants arises from the ability to reduce $\dot{G}$ or equivalently to increase $\dot{\rho}/\dot{G}$.

C. Inclusions

Fine, inert inclusions, as we observed earlier, exert a drag on the grain boundary and, if the drag is large enough, may pin the boundary. They appear not to have any significant influence on the diffusional transport of matter leading to densification. The effectiveness of inclusions can also be interpreted in terms of a decrease in $\dot{G}$ or an increase in $\dot{\rho}/\dot{G}$.

D. Uniformly Packed Fine Particles

Uniformly compacted fine powders have small pores with low coordination number (i.e., $N < N_c$). The densification rate for such a system is higher than for a similar system with heterogeneous packing. Furthermore, if the particle size distribution is narrow, the driving force for grain growth due to the curvature of the boundary is small. The effectiveness of this route can therefore be interpreted in terms of an increase in $\dot{\rho}/\dot{G}$.

E. Fast Firing

For powder systems in which the activation energy for densification is significantly greater than for grain growth (Fig. 9.40), fast heat-up to a high enough temperature where $\dot{\rho} \gg \dot{G}$ can provide an effective fabrication route. The process is usually referred to as *fast firing*. For MgO-doped Al_2O_3, the grain size versus density trajectory during fast firing is compared with those for hot pressing and sintering in Fig. 9.39. In this case, the benefits of fast firing over the more conventional sintering route is clear.

F. Liquid-Phase Sintering

A second phase that is liquid at the firing temperature can provide a fast diffusion path for densification. However, grain growth by the Ostwald

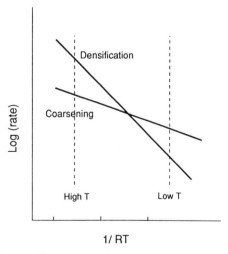

Figure 9.40 Under conditions where the densification mechanism has a higher activation energy than the coarsening mechanism, a fast heating rate to high firing temperatures (fast firing) can lead to the production of articles of high density.

Grain Growth and Microstructural Control

ripening process may also be enhanced. In this case, high density is normally accompanied by fairly appreciable grain growth. This commonly used fabrication approach is the subject of the next chapter.

9.14 CONCLUDING REMARKS

In this chapter we outlined the basic features of microstructural development in polycrystalline ceramics. Simple models allow the derivation of equations for the kinetics of normal grain growth. Many of the theories analyze an isolated grain boundary or a single grain and neglect the topological requirements of space filling. Computer simulation is playing an increasing role in exploring the interplay between the kinetics of grain growth and the topological requirements of space filling. Abnormal grain growth cannot be explained in terms of the reduction of a uniform grain boundary energy. It is believed to be due to local increases in the driving force or in the mobility of the grain boundary. A considerable gap exists between the theoretical understanding of microstructural evolution and the practical application to real powder systems. However, the simple models provide useful guidelines for the control of the microstructure during sintering. Excessive grain growth and abnormal grain growth especially must be prevented if high-density ceramics with controlled grain size are to be fabricated. The use of dopants that segregate to the grain boundary and fine inert inclusions can be very effective for reducing the grain boundary mobility. Improvements in the particle packing in the consolidated powder form also lead to benefits in densification. Pores that are large compared with the particle size are difficult to remove during sintering and therefore limit the final density achieved. The use of sintering with an externally applied pressure (e.g., hot pressing) and to a lesser extent fast firing also provide effective fabrication routes. Finally, if solid-state sintering is inadequate for achieving the desired endpoint density, then liquid-phase sintering, the subject of the next chapter, provides an effective fabrication route.

PROBLEMS

9.1 Compare the conditions that control the stability of a pore in a polycrystalline ceramic with those for a pore in a glass.

9.2 The average grain size of a dense ceramic after annealing for 120 minutes at 1200°C was found to be 5 μm. Annealing for 60 minutes at 1400°C gave an average grain size of 11 μm. If the average grain size at time = 0 was 2 μm, estimate what the average grain size will be after 30 minutes annealing at 1600°C.

9.3 Consider the grain boundary between two spherical particles with roughly the same radius. In one case, both grains are single crystalline. In the other case, one grain is polycrystalline while the second grain is single crystalline. Compare the grain growth phenomena in the two cases.

9.4 A dense Al_2O_3 ceramic (grain size of ≈ 1 μm) contains 10 vol% of fine ZrO_2 particles distributed fairly uniformly in the grain boundaries of the matrix. If the particle size of the ZrO_2 is ≈ 0.2 μm, estimate the maximum pinning force that the ZrO_2 particles can exert on unit area of the grain boundary. The grain boundary energy is ≈ 0.5 J/m^2. If the ceramic is annealed at a high enough temperature, discuss the grain growth phenomena.

9.5 Derive the equation in Table 9.1 for the pore mobility when pore migration is controlled by vapor transport.

9.6 Compare the densification of a homogeneous compact of 5 μm single crystal particles with that for a compact of 5 μm agglomerates of 0.5 μm single crystal particles. Assume that the chemical composition of the particles is the same.

9.7 According to Eq. (9.1), grain growth during sintering leads to a reduction in the densification rate. Can grain growth lead to densification of ceramic powder compacts? Explain.

9.8 Mercury porosimetry data reveal that an MgO powder compact (relative density ≈ 0.45) contains a bimodal pore size distribution in which the volume fraction of the small pores is approximately equal to that of the large pores. The average radii of the small pores and the large pores are one-half and four times the particle size of the MgO powder, respectively. Sketch and explain the possible variation of the isothermal densification rate of the compact as a function of its relative density.

REFERENCES

1. R. J. Brook, in *Ceramic Fabrication Processes* (*Treatise on Materials Science and Technology,* Vol. 9) (F. F. W. Wang, Ed.), Academic, New York, 1976, pp. 331–364.
2. J. E. Burke, Report No. 68-C-363, General Electric Research and Development Center, Schenectady, NY, October 1968.
3. W. D. Kingery, H. K. Bowen, and D. R. Uhlmann, *Introduction to Ceramics*, 2nd ed., Wiley, New York, 1976.

4. G. W. Greenwood, *Acta Metall.*, *4*:243 (1956).
5. C. Wagner, *Z. Electrochem.*, *65*:581 (1961).
6. I. M. Lifshitz and V. V. Slyozov, *Phys. Chem. Solids*, *19*:35 (1961).
7. C. S. Smith, *Metal Interfaces*, ASM, Cleveland, OH, 1952, pp. 65–113.
8. H. V. Atkinson, *Acta Metall.*, *36*:469 (1988).
9. J. E. Burke and D. Turnbull, *Prog. Metal Phys.*, *3*:220 (1952).
10. M. Hillert, *Acta Metall.*, *13*:227 (1965).
11. P. Feltham, *Acta Metall.*, *5*:97 (1957).
12. N. P. Louat, *Acta Metall.*, *22*:721 (1974).
13. F. N. Rhines and K. R. Craig, *Metall. Trans.*, *5A*:413 (1974).
14. M. P. Anderson, D. J. Srolovitz, G. S. Grest, and P. S. Sahni, *Acta Metall.*, *32*:783 (1984).
15. D. J. Srolovitz, M. P. Anderson, P. S. Sahni, and G. S. Grest, *Acta Metall.*, *32*:793 (1984).
16. D. J. Srolovitz, G. S. Grest, and M. P. Anderson, *Acta Metall.*, *33*:2233 (1985).
17. C. V. Thompson, H. J. Frost, and F. Spaepen, *Acta Metall.*, *35*:887 (1987).
18. R. L. Coble, Figure 10.11 in Ref. 3.
19. C. S. Smith, *Trans. AIME*, *175*:15 (1948).
20. J. W. Cahn, *Acta Metall.*, *10*:789 (1962).
21. M. F. Yan, R. M. Cannon, and H. K. Bowen, in *Ceramic Microstructures '76* (R. M. Fulrath and J. A. Pask, Eds.), Westview Press, Boulder, CO, 1977, pp. 276–307.
22. P. G. Shewmon, *Trans. AIME*, *230*:1134 (1964).
23. R. J. Brook, *J. Am. Ceram. Soc.*, *52*:56 (1969).
24. F. M. A. Carpay, in *Ceramic Microstructures '76* (R. M. Fulrath and J. A. Pask, Eds.), Westview Press, Boulder, CO, 1977, pp. 261–275.
25. W. D. Kingery and B. Francois, *J. Am. Ceram. Soc.*, *48*:546 (1965).
26. C. Greskovich and W. Lay, *J. Am. Ceram. Soc.*, *55*:142 (1972).
27. L. H. Edelson and A. M. Glaeser, *J. Am. Ceram. Soc.*, *71*:225 (1988).
28. J. W. Evans and L. C. De Jonghe, *The Production of Inorganic Materials*, Macmillan, New York, 1991, pp. 456–462.
29. S. C. Coleman and W. B. Beeré, *Phil. Mag.*, *31*:1403 (1975).
30. F. N. Rhines and R. T. DeHoff, in *Sintering and Heterogeneous Catalysis* (*Materials Science Research*, Vol. 16) (G. C. Kuczynski, A. E. Miller, and G. A. Sargent, Eds.), Plenum, New York, 1984, pp. 49–61.
31. W. D. Kingery and B. Francois, in *Sintering and Related Phenomena* (G. C. Kuczynski, N. A. Hooton, and C. F. Gibbon, Eds.), Gordon and Breach, New York, 1967, pp. 471–498.
32. T. K. Gupta, *J. Am. Ceram. Soc.*, *55*:276 (1972).
33. M. F. Yan, *Mater. Sci. Eng.*, *48*:53 (1981).
34. R. J. Brook, *Proc. Br. Ceram. Soc.*, *32*:7 (1982).
35. T. K. Gupta and R. L. Coble, *J. Am. Ceram. Soc.*, *51(9)*:521 (1968).
36. T. E. Clare, *J. Am. Ceram. Soc.*, *49(3)*:159 (1966).
37. R. L. Coble, *J. Appl. Phys.*, *32(5)*:787 (1961).

FURTHER READING

The following list of conference proceedings contains useful review papers on many aspects of ceramic microstructures.

R. M. Fulrath and J. A. Pask (Eds.), *Ceramic Microstructures: Their Analysis, Significance, and Production*, Wiley, New York (1968).

R. M. Fulrath and J. A. Pask (Eds.), *Ceramic Microstructures '76: With Emphasis on Energy Related Applications*, Westview Press, Boulder, CO (1977).

J. A. Pask and A. G. Evans, *Ceramic Microstructures '86: Role of Interfaces*, Materials Science Research Vol. 21, Plenum, New York (1987).

10
Liquid-Phase Sintering

10.1 INTRODUCTION

In the last few chapters, our discussion of the firing process has been concerned with solid-state sintering, in which the material remains entirely in the solid state. However, a wide variety of ceramic materials can also be produced by the process of liquid-phase sintering. In this process, the composition of the powder and the firing temperature are chosen such that a small amount of liquid forms between the grains. Rearrangement of the grains soon after formation of the liquid produces densification through more efficient packing. Transport of dissolved solid through the liquid leads to further densification and coarsening of the microstructure.

Liquid-phase sintering is important for systems that are difficult to densify by solid-state sintering. Examples of such systems are the ceramics that have a high degree of covalent bonding, such as SiC and Si_3N_4. The process is also important when the use of solid-state sintering is too expensive or requires too high a fabrication temperature. For example, the fabrication, by solid-state sintering, of Al_2O_3 and ZrO_2 materials of high density requires the use of fine deagglomerated powders or fairly high temperatures. A disadvantage is that the liquid phase used to promote sintering normally remains as a glassy grain boundary phase that may lead to a deterioration of properties.

Compared to solid-state sintering, the additional phase makes liquid-phase sintering more difficult to analyze. However, many features of the process are fairly well understood. The process is normally viewed in terms of three overlapping stages, with each stage described with respect to the dominant mechanism occurring in that stage. The mechanisms are rearrangement (Stage 1), solution-precipitation (Stage 2) and Ostwald ripening (Stage 3). Theoretical analysis and experiments show that the densification rate depends on a number of key parameters. Low contact angle, low dihedral angle, and high solubility of the solid in the liquid are essential for the achievement of high density. In addition, homogeneous packing of the particulate solid, homogeneous distribution of the liquid phase between the grains, and fairly fine particle size are the important processing requirements that must be met.

In our discussion of the basic mechanisms of liquid-phase sintering, we have not hesitated to draw on the extensive literature on metallic systems. Considerable work on the mechanisms of liquid-phase sintering has been performed with metallic systems, and this work has played a key role in our understanding of the process. The mechanisms of rearrangement and solution-precipitation are common to both metallic and ceramic systems. Observations also show clearly that many of the microstructural changes observed in metallic systems have some parallel in ceramic systems.

For consistency, we use the following nomenclature to describe the systems in liquid-phase sintering: the particulate solid forming the major component is written first, and the liquid-producing component is written in parentheses. In this nomenclature, silicon nitride sintered in the presence of a liquid phase produced by the addition of magnesium oxide is written $Si_3N_4(MgO)$. Some common systems in the liquid-phase sintering of ceramics and their application are summarized in Table 10.1.

Table 10.1 Some Liquid-Phase Sintering Systems for the Production of Ceramic Materials

Ceramic system	Additive content (wt %)	Application
$Si_3N_4(MgO)$	5–10	Structural ceramics
$MgO(CaO-SiO_2)$	<5	Refractories
$ZnO(Bi_2O_3)$	2–3	Electrical varistors
Al_2O_3(talc)	5	Electrical insulators
$UO_2(TiO_2)$	1	Nuclear ceramics
$BaTiO_3(TiO_2)$	<1	Dielectrics
$ZrO_2(CaO-SiO_2)$	<1	Ionic conductors
$MgO(LiF)$	<3	Refractories
$BaTiO_3(LiF)$	<3	Dielectrics

10.2 ELEMENTARY FEATURES OF LIQUID-PHASE SINTERING

Figure 10.1 shows a sketch of an idealized two-sphere model in which the microstructural aspects of liquid-phase sintering are compared with those of solid-state sintering. In liquid-phase sintering, if, as we assume, the liquid wets and spreads to cover the solid surfaces, the particles will be separated by a liquid bridge. The thickness of the liquid bridge is typically many times greater than the grain boundary thickness in solid-state sintering. Furthermore, since diffusion through a liquid is normally much faster than in a solid, the presence of the liquid bridge provides a path

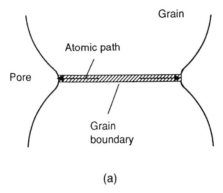

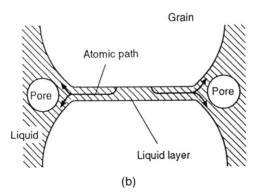

Figure 10.1 Sketch of an idealized two-sphere model comparing the microstructural aspects of solid-state sintering (a) with those of liquid-phase sintering (b).

for enhanced diffusion, providing that the solid is fairly soluble in the liquid.

A. Driving Force for Densification

Assuming as above that the liquid wets and spreads over the solid surfaces, the solid/vapor interface of the particulate system will be eliminated and pores will form in the liquid. The reduction of the liquid/vapor interfacial area provides a driving force for shrinkage and densification of the overall system. For a spherical pore of radius r in a liquid, the pressure difference across the curved surface is given by the equation of Young and Laplace

$$\Delta p = -\frac{2\gamma_{lv}}{r} \qquad (10.1)$$

where γ_{lv} is the surface energy of the liquid/vapor interface. The pressure in the liquid is lower than that in the pore, and this generates a compressive stress on the particles. This compressive stress due to the liquid is equivalent to placing the system under an external hydrostatic pressure, the magnitude of which is given by Eq. (10.1). Taking $\gamma_{lv} \approx 1$ J/m^2 and $r \approx 0.5$ μm gives $\Delta p \approx 4$ MPa. Pressures of this magnitude can provide a fairly appreciable driving force for sintering.

B. Enhancement of Densification

Compared to solid-state sintering, the presence of the liquid phase leads to enhanced densification through (i) enhanced rearrangement of the particulate solid and (ii) enhanced matter transport through the liquid.

Enhanced Rearrangement

The friction between the particles is significantly reduced, so they can rearrange more easily under the action of the compressive stress exerted by the liquid.

Enhanced Matter Transport

In solid-state sintering by, for example, grain boundary diffusion, an important parameter that controls the rate of diffusion is the product of the grain boundary diffusion coefficient D_{gb} and the grain boundary thickness δ_{gb}. In liquid-phase sintering, the corresponding parameter is the product of the diffusion coefficient D_L of the solute atoms in the liquid and the thickness of the liquid bridge, δ_L. As noted earlier, δ_L is typically many times greater than δ_{gb}, and diffusion through a liquid is much faster than in solids. The liquid therefore provides a path for enhanced matter transport.

C. Source of the Liquid Phase

The consolidated powder form for liquid-phase sintering is normally produced from a mixture of two powders: a major component and an additive phase. On heating, the additive melts or reacts with a small part of the major component to form a eutectic liquid. The production of the liquid phase by melting of the additive is fairly common in metallic systems, e.g., Fe(Cu) and W(Ni). In ceramic systems, the formation of a eutectic liquid is more common, e.g., MgO(CaO-SiO$_2$) and ZnO(Bi$_2$O$_3$). For systems that rely on the formation of a eutectic liquid, phase diagrams play a key role in the selection of the additive and in the choice of the firing conditions. The use of phase diagrams in liquid-phase sintering will be discussed later in this chapter.

D. Amount of Liquid Phase

For the production of most advanced ceramics by liquid-phase sintering, the amount of liquid phase produced at the firing temperature is kept typically below ≈5 vol %, although in a few cases it can be as high as ≈10 vol %. The volume of liquid is therefore insufficient to fill the interstices between the solid particles. However, many traditional ceramics are fabricated by a process in which a much higher volume of liquid is produced (≈25–35 vol %). In this case, the volume of liquid is sufficient to fill the interstices between the solid particles (after some rearrangement of the system), and no further densification by other mechanisms is required for the production of the final article. As we will recall from Chapter 1, this type of sintering in which the liquid volume is sufficient to fill up the interstices between the solid particles is referred to as *vitrification*.

E. Persistent and Transient Liquid-Phase Sintering

In most systems, the liquid persists throughout the firing process and its volume does not change appreciably. This situation is sometimes referred to as *persistent liquid-phase sintering*. On cooling, the liquid forms a glassy grain boundary phase, which, as outlined earlier, may lead to a deterioration in high-temperature mechanical properties. In a small number of systems, the liquid may be present over a major portion of the firing process but then disappears by (i) incorporation into the solid phase to produce a solid solution, e.g., Si$_3$N$_4$(Al$_2$O$_3$-AlN); (ii) crystallization of the liquid, e.g., Si$_3$N$_4$(Al$_2$O$_3$-Y$_2$O$_3$); or (iii) evaporation, e.g., the system BaTiO$_3$(LiF). The term *transient liquid-phase sintering* is used to describe the sintering in which the liquid phase disappears prior to the completion of firing. The interest in ceramic materials for mechanical engineering applications at high temperatures has led to the investigation of transient liquid-phase sintering in a few Si_3N_4 systems. However, the process is generally difficult to control and requires much further work if it is to be practiced successfully on a larger scale.

In this book, the term *liquid-phase sintering* will refer most generally to the case of a persistent liquid. A distinction between persistent and transient liquid-phase sintering will be made only when it is convenient.

F. Cohesion of the Particulate Solid During Firing

Despite the presence of a viscous liquid between the particles during liquid-phase sintering, the structure does not collapse unless the volume of liquid is very large. The capillary stress produced by the liquid holds the solid particles together. The creep (or shear) viscosity of the system is, however, much lower than that of a similar system without the liquid phase.

G. Advantages and Disadvantages of Liquid-Phase Sintering

The major advantages of liquid-phase sintering, as discussed earlier, are (i) the enhanced densification leading to the production of high density and (ii) the economic benefits arising from the use of a lower firing temperature than that required for solid-state sintering of the major component. However, the use of liquid-phase sintering is not without its disadvantages. The liquid formed during firing normally remains as a grain boundary phase on cooling (see Fig. 1.15). This grain boundary phase can cause a deterioration of the properties of the fabricated article. An important example is the production of ceramics for structural applications at high temperature. The grain boundary phase may soften prematurely, thereby causing a reduction in the creep resistance of the solid. In many cases, it is not easy to control the grain growth during liquid-phase sintering. The liquid may also enhance the coarsening process so that the achievement of a fine-grained microstructure may be difficult. Finally, compared to solid-state sintering, the presence of an additional phase may make the analysis and understanding of certain aspects of liquid phase sintering more difficult.

10.3 MICROSTRUCTURE PRODUCED BY LIQUID-PHASE SINTERING

In addition to any porosity that may be present, the microstructure of the product produced by liquid-phase sintering consists of two phases: (i) the grain boundary phase formed from the liquid phase on cooling and (ii) the grains. Unless it is crystallizable, the grain boundary phase normally has

Liquid-Phase Sintering

an amorphous structure. It may penetrate the grain boundaries completely so that the grains are separated from one another by a fairly thin ($\approx 1-5$ nm) grain boundary phase (Fig. 1.15), or it may only partially penetrate the grain boundaries. As discussed later, the extent to which the liquid penetrates the grain boundary is controlled by the ratio of the solid/solid interfacial tension to the solid/liquid interfacial tension (i.e., by the dihedral angle).

For the particulate solid, three common types of grain structures are observed. When the amount of liquid is moderate (say, roughly above ≈ 5 vol %), grains with a fairly rounded shape are observed (Fig. 10.2a). For higher liquid content, the grain shape becomes spheroidal. For low volume fraction of liquid ($<2-5$ vol %), the grains undergo considerable changes in shape and develop a morphology in which the contact regions between neighboring grains are fairly flat (Fig. 10.2b). The shape changes allow the grains to pack more efficiently, a phenomenon usually described as grain shape accommodation. Figure 10.2c shows a third type of structure that is commonly observed when the liquid content is fairly high [1]. The grains take up a prismatic shape (instead of the spheroidal shape discussed above), and there may be considerable solid/solid contact between neighboring grains. The growth rate of the prismatic grains is normally slower than that for systems with the same amount of liquid that develop spheroidal grains.

The difference between the grain structures in Figs. 10.2a and b may be explained in terms of geometric considerations. For a smaller amount of liquid, the grains must pack more efficiently for the pores to be filled up. However, the difference between the rounded grain structure and the prismatic grain structure of Fig. 10.2c must depend on the intrinsic parameters such as the relative magnitudes of the interfacial tensions.

Depending on the parameters of the system such as the composition of the liquid, the powder characteristics, and the firing temperature, Al_2O_3 grains develop in a complex way during sintering with a small amount of liquid. A variety of grain types ranging from equiaxial and elongated grains with curved sides to platelet and platelike grains with straight (faceted) sides have been observed [2].

10.4 THE STAGES IN LIQUID-PHASE SINTERING

As sketched in Fig. 10.3, liquid-phase sintering is usually considered to occur in three stages that overlap to a certain extent [3]. Compared to solid-state sintering, where each of the three stages is connected with a dominant microstructural change, each stage in liquid-phase sintering is identified with the dominant mechanism that occurs in that stage.

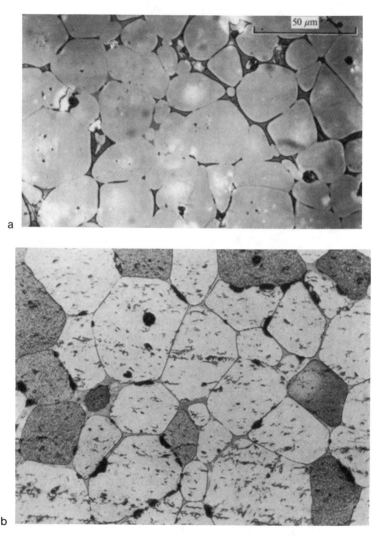

Figure 10.2 Commonly observed microstuctures of ceramics produced by liquid-phase sintering. (a) Rounded grains produced with a moderate liquid content (above ≈ 5 vol %). (From Ref. 30, with permission.); (b) grains with flat contact surfaces produced with a low volume fraction of liquid (<2–5 vol %). (From Ref. 29, with permission.); (c) relatively small prismatic grains with some solid–solid contact between neighboring grains produced when the liquid volume is moderate to high. (From Ref. 1, with permission.)

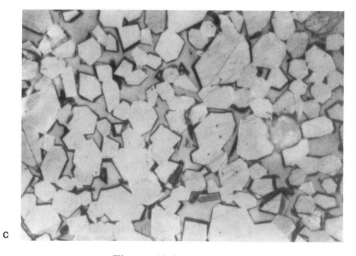

Figure 10.2 Continued.

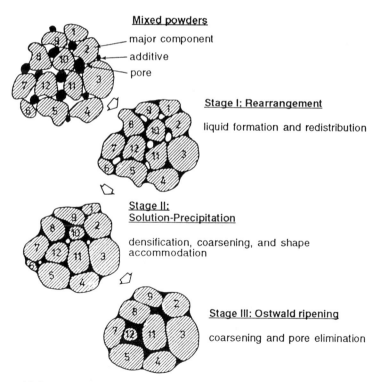

Figure 10.3 Schematic illustrating the three stages of liquid-phase sintering, with the dominant mechanism and processes occurring in each stage. (Adapted from Ref. 3.)

A. Stage 1: Rearrangement

As the temperature of the consolidated powder form is raised, liquid will form at some point. Assuming that there is good wetting between the liquid and the particulate solid, the particles will rearrange themselves under the action of the surface tension forces to produce a more stable packing. Capillary stresses will cause the liquid to redistribute itself between the particles and into the small pores, leading to further rearrangement. Assuming that the solid is soluble in the liquid, contact points between agglomerates will be dissolved (because this leads to a reduction in interfacial energy) and the fragments will also undergo rearrangement. Throughout the process, dissolution of sharp edges will make the particle surfaces smoother, thereby aiding in the rearrangement and packing of the system. Liquid redistribution and rearrangement will continue until the liquid has reached a relatively stable configuration, at which point the rearrangement process will have caused the particles to achieve a packing density close to that of dense random packing.

Rearrangement occurs fairly rapidly and may be dominant only over the first few minutes of the densification process. Depending on the amount of liquid present, a significant amount of densification can occur in the rearrangement stage. If the liquid volume is sufficient to fill up the pores within the particulate structure, full densification will be achieved by rearrangement alone. As outlined earlier, this process in which full density is achieved by rearrangement alone is widely used in the fabrication of traditional ceramics and is referred to as vitrification. For advanced ceramics, the amount of liquid present is fairly small, and other processes must contribute significantly to the achievement of a high final density.

B. Stage 2: Solution-Precipitation

As densification by rearrangement slows, assuming that the solid is soluble in the liquid, a process referred to as solution-diffusion-precipitation (or simply solution-precipitation) becomes dominant. The solid dissolves at the solid/liquid interfaces with a higher chemical potential, diffuses through the liquid, and precipitates on the particles at other sites with a lower chemical potential. One type of dissolution site is the wetted contact area between the particles, where the capillary stress due to the liquid or an externally applied stress leads to a higher chemical potential (Fig. 10.1b). Precipitation occurs at sites away from the contact area. For systems with a distribution of particle sizes, another type of dissolution site is the wetted surface of smaller particles where the smaller particle radius

Liquid-Phase Sintering

leads to a higher chemical potential. Precipitation will occur on the surfaces of the larger particles. This dissolution of the smaller grains and precipitation on the larger grains is the classic Ostwald ripening process considered in Chapter 9.

Densification by the solution-precipitation mechanism is accompanied by considerable coarsening (grain growth) and by changes in the shape of the grains. When the amount of liquid is fairly large, the grains normally take a rounded shape (Fig. 10.2a). However, for a small amount of liquid, the grains develop flat faces and assume the shape of a polyhedron. By taking the shape of a polyhedron, the grains can pack more efficiently, a process described as grain shape accommodation (see Fig. 10.2b).

C. Stage 3: Ostwald Ripening

Densification becomes very slow because of the large diffusion distances in the coarsened structure, and this stage normally takes most of the time. Microstructural coarsening by the solution-precipitation mechanism (Ostwald ripening) dominates the final stage. It is accompanied by grain shape accommodation, which allows more efficient packing of the grains. Liquid may be released from the more efficiently packed regions, and it may flow into the isolated pores, leading to densification. For systems with a dihedral angle greater than zero, solid–solid contacts between the grains develop, leading to a more rigid structure.

The three stages of liquid-phase sintering are sketched on a sintering curve in Fig. 10.4.

10.5 THE CONTROLLING KINETIC AND THERMODYNAMIC FACTORS

The discussion in the previous section indicated that the successful production of the desired density and microstructure in liquid-phase sintering depends on a number of kinetic and thermodynamic factors. In addition, several processing parameters influence the process. These processing parameters include the uniformity of the particle packing, the homogeneity of the liquid distribution, and the particle size characteristics of the major component and the additive. In this section, we examine how the key kinetic and thermodynamic factors influence liquid-phase sintering. The effect of the processing parameters will be described in Sections 10.7A–10.7C when we examine in more detail the processes occurring in liquid-phase sintering.

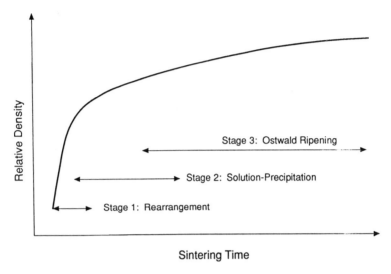

Figure 10.4 Schematic diagram illustrating the three stages of liquid-phase sintering on a typical sintering curve.

A. Wetting Behavior and the Contact Angle

It is generally found that liquids with a low surface tension readily wet most solids, giving a small contact angle, whereas liquids with a high surface tension show poor wetting, with a large contact angle (Fig. 10.5). At a molecular level, if the cohesion between the liquid molecules is greater than the adhesion between the liquid and the solid, the liquid will not show a tendency to wet the solid.

At equilibrium, the contact between a solid and a liquid is shown in Fig. 10.5d. If the specific energies of the liquid/vapor, solid/vapor, and solid/liquid interfaces are γ_{lv}, γ_{sv}, and γ_{sl}, respectively, then by the principle of virtual work,

$$\gamma_{sv} = \gamma_{sl} + \gamma_{lv} \cos \theta \tag{10.2}$$

This equation, derived by Young in 1805 and by Dupré in 1869, is what we would obtain by taking the horizontal components of the surface tension forces. The contact angle is an important parameter in liquid-phase sintering. We should have a small contact angle for good wetting of the solid. Furthermore, as we shall show in Section 10.5E, the contact angle has a significant effect on the magnitude and nature of the capillary forces exerted by the liquid on the solid grains.

Liquid-Phase Sintering

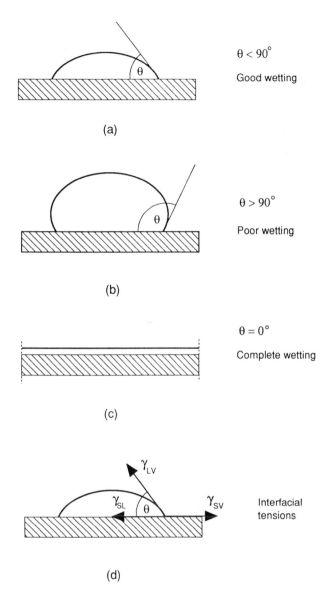

Figure 10.5 Wetting behavior between a liquid and a solid showing (a) good wetting, (b) poor wetting, (c) complete wetting, and (d) the balance between the interfacial tensions for a liquid with a contact angle of θ.

B. Spreading of the Liquid

Spreading is the process in which the liquid distributes itself to cover the surfaces of the particulate solid. It is expected to be important in the rearrangement stage soon after the formation of the liquid. For spreading to occur, the total interfacial energy must be reduced. For an infinitesimal change in the contact area between the solid and the liquid (Fig. 10.5d), spreading will occur if

$$\gamma_{lv} + \gamma_{sl} - \gamma_{sv} = 0 \tag{10.3}$$

Thus a spreading liquid has a contact angle equal to zero.

C. Dihedral Angle

Consider a liquid in contact with the corner of the grains as sketched in Fig. 10.6 for the two-dimensional situation. The dihedral angle is defined as the angle between the solid/liquid interfacial tensions. Applying a force balance, we obtain

$$2\gamma_{sl} \cos\left(\frac{\psi}{2}\right) = \gamma_{ss} \tag{10.4}$$

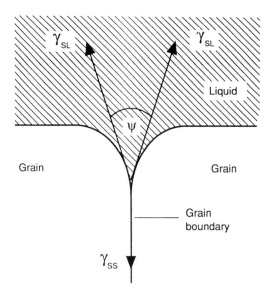

Figure 10.6 The definition of the dihedral angle ψ for a liquid in contact with the corners of the grains.

Liquid-Phase Sintering

After rearranging, we have

$$\cos\left(\frac{\psi}{2}\right) = \frac{\gamma_{ss}}{2\gamma_{sl}} \tag{10.5}$$

The solid/solid interfacial tension γ_{ss} is the same as the interfacial tension in the grain boundary, γ_{gb}, which was described earlier in the discussion of solid-state sintering.

Liquid Penetration of the Grain Boundary

The variation of the dihedral angle ψ as a function of the ratio γ_{ss}/γ_{sl} is sketched in Fig. 10.7. For $\gamma_{ss}/\gamma_{sl} < 2$, the dihedral angle takes values between 0° and 180° and the liquid does not completely penetrate the grain boundary. The condition $\gamma_{ss}/\gamma_{sl} = 2$ represents the limiting condition for complete penetration of the grain boundary by the liquid. For $\gamma_{ss}/\gamma_{sl} > 2$, no value of ψ satisfies Eq. (10.5); this represents the condition for complete penetration of the grain boundary. Basically, the condition $\gamma_{ss}/\gamma_{sl} > 2$ means that the sum of the specific energy associated with the two solid/liquid interfaces is less than that of the solid/solid interface, so

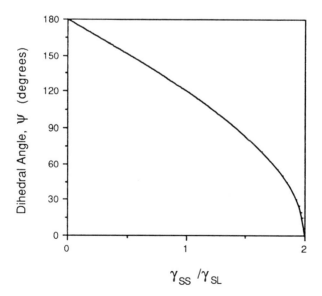

Figure 10.7 Variation of the dihedral angle ψ as a function of the ratio of the solid/solid interfacial tension γ_{ss} to the solid/liquid interfacial tension γ_{sl}.

penetration of the boundary will lead to an overall decrease in the energy of the system.

Shape of the Liquid and the Grains

In solid-state sintering, we found that the dihedral angle controlled the equilibrium shape of the pores. In liquid-phase sintering, the equilibrium shape of the liquid phase depends on the volume fraction of the liquid phase and on the dihedral angle. If we assume that no porosity exists in the structure, then the equilibrium shape of the liquid phase can be calculated [4,5]. Our discussion follows the work of Smith [6], who showed that the equilibrium distribution of second phases in a granular structure can be explained on the assumption that at three-grain junctions the interfacial tensions should be in a state of balance (Fig. 10.6). In two dimensions, Fig. 10.8 shows in an idealized form the shapes a small volume of liquid phase must have if it appears at the corners of three grains.

To obtain a more realistic idea of the shape of the liquid phase in the structure, we must consider the three-dimensional situation. For $\psi = 0$, the liquid completely penetrates the grain boundary and no solid–solid contact exists, as sketched in Fig. 10.9a. As ψ increases, the penetration of the liquid phase between the grains should decrease while the amount of solid–solid contacts (i.e., grain boundary area) should increase. However, for values of ψ up to 60°, the liquid should still be capable of penetrating indefinitely along the three-grain edges. The structure should therefore consist of two continuous interpenetrating phases (Fig. 10.9b). When ψ is greater than 60°, the liquid should form isolated pockets at the corners

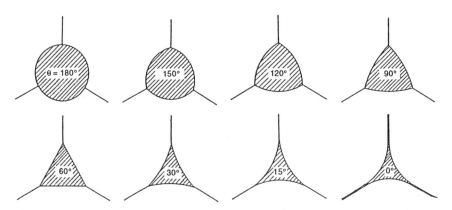

Figure 10.8 Effect of the dihedral angle on the idealized shape (in two dimensions) of the liquid phase at the corners of three grains. (From Ref. 6.)

Liquid-Phase Sintering

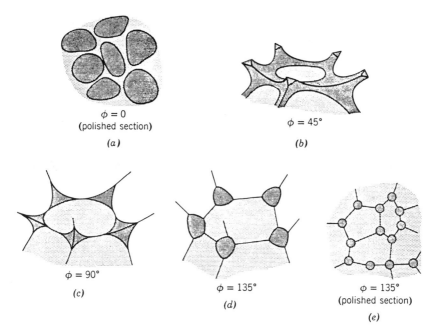

Figure 10.9 Idealized liquid-phase distribution (in three dimensions) for selected values of the dihedral angle. (From Ref. 29, with permission.)

of three (or four) grains, Figs. 10.9c and d. The microstructural features obtained for various values of the dihedral angle are summarized in Table 10.2. Finally, it should be remembered that these equilibrium liquid shapes represent idealizations. Local departures from them will occur in real systems.

Table 10.2 Microstructural Features Obtained for Two-Phase Ceramics Produced by Liquid-Phase Sintering for Various Ratios of the Interfacial Energies γ_{ss}/γ_{sl} and the Corresponding Dihedral Angles ψ

γ_{ss}/γ_{sl}	ψ	Microstructure
≥ 2	$0°$	All grains separated by liquid phase
$\sqrt{3}-2$	$0-60°$	Continuous liquid phase penetrating all three-grain junctions; partial penetration of the grain boundaries by the liquid
$1-\sqrt{3}$	$60-120°$	Isolated liquid phase partially penetrating the three-grain junctions
≤ 1	$\geq 120°$	Isolated liquid phase at four-grain junctions

D. Effect of Solubility

Two factors need to be distinguished: (i) the solubility of the liquid in the solid and (ii) the solid solubility in the liquid.

Liquid Solubility in the Solid

Solubility of the liquid in the solid, as noted earlier, can lead to transient liquid-phase sintering. If the liquid has a high solubility in the solid, it will be incorporated into the solid prior to complete densification. The fabrication of articles of high density will be difficult unless solid-state diffusion occurs. In the ideal situation, the liquid should be present in a sufficient quantity until the desired density is achieved, after which it can be incorporated into solid solution. Transient liquid-phase sintering has proved to be fairly successful in the fabrication of some silicon nitride solid solutions (e.g., a class of materials referred to as *sialons*). However, the process is generally difficult to control. A high liquid solubility in the solid should be avoided if high density in the fabricated article is a requirement.

Solid Solubility in the Liquid

Good solubility of the solid in the liquid is essential for the achievement of high density. In stage 1, dissolution of sharp grain edges and contact points between the grains enhance the efficiency of the rearrangement process. Good solid solubility in the liquid is essential for the solution-precipitation process in stages 2 and 3.

The size of the particle has a key effect on its solubility. As discussed in Chapter 9 (see Section 9.3), the concentration of the dissolved solid, C, immediately surrounding a spherical particle of radius R is given by

$$\ln\left(\frac{C}{C_0}\right) = \frac{2\gamma_{sl}\Omega}{kTR} \tag{10.6}$$

where C_0 is the equilibrium concentration of the solid in the liquid at a planar interface and Ω is the atomic volume. If the solubility is taken as the concentration, Eq. (10.6) becomes

$$\ln\left(\frac{S}{S_0}\right) = \frac{2\gamma_{sl}\Omega}{kTR} \tag{10.7}$$

According to this equation, the solubility increases with decreasing particle radius. Matter transport will therefore occur from surfaces with a smaller radius of curvature to those with a higher radius of curvature.

Liquid-Phase Sintering

Since asperities have a small radius of curvature, they tend to dissolve. Pits, crevices and necks between the particles have negative radii of curvature so solubility is diminished and precipitation is enhanced in those regions.

E. Capillary Forces

As observed earlier, for a liquid that completely wets the solid, the pressure deficit in the liquid, given by Eq. (10.1), leads to a fairly large compressive stress on the particles. In general, the magnitude and nature of the stress will depend on several factors such as the contact angle, the volume of liquid, the separation of the particles, and the particle size. To get a semiquantitative estimate of how these variables influence the capillary force exerted by the liquid, let us consider an idealized model consisting of two spheres of the same radius R separated by a distance d by a liquid with a contact angle θ. The geometrical parameters of the model are shown in Fig. 10.10.

The force acting on the two spheres is the sum of the contributions from (i) the pressure difference Δp across the liquid/vapor meniscus and (ii) the surface tension of the liquid/vapor meniscus, γ_{lv}. The equation for the force can be written

$$F = -\pi X^2 \Delta p + 2\pi X \gamma_{lv} \cos \beta \qquad (10.8)$$

where F is taken to be positive when the force is compressive. Some of the earlier papers in the literature include only the Δp term in Eq. (10.8).

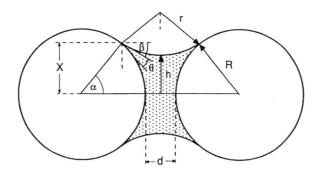

Figure 10.10 Geometrical parameters for an idealized model of two spheres separated by a liquid bridge.

However, Heady and Cahn [7] clearly showed that Eq. (10.8) is the correct expression for F. The pressure difference is given by

$$\Delta p = \gamma_{lv}\left(\frac{1}{r_1} - \frac{1}{r_2}\right) \tag{10.9}$$

where r_1 and r_2 are the two principal radii of curvature of the liquid/vapor meniscus. The shape of the meniscus is normally complex, and numerical methods must be used to determine r_1 and r_2 and to compute F. To simplify the mathematics, we shall make a rough approximation by assuming that the liquid/vapor meniscus forms a circle of radius r in one orthogonal direction. This approximation is referred to as the circle approximation. With this approximation, Δp is given by

$$\Delta p = \gamma_{lv}\left(\frac{1}{h} - \frac{1}{r}\right) \tag{10.9a}$$

where h and r are the radii of curvature shown in Fig. 10.10. With this approximation for Δp and putting $X = R \sin \alpha$, Eq. (10.8) becomes

$$F = -\pi R^2 \gamma_{lv}\left(\frac{1}{h} - \frac{1}{r}\right)\sin^2\alpha + 2\pi R \gamma_{lv} \sin \alpha \cos \beta \tag{10.10}$$

The distance of separation between the spheres is

$$d = 2[r \sin \beta - R(1 - \cos \alpha)] \tag{10.11}$$

and the angles are related by

$$\alpha + \beta + \theta = \frac{\pi}{2} \tag{10.12}$$

Substituting for β in Eq. (10.11) and rearranging gives

$$r = \frac{d + 2R(1 - \cos \alpha)}{2 \cos(\theta + \alpha)} \tag{10.13}$$

The positive radius of curvature of the meniscus is given by

$$h = R \sin \alpha - r[1 - \sin(\theta + \alpha)] \tag{10.14}$$

Substituting for r from Eq. (10.13) gives

$$h = R \sin \alpha - \left[\frac{d + 2R(1 - \cos \alpha)}{2 \cos(\theta + \alpha)}\right][1 - \sin(\theta + \alpha)] \tag{10.15}$$

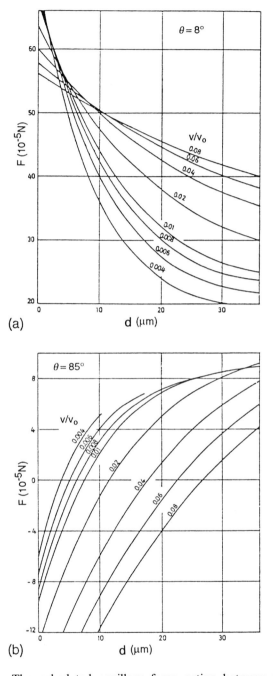

Figure 10.11 The calculated capillary force acting between two tungsten spheres separated by a liquid copper bridge as a function of the interparticle distance for two extreme values of the contact angle. (a) $\theta = 8°$; (b) $\theta = 85°$. V_O is the volume of a tungsten sphere. (From Ref. 8, reprinted with permission from Elsevier Science Ltd.)

The volume of the liquid bridge is

$$V = 2\pi(r^3 + r^2h)\left[\cos(\theta + \alpha) - \left[\frac{\pi}{2} - (\theta + \alpha)\right]\right]$$
$$+ \pi h^2 r \cos(\theta + \alpha)$$
(10.16)

As an example of the application of Eqs. (10.10) to (10.16), the results for the capillary force as a function of distance of separation of the spheres are shown Fig. 10.11 for two contact angles [8]. For small contact angle

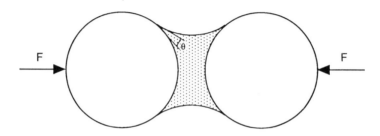

Small contact angle: compressive capillary force

(a)

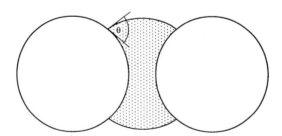

Large contact angle: equilibrium separation

(b)

Figure 10.12 The effect of two extreme values of the contact angle on the capillary force between two spherical particles separated by a liquid bridge. (a) Good wetting leading to a compressive capillary force; (b) poor wetting leading to an equilibrium separation of the particles.

($\theta = 8°$), the capillary force is compressive (Fig. 10.12a). When the contact angle is large ($\theta = 85°$), the capillary force is compressive for large separations between the spheres but repulsive for small separations. An equilibrium separation defined by the condition $F = 0$ will be attained (Fig. 10.12b).

An important result arising from these calculations is that very different rearrangement processes can be expected for low and for high contact angles. For efficient rearrangement and the achievement of high density, the contact angle must be kept low.

10.6 IS THE LIQUID SQUEEZED OUT FROM BETWEEN THE GRAINS?

The achievement of adequate densification by rearrangement and by the solution-precipitation mechanism, as we observed earlier, requires the presence of a liquid layer between the grains. However, in the previous section we found that a wetting liquid layer leads to the development of a compressive capillary force, which is equivalent to placing the system under a fairly large hydrostatic compression. Arguments have arisen from time to time as to whether the liquid layer can support a compressive stress and whether the liquid might not be squeezed out from between the grains.

After the rearrangement stage, we may envisage a situation where densification occurs by the solution-precipitation mechanism and the liquid layer separating the grains becomes progressively thinner with time. Assuming that there is no repulsion between the grain surfaces, we would eventually reach a stage where the liquid capillary becomes so narrow that liquid flow is very difficult. The solution-precipitation can still continue but at a reduced rate. A competition is therefore set up between the rate at which the liquid layer becomes thinner and the rate of densification. This type of argument appears to point to the development of a finite thickness of the liquid layer when densification is completed because liquid flow though the capillary becomes very slow. Observations by high-resolution electron microscopy often reveal an amorphous layer with a thickness of 1–5 nm between the grains in ceramics that have been fabricated by liquid-phase sintering (see Fig. 1.15).

Lange [9] applied a theory developed for plates separated by a liquid layer to the case of spherical particles undergoing liquid-phase sintering. His analysis showed that although the thickness of the liquid layer decreases progressively with time, a liquid layer of finite thickness always remains between the particles within any experimental time frame as long as the liquid completely wets the solid. The rate of approach of two plates separated by a Newtonian viscous liquid is [10]

$$\frac{ds}{dt} = -\frac{2\pi s^5}{3\eta A^2 s_0^2} F \qquad (10.17)$$

where s is the thickness of the liquid layer at time t, s_0 is the initial thickness of the layer, F is the compressive force exerted on the plates, η is the viscosity of the liquid, and A is the contact area between the liquid and the plate.

To apply Eq. (10.17) to the case of liquid-phase sintering, let us consider the model shown in Fig. 10.13 for two spheres held together by the capillary force due to a liquid bridge. Instead of asking whether the liquid will be squeezed out or not, it is more convenient to phrase the question another way: How long will it take for the liquid to be squeezed out? Putting $y = s/s_0$ in Eq. (10.17) and integrating, the time for y to approach zero is found to be

$$t_f = \frac{3\eta A^2}{8\pi s_0^2 F}\left(\frac{1}{y^4} - 1\right) \qquad (10.18)$$

This equation shows that $t_f = \infty$ for $y = 0$. Therefore within any experi-

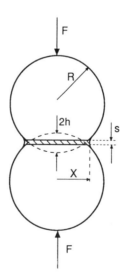

Figure 10.13 Parameters of an idealized two-sphere model separated by a liquid layer of thickness s, used to analyze the change in the thickness of the layer during sintering. (From Ref. 9, with permission.)

Liquid-Phase Sintering

mental time frame, some liquid is predicted to remain between the particles.

The problem can be analyzed further to take into account the effect of the approach of the particle centers due to densification. For the geometrical model shown in Fig. 10.13, for small changes in the center-to-center distance, the shrinkage $\Delta L/L_0 = -h/R$. The contact area $A = 2\pi X^2$ and, as found earlier for this geometry (see Fig. 8.10), $h \approx X^2/2R$. The contact area is therefore given by

$$A \approx 4\pi R^2 \left(\frac{\Delta L}{L_0}\right) \quad (10.19)$$

As described later, models for the solution-precipitation process predict shrinkage kinetics of the form

$$\frac{\Delta L}{L_0} = Kt^n \quad (10.20)$$

where K is a constant and the exponent n is equal to 1/2 or 1/3. Data for the shrinkage kinetics normally give n values between these two values. Combining Eq. (10.20) with Eqs. (10.18) and (10.19) gives

$$t_f = \left[\frac{6\pi\eta R^4 K(1 - 2n)}{s_0^2 F}\left(\frac{1}{y^4} - 1\right)\right]^{1/(1-2n)} \quad (10.21)$$

According to this equation, $t_f \to \infty$ as $y \to 0$ as long as $n \leq 1/2$. A small liquid layer is predicted to remain between the particles within any experimental time scale. The analysis of Lange shows that although the thickness of the liquid layer decreases with time, thereby reducing the rate of matter transport, a liquid layer of finite thickness will always remain between the grains as long as the liquid completely wets the solid.

Clarke et al. [11] have explored the possibility that an electrical double layer repulsion between the surfaces of the grains acts to stabilize an equilibrium thickness of the liquid layer in polycrystalline ceramics. The model has features that are parallel to those discussed in Chapter 4 for the stability of colloidal particles in a suspension. In this model, an equilibrium separation will be achieved when the sum of the compressive capillary force and the van der Waals force of attraction balances the repulsive force due to the electrical double layer. Although the magnitude of the calculated forces appears reasonable, the existence of an electrical double layer on the surfaces of crystalline particles that undergo liquid-phase sintering has not been established.

To summarize, qualitative arguments and theoretical considerations indicate that the liquid will not be squeezed out from between the grains as long as the liquid completely wets the solid. While the mechanism by which an equilibrium thickness of the liquid layer can be stabilized is not clear, it is likely that ordering of the liquid molecules at the surface may play an important role.

10.7 THE BASIC MECHANISMS OF LIQUID-PHASE SINTERING

Having discussed the general features of the process and the influence of the key thermodynamic and kinetic factors, we are now in a position to examine in more detail the basic mechanisms and processes occurring in liquid-phase sintering. For convenience, we divide our discussion into three main parts connected with the three stages of liquid-phase sintering. However, it should be remembered that there is some degree of overlap between the successive stages.

A. Stage 1. Rearrangement

Rearrangement, as we outlined earlier, is the dominant mechanism in stage 1. The capillary forces act to bring about physical movement of the grains and, for a wetting liquid, center-to-center approach occurs, leading to densification. However, a number of processes that have important consequences for liquid-phase sintering may accompany rearrangement. These processes include redistribution of the liquid, particle disintegration, and swelling of the powder compact.

Soon after the formation of the liquid, the particles rearrange quickly to produce a packing with some degree of stability. However, a further decrease in the energy of the system can come about by the redistribution of the liquid under the action of capillary forces. Prior to analyzing the rearrangement mechanism, we consider the effects of the liquid redistribution.

Liquid Redistribution

Experimental observations indicate that considerable redistribution of the liquid may occur during liquid-phase sintering. Perhaps the most dramatic evidence for such redistribution comes from the experiments of Kwon and Yoon [12], who investigated the sintering of fine tungsten powder containing coarse nickel particles that melt to form the liquid phase. As shown in Fig. 10.14, sequential filling of the pores occurs in such a way that the small pores are filled first and the larger pores later.

The redistribution of the liquid during liquid-phase sintering was analyzed by Shaw [13] using a two-dimensional model of circular particles. Shaw's approach was to determine the equilibrium distribution of the liquid in different packing arrangements of particles under the condition that

Liquid-Phase Sintering

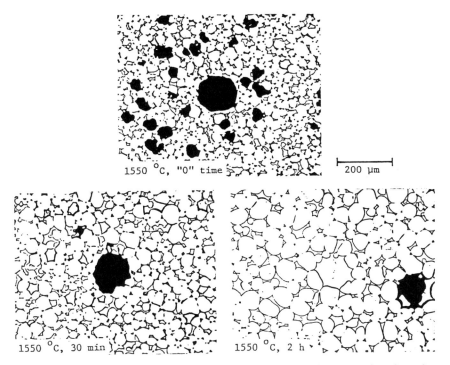

Figure 10.14 The change in microstructure during liquid-phase sintering of a mixture of fine (10 μm) tungsten powder, 2 wt % of 30 μm nickel spheres, and 2 wt % 125 μm nickel spheres, showing sequential filling of the pores. (From Ref. 12, with permission.)

the chemical potentials of the liquid in all the pores in a particle array are the same at equilibrium. The chemical potential of an atom under the surface of a liquid/vapor meniscus with an average radius of curvature r is

$$\mu = \mu_0 + \frac{\gamma_{lv}\Omega}{r} \tag{10.22}$$

where μ_0 is the chemical potential of an atom under a flat surface, γ_{lv} is the liquid/vapor surface energy, and Ω is the atomic volume. The condition that the chemical potential must be the same is therefore equivalent to the radius of the liquid menisci being the same.

Shaw's analysis provides considerable insight into how a liquid will redistribute within an array of particles, and we shall describe the main features of the results. For a regular array of circles (with threefold coordinated pores) in which no shrinkage occurs, Fig. 10.15 shows two possible ways in which the liquid will distribute itself. At small volume fractions, the liquid will be distributed evenly in the necks between the particles (Fig. 10.15A). Increasing the volume of liquid will cause the necks to fill with liquid until some critical volume fraction is reached. Above this critical volume fraction, instead of filling each pore evenly, the liquid prefers to adopt a distribution in which a certain fraction of the pores are completely filled and the remainder of the liquid is in isolated necks (Fig. 10.15B). This distribution of the liquid has the minimum surface area. Altering the amount of liquid in this regime has no effect on the amount of liquid situated at the necks but simply alters the fraction of pores that

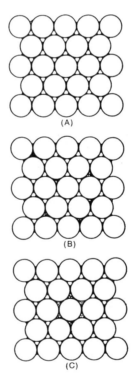

Figure 10.15 Possible equilibrium configurations that can be adopted by a liquid in a close-packed array of particles. (A) Isolated necks filled with liquid, (B) fraction of pores completely filled with liquid, and (C) same as (B) but with inhomogeneous liquid distribution. (From Ref. 13, with permission.)

Liquid-Phase Sintering

are filled. Increasing the contact angle decreases the range of liquid volume over which the liquid is evenly distributed in the necks.

An important finding of the analysis is the way in which an initially inhomogeneously distributed liquid will redistribute itself in the two-dimensional array. For low liquid content where the liquid is situated at isolated necks, there is always a force driving the liquid to redistribute itself homogeneously. In the second regime, where some of the pores are filled, it is immaterial which pores fill with liquid as long as the correct fraction of pores are filled. Consequently, once an inhomogeneous distribution such as that in Fig. 10.15C forms, there is no driving force to redistribute the liquid homogeneously.

Shaw also considered an array of circles containing pores with threefold and sixfold coordination (Fig. 10.16). While simple in geometry, the model can provide some insight into how a liquid will redistribute itself in the practical situation of an inhomogeneously packed powder system. The free energy calculations can be used to construct a diagram (Fig. 10.17) showing the distribution of the liquid in the array. The diagram shows the fraction of particles surrounding sixfold coordinated pores as a function of the volume fraction of the liquid. Four regions can be distinguished if we neglect the shaded region to the right of the diagram corresponding to a region in which the pores are completely filled.

To illustrate the significance of Fig. 10.17, let us consider a structure containing a fixed fraction of particles at sixfold coordinated pores and consider how the pores will fill as the volume of liquid is increased. For small volume fraction, the liquid is situated only at isolated necks between the particles. Each point in this region lies on a tie line representing liquid-filled necks at threefold coordinated pores at the bottom of the diagram and liquid-filled necks at sixfold coordinated pores at the top of the dia-

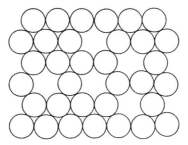

Figure 10.16 Example of a two-dimensional arrangement of pores that contains three- and sixfold coordinated pores. (From Ref. 13, with permission.)

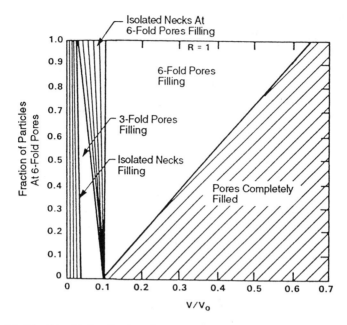

Figure 10.17 Liquid distribution diagram for a particle array that contains three- and sixfold coordinated pores. (From Ref. 13, with permission.)

gram. The tie lines connect systems in which the chemical potential of the liquid is the same in the two types of pores. As the volume fraction of liquid is increased, we enter the second region, in which it becomes energetically more favorable to fill threefold coordinated pores than to continue filling the necks between the particles. In this region, the liquid content between the necks remains constant and the increase in the liquid volume goes into filling the threefold coordinated pores. When the third region is reached, all the threefold coordinated pores are filled with liquid and an increase in liquid volume goes into filling the necks between particles surrounding the sixfold coordinated pores. Finally, in the fourth region, the sixfold coordinated pores become completely filled. Shaw extended his analysis to include the effect of shrinkage on the liquid distribution. The results show the general sequence of behavior described for the simple mode.

Shaw's analysis shows that the pores in the model fill sequentially. If we extend the results to the more complex situation of a powder compact with a distribution of pore sizes, the same behavior of sequential

Liquid-Phase Sintering

filling of the pores is expected to occur. The pores with the smallest coordination number will be the first to fill because such pores have a high surface to volume ratio so a given volume of liquid eliminates more solid/vapor interface. Only if there is sufficient liquid will the pores with higher coordination number start to fill. Increasing the liquid volume will cause the pores to fill sequentially in the order of increasing pore coordination number, as observed by Kwon and Yoon (Fig. 10.14).

To summarize at this stage, Shaw's analysis shows the importance of (i) controlling the homogeneity of the particle packing to produce pores with a narrow distribution of sizes and (ii) controlling the homogeneity of the mixing between the major component and the additive to produce a homogeneous distribution of the liquid phase in the powder compact. Heterogeneous packing leads to sequential filling of the pores so that the larger pores are filled later in the firing process, thereby producing regions that are enriched with the liquid composition. Inhomogeneous mixing leads to an inhomogeneous liquid distribution such that there is no driving force for redistribution of the liquid. Also, using large particles to create the liquid phase (as in Fig. 10.14) leaves huge voids when the particles melt and the liquid invades the smaller pores. The optimal situation is to start with coated particles in which the coating forms the liquid phase.

Particle Rearrangement

Two different approaches have been used to analyze rearrangement during liquid-phase sintering. Kingery [14] used an empirical approach in which the surface tension forces driving densification were balanced by the frictional forces resisting rearrangement. He derived a simple kinetic relationship for the variation of the shrinkage with time:

$$\frac{\Delta L}{L_0} \sim t^{1+y} \tag{10.23}$$

where ΔL is the change in length, L_0 is the original length, and the exponent $1 + y$ is slightly greater than unity. The experimental verification of Eq. (10.23) has not been convincing. Kingery's approach treats rearrangement during liquid-phase sintering as an isotropic and uniform process. It provides little insight into how the process is influenced by the key parameters (such as the contact angle) or how the microstructure develops as a result of the rearrangement process.

The second approach to rearrangement involves the analysis of the capillary forces between particles separated by a liquid bridge. The results of such an analysis were discussed earlier for a model consisting of two spheres (see Section 10.5E). Huppmann and Riegger [8] compared the results of their analysis with data for copper-coated tungsten spheres in

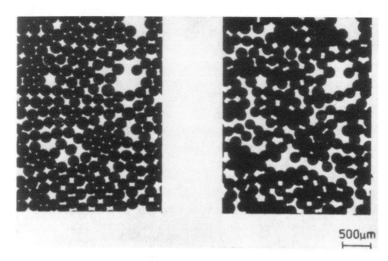

Figure 10.18 Planar array of copper-coated tungsten spheres before (left) and after (right) liquid-phase sintering. (From Ref. 8, reprinted with permission from Elsevier Science Ltd.)

which the copper melted to form the liquid phase. For close-packed planar arrangements of the spheres, good agreement between the calculations and the data was obtained. However, for randomly packed arrays, local densification occurred, leading to the opening of large pores (Fig. 10.18). Less densification than that predicted by the model was observed.

Because of their axial symmetry, two-particle models have severe limitations for describing the rearrangement of a randomly packed array. They cannot account for the presence of other forces, such as torques, that may act to close the angle between the particle contacts. The analysis of the rearrangement process of a randomly packed array of particles is a difficult problem. More complex models must be used, so the calculations become fairly difficult.

Primary and Secondary Rearrangement

For a system of polycrystalline particles, the overall rearrangement process can consist of two stages, referred to as primary rearrangement and secondary rearrangement. *Primary rearrangement* describes the rapid rearrangement, soon after the formation of the liquid, of the polycrystalline particles under the surface tension forces of the liquid bridge. However, as described earlier, if $\gamma_{ss}/\gamma_{sl} > 2$, the liquid can penetrate the grain boundaries between the particles in the polycrystals. Fragmentation of the polycrystalline particles will occur. *Secondary rearrangement* de-

Liquid-Phase Sintering

scribes the rearrangement of these fragmented particles. Since it depends on the rate at which the grain boundaries are dissolved away, secondary rearrangement occurs more slowly than primary rearrangement. The two types of rearrangement are sketched in Fig. 10.19.

Liquid Volume and Densification

Earlier, we mentioned the effect of the liquid volume on the capillary force for two particles separated by a liquid bridge. The liquid volume also has a significant effect on the densification resulting from the rearrangement process. Let us consider a consolidated powder form with a relative density of 60% (i.e., with a porosity of 40%), consisting of a major component and a liquid-producing additive. After melting of the additive, suppose that the particles of the major component rearrange to form 64% of the final product, i.e., a packing density close to that for dense random packing of monosize spheres. Prior to any further densification, we can calculate the residual porosity for a given volume fraction of liquid. If the

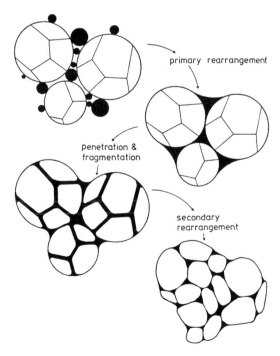

Figure 10.19 Schematic diagram illustrating fragmentation and rearrangement of polycrystalline particles. (From Ref. 3.)

volume fraction (in percent) of the liquid is f_L, the volume percent of the body consisting of solid and liquid is $64 + f_L$. The residual porosity (in percent) after the rearrangement is therefore $100 - (64 + f_L)$. The results are shown in Fig. 10.20. For a liquid volume fraction greater than 36%, full densification is achieved by rearrangement alone. When the liquid content is less than 36%, full densification requires the occurrence of additional processes such as solution-precipitation.

Compact Shrinkage and Compact Swelling

When the contact angle of the liquid is small and the density of the consolidated powder form is not too great, a shrinkage normally accompanies the rearrangement process as a result of the compressive surface tension forces. However, if the density of the consolidated powder form is too great, the flow of the liquid between the particles can lead to a swelling of the powder form. A high liquid solubility in the solid can also lead to a swelling of the powder form, as sketched in Fig. 10.21.

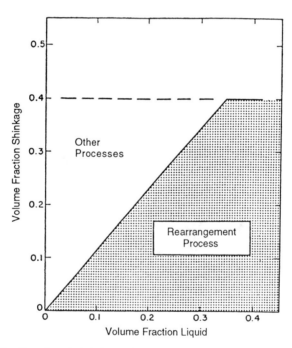

Figure 10.20 Fractional shrinkage due to the rearrangement process for different liquid contents. (From Ref. 14.)

Liquid-Phase Sintering

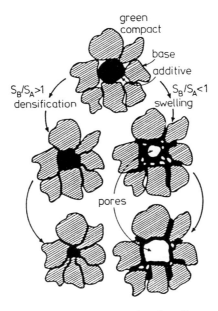

Figure 10.21 A schematic diagram contrasting the effects of solubility on densification or swelling during liquid-phase sintering. (From Ref. 3.)

B. Stage 2. Solution-Precipitation

In stage 2, rearrangement decreases considerably and the solution-precipitation mechanism becomes dominant. The major processes that occur by the solution-precipitation mechanism are densification and coarsening (Ostwald ripening). They occur concurrently and may be accompanied by grain shape accommodation if the liquid volume is fairly small. Two models have been put forward to account for the densification process. We refer to these models as (i) densification by contact flattening and (ii) densification accompanied by Ostwald ripening.

Densification by Contact Flattening
Densification by contact flattening was described by Kingery [14]. As a result of the compressive capillary force of a wetting liquid, the solubility at the contact points between the particles is higher than that at other solid surfaces. This difference in solubility (or chemical potential) leads to matter transport away from the contact points, thereby allowing center-to-center approach under the action of the surface tension forces. The

contact regions between the particles also become flat, as sketched in Fig. 10.22.

Kingery made quantitative estimates of the densification rate using a model consisting of two spherical particles of the same radius. If the radius of the sphere is R and each sphere is dissolved away along the center-to-center line by a distance h to give a circular contact area of radius X, we have

$$h = X^2/2R \tag{10.24}$$

The volume of material removed from each sphere is given by $V \approx \pi X^2 h/2$. Using Eq. (10.24), the equation for V can be written

$$V = \pi R h^2 \tag{10.25}$$

Two different mechanisms can control the rate of matter transport: (i) diffusion through the liquid and (ii) the interface reaction leading to solution (or precipitation).

Rate Control by Diffusion Through the Liquid. Kingery used the same diffusional flux equation assumed by Coble in his model for the intermediate stage of solid-state sintering (see Section 8.5F). The flux from the boundary per unit thickness is given by Eq. (8.46). That is,

$$J = 4\pi D_L \Delta C \tag{10.26}$$

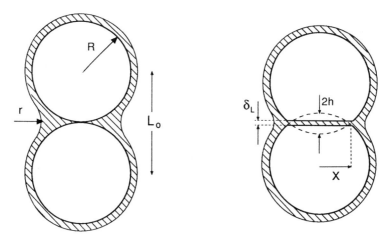

Figure 10.22 Idealized two-sphere model for densification by contact flattening.

Liquid-Phase Sintering

where D_L is the diffusion coefficient for the solute atom in the liquid and ΔC is the solute concentration difference between that at the contact area, C, and that at a flat, stress-free surface, C_0. If the thickness of the liquid bridge is δ_L, the rate of removal of the solid is

$$\frac{dV}{dt} = \delta_L J = 4\pi D_L \delta_L (C - C_0) \tag{10.27}$$

As derived in Chapter 7, the chemical potential of an atom is related to its concentration and the local stress p by

$$\mu_a - \mu_0 = kT \ln\left(\frac{C}{C_0}\right) = p\Omega \tag{10.28}$$

If the concentration difference $C - C_0$ is small, we can write

$$\frac{\Delta C}{C_0} = \frac{p\Omega}{kT} \tag{10.29}$$

where k is the Boltzmann constant and T is the absolute temperature. The capillary pressure due to the spherical pore in the liquid is given by $p_0 = 2\gamma_{lv}/r$. As discussed earlier, this capillary pressure is equivalent to an externally applied hydrostatic pressure that would have to be applied to the particulate system in order for the resulting interparticle force to be equal to that set up by the liquid bridge. Because the contact area is less than that of the external area, the pressure is magnified at the contact area. The pressure at the contact area, p, is found by a simple force balance as described earlier in our discussion of the stress intensification factor in solid-state sintering, i.e.,

$$p = k_2 \frac{R^2}{X^2} p_0 = k_2 \frac{R^2}{X^2}\left(\frac{2\gamma_{lv}}{r}\right) \tag{10.30}$$

where k_2 is a geometrical constant. Substituting for X from Eq. (10.24), we obtain

$$p = k_2 \frac{\gamma_{lv} R}{rh} \tag{10.31}$$

Assuming that the radius of the pore is proportional to the sphere radius, i.e., $r \sim k_1 R$, where k_1 is assumed to remain constant during sintering, Eq. (10.31) becomes

$$p = \frac{k_2}{k_1}\left(\frac{\gamma_{lv}}{h}\right) \tag{10.32}$$

Substituting Eq. (10.29) into Eq. (10.27) and using the expression for p given by Eq. (10.32), after some rearranging we obtain

$$\frac{dV}{dt} = \frac{4\pi k_2 D_L \delta_L C_0 \Omega \gamma_{lv}}{k_1 hkT} \tag{10.33}$$

But dV/dt is also equal to $2\pi Rh(dh/dt)$ from Eq. (10.25). Equating these two expressions for dV/dt and rearranging, we obtain

$$h^2 dh = \frac{2k_2 D_L \delta_L C_0 \Omega \gamma_{lv}}{k_1 kTR} dt \tag{10.34}$$

Integrating and applying the boundary condition that $h = 0$ at $t = 0$, Eq. (10.34) gives

$$h = \left(\frac{6k_2 D_L \delta_L C_0 \Omega \gamma_{lv}}{k_1 kT}\right)^{1/3} \left(\frac{t^{1/3}}{R^{1/3}}\right) \tag{10.35}$$

Since $h/R = -\Delta L/L_0 = -(1/3)\Delta V/V_0$ (for small $\Delta L/L_0$), where $\Delta L/L_0$ and $\Delta V/V_0$ are the linear shrinkage and the volumetric shrinkage, respectively, of the powder compact, we can write

$$-\frac{\Delta L}{L_0} = -\frac{1}{3}\left(\frac{\Delta V}{V_0}\right) = \left(\frac{6k_2 D_L \delta_L C_0 \Omega \gamma_{lv}}{k_1 kT}\right)^{1/3} \left(\frac{t^{1/3}}{R^{4/3}}\right) \tag{10.36}$$

According to this equation, when diffusion through the liquid is the rate-controlling mechanism, the shrinkage is proportional to the one-third power of the time and inversely proportional to the four-thirds power of the initial particle size.

Rate Control by Phase Boundary Reaction. When the phase boundary reaction leading to dissolution of the solid into the liquid is the rate-controlling mechanism of sintering, the volumetric rate of material transfer is assumed to be directly proportional to the contact area times a rate constant for the phase boundary reaction times the increase in the activity of the solid at the contact area due to the capillary pressure, that is,

$$\frac{dV}{dt} = k_3 \pi X^2 (a - a_0) = 2\pi k_3 hR(C - C_0) \tag{10.37}$$

where k_3 is the reaction rate constant and the activities are taken to be equal to the concentrations. Following the steps outlined above for the case of diffusion control, the shrinkage is given by

$$-\frac{\Delta L}{L_0} = -\frac{1}{3}\left(\frac{\Delta V}{V_0}\right) = \left(\frac{2k_3 k_2 C_0 \Omega \gamma_{lv}}{k_1 kT}\right)^{1/2} \left(\frac{t^{1/2}}{R}\right) \tag{10.38}$$

According to this equation, the shrinkage is predicted to be proportional to the square root of the time and inversely proportional to the initial particle radius.

To summarize at this stage, the mechanism of contact flattening as described by Kingery allows quantitative estimates of the shrinkage rates. However, we must remember the drastic assumptions made in the derivation of the equations and the idealized geometry of the model. Specifically, the equations apply to particles of the same size where coarsening is absent. A few studies have shown good agreement between the exponents determined from data and those predicted by Kingery's model. However, in most of these studies the fitting of the data to determine the exponents appears somewhat arbitrary. The value of the exponent is dependent on the time at which stage 2 is assumed to begin after the rearrangement stage (see Fig. 10.4). In practice, the shrinkage data can normally be fitted with a curve having a smoothly varying slope instead of a line having a fixed slope equal to the predicted value.

Kingery's model can account for the phenomenon of grain shape accommodation. However, observations with real powder systems, which normally have a distribution of particle sizes, show that densification is accompanied by significant coarsening of the grains. The contact flattening mechanism described by Kingery cannot account for the observed coarsening in real powder systems.

Densification Accompanied by Ostwald Ripening

The second mechanism of densification in stage 2 has been described by Yoon and Huppmann [15], based on their observations of the liquid-phase sintering of W(Ni) powder mixtures. In the model, the smaller grains dissolve and precipitate on the larger grains. This solution-precipitation process is the classic Ostwald ripening process we considered in Chapter 9. As sketched in Fig. 10.23, dissolution of the small grains and precipitation away from the contact points lead to center-to-center approach of the larger grains (i.e., to shrinkage). In this way, Ostwald ripening accompanies the densification process.

Theoretical estimates of the shrinkage rates due to the Ostwald ripening mechanism are difficult. However, experimental observations of real powder systems show that this mechanism is the dominant one in the third stage. In a classic set of experiments, Yoon and Huppmann [15] studied the sintering of a mixture of coarse, single-crystal, spherical tungsten particles (200–250 μm in diameter), fine tungsten particles (average size ≈ 10 μm), and fine nickel powder. The nickel melted to produce the liquid phase. When the tungsten dissolves in the nickel and precipitates out, the precipitated material is not pure tungsten but a solid solution

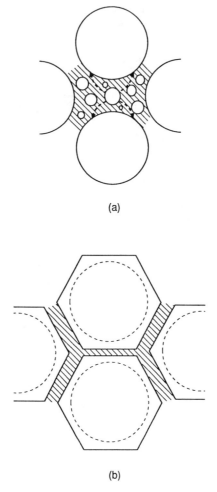

Figure 10.23 Schematic diagram illustrating densification accompanied by Ostwald ripening. The process is accompanied by grain shape accommodation.

containing a small amount of nickel (≈0.15 wt %). By etching in Murakami's solution, the precipitated tungsten can be distinguished from the pure tungsten. Figure 10.24 shows the microstructure, after sintering at 1670°C for 3, 20, 120 and 360 min, of a powder mixture that contained 48 wt % large tungsten spheres, 48 wt % fine tungsten particles, and 4 wt % nickel. The porosity and the fraction of fine tungsten particles are given

Liquid-Phase Sintering

in Table 10.3. The results show clearly that the fine tungsten particles dissolve and precipitate on the coarse tungsten spheres. However, this observation does not prove that Ostwald ripening causes densification, only that it accompanies densification. The growth of the precipitated tungsten does not occur uniformly around the large tungsten spheres; it

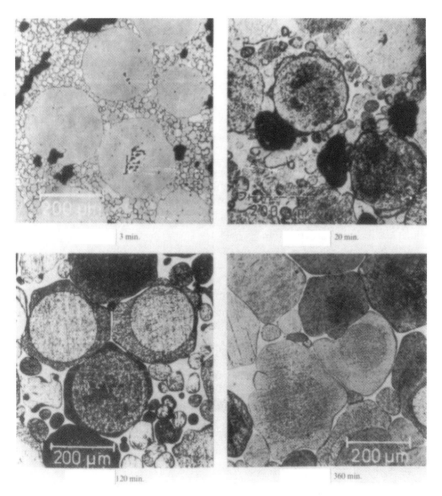

Figure 10.24 The microstructures of a mixture of 48 wt % large tungsten spheres, 48 wt % fine tungsten powder, and 4 wt % nickel, after sintering for various times at 1670°C. (From Ref. 15, reprinted with permission from Elsevier Science Ltd.)

Table 10.3 Porosity and Fraction of Small Grains in the Mixture of 48 wt % Large Tungsten Spheres, 48 wt % Fine Tungsten Powder, and 4 wt % Nickel[a]

Sintering time (min)	3	20	120	360
Porosity (vol %)	10	6	2	2
Fraction of small grains (wt %)	48	38	27	15

[a] See Fig. 10.24.
Source: Ref. 15.

occurs preferentially in the regions where no neighboring large grains impede it. The coarsened large tungsten spheres develop flat faces and assume the shape of a polyhedron; i.e., the densification and coarsening processes for this powder composition are accompanied by grain shape accommodation. The occurrence of grain shape accommodation indicates that the contact flattening mechanism may also be active. However, it does not indicate which mechanism causes shrinkage.

To summarize, the experiments of Yoon and Huppmann indicate that densification during liquid-phase sintering is significant when the conditions promote Ostwald ripening. Many observations on real powder systems have shown that densification after the rearrangement stage is slight unless significant coarsening occurs.

Grain Shape Accommodation

Grain shape accommodation, as discussed earlier, is sometimes observed in the microstructure of materials fabricated by liquid-phase sintering. It is clearly observed in the micrographs of Fig. 10.24 for the W(Ni) system containing 4 wt % Ni. However, many microstructures of materials produced by liquid-phase sintering also show rounded (spheroidal) grains. As an example, Fig. 10.25 shows the microstructure for the same W(Ni) system described in Fig. 10.24 but with a higher proportion of liquid (14 wt % Ni). It is clear from Figs. 10.24 and 10.25 that the volume fraction of liquid is a key factor in determining whether or not grain shape accommodation will occur.

The occurrence of grain shape accommodation is normally favored when the amount of liquid is relatively small. In this case, the liquid is insufficient to completely fill the voids between the grains as long as the grains assume a rounded or spheroidal shape. Grain shape accommodation produces a polyhedral grain with flat contact surfaces, and this leads to more efficient packing of the grains. Liquid released from the well-packed regions can flow into the pores.

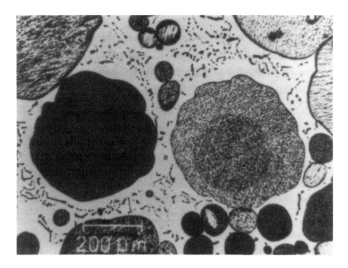

Figure 10.25 The microstructure of a mixture of 43 wt % large tungsten spheres, 43 wt % fine tungsten powder, and 14 wt % nickel, after sintering for 60 min at 1670°C. (From Ref. 15, reprinted with permission from Elsevier Science Ltd.)

Grain shape accommodation will occur if it leads to a decrease in the energy of the system. Compared to a sphere with the same volume, a polyhedral grain has a higher surface area. If grain shape accommodation is to occur, the decrease in the interfacial energy associated with the filling of the pores must overcome the increase in the interfacial energy associated with the development of the polyhedral grain shape. For a fairly large amount of liquid, the driving force for shape accommodation is absent and the spheroidal shape is maintained, as shown in Fig. 10.25.

Coarsening

As densification slows, coarsening starts to become dominant toward the latter part of stage 2 and continues into stage 3 to become the clearly dominant process. Because of its dominant role, we shall consider coarsening in detail in the next section, which is devoted to stage 3 mechanisms. Here, we consider some further issues associated with coarsening in stage 2.

For $\gamma_{ss}/\gamma_{sl} > 2$, the liquid completely penetrates the grain boundaries between the grains. Coarsening occurs by the Ostwald ripening mechanism. However, for $\gamma_{ss}/\gamma_{sl} < 2$, the dihedral angle is greater than zero and solid–solid contacts develop between the grains. These solid–solid contacts start to become noticeable late in stage 2 and continue into stage

3, during which some equilibrium distribution of the solid and liquid phases is developed.

Coalescence. The development of solid/solid contacts opens up the possibility that coarsening may occur by coalescence of the grains. As sketched in Fig. 10.26, a possible mechanism for coalescence consists of contact between the grains, neck growth, and migration of the grain boundary by solid-state diffusion. The occurrence of this mechanism depends on whether the geometry permits it. For an incremental movement of the boundary, as we discussed earlier for grain growth in porous ceramics

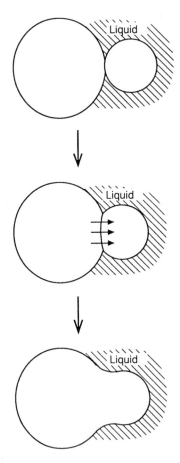

Figure 10.26 Schematic diagram illustrating coalescence by solid-state migration of the grain boundary.

Liquid-Phase Sintering

(see Section 9.9), the energy of the system must decrease. When the dihedral angle is small, the liquid partially penetrates the grain boundary. Movement of the grain boundary will involve initially an increase in the grain boundary energy. Coalescence is therefore not expected to be favorable for small dihedral angles. For larger dihedral angles, where the penetration of the boundary is significantly reduced or the liquid exists in isolated pockets at the grain corners, coalescence may become energetically favorable. The process is enhanced when the difference between the particle sizes becomes greater. The formation of necks with large dihedral angles and coalescence of the grains by migration of the boundary have been reported in a few metallic systems [16].

In systems where no solid/solid contact exists, coalescence may be possible by migration of the liquid film separating the grains (Fig. 10.27). This mechanism has been found to occur in a few metallic systems and is normally referred to as *directional grain growth*. The coalescence process is not driven by a reduction in interfacial energy but by a reduction in chemical energy. Yoon and Huppmann [17] observed that when single-crystal tungsten spheres are sintered in the presence of liquid nickel, one tungsten sphere grows at the expense of its neighbor (Fig. 10.28a). An electron microprobe analysis shows that the shrinking grain consists of pure tungsten whereas the precipitated material on the growing grain is a solid solution of tungsten containing 0.15 wt % nickel (Fig. 10.28b). This compositional difference between the pure tungsten and the solid solution is responsible for a large decrease in chemical energy, which more than offsets the increase in interfacial energy.

C. Stage 3. Ostwald Ripening

In stage 3, densification slows considerably and microstructural coarsening clearly becomes the dominant process. For $\gamma_{ss}/\gamma_{sl} > 2$, the grains are

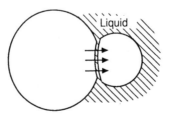

Figure 10.27 Schematic diagram illustrating coalescence by migration of the liquid film separating the grains.

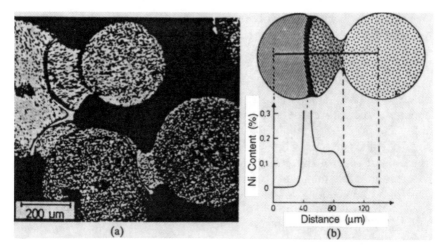

Figure 10.28 Directional grain growth during liquid-phase sintering of single-crystal tungsten spheres with nickel at 1640°C, showing (a) the microstructure and (b) the microprobe analysis. (From Ref. 17, reprinted with permission from Elsevier Science Ltd.)

completely separated by a liquid layer, the thickness of which, as discussed earlier, may decrease asymptotically. When the dihedral angle is greater than zero, solid/solid contacts lead increasingly to the formation of a rigid skeleton. With the development of the solid/solid contacts, matter transport can occur by the solution-precipitation mechanism, and by solid-state diffusion.

Densification

When the volume of the liquid phase is fairly small, solution-precipitation and grain shape accommodation lead to a slow, continuous elimination of the isolated porosity. However, for a larger volume of liquid, Kang et al. [18] have shown that the filling of the isolated pores may occur in a discontinuous manner. Kang et al. studied the shape change of grains surrounding large isolated pores in the Mo(Ni) system. The sample was heated for 30 min at 1460°C and cooled. By repeating the sintering process three times, the shape of the growing grains after each heating cycle was revealed by ghost boundaries formed within the grains as a result of strong etching. As shown in Fig. 10.29a, since the grains surrounding the pore grow laterally along the pore surface (see, e.g., the grains labeled A and B), it is clear that the pore is not filled continuously by matter deposited on its surface. The pore can remain roughly unchanged for an extended

Liquid-Phase Sintering

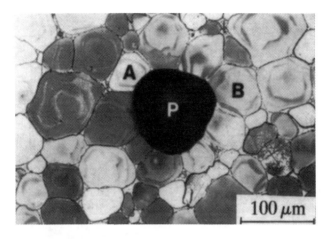

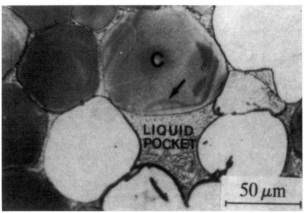

Figure 10.29 Microstructures of a Mo(4 wt % Ni) system showing (top) lateral growth of grains (such as A and B) around a pore (P) for a specimen sintered in three cycles (for 30 min in each cycle) at 1460°C; (bottom) after pore filling, preferential growth of a grain (C) into a liquid pocket formed at a pore site, resulting in a more homogeneous microstructure in a specimen sintered in three cycles (60, 30, and 30 min) at 1460°C. (From Ref. 18, with permission.)

period of sintering. However, after this extended period of sintering, the pore can be filled almost instantaneously by the liquid. After filling of the pore (Fig. 10.29b), the etch boundaries of the grain labelled C reveal that the grains grow preferentially into the pocket of liquid, thereby leading to a more rounded grain shape. The growth of the grains into the liquid pocket also has the effect of improving the homogeneity of the microstructure because the size of the heterogeneity (i.e., the liquid pocket) is reduced.

A subsequent theoretical analysis by Park and coworkers [19,20] revealed that the filling of the large pore occurs after a certain critical grain size is reached. A model of the process is sketched in Fig. 10.30. According to Park et al., when the filling of the pore occurs, it is accompanied by an almost instantaneous drop in the capillary pressure of the liquid, which in turn causes a gradual accommodation in the shape of the grains. The grain shape accommodation causes the sample to shrink gradually. At the same time, the rounding of the grains leads to a gradual homogenization of the microstructure. While the physical basis of the model is not clear, the results of Park et al. indicate that grain growth controls the filling of the large isolated pores (i.e., densification) in the final stages of liquid-phase sintering.

Coarsening

Microstructural coarsening by the Ostwald ripening mechanism was suggested as early as 1938 to account for the significant grain growth observed during the liquid-phase sintering of heavy metals. Since the theoretical analysis of Lifshitz, Slyozov, and Wagner, referred to as the LSW theory (see Section 9.3), many studies have been made to compare coarsening data obtained in liquid-phase sintering with the theory. We will recall that the LSW theory is strictly applicable to fairly dilute precipitates in a matrix i.e., to very high volume fraction of liquid. The growth of the average size of the precipitate with time is given by

$$R^m - R_0^m = Kt \qquad (10.39)$$

where R_0 is the initial value of R, K is a temperature-dependent parameter, and the exponent m is dependent on the rate-controlling mechanism. When diffusion through the matrix is rate-controlling, $m = 3$, whereas when the interface reaction is rate-controlling, $m = 2$.

Ardell [21] extended the LSW theory to account for a varying volume fraction ϕ of the precipitates (particles). For the case of diffusion through the matrix as the rate-controlling mechanism, the exponent $m = 3$ is still valid. However, for $\Phi > 0.2$, the size distribution of the precipitates is nearly identical to that predicted by the LSW theory when the interface reaction is the rate-controlling mechanism.

Grain Growth Law

The grain growth exponent m in Eq. (10.39) is observed to be close to 3 for most ceramic and metallic systems. As outlined earlier, the exponent

Liquid-Phase Sintering

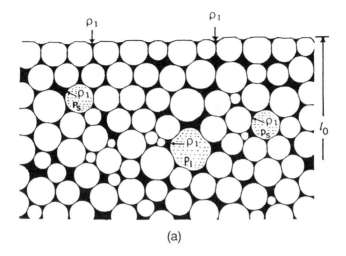

(a)

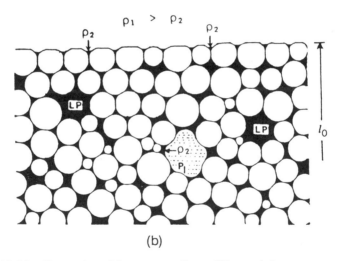

(b)

Figure 10.30 Illustration of the process of pore filling and shape accommodation during the final stage of liquid-phase sintering. (a) Just before the liquid filling of small pores (P_s) (at a critical condition); (b) right after filling of small pores; (c) grain shape accommodation by grain growth and homogenization of the microstructure around the liquid pockets formed at the pore sites during prolonged sintering; (d) just before liquid filling of a large pore (P_l). For simplicity, the equilibrium shape of the grains is assumed to be spherical. In (c), in order to show grain shape accommodation during grain growth, the grains are drawn extremely anhedral. ρ_i is the radius of curvature of the liquid meniscus, l_i is the specimen length, and LP is the liquid pocket. (From Ref. 18, with permission.)

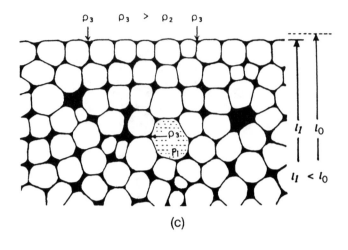

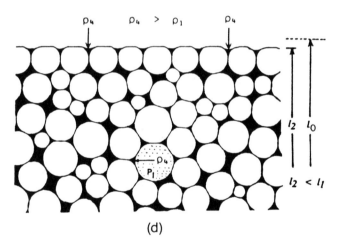

Figure 10.30 Continued.

$m = 3$ corresponds to the situation when diffusion through the liquid is the rate-controlling mechanism. However, grain growth controlled by the interface reaction has also been reported in a few studies.

In a detailed set of experiments, Buist et al. [22] investigated the growth of periclase (MgO), lime (CaO), and corundum (Al_2O_3) grains in a variety of liquid-phase compositions for liquid content of 10–15 vol %. Typical microstructures of the fired samples are shown in Fig. 10.31. In

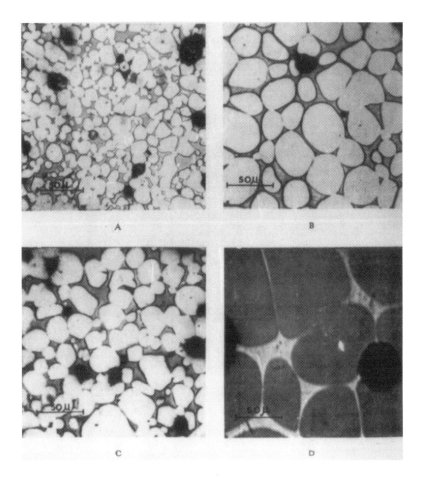

Figure 10.31 Microstructures of (A) composition 3 (see Table 10.4) after firing for 0.5 h at 1550°C, showing rounded grains in a silicate matrix; (B) composition 3 after firing for 8 h at 1550°C; (C) composition 2 (see Table 10.4) after firing for 8 h at 1550°C; (D) a composition of 85 wt % CaO(15 wt % $Ca_2Fe_2O_5$) after firing for 8 h at 1550°C, showing rounded grains in a ferrite matrix. (The very dark areas are pores, some of which contain Araldite.) (From Ref. 22.)

all cases, a grain growth exponent very close to 3 was observed. Figure 10.32 shows the data for the growth of periclase grains in four different liquids corresponding to systems with the compositions given in Table 10.4. Despite the difference in the composition of the liquids, the results show good agreement with the cubic growth law.

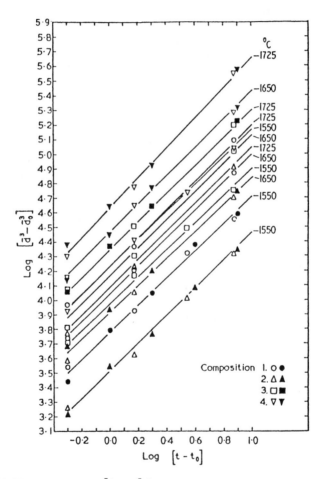

Figure 10.32 Plots of $\log(\bar{d}^3 - \bar{d}_0^3)$ against $\log(t - t_0)$ for periclase grains. (See Table 10.4 for compositions 1–4.) $\bar{d}$ and $\bar{d}_0$ are the mean grain diameters (in μm) at times t and t_0 (in hours). Firing temperatures indicated alongside the plots. Open symbols are for $t_0 = 0.5$ h. Closed symbols are for $t_0 = 0$, with $\bar{d}_0$ assumed to be negligible. (From Ref. 22.)

Table 10.4 Parameters for the Liquid-Phase Sintering of MgO in Four Different Liquids at 1550 and 1725°C[a]

Number	Composition (wt %)	Dihedral angle		Vol % liquid	
		1550°C	1725°C	1550°C	1725°C
1	85 MgO(15 CaMgSiO$_4$)	25	25	15	15
2	80 MgO(15 CaMgSiO$_4$·5Cr$_2$O$_3$)	40	30	16	15
3	80 MgO(15 CaMgSiO$_4$·5Fe$_2$O$_3$)	20	20	16	16
4	85 MgO(15 Ca$_2$Fe$_2$O$_5$)	15	15	11	12

[a] See Figs. 10.31–10.33.
Source: Ref. 22.

Effect of Key Parameters on Coarsening

Coarsening in stage 3 appears to be understood fairly well. The key parameters that control coarsening rate are the temperature, the dihedral angle, and the volume fraction of liquid (or solid).

Effect of Temperature. The solubility of the solid in the liquid and the diffusion of the solute atoms through the liquid increase with temperature in accordance with the Arrhenius relationship. An enhancement of the grain growth rate with increasing temperature can be observed from the data of Buist et al. shown in Fig. 10.32.

Effect of Dihedral Angle. An increase in the dihedral angle reduces the area of the solid in contact with the liquid. This decrease in the solid–liquid contact area will reduce the rate of matter transport by the solution-precipitation mechanism and is therefore expected to reduce the grain growth rate. For the powder compositions investigated by Buist et al. [22] (Table 10.4), it is known that the addition of Cr$_2$O$_3$ increases the dihedral angle in MgO(CaMgSiO$_4$) mixtures whereas the addition of Fe$_2$O$_3$ decreases it. Furthermore, the system MgO(2CaO·Fe$_2$O$_3$) has the smallest dihedral angle. Figure 10.33 shows the results of Buist et al. [22] for the dependence of the grain growth rate constant, i.e., K in Eq. (10.39), as a function of the dihedral angle. An increase in the dihedral angle is observed to lead to a decrease in the grain growth rate constant. The microstructures of the samples with the Fe$_2$O$_3$ and Cr$_2$O$_3$ additions should also be compared in Fig. 10.31. Under nearly identical conditions, the sample with the Fe$_2$O$_3$ addition (smaller dihedral angle) has a larger grain size and less solid–solid contact (Fig. 10.31B) than the sample with the Cr$_2$O$_3$ addition (Fig. 10.31C).

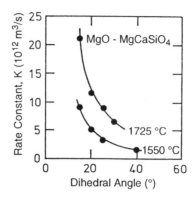

Figure 10.33 Effect of dihedral angle on the grain growth rate constant. (From Ref. 22.)

Effect of Liquid (or Solid) Volume Fraction. Kang and Yoon [23] made a detailed study of the coarsening behavior of the W(Ni) system containing different amounts of liquid phase (1, 7, and 30 wt %). For the system containing 30 wt % nickel, the tungsten grains developed a spherical shape. The grain growth exponent m was found to be 3, as predicted for diffusion control. Comparison of the observed grain size distribution with the LSW theory is not easy. However, Kang and Yoon found that the distribution measured from linear grain intercepts agreed more closely with the LSW theory for reaction control (Fig. 10.34). For the system containing 30 wt % nickel, the data are therefore consistent with Ardell's modification of the LSW theory for diffusion control.

For the systems with the lower nickel content (1 and 7 wt %), flattening of the contact region between the grains occurred as a result of grain shape accommodation. This observation is consistent with our earlier observation that the occurrence of grain shape accommodation is more favorable at fairly small liquid volume fraction. Because of the flattening of the grain contacts, an accurate comparison of the grain size distribution with the LSW theory was not possible. The grain growth exponent m decreased and the grain growth rate increased with decreasing nickel content (decreasing liquid volume). The increase in the grain growth rate with decreasing nickel content is consistent with diffusion through the liquid being the rate-controlling mechanism. With decreasing liquid volume, the diffusion distance decreases; matter transport is therefore faster, leading to an increase in the growth rate.

Liquid-Phase Sintering 569

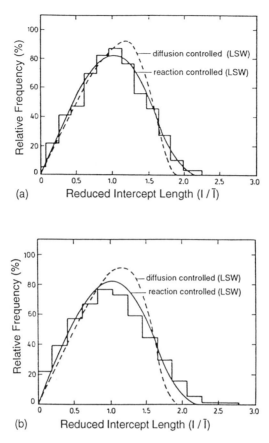

Figure 10.34 The linear intercept distribution in a W(30 wt % Ni) alloy annealed at 1540°C for (a) 30 min and (b) 15 h. (From Ref. 23.)

10.8 HOT PRESSING WITH A LIQUID PHASE

Densification in the presence of a liquid phase can also be used with hot pressing. In parallel with the hot pressing of solid-state systems discussed in Chapter 8, the chemical potential of the atoms under the contact surfaces increases with the applied stress. Matter transport from the contact regions to the pores is enhanced, thereby leading to an increase in the densification rate.

Hot pressing with a liquid phase was used by Bowen et al. [24] to densify Si_3N_4 powder containing MgO as the liquid-producing additive. The MgO reacts with the silica-rich surface layer on the Si_3N_4 powder to produce a eutectic liquid at $\approx 1550°C$. Bowen et al. showed that the densification kinetics were consistent with a solution-precipitation mechanism in which diffusion through the liquid was rate-controlling. The densification kinetics obeyed an equation similar to that derived by Coble for grain boundary diffusion (see Table 8.6) provided that (i) the grain boundary thickness δ_{gb} is replaced by the thickness of the liquid layer δ_L; (ii) the grain boundary diffusion coefficient D_{gb} is replaced by the diffusion coefficient for the solute in the liquid, D_L; and (iii) the solid/vapor interfacial energy γ_{sv} is replaced by the liquid/vapor interfacial energy γ_{lv}. When the applied pressure p_a is much greater than the pressure due to the liquid meniscus, the densification rate can be written

$$\frac{1}{\rho}\frac{d\rho}{dt} = \frac{AD_L \delta_L \Omega}{G^3 kT} p_a \phi \tag{10.40}$$

where A is a geometrical constant, Ω is the atomic volume of the rate-controlling species, G is the grain size, k is the Boltzmann constant, T is the absolute temperature, and ϕ is the stress intensification factor.

10.9 PHASE DIAGRAMS AND THEIR USE IN LIQUID-PHASE SINTERING

Phase equilibrium diagrams (or simply phase diagrams) play a useful role in the selection of powder compositions and firing conditions. However, it must be remembered that the diagrams give the phases present under equilibrium conditions. The kinetics of the reactions may not be fast enough for equilibrium to be achieved during the firing schedule in sintering. The phase diagrams should therefore serve only as a guide.

For a powder composition consisting of a major component and a liquid-producing additive, the first task is to determine which additives will form a liquid phase with the major component and under what conditions. For most systems, phase diagrams will be available in reference books [e.g., 25]. If the desired diagram or region of the diagram is not available, it is useful to spend some time to construct it.

For any major component, there will usually be a variety of additives that will produce a liquid phase under different conditions. For example, the temperatures for the formation of a eutectic liquid between ZnO and a few selected additives are as follows: Bi_2O_3, 750°C; TiO_2, 1420°C; SiO_2, 1425°C; and Al_2O_3, 1700°C. On cooling of the sample after firing, the liquid will form a second phase (or more than one second phase). The nature and composition of the second phase control the properties. The choice

Liquid-Phase Sintering

of additive is therefore largely governed by the desired properties of the fabricated article. In the case of ZnO, it is fortuitous that the additive Bi_2O_3 produces a liquid with a fairly low eutectic temperature as well as useful properties for varistor applications. In practice, the desired properties are achieved with a mixture of Bi_2O_3 and very small additions of various oxides such as MnO, CoO, Sb_2O_3, and SnO.

For a given system, the amount of additive must be determined so that (i) the composition falls in a phase field where a liquid phase is present and (ii) a sufficient amount of liquid is present at the firing temperature to achieve the desired density of the fabricated article. Assuming that equilibrium is achieved, the amount of liquid formed can be determined from the phase diagram by the lever rule.

As an illustration of the use of phase diagrams in liquid-phase sintering, let us consider the fabrication of silicon nitride.

Silicon Nitride

Silicon nitride is one of the ceramics best suited for mechanical engineering applications at high temperatures. The crystal structure of Si_3N_4 contains a high degree of covalent bonding. Because of the strong covalent bonds, matter transport by solid-state diffusion is very slow. Under normally available conditions, polycrystalline Si_3N_4 in the form of a bulk solid is fabricated by liquid-phase sintering. Much of the progress made in its fabrication for high-temperature applications can be traced to careful work in the construction and use of phase diagrams.

The role of the liquid-producing additive and the polyphase nature of the fabricated Si_3N_4 can be summarized as [26]

$$Si_3N_4 + \text{additive} + SiO_2 \xrightarrow{\text{Firing}} Si_3N_4 \text{(solid)} + \text{liquid}$$
$$\xrightarrow{\text{Cooling}} Si_3N_4 + \text{second phase(s)}$$

The Si_3N_4 powder normally contains 1–5 wt % SiO_2 as an oxidation layer on its surface. At the firing temperature the components react to form a liquid. On cooling after densification, the liquid forms a second phase or more than one second phase.

Magnesium oxide was one of the first successful additives used in the fabrication of Si_3N_4. It forms a eutectic liquid with SiO_2 at $\approx 1550°C$. Normally 5–10 wt % of MgO is used, and the firing temperature can be anywhere between 100 and 300°C higher than the eutectic temperature. Figure 10.35a shows the identified tie lines in the subsolidus system Si_3N_4-SiO_2-MgO and the compositions used in their identification [27]. If equilibrium could be achieved during cooling, the fabricated materials would contain two phases if the composition lies on the tie lines or three phases

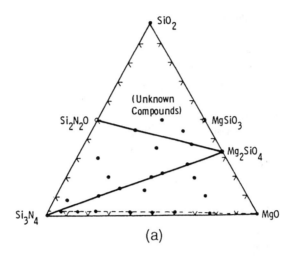

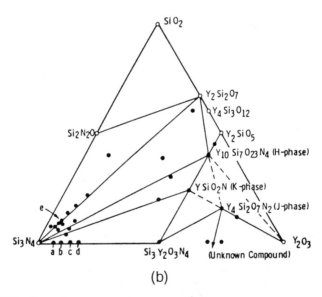

Figure 10.35 Experimental phase relations in (a) the system Si_3N_4-SiO_2-MgO determined from specimens hot pressed between 1500 and 1750°C and (b) the system Si_3N_4-SiO_2-Y_2O_3 determined from specimens hot pressed between 1600 and 1750°C. The filled circles represent the compositions examined. (From Refs. 27 and 28, with permission.)

Liquid-Phase Sintering

if the composition lies within one of the compatibility triangles Si_3N_4 + MgO + Mg_2SiO_4 or Si_3N_4 + Si_2N_2O + Mg_2SiO_4. However, experiments show that all of the magnesium is incorporated into the continuous glassy phase shown in Fig. 10.36A. The continuous glassy phase has a relatively low softening temperature, and this causes a severe reduction in the high-temperature creep resistance of the material.

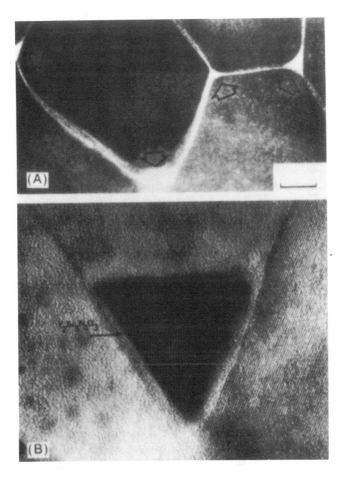

Figure 10.36 Transmission electron micrographs of (A) commercial sample of hot pressed Si_3N_4(MgO), showing continuous glassy phase between β-Si_3N_4 grains (dark field, bar = 50 nm) and (B) $Y_2Si_3O_3N_4$ grain surrounded by β-Si_3N_4 grains, with a glass phase between the crystalline phases (lattice fringe spacing for the β-Si_3N_4 grain at lower left is 0.663 nm). (Courtesy of D. R. Clarke.)

The use of Y_2O_3 as the additive instead of MgO leads to the formation of a eutectic liquid at a higher temperature ($\approx 1660°C$). Higher firing temperatures therefore have to be used. However, an improvement in the high-temperature creep resistance of the fabricated material is also observed. The phase relations for the Si_3N_4-SiO_2-Y_2O_3 system are shown in Fig. 10.35b along with the compositions used in their identification [28]. The system contains three quaternary Y-Si-O-N compounds compatible with Si_3N_4 that crystallize on cooling. Figure 10.36B shows the $Y_2Si_3O_3N_4$ phase that crystallized on cooling at a three- (or four-) grain junction for a material with a composition on the Si_3N_4-$Y_2Si_3O_3N_4$ tie line. It is separated from the surrounding Si_3N_4 grains by a thin residual glass film. The conversion of the glassy phase to a crystalline phase leads to a significant improvement in the creep resistance. Many Si_3N_4 ceramics are now sintered with an additive consisting of a combination of Y_2O_3 and Al_2O_3. On cooling, the liquid phase crystallizes to an yttrium-aluminum garnet phase. The fabricated Si_3N_4 shows improved high-temperature creep resistance as well as better oxidation resistance compared to the material fabricated with Y_2O_3 as the additive.

10.10 VITRIFICATION

The expression "to vitrify" means to make glasslike. Vitrification is used specifically to describe liquid-phase sintering when densification is achieved by the viscous flow of a sufficient amount of liquid to fill the pore spaces between the solid grains. It is the common firing method in the fabrication of many silicate systems. *Vitrification* is distinguished from the normal use of the term *liquid-phase sintering* in that densification by the solution-precipitation mechanism, if present, plays an insignificant role. Vitrification should not be confused with the term *devitrification*, which refers to the crystallization of a glassy phase.

As we discussed earlier (Section 10.7A), the amount of liquid required to produce full densification by vitrification depends on the packing density the solid grains can achieve after rearrangement. For example, if the grains rearrange to achieve a packing density of 64% (i.e., a packing density equal to that for dense random packing of monosize spheres), then 36 vol % of liquid is required to produce full densification by vitrification (Fig. 10.20). In real powder systems, the required liquid volume is normally somewhat less than 36 vol % because (i) the particles have a distribution in sizes that leads to a higher packing density and (ii) some densification occurs by the solution-precipitation mechanism during the time required for complete densification by viscous flow.

A. The Controlling Parameters

In vitrification, the amount and viscosity of the liquid must be such that the desired density (normally full density) is achieved within a reasonable time without the sample deforming under the force of gravity. We require a high enough densification rate that vitrification is complete within a reasonable time (less than a few hours) and a high ratio of densification rate to deformation rate so that densification is achieved without significant deformation of the article. These requirements determine, to a large extent, the firing temperature and the composition of the powder mixture.

Sintering of glasses by a viscous flow mechanism was described in Chapter 8. The models predict that the densification rate depends on three major variables: the surface tension of the of the glass, γ_{sv}; the viscosity of the glass, η; and the pore radius r [see Eq. (8.69)]. Assuming that the pore radius is proportional to the particle radius R, the dependence of the densification rate on these parameters can be written

$$\dot{\rho} = \frac{d\rho}{dt} \sim \frac{\gamma_{sv}}{\eta R} \qquad (10.41)$$

In many silicate systems, the surface tension of the glassy phase does not change significantly with composition. Furthermore, the change in surface tension with temperature is fairly small in the temperature range used for the firing of a given system. The particle size has a significant effect: the densification rate increases inversely as the particle size. For example, a decrease in the particle size from 10 μm to 1 μm increases the densification rate by a factor of 10. However, by far the most important variable is the viscosity. As we discussed in Chapter 8, the dependence of the viscosity of a glass on temperature is described by the Vogel–Fulcher equation [Eq. (8.17)]. This dependence is very strong; for example, the viscosity of soda-lime glass can decrease typically by a factor of ≈ 1000 for an increase in temperature of only $\approx 100°C$. The viscosities of glasses also change significantly with composition. The rate of densification can therefore be increased significantly by changing the composition, increasing the temperature, or some combination of these to reduce the viscosity. However, an appreciable amount of liquid is present during vitrification, and if the viscosity is too low the sample deforms easily under the force of gravity. We must consider not only the absolute rate of densification but also the rate of densification relative to the rate of deformation of the system.

Basically, if the ratio of the densification rate to the deformation rate is large, densification without significant deformation will be achieved. According to Eq. (10.41), the densification rate varies as $1/\eta R$. The deformation rate is related to the applied stress σ and the viscosity by the expression

$$\dot{\epsilon} = \sigma/\eta \tag{10.42}$$

The force due to gravity exerted on a particle of mass m is $W = mg$, where g is the acceleration due to gravity. The mass of the particle varies as R^3, and the area over which the force acts is proportional to the area of the particle, i.e., to R^2. The stress therefore varies as R. The ratio of the densification rate to the deformation rate varies according to

$$\frac{\dot{\rho}}{\dot{\epsilon}} \sim \frac{1}{\eta R}\left(\frac{\eta}{R}\right) \sim \frac{1}{R^2} \tag{10.43}$$

According to this equation, the ratio of densification rate to deformation rate of the system is enhanced as the particle size decreases. Therefore, the best way to achieve high densification without significant deformation is to use fine particles. Many successful silicate systems satisfy this requirement; the compositions contain a substantial amount of clays that are naturally fine-grained.

As noted earlier, the amount of liquid formed at the firing temperature is also an important parameter. A sufficient amount of liquid must be formed to fill the pore spaces between the grains. However, the formation of the liquid must be controlled to prevent the sudden formation of a large amount of liquid that will lead to warping or slumping of the sample under the force of gravity. Control of the amount of liquid formed is best illustrated by considering a binary phase diagram. As sketched in Fig. 10.37, suppose two components, A and B, of the powder composition react to form a liquid at a eutectic temperature T_1. The mixture with the composition labeled C_1 is better than that labeled C_2 because less liquid is formed at T_1, thereby reducing the potential for collapse of the sample. Composition C_1 is also beneficial for another reason. The quantity of liquid formed changes less rapidly as the temperature is raised above T_1, giving better flexibility in the choice of the firing temperature.

B. Vitrification of Silicate Systems

Vitrification is the common firing method in the fabrication of many silicate systems. A viscous silicate glass forms at the firing temperature that fills the pore spaces and glues the particles together. On cooling, a dense solid product is produced.

As an example, consider the vitrification of triaxial porcelain compositions. The materials are produced from powder mixtures containing three components with a typical composition being [29]

Liquid-Phase Sintering

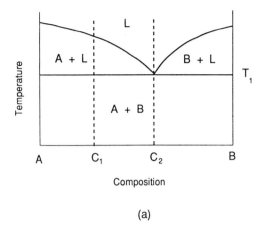

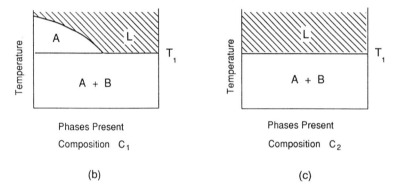

Figure 10.37 Sketch of (a) the phase diagram for a binary system showing the phases present for two compositions C_1 (b) and C_2 (c).

50 wt % clay. For a clay consisting of predominantly the mineral kaolinite, with a formula $Al_2(Si_2O_5)(OH)_4$, the composition of the clay corresponds to $\approx$45 wt % Al_2O_3 and $\approx$55 wt % SiO_2.

25 wt % feldspar. Feldspar is an alkali-containing mineral that acts as a flux, i.e., the alkali (often K^+) serves to lower the temperature at which the viscous liquid forms. A common feldspar has the formula $KAlSi_3O_8$.

25 wt % silica, present as flint.

This and similar compositions lie in the primary mullite phase field in the ternary K_2O-SiO_2-Al_2O_3 phase diagram (Fig. 10.38). Typical firing conditions are in the range 1200–1400°C. Between 1200 and 1600°C, the equilibrium phases are mullite and liquid. An isothermal section of the phase diagram (Fig. 10.39) shows the equilibrium phases at 1200°C, i.e., the lower end of the temperature range used in firing. At this temperature, the liquid has the composition of approximately 75 wt % SiO_2, 12.5 wt % K_2O, and 12.5 wt % Al_2O_3. (The letters K, A, and S in Fig. 10.39 represent K_2O, Al_2O_3 and SiO_2, respectively.)

In practice, the approach to equilibrium is incomplete, and only a small part of the SiO_2 present as flint enters the liquid phase. However, the amount of SiO_2 that dissolves does not have a strong effect on the amount and composition of the liquid. The cooled material contains mull-

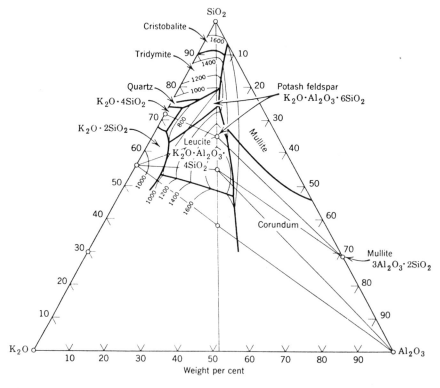

Figure 10.38 The ternary system K_2O-Al_2O_3-SiO_2. [From J. F. Schairer and N. L. Bowen, *Am. J. Sci.*, 245:199 (1947).]

Liquid-Phase Sintering

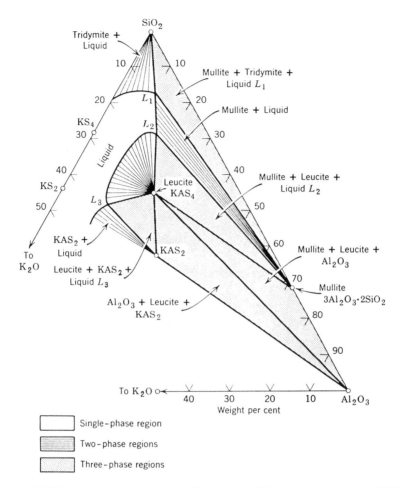

Figure 10.39 Isothermal cut in the K_2O-Al_2O_3-SiO_2 phase diagram at 1200°C. (From Ref. 29, with permission.)

ite grains, a glass, and SiO_2 grains. Triaxial porcelains are white, dense, and translucent and are used for the production of pottery, tiles, and insulators.

10.11 CONCLUDING REMARKS

Liquid-phase sintering is a common firing process used in the production of a wide variety of ceramic materials. The liquid phase enhances densification through easier rearrangement of the grains and faster matter transport through the liquid. However, the liquid normally remains as a glassy grain boundary phase in the fabricated article, which may lead to a deterioration of properties. The overall process of liquid-phase sintering is divided into three overlapping stages, which are defined in terms of the dominant mechanism that occurs in each stage. The dominant mechanism in stage 1 is the rearrangement of the grains, aided by the presence of the lubricating liquid. Extensive redistribution of the liquid may also take place in this stage. The dominant mechanism in stage 2 is solution of the grains at regions with higher chemical potential and precipitation at regions with lower chemical potential.

Two models have been proposed to account for the densification process: (i) contact flattening, which is more appropriate to systems that undergo very limited or no coarsening, and (ii) densification accompanied by Ostwald ripening. While the contact flattening model allows quantitative estimates of the shrinkage rates, experimental support for it is not very convincing. In the vast majority of real systems, observations support the model in which Ostwald ripening accompanies the densification. However, it is difficult to make quantitative estimates of the shrinkage rates. In stage 3, the dominant mechanism is Ostwald ripening, which leads to extensive coarsening of the microstructure. Successful exploitation of liquid-phase sintering depends on the control of a number of key parameters. Small contact angle, small dihedral angle, and high solubility of the solid in the liquid are essential material parameters, while homogeneous packing of the particulate solid, homogeneous distribution of the liquid-producing additive in the powder system, and fairly fine particle size are important processing parameters for the achievement of high density.

Vitrification is a type of liquid-phase sintering used for the fabrication of many silicate systems; a viscous glassy phase formed in sufficient quantity at the firing temperature fills the pores and glues the particles together. The size of the solid grains and the viscosity of the glassy phase are the important parameters in vitrication that must be controlled to achieve high density without significant deformation of the fabricated article. Finally, phase diagrams play a useful role in the selection of the powder compositions and the firing temperatures; we illustrated the use of phase diagrams in the liquid-phase sintering of silicon nitride and the vitrification of a triaxial porcelain composition.

PROBLEMS

10.1 For each of the stages associated with classic liquid phase sintering, give (a) the name of the stage and (b) the main microstructural changes.

Liquid-Phase Sintering

10.2 Given the following interfacial energies (in units of J/m^2) for liquid phase sintering: $\gamma_{sv} = 1.0$, $\gamma_{lv} = 0.8$, $\gamma_{ss} = 0.75$, and $\gamma_{sl} = 0.35$, determine (a) the solid/liquid contact angle and (b) whether the liquid will penetrate the grain boundaries.

10.3 Consider two spheres with the same radius R in contact in a liquid. Derive an expression for the equilibrium radius of the neck, X, between the spheres when the dihedral angle is ψ. Plot a graph of the ratio X/R versus γ_{ss}/γ_{sl}, where γ_{ss} and γ_{sl} are the interfacial tensions for the solid/solid and solid/liquid interfaces, respectively.

10.4 Discuss whether each of the following factors is likely to promote shrinkage or swelling during the initial stage of liquid phase sintering: (a) high-solid solubility in the liquid, (b) low-liquid solubility in the solid, (c) large contact angle, (d) small dihedral angle, and (e) large particle size of the liquid-producing additive.

10.5 Estimate the capillary force acting on two spherical particles (diameter 10 μm) as a function of distance of separation when the liquid phase content is 5 vol% and the solid/liquid contact angle is (a) 5° and (b) 120°. State any assumptions that you make. Plot the results on a graph.

10.6 Derive Eq. (10.38).

10.7 Discuss the effects of each of the following factors on the densification and microstructural evolution during liquid phase sintering of silicon nitride: (a) composition of the liquid-producing additive, (b) volume fraction of the additive, (c) sintering temperature, (d) particle size of the silicon nitride powder, and (d) applied pressure.

REFERENCES

1. W. D. Kingery, E. Niki, and M. D. Narasimhan, *J. Am. Ceram. Soc., 44*: 29 (1961).
2. H. Song and R. L. Coble, *J. Am. Ceram. Soc., 73*:2077 (1990).
3. R. M. German, *Liquid Phase Sintering*, Plenum, New York, 1985.
4. W. Beeré, *Acta Metall., 23*:131 (1975).
5. H. H. Park and D. N. Yoon, *Met. Trans. A, 16*: 923 (1985).
6. C. S. Smith, *Trans. AIME, 175*:15 (1948).
7. R. B. Heady and J. W. Cahn, *Met. Trans., 1*:185 (1970).
8. W. J. Huppmann and H. Riegger, *Acta Metall., 23*:965 (1975).
9. F. F. Lange, *J. Am. Ceram. Soc., 65*:C-23 (1982).
10. D. D. Eley, *Adhesion*, Oxford Univ. Press, New York, 1961, pp. 118–120.
11. D. R. Clarke, T. M. Shaw, A. P. Philipse, and R. G. Horn, *J. Am. Ceram. Soc., 76*:1201 (1993).
12. O.-J. Kwon and D. N. Yoon, *Int. J. Powder Metall. Powder Technol., 17*: 127 (1981).

13. T. M. Shaw, *J. Am. Ceram. Soc., 69*:27 (1986).
14. W. D. Kingery, *J. Appl. Phys., 30*:301 (1959).
15. D. N. Yoon and W. J. Huppmann, *Acta Metall., 27*: 693 (1979).
16. W. J. Huppmann and G. Petzow, in *Sintering Processes* (G. C. Kuczynski, Ed.) (*Mater. Sci. Res.,* Vol. 13), Plenum, New York, 1980, pp. 189–201.
17. D. N. Yoon and W. J. Huppmann, *Acta Metall., 27*:973 (1979).
18. S.-J. L. Kang, K.-H. Kim, and D. N. Yoon, *J. Am. Ceram. Soc., 74*:425 (1991).
19. H. H. Park, S. J. Cho, and D. N. Yoon, *Met. Trans. A, 15*:1075 (1984).
20. H. H. Park, O. J. Kwon, and D. N. Yoon, *Met. Trans. A, 17*:1915 (1986).
21. A. J. Ardell, *Acta Metall., 20*:61 (1972).
22. D. S. Buist, B. Jackson, I. M. Stephenson, W. F. Ford, and J. White, *Trans. Br. Ceram. Soc., 64*:173 (1965).
23. T.-K. Kang and D. N. Yoon, *Met. Trans. A, 9*:433 (1978).
24. L. J. Bowen, R. J. Weston, T. G. Carruthers, and R. J. Brook, *J. Mater. Sci., 13*:341 (1978).
25. E. M. Levin, C. R. Robbins, and H. W. McMurdie, *Phase Diagrams for Ceramists*, American Ceramic Society, Columbus, OH, 1964.
26. F. F. Lange, *Am. Ceram. Soc. Bull., 62*:1369 (1983).
27. F. F. Lange, *J. Am. Ceram. Soc., 61*:53 (1978).
28. F. F. Lange, S. C. Singhal, and R. C. Kuznicki, *J. Am. Ceram. Soc., 60*: 249 (1977).
29. W. D. Kingery, H. K. Bowen, and D. R. Uhlmann, *Introduction to Ceramics*, 2nd ed., Wiley, New York, 1976, pp. 490–497.
30. R. W. Davidge, *Mechanical Behavior of Ceramics*, Cambridge Univ. Press, New York, 1979.

FURTHER READING

R. M. German, *Liquid Phase Sintering*, Plenum, New York, 1985.
V. N. Eremenko, Y. V. Naidich and I. A. Lavrinenko, *Liquid Phase Sintering*, Consultants Bureau, New York, 1970.
J. Marion, C.-H. Hsueh and A. G. Evans, J. Am. Ceram. Soc., *70*:(10) 708 (1987).
D. R. Clarke, J. Am. Ceram. Soc., *70*:(1) 15 (1987).

11
Problems of Sintering

11.1 INTRODUCTION

The effects of variables such as grain size, temperature, and applied pressure on sintering are well understood. For powder systems that approach the assumptions of the theoretical models (i.e., homogeneously packed spherical particles with a narrow size distribution), the sintering kinetics can be well described by the models. However, some important issues remain, and these need to be examined in terms of their ability to limit the attainment of high density with controlled microstructure.

The structure of real powder compacts, as we discussed earlier in this book, is never homogeneous. In general, we can expect various types of inhomogeneities, such as variations in packing density, a distribution of pore sizes, and a distribution of particle sizes. Inhomogeneities lead to differential densification during sintering where different regions of the powder compact sinter at different rates. Differential densification, in turn, leads to the generation of stresses that reduce the densification rate and to the creation of microstructural flaws that limit the ability to reach high density. A normal feature of inhomogeneities is that their effects become magnified during firing. Major issues that need to be examined include the quantification and control of inhomogeneities in the powder compact. Equally important is the examination of techniques that might allow some correction of inhomogeneities during sintering.

Inhomogeneities are a necessary feature of the microstructural design of composites. Rigid inclusions such as particles, platelets, or fibers are incorporated into the material to improve its properties. However, they invariably create severe problems for sintering. Densification is retarded because the matrix surrounding the inclusions is unable to shrink freely. The retardation is observed to be much more significant in polycrystalline matrices than in glass matrices. The mechanisms responsible for the reduced densification must be determined so that suitable methods can be developed to alleviate the problems. Transient stresses due to differential densification between the matrix and the inclusions provide one mechanism for the retardation of densification. Processing problems such as the inhomogeneous packing of the powder system may be enhanced in the composites due to the presence of the inclusions. New problems may also be introduced. One such problem is the formation of a percolating network of the inclusion phase that may resist densification.

The sintering of films attached to a rigid substrate is required in several important applications. The substrate provides an external constraint, preventing shrinkage in the plane of the substrate; all of the shrinkage occurs in the direction perpendicular to the plane of the film. The inhibition of the shrinkage in the plane of the substrate leads to stresses in the film that influence the sintering kinetics. The sintering rate of the constrained film relative to that of a free film is an important parameter that must be quantified.

The use of solid solution additives or dopants, as we outlined in Chapter 9, provides an effective approach for the fabrication, by sintering, of ceramics with high densities and controlled grain size. However, except for a few well-recognized systems, the role of the dopant in sintering is not understood. This gap in understanding limits the applicability of the approach. The major roles of the dopant and the parameters that control the effectiveness of a dopant in a given role form important issues in the solid solution additive approach. Addition of MgO to Al_2O_3 is perhaps the best understood system. The present understanding of the dopant role in this system and the applicability of this understanding to other systems may serve as a useful starting point for the investigation of the dopant role in other systems.

Reaction sintering, in which the chemical reaction of the starting materials and the densification of the powder compact are achieved in a single heat treatment step, may provide an alternative route to conventional sintering for the production of single-phase ceramics and composites. Its applicability is, however, severely limited because of problems in controlling the microstructure of the product, particularly in the fabrication of single-phase ceramics. The parameters that control the separate processes of densification and reaction and the interaction of the two processes need to be examined to provide a basis for selecting the appropriate processing conditions for reaction sintering.

Techniques such as sol-gel processing, discussed earlier in this book, can be used to prepare porous materials (films or bulk solids) or powders that are normally amorphous in structure. The porous material or the compacted powder must be fired to produce a dense article. Crystallization prior to full densification has been observed to cause the retention of a considerable amount of porosity in the final article. A major issue is the quantification of the effect of crystallization on densification in order to determine its significance. Another major issue is understanding the influence of key variables on the processes of densification and crystallization in order to determine whether the two processes can be decoupled. If decoupling can be achieved, it would be possible to alter the processing parameters to achieve the desired goal of densification prior to crystallization.

11.2 INHOMOGENEITIES AND THEIR EFFECTS ON SINTERING

The structure of real powder compacts is never completely homogeneous. The characteristics of the powder and the forming method control the extent of the structural inhomogeneities in the consolidated powder form. In the case of monodisperse particles that have been consolidated from stabilized suspensions by colloidal methods, local packing variations exist between regions (domains) of uniformly packed particles. In general, we can expect various types of structural inhomogeneity such as variations in packing density, a distribution in pore sizes, and a distribution in particle sizes. If the powder contains impurities, chemical inhomogeneity in the form of variations in composition will also be present. Inhomogeneities can limit the ability to reach high density. To make the most effective use of the sintering process, we must clarify the effects of inhomogeneities on sintering and how these effects can be controlled.

A. Differential Densification

Structural inhomogeneities lead to some regions of the powder compact densifying at different rates from other regions, a process referred to as *differential densification*. The regions densifying at different rates interact with each other, and this interaction leads to the development of transient stresses during sintering. The stresses are said to be transient because they diminish rapidly when the densification is complete or when the sample is cooled to temperatures where no densification takes place. However, in many cases, some residual stress remains after densification. The development of transient stresses during sintering is analogous to the development of thermal stresses in materials that differ in thermal expansion coefficients (e.g., a bimetallic strip in which the two metals have different thermal expansion coefficients). In sintering, the volumetric strain rate takes the place of the thermal strain.

Stresses due to differential densification have been analyzed for a model consisting of a spherical inhomogeneity surrounded by a uniform powder matrix [1,2]. The stress system is analogous to that of a thermal stress problem in which a spherical core is surrounded by a cladding with a different thermal expansion coefficient. If the inhomogeneity (e.g., a hard, dense agglomerate) shrinks more slowly than the surrounding matrix, a hydrostatic backstress is generated in the matrix. This backstress opposes the sintering stress and leads to a reduction in the densification rate of the powder matrix (Fig. 11.1a). Differential densification may also generate microstructural flaws in the sample. For the case of the spherical inhomogeneity that shrinks more slowly than the powder matrix, a circumferential (or hoop) stress is also set up in the matrix. If this stress is large enough, it can produce cracklike voids in the sample (Fig. 11.1b).

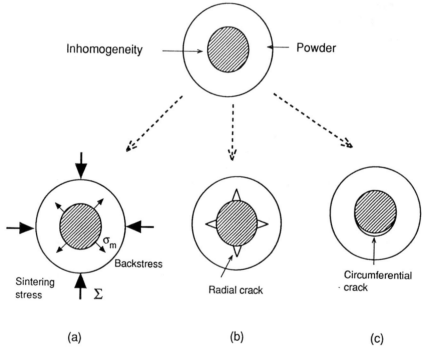

Figure 11.1 Schematic diagram illustrating the effects of inhomogeneities in a powder compact. Reduced densification (a) and radial flaws (b) are produced by an inhomogeneity that sinters at a slower rate than the surrounding region whereas a circumferential flaw (c) is produced when the inhomogeneity sinters at a faster rate than the surrounding region.

Problems of Sintering

Microstructural flaws can also be produced if the inhomogeneity shrinks faster than the surrounding matrix. In this case, the inhomogeneity can shrink away from the matrix, thereby producing a circumferential void (Fig. 11.1c). Microstructural flaws such as those sketched in Fig. 11.1b and c have been observed in some sintered articles produced from heterogeneous powder compacts [3].

Another type of microstructural flaw produced by differential densification has been observed in irregular two-dimensional arrays of monosize copper spheres [4]. An example of the microstructural evolution in such arrays is shown in Fig. 11.2. A detailed statistical analysis of the evolution of a similar type of array revealed that rearrangement of the particles occurred as a result of differential densification [5]. The regions that densify faster exert tensile stresses on the neighboring regions that are densifying more slowly. If the stresses are greater than, and opposed to, the sintering stress, pores at that location will grow rather than shrink despite the fact that the compact undergoes an overall shrinkage. In three dimensions, movement is more constrained (because of the higher coordination of the particles) so that particle rearrangement due to differential densification is expected to be less dominant than that observed in the two-dimensional arrays.

To summarize, differential densification leads to the development of microstructural flaws such as large pores and cracklike voids. Since these flaws cannot normally be removed during sintering, the overall effect is a reduction of the densification rate and a limitation of the density that can be attained.

B. Control of Inhomogeneities

It is clear from Figs. 11.1 and 11.2 that differential densification can have a significant effect on densification and microstructural evolution. To reduce the undesirable effects of densification, we must reduce the extent of inhomogeneity present in the powder compact. In seeking the origins of inhomogeneities, it should be considered that the powder quality influences both the particle size distribution and, assuming no impurities are incorporated in the subsequent processing steps, the variation in chemical composition. The powder characteristics and the forming method influence the density variations in the powder compact. A common feature of such variations, as discussed earlier in this book, is that they become exaggerated during firing.

Important issues in the control of inhomogeneities in the consolidated powder form are (i) the quantification of the extent of inhomogeneity in

Figure 11.2 Evolution of the structure of two planar arrays of copper spheres during sintering. Two different starting densities, expressed as x/a, are shown; x is the neck radius between particles of radius a. (From Ref. 4, reprinted with permission of Freund Publishing House, Ltd.)

a powder compact, (ii) methods for reducing or averting inhomogeneities, and (iii) mechanisms for reversing structural inhomogeneities. The quantification of inhomogeneities may allow us to determine what level of inhomogeneity can be tolerated in a given process. However, such quantification is difficult and has been discussed only at a very qualitative level. The inhomogeneity depends on the scale of observation: the structure appears more uniform as the scale of observation becomes coarser.

One approach to reducing structural inhomogeneities in a powder compact involves the application of pressure, either during compaction (e.g., cold isostatic pressing) or during firing (e.g., hot pressing). Another approach is to attempt to attain a uniform pore size that is smaller than

Problems of Sintering

the grain size. As discussed earlier, it is thermodynamically feasible for such pores to shrink. This approach has shown clear benefits for the achievement of high density and small grain size. A good example of such an approach is the work of Yeh and Sacks [6] in which the effect of particle packing on the sintering of Al_2O_3 powder compacts was investigated. Using the same powder, which had a fairly narrow size distribution, compacts were prepared by slip casting of a well-stabilized suspension (pH 4) and a flocculated suspension (pH 9). The structural variation in the two types of powder compacts is shown in Fig. 11.3. During firing under the same conditions (1340°C), the more homogeneous compact with the higher packing density (pH 4) had a higher densification rate and reached a higher density than the less homogeneous compact (Fig. 11.4).

To summarize, a reduction in the extent of inhomogeneities can be achieved by (i) reducing the structural variations in the powder compact, (ii) increasing the packing density, and (iii) maximizing the fraction of pores that are smaller than the grain size.

C. Correction of Inhomogeneities

The structure of real powder compacts is never homogeneous. Faced with this problem, it is worth examining (i) mechanisms that may cause a reversal of structural variations already present in a powder compact and (ii) the extent to which the inhomogeneity can be reversed by each mechanism.

A common type of inhomogeneity present in powder compacts consists of agglomerates. They lead to density variations and to a distribution in pore sizes. At one extreme of the pore size range, we have pores that are large compared to the grain size of the powder system. These large pores would normally be characteristic of the pores between the agglomerates. They are surrounded by a large number of grains. The pore coordination number N of such pores (i.e., the number of grains surrounding a pore) is large. At the other extreme, small pores with a low coordination number N will exist. Such small pores would be characteristic of the pores in the well-packed regions of the agglomerates. Following our discussion in Chapter 9 of the thermodynamics of pore closure, the reader will recall that pores with N greater than some critical value N_C will grow whereas those with $N < N_C$ will shrink. An important question relating to the correction of inhomogeneity in the powder system is whether the growth of the large pores can be reversed, thereby reversing the progress to a more heterogeneous structure.

As sketched in Fig. 11.5, assuming that the grains grow faster than the pore, grain growth can eventually lead to a situation where the coordination number N of the large pore becomes less than N_C. When this

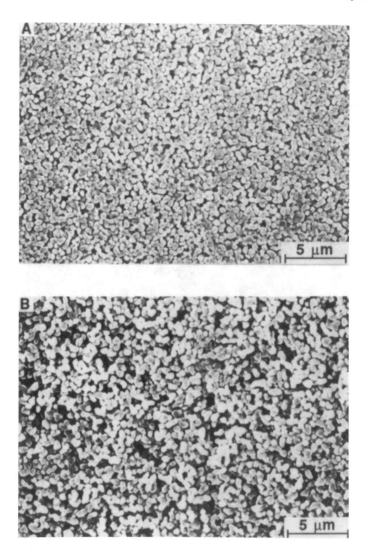

Figure 11.3 Scanning electron micrographs of the top surfaces of slip-cast samples of Al_2O_3 formed from (A) well-dispersed (pH 4) and (B) flocculated (pH 9) suspensions. Note the difference in packing homogeneity between the samples. (From Ref. 6, reprinted with permission from the American Ceramic Society.)

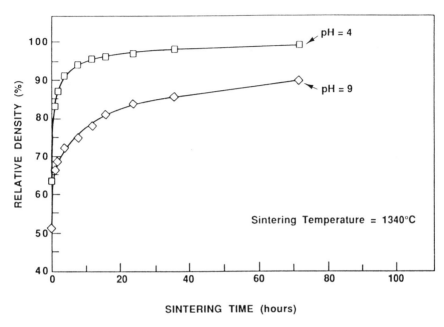

Figure 11.4 Relative density versus sintering time at 1340°C for the two samples described in Fig. 11.3. The sample with the more homogeneous packing (prepared from the suspension at pH 4) has higher density at any given time. (From Ref. 6, reprinted with permission of the American Ceramic Society.)

situation occurs, the large pore will shrink, and the progress toward a wide distribution in pore sizes (i.e., increased inhomogeneity) will be reversed. Based on reasoning of this nature, it has been suggested that normal grain growth might be beneficial for the sintering of heterogeneous powder compacts because it leads to a reduction in the pore coordination number [7].

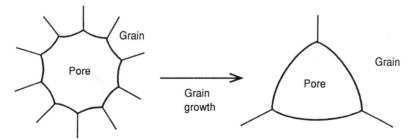

Figure 11.5 Schematic diagram illustrating the reduction in the pore coordination number as a result of normal grain growth.

592 *Chapter 11*

(Abnormal grain growth should still be avoided in this concept.) This suggestion that normal grain growth might be beneficial for sintering has been the subject of some debate. Although grain growth can indeed lead to a reduction in the pore coordination number, we must also consider the kinetics of achieving the desired density. An analysis indicates that the kinetics of densification can be decreased significantly by the grain growth so that no long-term benefit for densification is achieved [8].

Although the role of grain growth for the attainment of high density may be controversial, the ability of a limited amount of grain growth to homogenize a microstructure has been clearly demonstrated [9]. Figure 11.6 shows a sequence of microstructures of an MgO powder compact sintered at 1250°C to various densities. The initial compact contains clearly identifiable large pores in a fine-grained matrix. As sintering pro-

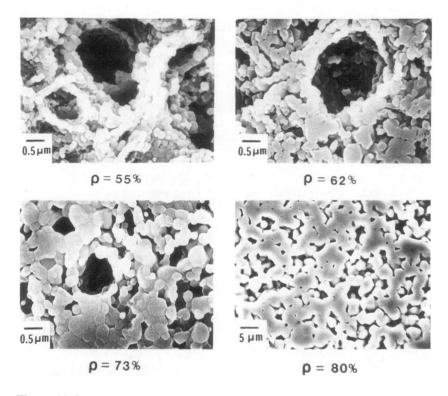

Figure 11.6 Microstructural evolution of an inhomogeneous MgO powder compact sintered to various densities at 1250°C. Note the trend toward homogenization of the microstructure. (From Ref. 9.)

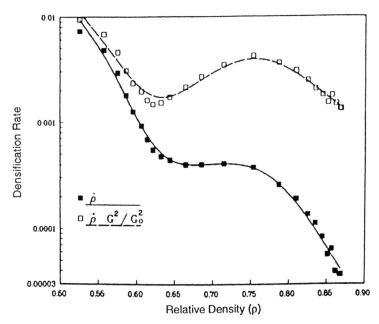

Figure 11.7 Observed densification rate $\dot{\rho}$ and the grain size compensated densification rate, $\dot{\rho}G^2/G_0^2$, versus relative density for the powder compact described in Fig. 11.6. G is the grain size at a given time (or density) and G_0 is the initial grain size. The grain size compensated densification rate is a measure of the densification rate at roughly the same grain size (equal to G_0). When a sufficient amount of grain growth and microstructural homogenization have occurred (i.e., at $\rho \approx 0.65$), the large pores originally present in the powder compact can shrink, leading to an increase in the densification rate. (From Ref. 9.)

ceeds, grain growth and pore coalescence cause the microstructure to become more homogeneous. At a relative density of ≈0.65, grain growth has proceeded to such an extent that it becomes thermodynamically feasible for the large pores in the original compact to shrink. The onset of shrinkage of the large pores is observed as an increase in the measured densification rate of the compact (Fig. 11.7). A limited amount of precoarsening (i.e., coarsening at temperatures lower than the onset of densification) has also been observed to lead to an improvement in the microstructural homogeneity of the powder compact, which in turn leads to a more uniform fired microstructure [10]. Such precoarsening treatments have been successfully applied to the sintering of MgO, ZnO, and Al_2O_3. However, the generalization of the method to other powder systems is not clear.

Another example where some homogenization of the microstructure has been observed comes from the sintering of compacts formed from mixtures of powders with particles of two different sizes [11]. The powder mixture is said to have a bimodal particle size distribution. As a first consideration, we might assume that each fraction in the mixture densifies in the same way as it would independently of the other. In this case we may expect the densification of the mixture to obey a simple rule of mixtures, i.e., the densification of the mixture is some weighted average of the densification of the separate powder fractions. For Al_2O_3 powder in which the particle size of the coarse fraction was ≈ 10 times that of the fine fraction, the densification behavior was indeed found to approximately obey a rule of mixtures. However, an interesting observation was that grain growth occurred only in the fine powder fraction. Thus the microstructure became more homogeneous than that of the initial powder compact (Fig. 11.8).

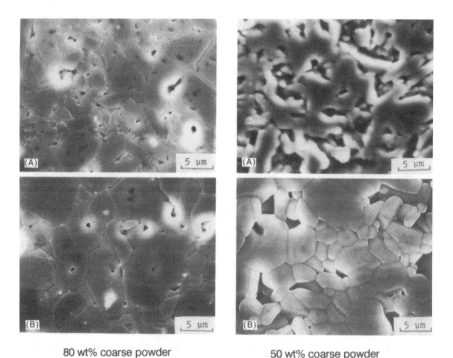

Figure 11.8 Scanning electron micrographs of Al_2O_3 powder compacts formed from a mixture with two discrete particle sizes (coarse and fine) after sintering for (A) 60 min and (B) 360 min at 1600°C. The compacts consisted of 80 wt % coarse powder (left pair) and 50 wt % coarse powder (right pair). (From Ref. 11, with permission.)

Problems of Sintering

To summarize, the ability of a limited amount of normal grain growth to improve the microstructural homogeneity has been clearly demonstrated for compacts with a wide pore size distribution or with a bimodal grain size distribution. As we will describe later in this chapter, the use of solid solution additives (dopants) provides one of the most effective methods for the homogenization of a microstructure.

11.3 CONSTRAINED SINTERING: I. RIGID INCLUSIONS

Every system is constrained to some extent. For example, the sintering of a pure single-phase powder is constrained by inhomogeneities (e.g., agglomerates) present in the powder compact. However, the term *constrained sintering* is usually taken to refer to sintering in which the constraint is deliberately imposed and is a necessary feature of the system. Compared to the constrained system, an identical system in which the constraint is absent is taken to be the free or unconstrained system. The constraint may be either internal or external to the sintering system. Ceramic matrix composites are a well-known system in which the constraint is internal (Fig. 11.9a). A dense, rigid second phase in the form of particles,

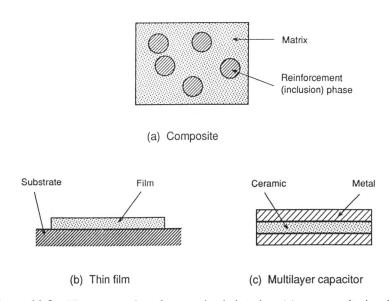

Figure 11.9 Three examples of constrained sintering: (a) a composite in which inclusions prevent the powder matrix form sintering freely; (b) a thin film attached to a rigid substrate where the substrate inhibits the sintering in the plane of the film; (c) a multilayer in which the layers sinter at different rates.

whiskers, platelets, or continuous fibers is incorporated into a glass or polycrystalline matrix to modify its properties. The sintering of a thin film that adheres to a rigid substrate is an important example in which the constraint is external to the system (Fig. 11.9b). The film is constrained from shrinking in the plane of the substrate. Another example of an externally constrained system is the sintering of laminated substrates for electronic applications (Fig. 11.9c). Different layers normally have different sintering characteristics and constrain the sintering of neighboring layers.

We start our discussion of the sintering of constrained systems by considering the case of sintering of composites. In the fabrication of ceramic matrix composites, the reinforcing phase is normally incorporated into the powder matrix to form the shaped article, which is then densified to produce the final article. The effects of the reinforcing phase on the sintering of the powder matrix must be understood in order to achieve the desired density and microstructure in the composite.

A reinforcing phase has been observed to produce a drastic reduction in the densification rate of a ceramic powder matrix. We consider the simple system in which the reinforcing phase is in the form of dense, rigid (i.e., undeformable) inert particulate inclusions.

A. Volume Fraction of Inclusions

The volume fraction of the inclusions is an important parameter. It is defined as the volume of the inclusions divided by the total volume of the composite (solid plus pores). Let us consider a shaped composite consisting of a porous powder matrix with a relative density ρ_m (i.e., relative to the fully dense powder) and rigid, fully dense inclusions (i.e., the relative density of the inclusions, ρ_i, is equal to 1). As the sample densifies (Fig. 11.10), the total volume decreases so the volume fraction of the inclusions increases. The volume fraction of the inclusions, v_f, is normally determined on the basis of the fully dense composite, i.e.,

$$v_f + v_{m\infty} = 1 \tag{11.1}$$

where $v_{m\infty}$ is the volume fraction of the matrix in the fully dense composite. If v_i is the volume fraction of the inclusions when the relative density of the matrix is ρ_m (<1), then v_i is related to v_f by

$$v_i = \frac{\rho_m}{\rho_m + (1 - v_f)/v_f} \tag{11.2}$$

Therefore, knowing v_f we can determine v_i at any matrix relative density ρ_m.

Problems of Sintering

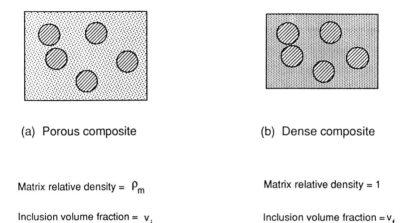

(a) Porous composite   (b) Dense composite

Matrix relative density = ρ_m Matrix relative density = 1

Inclusion volume fraction = v_i Inclusion volume fraction = v_f

Figure 11.10 Matrix and inclusion parameters in the sintering of composites.

B. Densification Rate of the Composite and the Matrix

Normally we measure the density of the composite. However, in many cases we need to determine the density of the matrix of the composite in order to compare its densification kinetics with those of the free matrix. If D_c and D_m are the densities (i.e., mass/volume) of the composite and the matrix, respectively, then by simple geometry we can show that

$$D_m = D_c \frac{D_{c0} - v_{i0} D_i}{D_{c0} - v_{i0} D_c} \tag{11.3}$$

where D_i is the density of the inclusion, and D_{c0} and v_{i0} are the initial values of D_c and v_i, respectively. Differentiating Eq. (11.3) followed by some manipulation gives

$$\frac{1}{D_m}\frac{dD_m}{dt} = \frac{1}{D_c}\frac{dD_c}{dt}\left(\frac{D_{c0}}{D_{c0} - v_{i0} D_c}\right) \tag{11.4}$$

If ρ_c and ρ_m are the relative densities of the composite and the matrix, respectively, this equation becomes

$$\frac{1}{\rho_m}\frac{d\rho_m}{dt} = \frac{1}{\rho_c}\frac{d\rho_c}{dt}\left(\frac{\rho_{c0}}{\rho_{c0} - v_{i0}\rho_c}\right) \tag{11.5}$$

It is left as an exercise for the reader to derive Eqs. (11.2)–(11.4) from first principles.

C. The Rule of Mixtures

The rule of mixtures assumes that the densification of the composite is a weighted average of the independent densification rates of the matrix and the inclusions; that is, it assumes that in the composite each phase densifies in the same way as it would independently by itself. If, for example, we consider the linear densification rate $\dot{\epsilon}$ defined as $(1/3)(1/\rho)d\rho/dt$, where $(1/\rho)d\rho/dt$ is the volumetric densification rate, according to the rule of mixtures,

$$\dot{\epsilon}_c^{rm} = \dot{\epsilon}_{fm}(1 - v_i) + \dot{\epsilon}_i v_i = \dot{\epsilon}_{fm}(1 - v_i) \tag{11.6}$$

In this equation, $\dot{\epsilon}_c^{rm}$ is the composite densification rate predicted by the rule of mixtures, $\dot{\epsilon}_{fm}$ is the densification rate of the free matrix (i.e., containing no inclusions), $\dot{\epsilon}_i$ is the densification rate of the inclusions, taken to be zero, and v_i is the volume fraction of the inclusions. According to Eq. (11.6), the ratio $\dot{\epsilon}_c^{rm}/\dot{\epsilon}_{fm}$ varies linearly as $1 - v_i$.

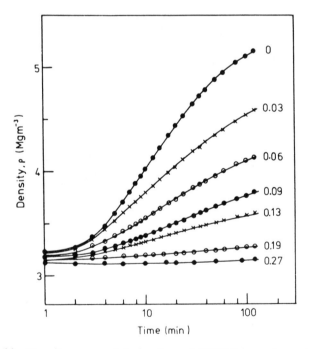

Figure 11.11 Density versus sintering time at 700°C for a polycrystalline ZnO powder containing different volume fractions of rigid SiC inclusions. (From Ref. 12.)

D. Experimental Observations of Sintering with Rigid Inclusions

Data for the sintering of a polycrystalline ZnO powder matrix with controlled amounts of rigid SiC particulate inclusions are shown in Fig. 11.11. In the experiments [12], fine-grained ZnO powder (particle size ≈0.4 μm) was mixed with coarse, dense SiC particles (size ≈14 μm) to produce mixtures with the required volume fraction of inclusions. After compaction to roughly the same matrix density, the composites were sintered at 700°C in a dilatometer. At the sintering temperature, no significant reaction was detected at the ZnO/SiC interface, so the inclusions can be considered to be inert. The inclusions produce a drastic reduction in the densification of the composite relative to that of the free ZnO matrix. For inclusion contents greater than ≈20 vol %, densification is almost completely inhibited.

Figure 11.12 shows data [13] for the sintering of a soda-lime glass powder (particle size ≈4 μm) containing controlled amounts of coarse, rigid SiC inclusions (particle size ≈35 μm). At the sintering temperature (600°C), the SiC particles did not react with the glass. For the glass matrix composite, the effect of the inclusions on the densification is considerably weaker than that observed above for the polycrystalline ZnO matrix.

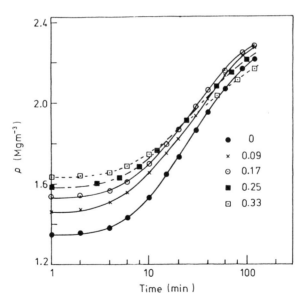

Figure 11.12 Density versus sintering time at 600°C for a soda-lime glass powder containing different volume factions of rigid SiC inclusions. (From Ref. 13.)

E. Comparison of Sintering Data with the Rule of Mixtures

The data of Figs. 11.11 and 11.12 can be manipulated to produce results suitable for comparison with the predictions of the rule of mixtures. The procedure is to fit smooth curves to the data and differentiate to find the densification rate at any time. Figure 11.13 shows the results of the comparison. According to the rule of mixtures, the ratio $\dot{\epsilon}_c/\dot{\epsilon}_{fm}$ should decrease linearly with increasing v_i. However, large deviations are observed for the ZnO matrix composite for v_i as low as 5 vol %. The glass matrix composite shows good agreement with the predictions of the rule of mixtures for v_i less than ≈ 20 vol %, but the deviations become increasingly large at higher values of v_i.

As a first approach, it may be argued that the rule of mixtures is inadequate to describe the data because the assumptions are too simplistic. It ignores key factors such as the following:

1. Transient stresses may exist due to differential sintering between the

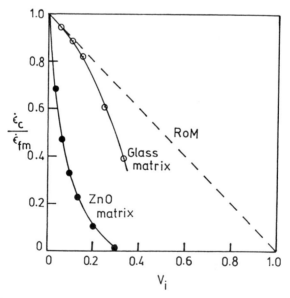

Figure 11.13 Comparison of the sintering data for the composites described in Figs. 11.11 and 11.12 with the predictions of the rule of mixtures (RoM). The densification rate of the composite relative to that for the free matrix is plotted versus the volume fraction of inclusions. Note the strong deviation from the predictions of the rule of mixtures for the ZnO matrix even at fairly small inclusion content.

Problems of Sintering

inclusions and the matrix. As discussed qualitatively earlier in this chapter, if the stresses are large enough they can reduce the densification and cause microstructural damage.
2. The inclusions will interfere and eventually form a touching network at some critical value of v_i. The formation of a network will inhibit densification, and if the network is rigid enough densification may stop completely.
3. The presence of the inclusions may disrupt the packing of the matrix phase, especially in the regions immediately surrounding the inclusions. Packing inhomogeneities, as we observed earlier, can severely retard the densification of a polycrystalline matrix.

We must analyze the effects of these and other possible factors to determine whether they can explain the observed deviations from the rule of mixtures.

F. Transient Stresses During Sintering

When one region of a sample (e.g., the matrix) shrinks at a different rate from a neighboring region (e.g., the inclusion), transient stresses are generated. However, creep or viscous flow will always seek to relieve the stresses. The calculation of the stresses therefore requires a time-dependent viscoelastic solution.

A viscoelastic material is one whose mechanical response to a stress (or strain) is a combination of a time-independent elastic response and a time-dependent viscous response. The reader is familiar with the two types of ideal behavior: the elastic solid and the viscous liquid. For an elastic solid, if the stress is linearly proportional to the strain, the behavior is described by Hooke's law, and the solid is said to be linearly elastic. For a liquid, if the shear stress is linearly proportional to the shear rate (the rate of change of the shear strain with time), the behavior obeys Newton's law of viscosity, and the liquid is said to be linearly viscous. A simple constitutive equation for the mechanical response of a linear viscoelastic solid is obtained by combining Hooke's law with Newton's law:

$$\sigma = Ge + \eta \frac{de}{dt} \tag{11.7}$$

where σ is the shear stress, G is the shear modulus, η is the viscosity, e is the shear strain, and t is the time. Both G and η are constant, independent of the strain or the strain rate. Hooke's and Newton's laws are normally obeyed for small strain or strain rate. When one or both of these laws are not obeyed, the material is no longer linearly viscoelastic.

Geometrical Model: The Composite Sphere

To determine the stresses, a geometrical model must be assumed. A model that has been commonly used is the composite sphere, in which the core represents the inclusion and the cladding represents the porous matrix. As sketched in Fig. 11.14, the composite containing a volume fraction v_i of inclusions is conceptually divided into composite spheres, each having a volume fraction v_i of core. We can consider a single composite sphere, with the assumption that the properties of the composite sphere are representative of the whole composite. When the inclusion is much larger than the particle size of the matrix, as in most practical composites, the matrix can be regarded, to a good approximation, as a continuum. With this assumption, phenomenological constitutive equations can be used for the matrix and microscopic considerations can be neglected. It should be recognized at this stage that the use of the composite sphere model means that the inclusions are assumed to be spherical.

The Stress Components

A shrinking cladding (matrix) around a core (inclusion) gives rise to (i) compressive stresses within the core and (ii) a compressive radial stress and tangential tensile stresses within the cladding [14]. If σ_r, σ_θ, and σ_ϕ are the components of the stress in spherical coordinates (Fig. 11.15), all

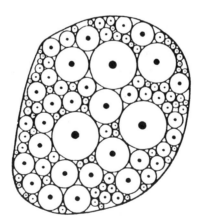

Figure 11.14 The composite sphere model. A composite containing spherical inclusions is conceptually divided into composite spheres (cross-sectional view). The core of each sphere is an inclusion, and the outer radius of the cladding of the sphere is chosen such that each sphere has the same volume fraction of inclusion as the whole body. (From Ref. 20, with permission.)

Problems of Sintering

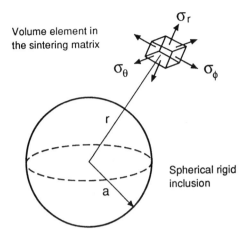

Figure 11.15 The stress components in a volume element at a distance r in the sintering matrix of a composite sphere. The radius of the inclusion is a, and the outer radius of the cladding (the matrix) is b.

three stress components in the inclusion are equal:

$$\sigma_{ri} = \sigma_{\theta i} = \sigma_{\phi i} = \sigma_i \quad (11.8)$$

where σ_i is the mean hydrostatic stress in the inclusion. Therefore, the inclusion is under purely hydrostatic stresses. In the convention that compressive stresses and strains are negative quantities, σ_i is negative. The stresses in the matrix are given by [14]

$$\sigma_{rm} = -\frac{v_i \sigma_i}{1-v_i}\left(1 - \frac{b^3}{r^3}\right) \quad (11.9)$$

$$\sigma_{\theta m} = \sigma_{\phi m} = -\frac{v_i \sigma_i}{1-v_i}\left(1 + \frac{b^3}{r^3}\right) \quad (11.10)$$

where a and b are the inner and outer radii, respectively, of the composite sphere and v_i is the volume fraction of inclusions, given by

$$v_i = a^3/b^3 \quad (11.11)$$

The stresses in the matrix have their maximum values at the inclusion/matrix interface (i.e., $r = a$) and decrease as $1/r^3$. If $\sigma_{rm}(a)$ is the radial stress and $\sigma_{\theta m}(a)$ is the tangential (hoop) stress in the matrix at the interface, then from Eqs. (11.9–11.11) we obtain

$$\sigma_{rm}(a) = \sigma_i \tag{11.12}$$

$$\sigma_{\theta m}(a) = -\sigma_i \frac{1 + 2v_i}{2(1 - v_i)} \tag{11.13}$$

The mean hydrostatic stress in the matrix is

$$\sigma_m = \frac{1}{3}(\sigma_{rm} + \sigma_{\theta m} + \sigma_{\phi m}) = -\sigma_i \left(\frac{v_i}{1 - v_i}\right) \tag{11.14}$$

According to this equation, the mean hydrostatic stress in the matrix is uniform, i.e., it is independent of r. Furthermore, the matrix is under both hydrostatic and tangential stresses. Since σ_i is compressive (i.e., negative), σ_m is tensile (positive).

Transient Stresses: Effect on Sintering

The densification rate of the matrix is affected by the hydrostatic component of the stress in the matrix. For a free matrix (i.e., containing no inclusions), the linear densification rate (or strain rate) can be written [see Eq. (8.96)]

$$\dot{\epsilon}_{fm} = \Sigma/K_m \tag{11.15}$$

where Σ is the sintering stress and K_m is the densification or bulk viscosity. Both $\dot{\epsilon}_{fm}$ and Σ are negative quantities. For the matrix phase of the composite, the linear densification rate is

$$\dot{\epsilon}_m = \frac{\Sigma + \sigma_m}{K_m} = \frac{1}{K_m}\left(\Sigma - \sigma_i \frac{v_i}{1 - v_i}\right) \tag{11.16}$$

Since σ_i is compressive (negative), its effect is to reduce the densification rate of the matrix. The tensile hoop stress $\sigma_{\theta m}$ may influence the development of radial cracks in the matrix (see Fig. 11.1b).

To summarize, if the transient stresses generated by the presence of the inclusions were the only significant factor, the important effects in the sintering of composites would be determined by the value of σ_i. An important issue, therefore, is the calculation of σ_i.

G. Calculation of the Stresses and Strain Rates

Several models have been used to determine analytically the transient stresses and their effects on sintering. A common approach is to determine the elastic solution to the stress problem and then transform it into the time-dependent viscoelastic solution. If the material is linearly viscoelas-

Problems of Sintering

tic, the calculation of the stresses is greatly facilitated by the viscoelastic analogy, discussed in detail elsewhere [15,16]. In the viscoelastic analogy, the general time-dependent solutions are obtained by taking the Laplace transform of the elastic solutions and then inverting the solutions into real-time space. The equations for the viscoelastic stress exhibit the general form [18]

$$\sigma(t) = \int_0^t \dot{\epsilon}(u) \, F(G, K, \eta, t - u) \, du \tag{11.17}$$

where $\dot{\epsilon}$ is the linear strain rate; F is a stress relaxation function; G, K, and η are the shear modulus, bulk modulus, and viscosity, respectively, of the matrix; and t is the time. Determination of the stresses is dependent on choosing the form of F. Below, we outline two models that have employed this approach.

The Model of Raj and Bordia

An analysis by Raj and Bordia [2] assumed explicit spring–dashpot elements to represent the constitutive properties of a porous material. Such spring–dashpot elements are commonly used to analyze the stress–strain response of polymeric materials, an important class of viscoelastic materials [17]. The viscoelastic analogy was used to calculate the time-dependent stresses from the elastic solutions. The shear or creep response was represented by a Maxwell element (Fig. 11.16), where G_0 is the shear modulus

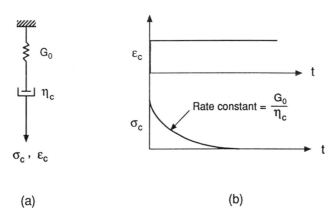

Figure 11.16 (a) The Maxwell element used in the model of Raj and Bordia [2] to represent the creep or shear response of a porous solid; (b) the response of the Maxwell element to a step strain.

and η_c is the shear viscosity of the matrix. If a step strain ϵ is applied to the Maxwell element, the stress decays according to

$$\sigma_c = \sigma_0 \exp(-t/\tau_c) \tag{11.18}$$

where σ_0 is the initial value of the stress and τ_c is a relaxation time given by

$$\tau_c = \eta_c/G_0 \tag{11.19}$$

Raj and Bordia also assumed that the densification response of the matrix can be represented by a Kelvin–Voigt element, shown in Fig. 11.17, where K_0 is the bulk modulus and η_b is the bulk viscosity of the matrix. The use of the Kelvin–Voigt element is purely phenomenological in that typical

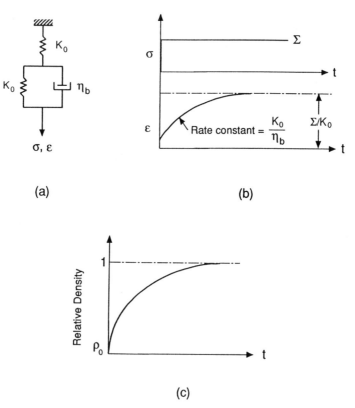

Figure 11.17 (a) The Kelvin–Voigt element used in the model of Raj and Bordia [2] to represent the densification response; (b) the strain as a function of time for the element under a step load Σ; (c) a typical densification curve.

Problems of Sintering

isothermal sintering curves have a shape similar to the response of the Kelvin–Voigt element to a step load. If a constant stress, taken to be the sintering stress Σ, is applied to the element, the strain increases according to

$$\epsilon = \frac{\Sigma}{K_0}\left[1 - \exp\left(-\frac{t}{\tau_b}\right)\right] \tag{11.20}$$

where τ_b is a relaxation time given by

$$\tau_b = \eta_b/K_0 \tag{11.21}$$

A nondimensional parameter $\beta = \tau_b/\tau_c$ characterizes the relative rates of creep and densification. Raj and Bordia derived all of their results in terms of β. Stress development is seen as a balance between stress generation by differential densification and stress relaxation by creep of the matrix. For $\beta \gg 1$, stress relaxation is significant, so the buildup of the transient stresses is small. Conversely, if $\beta \ll 1$, the stress relaxation is insignificant and large stresses build up.

As outlined earlier, the greatest interest from the point of view of sintering is the magnitude of σ_i, equal to $\sigma_{rm}(a)$. According to the model of Raj and Bordia, the maximum interfacial stress is given by

$$\left[\frac{\sigma_{rm}(a)}{\Sigma}\right]_{max} = \frac{4(1 - v_i)}{4v_i + 9\beta} \tag{11.22}$$

A plot of the left-hand side of this equation as a function of β for various values of v_i is shown in Fig. 11.18. Large values of the interfacial stress are predicted for small values of β and v_i. Using Eqs. (11.14) and (11.22), we obtain

$$\left[\frac{\sigma_m}{\Sigma}\right]_{max} = -\frac{1}{1 + 9\beta/4v_i} \tag{11.23}$$

According to this equation, the hydrostatic tension in the matrix is predicted to be comparable to the sintering stress when β is small. In this case, a drastic reduction in the densification rate is predicted for even small values of v_i.

In their model, Raj and Bordia assumed that the elastic moduli (G_0 and K_0) and the viscosities (η_c and η_b) are constants. The assumption of constant values for these parameters means that the material is linearly viscoelastic and the viscoelastic analogy can be correctly used to determine the time-dependent stresses. However, the assumption is physically unreasonable because it is known that the modulus and viscosity vary significantly during sintering.

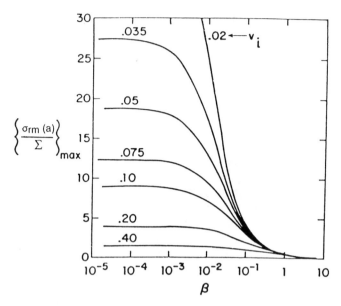

Figure 11.18 A plot of the value of the maximum interfacial stress $\sigma_{rm}(a)$ normalized by the sintering stress Σ versus β for different values of the inclusion volume fraction v_i. (From Ref. 2, reprinted with permission of Elsevier Science Ltd.)

The Model of Hsueh et al.

As in Raj and Bordia's model, Hsueh et al. [18] used the viscoelastic analogy and assumed that the creep response of the matrix can be represented by a Maxwell element. However, they used constitutive laws derived from experimental data to describe the density (or time) dependence of the densification rate, the bulk and shear modulus, and the viscosity. While the use of density-dependent parameters is physically attractive, it creates the problem that the material is no longer *linearly* viscoelastic and makes the use of the viscoelastic analogy invalid. Nevertheless, the model predicted large values for the interfacial stresses $\sigma_{rm}(a)$, which were typically more than 100 times the sintering stress Σ. As discussed earlier, interfacial stresses of this magnitude can severely reduce the densification rate of the matrix and cause substantial sintering damage.

Applicability of the Models

In a series of papers, Bordia and Scherer [19] reexamined the models of Raj and Bordia and Hsueh et al. They showed that the large stresses predicted by these models result from incompatible expressions for the parameters that yield negative values for Poisson's ratio. While a negative value for Poisson's ratio is not thermodynamically forbidden, it is not observed experimentally when porous solids are sintered under an applied stress.

As discussed earlier, the model of Raj and Bordia predicted large values for the interfacial stresses when β is small. Bordia and Scherer showed that Poisson's ratio becomes negative for $\beta < n$, where n is given by

$$n = \frac{2G_0}{3K_0} = \frac{1 - 2\nu_0}{1 + \nu_0} \tag{11.24}$$

In this equation, ν_0 is the elastic Poisson's ratio, as determined by the elastic moduli. Since the stresses increase as β decreases, the highest possible stress is obtained for the smallest allowed value of β, i.e., $\beta = n$. In their calculations, Raj and Bordia assumed that $\nu_0 = 0.35$ for the porous matrix, which corresponds to $n = 2/9$. If this value for β is substituted into Eq. (11.22), we find that $\sigma_{rm}(a)/\Sigma = 2$ when $\nu_i = 0$. Therefore, for $\nu_0 = 0.35$, the maximum allowable interfacial stress is only twice the sintering stress.

In the model of Hsueh et al., two different sets of experimental data were used to determine η_c and η_b. The samples used in the two experiments differed significantly in grain size and microstructure. The difference in grain size was not compensated for, which resulted in an overestimation in the ratio η_c/η_b and a negative Poisson's ratio.

To summarize at this stage, the large interfacial stresses predicted by the models of Raj and Bordia and Hsueh et al. result from incompatible expressions for the parameters that lead to a negative Poisson's ratio. In the next section, we describe a model for which the constitutive equations are derived in a consistent manner. As we shall see, the model leads to positive values of Poisson's ratio and predicts stresses that are significantly lower than those predicted by the two models considered above.

H. Scherer's Model for Sintering with Rigid Inclusions

It is reasonable to assume that a porous sintering material is viscoelastic because the response to an external stress consists of an elastic strain and a time-dependent deformation due to creep or viscous flow. However, the elastic strain is very much smaller than the strains observed during densification, so the observed deformation results almost entirely from creep or viscous flow. Even if the viscoelastic analogy were applicable, its use would be unnecessary. We are justified in neglecting the elastic response and considering only the viscous deformation. Instead of the viscoelastic analogy, we can now use the simpler elastic–viscous analogy, where the equations for the time-dependent viscous deformation are found from those for the elastic solution. This approach, based on a consideration of the purely viscous deformation of the porous sintering material, was adopted by Scherer [20]. Scherer's analysis is discussed in some detail below because of the significant role it played in the development of our present understanding of sintering with rigid inclusions.

The Geometrical Model

Two models were considered by Scherer: the composite sphere model described earlier and a self-consistent model [21]. In the self-consistent model, a microscopic region in the matrix is regarded as an island of sintering material in a continuum (the composite) that contracts at a slower rate. The mismatch in sintering rates causes stresses that influence the densification rate of each region. Compared to the composite sphere, the self-consistent model is expected to be valid over a larger range of inclusion volume fraction v_i. However, for v_i less than ≈ 20 vol %, a useful range in practice, the predictions of the two models are indistinguishable. Because of the limited space available in this chapter, we will describe the analysis based on the composite sphere model only. However, as outlined later, the equations for the stresses and strains in the self-consistent model differ only in that the shear viscosity of the *matrix* in the composite sphere model is replaced by the shear viscosity of the *composite*.

Calculation of the Stresses and Strain Rates

The first stage in the calculation of the stresses and strains in the sintering material is to choose a constitutive equation that relates the applied stresses to the resulting strains. For an elastic material, the behavior is described by two independent elastic constants such as the shear modulus G and the bulk modulus K. For an isotropic linear elastic solid, the constitutive equation has the form [14]

$$\epsilon_x = \epsilon_f + \frac{1}{E}[\sigma_x - \nu(\sigma_y + \sigma_z)] \qquad (11.25)$$

where ϵ_x is the strain in the x direction; σ_x, σ_y, and σ_z are the stresses in the x, y, and z directions, respectively; E is Young's modulus; ν is Poisson's ratio; and ϵ_f is the free strain, i.e., the strain that would be produced in the absence of local stresses. Young's modulus E and Poisson's ratio ν are an alternative pair of elastic constants that are related to G and K by

$$G = \frac{E}{2(1+\nu)}, \qquad K = \frac{E}{3(1-2\nu)} \qquad (11.26)$$

Problems of Sintering

For an isotropic, linearly viscous, incompressible material, the constitutive equation is easily obtained from Eq. (11.25) by invoking the elastic-viscous analogy: the strain is replaced by the strain rate, E is replaced by three times the shear viscosity η, and v becomes 1/2 (for an incompressible material), giving

$$\dot{\epsilon}_x = \dot{\epsilon}_f + \frac{1}{3\eta}\left[\sigma_x - \frac{1}{2}(\sigma_y + \sigma_z)\right] \tag{11.27}$$

where the dot denotes the derivative with respect to time. For a porous material, Eq. (11.27) must be modified to allow for the compressibility of the pores; that is,

$$\dot{\epsilon}_x = \dot{\epsilon}_f + \frac{1}{E_m}[\sigma_x - v_m(\sigma_y + \sigma_z)] \tag{11.28}$$

where E_m is the uniaxial viscosity of the matrix, which varies from 0 to 3η as the relative density of the matrix ρ_m goes from 0 to 1, and v_m is the Poisson's ratio of the matrix, which varies from 0 to 1/2 as ρ_m goes from 0 to 1. As ρ_m approaches 1, Eq. (11.28) becomes identical to Eq. (11.27). By analogy with Eq. (11.25), the shear viscosity G_m and the bulk viscosity K_m are related by

$$G_m = \frac{E_m}{2(1 + v_m)}, \quad K_m = \frac{E_m}{3(1 - 2v_m)} \tag{11.29}$$

Examination of Eqs. (11.28) and (11.29) indicates that the solutions for the sintering material are obtained from those for the elastic material, Eqs. (11.25) and (11.26), by replacing G and K by G_m and K_m and replacing the strains by the strain rates.

As outlined earlier, the sintering problem of mismatched shrinkage rates is analogous to the problem of thermal expansion mismatch in a composite sphere. By adopting the elastic solution for the thermal stress problem and using the elastic-viscous analogy to transform it to the time-dependent solution for the sintering problem, Scherer derived the following equations for the stresses and strain rates:

$$\sigma_i = (1 - v_i)K_{cs}\dot{\epsilon}_{fm} \tag{11.30}$$

$$\sigma_{rm} = \left[\left(\frac{a}{r}\right)^3 - v_i\right]K_{cs}\dot{\epsilon}_{fm} \tag{11.31}$$

$$\sigma_{\theta m} = -\left[\frac{1}{2}\left(\frac{a}{r}\right)^3 + v_i\right]K_{cs}\dot{\epsilon}_{fm} \tag{11.32}$$

where a is the inner radius of the composite sphere, and K_{cs} is given by

$$K_{cs} = \frac{1}{1/4G_m + v_i/3K_m} \tag{11.33}$$

The free strain rate (i.e., the linear densification rate) is given by

$$\dot{\epsilon}_{fm} = \frac{1}{3}\left(\frac{\Sigma}{K_m}\right) \tag{11.34}$$

where Σ is the sintering stress of the matrix. The hydrostatic stress, defined by Eq. (11.14), can be calculated from Eqs. (11.30), (11.33), and (11.34), giving

$$\frac{\sigma_m}{\Sigma} = -\frac{v_i}{v_i + 3K_m/4G_m} \tag{11.35}$$

According to this equation, as v_i approaches zero, so does σ_m. Furthermore, K_m and G_m are positive, so σ_m is always smaller than Σ.

The stress in the inclusion can be compared with the sintering stress by using Eqs. (11.30), (11.33), and (11.34). We obtain

$$\frac{\sigma_i}{\Sigma} = \frac{1 - v_i}{v_i + 3K_m/4G_m} \tag{11.36}$$

Using Eq. (11.29), as v_i approaches zero, the stress in an isolated inclusion is

$$\frac{\sigma_i}{\Sigma} = \frac{2(1 - v_m)}{1 + v_m} \tag{11.37}$$

This equation shows that σ_i/Σ can be greater than 2 only when $v_m < 0$. As outlined earlier, negative values of Poisson's ratio have not been observed experimentally. The absolute *maximum* value of σ_i predicted by Scherer's model is therefore 2Σ.

The maximum stresses in the matrix occur at the interface between the matrix and the inclusion. The radial stress is $\sigma_{rm}(a) = \sigma_i$, so it is also predicted to be less than 2Σ. The circumferential or hoop stress is found from Eqs. (11.32)–(11.34) to be

$$\frac{\sigma_{\theta m}(a)}{\Sigma} = -\frac{1/2 + v_i}{v_i + 3K_m/4G_m} \tag{11.38}$$

When v_i approaches zero, Eq. (11.38) becomes

$$\frac{\sigma_{\theta m}(a)}{\Sigma} = -\frac{1 - 2v_m}{1 + v_m} \tag{11.39}$$

Problems of Sintering

This equation shows that $\sigma_{\theta m}(a)$ can be no greater in magnitude than Σ unless $v_m < 0$. The analysis by Scherer is seen to predict transient stresses that are considerably smaller than those calculated by Raj and Bordia and by Hsueh et al.

The effect of the stresses on the densification rate of the composite can also be calculated from Scherer's model. The linear densification rate of the composite is predicted to be

$$\dot{\epsilon}_c = \frac{(1 - v_i)K_{cs}\dot{\epsilon}_{fm}}{4G_m} \tag{11.40}$$

According to the rule of mixtures, the linear densification rate of the composite, $\dot{\epsilon}_c^{rm}$, is given by Eq. (11.6). Using Eqs. (11.6), (11.33), and (11.40), we find that

$$\frac{\dot{\epsilon}_c}{\dot{\epsilon}_c^{rm}} = \frac{1}{1 + v_i\,(4G_m/3K_m)} \tag{11.41}$$

Since the denominator is always greater than 1, this equation predicts that the linear densification rate is lower than that predicted by the rule of mixtures. From Eq. (11.29), the ratio $4G_m/3K_m$ is given by

$$\frac{4G_m}{3K_m} = \frac{2(1 - 2v_m)}{1 + v_m} \tag{11.42}$$

The maximum value of this ratio is 2, which occurs when $v_m = 0$. For v_i less than ≈ 0.05, the maximum deviation of the composite densification rate from that predicted by the rule of mixtures is $\approx 10\%$. Returning to the data shown in Fig. 11.13, this predicted maximum deviation is still considerably smaller than that observed for polycrystalline matrix composites.

The equations for the self-consistent model, as we outlined earlier, can be obtained from those for the composite sphere model simply by replacing G_m with G_c, the shear viscosity of the composite. Taking Eq. (11.41), the corresponding equation for the self-consistent model is

$$\frac{\dot{\epsilon}_c}{\dot{\epsilon}_c^{rm}} = \frac{1}{1 + v_i\,(4G_c/3K_m)} \tag{11.43}$$

An approximation to G_c can be obtained from the Hashin–Shtrikman equation, which, for a viscous matrix, is given by [20]

$$G_c = G_m\left[1 + \frac{15}{2}\left(\frac{v_i}{1 - v_i}\right)\left(\frac{1 - v_m}{4 - 5v_m}\right)\right] \tag{11.44}$$

Viscous Sintering with Rigid Inclusions

In the case of viscous sintering where the matrix phase of the composite is a glass, explicit expressions can be derived for the moduli and Poisson's ratio [22]. The microstructure of the matrix is approximated by a model (described in Chapter 8) consisting of cylinders that intersect at right angles. Poisson's ratio is given to a good approximation by

$$v_m = \frac{1}{2}\left(\frac{\rho_m}{3 - 2\rho_m}\right)^{1/2} \tag{11.45}$$

where ρ_m is the relative density of the matrix. The uniaxial viscosity is given by

$$E_m = \frac{3\eta\rho_m}{3 - 2\rho_m} \tag{11.46}$$

Using Eqs. (11.42) and (11.45), the ratio $4G_m/3K_m$ can be calculated in terms of the relative density of the matrix. This ratio can then be substi-

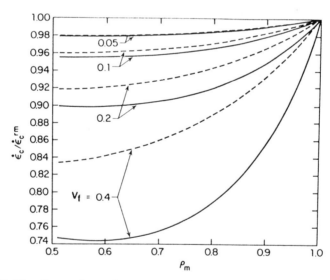

Figure 11.19 Comparison of the predictions of Scherer's theory of sintering with rigid inclusions with the predictions of the rule of mixtures. The linear strain rate of the composite normalized by the strain rate from the rule of mixtures [Eq. (11.6)], $\dot{\epsilon}_c/\dot{\epsilon}_c^{rm}$, is plotted versus the relative density of the matrix, ρ_m, for the indicated volume fraction of inclusions, v_f. The dashed curves represent the composite sphere model [Eq. (11.41)], and solid curves the self-consistent model [Eq. (11.43)]. (From Ref. 20, with permission.)

Problems of Sintering

tuted into Eqs. (11.36), (11.37), and (11.41) to calculate the stresses and densification rates for the composite sphere model. Alternatively, for the self-consistent model, G_c can be found from Eq. (11.44) and the same procedure repeated to determine the stresses and strain rates. Figure 11.19 shows the predicted values for the linear densification rate of the composite relative to that predicted by the rule of mixtures for the composite sphere and self-consistent models. For v_i less than ≈ 20 vol %, the predictions are almost identical; however, the deviations increase for higher values of v_i. Furthermore, for v_i less than ≈ 10–15 vol %, the predicted values of $\dot{\epsilon}_c$ are not very different from those of $\dot{\epsilon}_c^{rm}$, so the rule of mixtures is accurate enough for these values of v_i.

An experimental test of Scherer's theory of viscous sintering with rigid inclusions was made using the data described earlier in Fig. 11.12. As shown in Fig. 11.20 for the ratio $\dot{\epsilon}_c/\dot{\epsilon}_c^{rm}$, the theory performs well for v_i less than ≈ 15 vol % but overestimates the sintering rate at higher inclusion content.

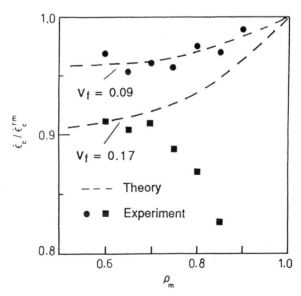

Figure 11.20 Comparison of the sintering data for the glass matrix composites described in Fig. 11.12 with the predictions of Scherer's theory for sintering with rigid inclusions [Eq. (11.43)]. The linear strain rate of the composite normalized by the strain rate from the rule of mixtures [Eq. (11.6)], $\dot{\epsilon}_c/\dot{\epsilon}_c^{rm}$, is plotted versus the relative density of the matrix, ρ_m, for the indicated volume fraction of inclusions, v_f. Note the deviation starting near $\rho_m = 0.75$ for the composite with $v_f = 0.17$. (From Ref. 13.)

To summarize, Scherer's model predicts that the transient stresses generated in sintering with rigid inclusions are fairly small. The model accounts very well for the observed densification rates in glass matrix composites when v_i is less than ≈ 15 vol %. However, a consideration of the transient stresses alone cannot account for the sintering of glass matrix composites at higher values of v_i or the sintering of polycrystalline matrix composites even at fairly low values of v_i (e.g., less than ≈ 5 vol %). Other factors must also play a role in hindering the sintering rates. As outlined earlier, a possible factor is that the inclusions may interfere and eventually form a network that resists shrinkage.

I. Percolation and Network Formation

As the volume fraction of inclusions increases, we reach a stage in which the inclusions become so numerous that they form enough inclusion–inclusion contacts to produce a continuous network extending all the way across the sample. We can say that the network of inclusions percolates through the sample, rather like water percolating through ground coffee in a coffeemaker. The stage where the inclusions first form a percolating network is called the *percolation threshold*.

The formation of a continuous network of inclusions has important consequences for sintering as well as for other properties of the system. If the inclusions are electrically conducting, the structure can carry an electric current as soon as the percolating network forms (Fig. 11.21). If the contacts between the inclusions are rigid, the structure will be mechanically rigid. The rigidity of the structure is important for sintering. Increased stiffness of the structure will retard sintering. If the structure is completely rigid, sintering will stop. The models considered so far do not take into account the formation of a percolating network of inclusions.

The Concept of Percolation

Percolation has been considered in some detail by Zallen [23]. The concept of percolation is illustrated in Fig. 11.22 using a triangular two-dimensional lattice [24]. If we place particles on sites at random, larger and larger clusters of adjoining particles will be formed as the number of particles increases. Eventually, one of these clusters, referred to as the percolating cluster or the spanning cluster, becomes large enough to extend all the way across the lattice. The fraction of sites that must be filled before the percolating cluster appears is called the *percolation threshold* p_c.

The percolation threshold depends on the shape and dimensionality of the lattice. For the triangular lattice shown in Fig. 11.22, the percolation threshold is reached when half of the sites are occupied, i.e., $p_c = 0.5$.

Problems of Sintering

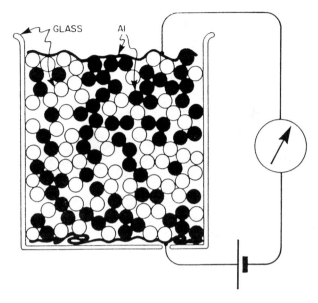

Figure 11.21 Sketch of an experiment to illustrate some elementary aspects of percolation in a mixture of aluminum and glass balls. When the percolation threshold for the aluminum balls is reached, the system will conduct an electric current. (From Ref. 23, with permission.)

For a square two-dimensional lattice, $p_c = 0.593$, while for a cubic lattice, $p_c = 0.311$. For a powder system undergoing sintering, no lattice is present. However, it has been shown that the percolation threshold occurs at a certain volume fraction (or area fraction in two dimensions) regardless of the nature of the lattice. In three dimensions, the percolation threshold is reached when the volume fraction of particles is ≈16 vol % even when no lattice is present. If, for example, glass balls are mixed with aluminum balls as sketched in Fig. 11.21, the structure will become electrically conducting when the aluminum balls occupy ≈16 vol % of the space. If the glass and aluminum balls are of the same size, percolation will occur when the number fraction of the aluminum balls is ≈0.27. This is because the particles occupy ≈64% of the total volume of the structure (i.e., the packing density for dense random packing) and the volume fraction of the aluminum balls (0.27 x 0.64) must equal 0.16. If glass and aluminum balls of the same size were placed at random on a simple cubic lattice, percolation would occur when the number fraction of the aluminum balls was ≈0.311. In this case, the packing density for the simple cubic lattice is 0.52, and 0.52 x 0.311 is approximately 0.16.

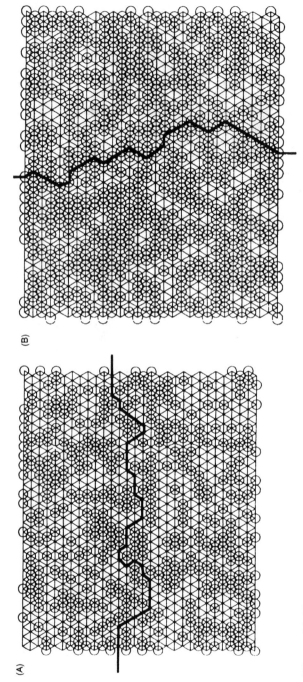

Figure 11.22 Illustration of percolation on a triangular lattice. When 40% of the sites are filled, no single cluster of particles extends all the way across the lattice, as indicated by the tick line in (A). When >50% of the sites are filled, a "spanning cluster" appears; then, as indicated by the thick line in (B), there is a path of occupied sites completely crossing the lattice. (From Ref. 24, reprinted by permission of the American Ceramic Society.)

Problems of Sintering

The percolation threshold of ≈0.16 applies to equiaxial particles in a random arrangement. This value will change if the particles agglomerate or repel one another. The percolation threshold also depends on the aspect ratio and orientation of the particles: p_c decreases with increasing aspect ratio but increases if the particles become aligned.

Effect of Percolation on Sintering
The effect of percolation on the rigidity of the composite is an important factor because it will influence the sintering kinetics. If the inclusions bond together on contact, the rigidity of the composite will increase dramatically near the percolation threshold and the sintering rate will show a corresponding decrease. Even before the percolation threshold, large clusters of inclusions will form, and these will have a significant effect locally, even before the overall stiffening produced by the percolating cluster.

As the results of Fig. 11.12 indicate, it is not impossible to densify composites with an inclusion volume fraction greater than the percolation threshold (≈16 vol %). In fact, some practical composites with an inclusion content significantly greater than the percolation threshold can be sintered to almost full density. The reason is that the inclusions do not bond together to form a rigid network. In many cases, particularly for glass matrices and systems with a significant amount of liquid phase at the sintering temperature, the glass or the liquid phase wets the contacts between the inclusions, providing a lubricating layer. When the inclusion–inclusion contacts are wetted, the stage where the system becomes fairly rigid, referred to as the *rigidity threshold*, is expected to occur at inclusion contents significantly higher than the percolation threshold.

To summarize at this stage, the nature of the interactions between the inclusions will have a significant effect on sintering. If the inclusions form strong bonds on contact (e.g., if they are not wetted by the matrix) or interlock (e.g., because of surface roughness), the densification rate will be reduced significantly near the percolation threshold. However, if the contacts between the inclusions are lubricated, higher inclusion contents can be accommodated without a drastic reduction in the densification rate.

Numerical Simulations
Numerical simulations can be extremely valuable in exploring the effects of particle interactions on the sintering of composites. Work in this area has been pioneered by Scherer and Jagota [24–26], who used a finite-element method to simulate the sintering of viscous matrices containing rigid inclusions. In a three-dimensional packing of spherical particles, ran-

domly selected particles are assigned to be inclusions and the rest are chosen to be matrix particles with the same size and surface tension as the inclusions. As sketched in Fig. 11.23, the composite contains three types of contacts: inclusion–inclusion (i–i), matrix–matrix (m–m), and inclusion–matrix (i–m). The results of the finite-element simulation are found to depend on (i) the degree of bonding between the inclusions and (ii) the ratio N of the viscosity of the inclusions to the viscosity of the matrix particles. For rigid inclusions ($N = 10^6$) that form strong inclusion–inclusion bonds, and for wetted inclusion–matrix contacts, Fig. 11.24 shows the results of a finite-element simulation plotted in the form of the ratio $4G_c/3K_m$ versus the volume fraction of inclusions. Near the percolation threshold, the viscosity rises sharply by a factor of $\approx 10^6$ characteristic of the inclusions themselves. Also shown for comparison are the predictions of two other models that were described earlier: the Hashin–Shtrikman model [Eq. (11.44)] and the self-consistent model. These two models do not account for the rapid rise in the viscosity near the percolation threshold. They do show a rapid rise in the viscosity reminiscent of a percolation threshold, but it appears at too high a volume fraction.

As discussed earlier, models that neglect the interactions between the inclusions overestimate the sintering rate of glass matrix composites near or above the percolation threshold (see Fig. 11.20). In one attempt to incorporate the effects of percolation, Dutton and Rahaman [27] measured the viscosity of a soda-lime glass matrix composite and used the data to provide an estimate of G_c in the model equation for the sintering rate, e.g., Eq. (11.43). The use of the measured values is expected to provide a more realistic estimate of G_c than that predicted by the Hashin–Shtrikman equation, especially near and above the percolation threshold. With the measured values of G_c, Eq. (11.43) was indeed found to provide a better fit to the sintering data than when the Hashin–Shtrikman expression for G_c was used.

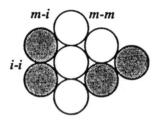

Figure 11.23 Contacts between inclusions (shaded) and matrix particles (white). (From Ref. 25, with permission.)

Problems of Sintering

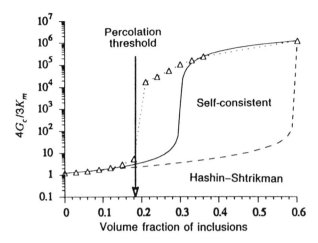

Figure 11.24 Shear viscosity of the composite, G_c, normalized by the bulk viscosity of the matrix, K_m, as a function of the volume fraction of inclusions (0.6 times the number fraction). Calculations by the finite-element method ($\triangle$), self-consistent model (solid line) and Hashin–Shtrikman lower bound (dashed line) are shown. (From Ref. 25, with permission.)

To summarize, when the effects of transient stresses and network formation between the inclusions are taken into account, the models appear to provide a reasonably good description of viscous sintering with rigid inclusions. The explanation of the sintering of polycrystalline matrix composites has proved considerably more difficult, and we consider the problem separately in the next section.

J. Sintering of Polycrystalline Ceramic Matrix Composites

Polycrystalline ceramic matrix composites, as discussed earlier, show severely reduced sintering rates even at fairly low inclusion volume fractions (less than ≈ 5 vol %). The effect cannot be explained in terms of the transient stresses generated during sintering because these stresses have been shown to be small. Another explanation, based on the development of cracklike voids resulting from the circumferential or hoop stress at the inclusion–matrix boundary (see Fig. 11.1b), is also considered to be inappropriate on account of the small transient stresses. Whereas some sintering damage has been observed in a few cases, it is not a significant factor.

The formation of a percolating network of inclusions, as noted earlier, can have a significant factor on sintering. The problem with polycrystalline matrix composites is that the drastically reduced sintering rates are observed far below the percolation limit. One of the earliest models for the effect of a constraining network of inclusions on sintering was put forward by Lange [28]. The model is purely geometric in origin and should therefore be equally applicable to glass and polycrystalline matrix composites. However, as Fig. 11.13 shows, there is a great difference in the sintering behavior of the two types of composites. Later a more complex model was proposed by Lange and coworkers [29,30], based on observations of the microstructural evolution of Al_2O_3 matrix composites containing ZrO_2 inclusions. Figure 11.25 shows a sketch of the important features of the model. The stress field due to differential densification leads to premature densification in certain regions between the inclusions, and these regions support grain growth (Fig. 11.25a). For a number of inclusions (Fig. 11.25b), the premature densification and associated grain growth lead to

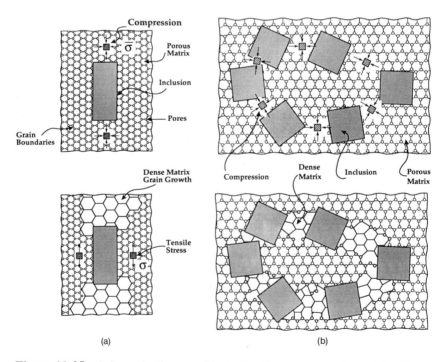

Figure 11.25 Schematic diagrams illustrating the sequence of events leading to the retardation of sintering and damage in bodies containing inclusions. (a) Single inclusion; (b) multiple inclusions. (According to Ref. 30, with permission.)

Problems of Sintering

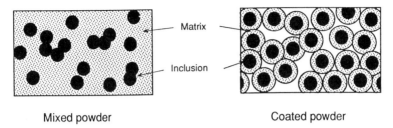

Mixed powder   Coated powder

Figure 11.26 Schematic diagrams illustrating the distribution of inclusions (cross-sectional view) in composites formed from (a) a mixture of the inclusions and powder matrix prepared by ball milling and (b) coated inclusions prepared by chemical precipitation.

the development of a dense annulus that resists densification. Severely reduced densification rates can therefore occur even for a fairly sparse network of nontouching inclusions (i.e., low inclusion volume fraction).

The importance of network formation between the inclusions and packing inhomogeneities in the powder matrix for retarding the sintering

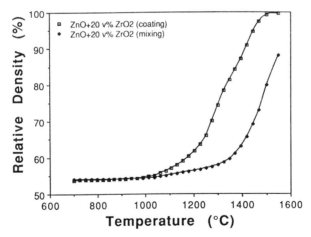

Figure 11.27 Data for relative density versus temperature during sintering at a constant heating rate (5°C/min) for composites formed from coated powders and from mechanically mixed powders. The system consisted of a ZnO matrix and 20 vol % ZrO_2 inclusions. Note the significantly higher density of the composite formed from the coated powders. (From Ref. 31.)

of polycrystalline matrix composites has been clearly demonstrated by the work Hu and Rahaman [31]. In these experiments, composites consisting of a polycrystalline ZnO matrix and inert, rigid inclusions of ZrO_2 were prepared by two separate methods. In one method, the matrix and inclusions were mixed mechanically in a ball mill, and in the other, individual ZrO_2 inclusions were coated with the required thickness of ZnO powder by chemical precipitation from solution. The mechanically mixed powder and the coated powder were consolidated and sintered under nearly identical conditions. A sketch of the two types of powders is shown in Fig. 11.26. The significance of the experiments is that for the coated powders (i) the inclusion particles are separated from one another by a layer of matrix so network formation is minimized and (ii) each inclusion is surrounded by a homogeneous layer of matrix so packing inhomogeneities, especially in the matrix immediately surrounding the inclusions, are significantly reduced. The sintering kinetics of the coated powders were observed to be similar to those of the unreinforced ZnO and to be significantly greater than those for the mechanically mixed powder (Fig. 11.27). The difference in microstructure between the composites fabricated from the coated powder and from the mechanically mixed powder is revealed in Fig. 11.28. Compared to the composite fabricated from the coated powder, which is almost fully dense, the composite fabricated from the mechanically mixed powder has clearly identifiable regions of porosity immediately surrounding the inclusions. In further work, the benefits of using coated powders for the fabrication, by sintering, of dense composites with up to $\approx 35-40$ vol % of particulate, whisker, or platelet reinforcement were clearly demonstrated [31].

To summarize, the sintering of polycrystalline matrix composites is less clearly understood than that of glass matrix composites. Experiments demonstrate that network formation between the inclusions and packing inhomogeneities of the matrix, particularly in regions immediately surrounding the inclusion, are the significant factors that retard the sintering of polycrystalline matrix composites. When these factors are alleviated, as by the use of coated powders, dense polycrystalline matrix composites containing an appreciable volume fraction of inclusions can be fabricated by solid-state sintering.

11.4 CONSTRAINED SINTERING: II. THIN FILMS

A variety of applications, particularly in the areas of electronic and optical ceramics, require the sintering of porous thin films on a substrate. Normally the film adheres to the substrate but is too thin to cause it to deform. The substrate can therefore be considered to be rigid. If the film remains

Problems of Sintering

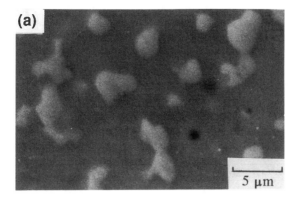

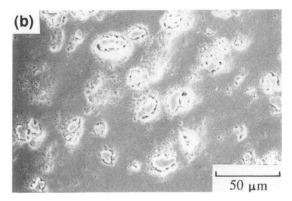

Figure 11.28 Scanning electron micrographs of sintered composites formed from (a) coated powders and (b) mechanically mixed powders. Note the dense matrix in (a) and the porosity, especially in the regions immediately surrounding the inclusions, in (b).

attached to the substrate during sintering and does not crack, shrinkage in the plane of the substrate is inhibited. All of the shrinkage occurs in the direction perpendicular to the plane of the film (Fig. 11.29). The inhibition of the shrinkage in the plane of the substrate leads to the generation of transient stresses that influence the sintering kinetics. In the case of sintering with rigid inclusions, we observed that the stress field in the composite sphere model is analogous to that caused by thermal expansion mismatch between the core and the cladding. The stresses that arise in the sintering of constrained films are analogous to those in a sandwich seal caused by mismatch in the thermal expansion coefficients in the layers.

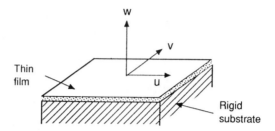

Figure 11.29 Geometry of a thin film attached to a rigid substrate. No shrinkage occurs in the plane of the film (uv plane); all of the shrinkage occurs in the direction perpendicular to the plane of the film (the w direction).

A. Models for Constrained Sintering of Thin Films

As in the approach used earlier for sintering with rigid inclusions, we must calculate the stresses and determine how they influence the sintering rate of the film. Two major models taken from the ceramics literature are discussed briefly. As we shall see, they incorporate some features that parallel those discussed earlier for sintering with rigid inclusions. Further studies on the sintering of thin films are given in Refs. 73 and 74.

The Model of Scherer and Garino

Scherer and Garino [32] assumed that the sintering rate of the film can be described by a constitutive relation for a porous, viscous body. By analogy with Eq. (11.28), the equation for the sintering rate along the orthogonal u, v, and w directions (see Fig. 11.29) can be written

$$\dot{\epsilon}_u = \dot{\epsilon}_f + \frac{1}{E_p}[\sigma_u - \nu_p(\sigma_v + \sigma_w)] \tag{11.47a}$$

$$\dot{\epsilon}_v = \dot{\epsilon}_f + \frac{1}{E_p}[\sigma_v - \nu_p(\sigma_u + \sigma_w)] \tag{11.47b}$$

$$\dot{\epsilon}_w = \dot{\epsilon}_f + \frac{1}{E_p}[\sigma_w - \nu_p(\sigma_u + \sigma_v)] \tag{11.47c}$$

where $\dot{\epsilon}_u$, $\dot{\epsilon}_v$, $\dot{\epsilon}_w$ and σ_u, σ_v, σ_w are the linear densification rates and stresses in the u, v, and w directions, respectively; $\dot{\epsilon}_f$ is the linear densification rate of the free or unconstrained film; E_p is the uniaxial viscosity; and ν_p is the Poisson's ratio of the porous film. As written, Eq. (11.47) is not based on any microstructural model, but a model must be chosen so that $\dot{\epsilon}_f$, E_p, and ν_p can be specified. Since no deformation of the film

Problems of Sintering

occurs in the u and v directions, $\epsilon_u = \epsilon_v = 0$. Also, there is no constraint in the w direction so that $\sigma_w = 0$. Putting $\sigma_u = \sigma_v = \sigma$, Eq. (11.47) gives

$$\sigma = -\frac{E_p \dot{\epsilon}_f}{1 - v_p} \tag{11.48}$$

and

$$\dot{\epsilon}_w = \left(\frac{1 + v_p}{1 - v_p}\right) \dot{\epsilon}_f \tag{11.49}$$

The volumetric densification rate is given by

$$\frac{\dot{\rho}}{\rho} = -\frac{\dot{V}}{V} = \dot{\epsilon}_u + \dot{\epsilon}_v + \dot{\epsilon}_w \tag{11.50}$$

where ρ is the relative density, V the volume, and the dot denotes the derivative with respect to time. Using Eq. (11.49) and putting $\dot{\epsilon}_u = \dot{\epsilon}_v = 0$, the densification rate of the constrained film is

$$\left(\frac{\dot{\rho}}{\rho}\right)_c = -\left[\frac{1 + v_p}{3(1 - v_p)}\right] 3\dot{\epsilon}_f \tag{11.51}$$

The densification rate of the unconstrained material is

$$\left(\frac{\dot{\rho}}{\rho}\right)_u = -3\dot{\epsilon}_f \tag{11.52}$$

The function of v_p in the square brackets of Eq. (11.51) represents the amount by which the densification rate of the film is retarded by the substrate.

For amorphous films where sintering occurs by a viscous flow mechanism, Scherer's cylinder model described in Chapter 8 can be used to determine the terms E_p, $\dot{\epsilon}_f$, and v_p in Eqs. (11.48), (11.51), and (11.52). The procedure is described in Ref. 22. The term in the square brackets of Eq. (11.51), i.e., $(1 + v_p)/3(1 - v_p)$, is always < 1 for $\rho < 1$, so that the constrained film is predicted to sinter at a slower rate than the unconstrained film. After integration, the results can be plotted as relative density versus dimensionless time (Fig. 11.30). Starting from the same initial density, ρ_0, the constrained film takes a longer time to reach a given density than does an unconstrained film of the same material. For example, for $\rho_0 = 0.5$, the constrained film takes $\approx 25\%$ longer than the unconstrained film to reach theoretical density.

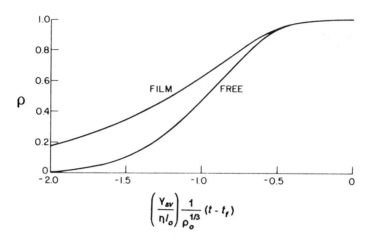

Figure 11.30 Predictions of the model of Scherer and Garino for the sintering of a constrained film and a free film. The relative density is plotted versus dimensionless time. The constrained film is predicted to sinter more slowly than the free film. (From Ref. 32, with permission.)

The stress in the film during sintering can be found from Eq. (11.48) and the equations for E_p, $\dot{\epsilon}_f$, and v_p. The equation can be written

$$\sigma = \frac{\gamma_{sv}}{l_0 \rho_0^{1/3}} f_1(\rho) \tag{11.53}$$

where γ_{sv} is the surface tension of the glass, l_0 is the length of the unit cell for Scherer's cylinder model (see Fig. 8.15), ρ_o is the initial density, and $f_1(\rho)$ is a function of the density of the film. This function reaches a maximum of ≈ 1 at $\rho \approx 0.81$. The maximum stress is therefore

$$\sigma_{max} = \gamma_{sv}/l_0 \rho_0^{1/3} \tag{11.54}$$

For a polymeric silica gel prepared by the hydrolysis of silicon tetraethoxide, $\gamma_{sv} \approx 0.25$ J/m^2, $l_0 \approx 10$ nm, and $\rho_o \approx 0.50$, so that $\sigma_{max} \approx 30$ MPa. This value of σ_{max} is on the order of the stresses used in hot pressing (but it is tensile in nature). If the film is not strong enough to withstand such high tensile stresses, cracking will result. For a colloidal silica gel, the corresponding values are $\gamma_{SV} \approx 0.25$ J/m^2, $l_0 \approx 100$ nm, $\rho_0 \approx 0.25$, and $\sigma_{max} \approx 3$ MPa; that is, the maximum stress is considerably smaller.

The Model of Bordia and Raj

Bordia and Raj [33] used an approach similar to that described earlier for their model of sintering with inclusions (see Section 11.3G). The stresses

Problems of Sintering

in the film and the densification behavior are derived in terms of the dimensionless parameter β. Figure 11.31 shows the results for the densification of the film as a function of the normalized time for various values of β. The quantity shown in the plot is $\Delta\rho/\Delta\rho_0$, where $\Delta\rho = \rho - \rho_i$, $\Delta\rho_0 = 1 - \rho_i$, and ρ_i is the initial density. The parameter $\Delta\rho/\Delta\rho_0$ is therefore the change in density normalized to the maximum possible change in density. For $\beta \gg 1$, the densification rate of the constrained film is predicted to be roughly similar to that for the unconstrained film, whereas for $\beta \ll 1$, the densification rate of the constrained film can be severely retarded. As discussed earlier for the case of sintering with rigid inclusions, a major flaw of the analysis is that values of $\beta \ll 1$ are physically unrealistic.

B. Experimental Observations of Sintering of Thin Films

Garino and Bowen [34] studied the sintering of constrained and free films of soda-lime-silica glass, polycrystalline ZnO, and polycrystalline Al_2O_3. The data were compared with the predictions of the model of Scherer and Garino discussed earlier. Constrained films (100–200 μm thick) were prepared by casting powder slurries onto a substrate. After drying, the shrinkage was measured using the laser reflectance apparatus sketched

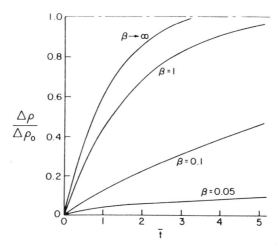

Figure 11.31 Predictions of the model of Bordia and Raj for the densification of a constrained film. The change in density divided by the maximum possible change in density ($\Delta\rho/\Delta\rho_0$) is shown as a function of normalized time for various values of the parameter β. The plot shows that the constraint can severely retard densification if β is small. (From Ref. 33, with permission.)

in Fig. 11.32. Free films were obtained by breaking off small pieces of the dried films from the substrate. The free shrinkage of these films was measured from scanning electron micrographs taken at various stages of the sintering process.

Figure 11.33 shows the data for the volumetric shrinkage of the free film and the constrained film of the glass powder at 650°C. The data for the free film were well fitted by Scherer's cylinder model for the viscous sintering of glass. The constrained sintering model of Scherer and Garino gave a good fit to the data for the constrained film for shrinkages up to ≈25% but overestimated the shrinkage at higher values. Sintering of the constrained film for times longer than those shown in the figure revealed that the density increased steadily and eventually reached the same density as the free film.

The data for the sintering of the polycrystalline ZnO films are shown in Fig. 11.34. The constrained film sintered considerably more slowly than the free film and, even after sintering for longer times, did not reach the

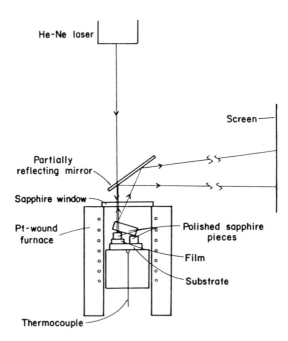

Figure 11.32 Schematic diagram of a laser reflectance apparatus used to measure the in situ shrinkage of a film during sintering. (From Ref. 34, with permission.)

Problems of Sintering

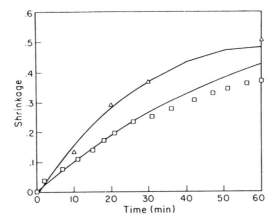

Figure 11.33 Isothermal shrinkage data for constrained (□) and free (△) glass films sintered at 650°C. The curve through the data for the free film represents the best fit to the data using Scherer's model for viscous sintering. The curve through the data for the constrained film represents the fit to the data by the thin-film model of Scherer and Garino using the fitting constants derived from the fit to the free-film data. (From Ref. 34, with permission.)

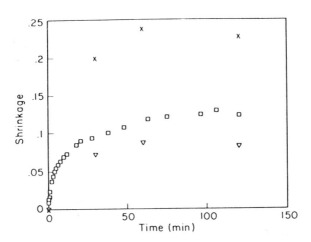

Figure 11.34 Shrinkage data for constrained and free films of ZnO during sintering at 778°C. Both the volumetric (x) and linear (▽) shrinkages of the free film are shown with the shrinkage of the constrained film (□). (From Ref. 34, with permission.)

same endpoint density as the free film. Instead, the shrinkage reached a steady value that was considerably lower than that reached by the free film. The data for the Al_2O_3 films showed trends similar to those for the ZnO films.

The data of Figs. 11.33 and 11.34 indicate that the constrained sintering of polycrystalline films is qualitatively different from that of glass films. The constrained polycrystalline films are difficult to densify. They reach a limiting density that is significantly lower than the endpoint density of the free films. The reason for this difference is not clear. However, two factors operating individually or in combination may play a role. The first factor is coarsening, which plays no part in the sintering of glassy films. It may be argued that since the constrained film takes a longer time to reach the same density as the free film, coarsening plays a larger role in reducing the sintering rate of the constrained film. Garino and Bowen attempted to incorporate the effects of grain growth into the constrained sintering model by assuming that the linear densification rate is described by an equation of the form [see Eq. (8.99)]

$$\dot{\epsilon} = \Sigma/\eta_\rho \qquad (11.55)$$

where Σ is the sintering stress and η_ρ is the effective viscosity for densification. Assuming that $\Sigma \sim 1/G$ and $\eta_\rho \sim 1/G^m$, where G is the grain size and m is an exponent that depends on the mechanism of sintering (i.e., $m = 2$ for lattice diffusion and $m = 3$ for grain boundary diffusion), the effects of grain growth can be examined. Even after compensating for grain growth, the constrained sintering model of Scherer and Garino failed to give an adequate representation of the data. The second factor is differential densification. Microstructural observations of the Al_2O_3 films revealed the presence of large pores surrounded by dense regions, i.e., features characteristic of differential densification. However, differential densification in a film is possible only if the adhesion fails [73,74].

C. Breakup Phenomena in Polycrystalline Thin Films

Ceramic thin films, as we discussed in Chapter 5, can be deposited onto substrates by dip coating or spin coating of liquid precursors such as solutions of alkoxides, acetates, or citrates. The films prepared by these methods can be fairly thin, typically less than a few tenths of a micrometer. It is normally found that these films (initially amorphous in structure) can be sintered readily to produce dense polycrystalline films that remain

Problems of Sintering

attached to the substrate. However, many experiments reveal that when subjected to further heat treatment for longer times or at higher temperatures, the films can break up into an interconnected network of islands, thereby uncovering the substrate. This instability of thin films can create severe problems in many applications; for example, it can cause failure of the overall device.

Miller et al. [35] made a detailed study of the breakup phenomena in thin polycrystalline films of ZrO_2 (containing 3 and 8 mol % Y_2O_3) attached to single-crystal Al_2O_3 substrates. The compositions were chosen such that in one case (3 mol % Y_2O_3) the grain growth of the ZrO_2 was inhibited and in the other case (8 mol % Y_2O_3) no inhibition of the grain growth took place. The films were prepared by spin coating mixed solutions of zirconium acetate and yttrium nitrate. After pyrolysis at 1000°C to decompose the metal salts, the films were heated to higher temperatures to produce densification. The grain size of the films increased rapidly during densification, and nearly fully dense films were obtained at 1400°C. Because grain growth was inhibited in the composition with 3 mol % Y_2O_3, the grain size of the films with this composition was smaller than that of the films with 8 mol % Y_2O_3.

When the dense films were further heated at 1400°C, regardless of the composition, they started to break up into islands, thereby uncovering the substrate. The onset of breakup occurred after grain growth had occurred to such an extent that the grain size was greater than the thickness of the film. However, the breakup mechanism was different for the two compositions. For the film containing 3 mol % Y_2O_3, the smallest grains disappeared (possibly through evaporation and condensation), leaving exposed regions of the substrate. The breakup of the film containing 8 mol % Y_2O_3 is shown in Fig. 11.35. In this case, larger grains grew at the expense of the smaller ones. The enlarged grains then spheroidized to uncover the substrate at the triple points of the grains.

Model for the Breakup Phenomena

Miller et al. analyzed the breakup phenomena observed in Fig. 11.35 by considering the free energy changes associated with the spheroidization of the grains. The geometrical parameters are sketched in Fig. 11.36 for a two-dimensional model. The film is assumed to consist of uniform grains with an initial size of D and thickness t. The grains are allowed to spheroidize at constant volume and the grain centers are assumed to remain fixed by the substrate. Initially, the change in configuration of the grains can be described by the angle ψ, which is allowed to decrease from its initial value of π. As ψ decreases, a point is reached where the grain

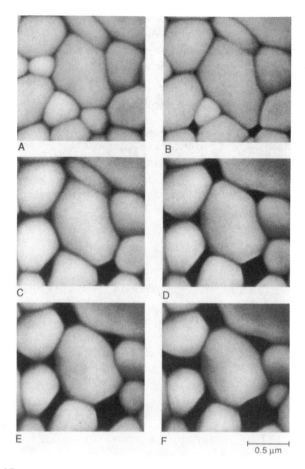

Figure 11.35 Microstructural evolution of a ZrO_2 (8 mol % Y_2O_3) thin film during heat treatment at 1400°C for the following times in hours: (A) 1, (B) 2, (C) 3, (D) 4, (E) 5, (F) 6. (From Ref. 35.)

boundary just disappears. The critical value of ψ when this occurs is related to D/t by

$$\frac{D}{t} = \frac{4(1 + \cos \psi_c)}{\pi - \psi_c - \sin \psi_c} \tag{11.56}$$

Further spheroidization of the particles is assumed to result in the uncovering of the substrate. When this occurs (Fig. 11.36), the geometry can be described in terms of the angle θ, which is related to ψ by the relation

Problems of Sintering

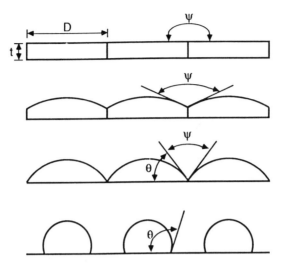

Figure 11.36 Model parameters for the spheroidization of a uniform thin film with initial thickness t and grain size D. (From Ref. 35.)

$\theta + \psi/2 = \pi/2$. The change in configuration of the model can therefore be described in terms of the angle ψ alone, where ψ varies in the range $-\pi \leq \psi \leq \pi$.

The total energy of the surfaces and interfaces can be expressed by

$$E = A_{sv}\gamma_{sv} + A_{gb}\gamma_{gb} + A_i\gamma_i + A_{sub}\gamma_{sub} \tag{11.57}$$

where A_{sv} is the solid/vapor interfacial area of the film, A_{gb} is the grain boundary area, A_i is the film/substrate interfacial area, A_{sub} is the substrate/vapor interfacial area, and γ_{sv}, γ_{gb}, γ_i, γ_{sub} are the corresponding interfacial energies. The areas in Eq. (11.57) depend on D/t and ψ only. Furthermore, the interfacial energies can be expressed in terms of the equilibrium dihedral angle ψ_e and the equilibrium wetting angle θ_e:

$$\frac{\gamma_{gb}}{\gamma_{sv}} = 2\cos\left(\frac{\psi_e}{2}\right) \tag{11.58a}$$

$$\frac{\gamma_{gb}}{\gamma_i} = 2\cos\left(\frac{\phi_e}{2}\right) \tag{11.58b}$$

$$\frac{\gamma_{sub} - \gamma_i}{\gamma_{sv}} = 2\cos\theta_e \tag{11.58c}$$

The free energy function E normalized to its initial value E_0 can be evaluated to determine the energy minimum as a function of the configuration of the film. This minimum determines the most stable configuration. The results of such calculations can be represented in terms of a diagram, referred to as an equilibrium configuration diagram, showing the minimum energy configuration for any desired values of D/t, ψ_e, and θ_e. As an example, the results of the minimum energy configuration are shown in Fig. 11.37 for a given value of $\psi_e = 120°$. The diagram is divided into three regions: the completely covered substrate, the uncovered substrate, and the partially connected film (corresponding to the situation where the grain boundary just disappears). When $D/t < 8/\pi$, grain boundaries will always exist and the covered substrate is the most stable configuration. For $2\theta_e > \pi - \psi_e$, two stability regions exist: the completely covered film and the uncovered substrate. As θ_e decreases, the D/t values for the substrate to remain completely covered increases. For $2\theta_e < \pi - \psi_e$, the three stability regions can exist, depending on the value of D/t. The transition from the completely covered film to the partially connected film is found by putting $\psi_c = \psi_e$ in Eq. (11.56). The transition from the partially connected film to the uncovered substrate is given by

$$\frac{D}{t} = \frac{4(1 - \cos 2\theta_e)}{2\theta_e - \sin 2\theta_e} \tag{11.59}$$

The analysis becomes more complicated for the three-dimensional case. However, trends similar to those described here are predicted.

To summarize, the model of Miller et al. predicts the conditions that

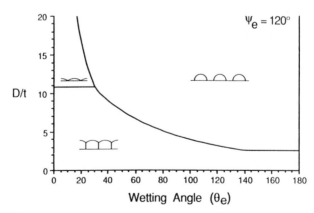

Figure 11.37 Two-dimensional equilibrium configuration diagram for the sintering of a thin film, plotted for an equilibrium dihedral angle $\psi_e = 120°$. (From Ref. 35.)

govern the thermal stability of polycrystalline thin films. The calculations allow us to predict under what conditions breakup of the films will occur and to determine what steps might be taken to prevent breakup.

11.5 SOLID SOLUTION ADDITIVES AND THE SINTERING OF CERAMICS

The use of additives, as we discussed earlier in this book, provides a very effective approach for the fabrication, by sintering, of ceramics with high density and controlled grain size. Additives that aid sintering are generally grouped into two classes: (i) those that form a second phase at the grain boundary and (ii) those that go into solid solution into the host material. As a second phase, an additive may provide a high-diffusivity path for matter transport; a common example of this function is liquid-phase sintering, which we considered in Chapter 10. An additive may also exist as discrete second-phase particles that inhibit grain boundary migration by pinning (i.e., the Zener mechanism discussed in Chapter 9). In this section, we confine our discussion to additives that go into solid solution; these additives are also generally known as dopants. It must be remembered, however, that most dopants have only a limited solid solubility in the host. If the solid solubility limit is exceeded, a second phase will form, and its effects must be taken into account.

The most celebrated example of the use of the solid solution additive approach was reported by Coble [36,37], who showed in 1961 that small additions of MgO (0.25 wt %) to Al_2O_3 produced polycrystalline translucent alumina with theoretical density (Lucalox). Figure 11.38 illustrates the effect of the MgO additive. Since Coble's work, many examples of the effectiveness of the solid solution approach can be found in the ceramics literature [38]. Table 11.1 shows examples of dopants used in some com-

Table 11.1 Examples of Dopants Used in the Sintering of Some Common Ceramics

Host	Dopant	Concentration (at %)
Al_2O_3	Mg	0.025
$BaTiO_3$	Nb, Co, La, W	0.5–1.0
CeO_2	Y, Nd, Ca	3–5
SiC	(B + C)	0.1 (B); 0.1 (C)
Y_2O_3	Th	5
ZnO	Al	0.02
ZrO_2	Y	6–10

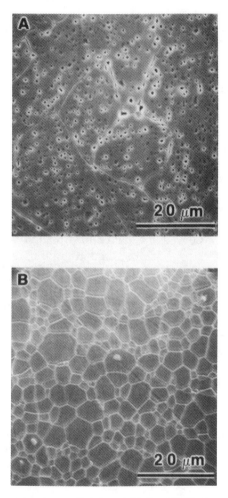

Figure 11.38 Microstructures of sintered Al_2O_3. (A) Undoped material showing pore–grain boundary separation and abnormal grain growth; (B) MgO-doped material showing high density and equiaxial grain structure. (From Ref. 51, reprinted by permission of the American Ceramic Society.)

mon ceramics. However, a major problem that limits the wider applicability of the approach is that apart from a few recognized systems (e.g., MgO-doped Al_2O_3), the role of the dopant is not understood very well. The main reason for this gap in our understanding is that dopants can display a variety of functions, which makes understanding of the dopant role difficult.

A. Solid Solution Additive Functions

An additive can influence both the kinetic and thermodynamic factors in sintering. As discussed earlier, an additive can alter the defect chemistry of the host, thereby changing the diffusion coefficient for transport of ions through the lattice, D_l. Segregation of the additive can alter the structure and composition of surfaces and interfaces. The additive can therefore alter the grain boundary diffusion coefficient D_b, the surface diffusion coefficient D_s, and the diffusion coefficient for the vapor phase D_g (i.e., the evaporation/condensation process). Segregation can also alter the interfacial energies. An additive can therefore also act thermodynamically to change the surface energy γ_{sv} and the grain boundary energy γ_{gb}. Another consequence of segregation is that the additive can alter the intrinsic grain boundary mobility M_b.

In principle, an additive will influence each of these factors to a certain extent. In a general sense, we can say that a good additive is one that alters many phenomena in a favorable way but few phenomena in an unfavorable way. This multiplicity of functions is the major reason that understanding the role of additives has proved difficult. As a result, the selection of additives has remained largely empirical.

B. Effect of Additives on the Kinetic Factors

As discussed above, an additive can alter each of the diffusion coefficients for matter transport, i.e., D_l, D_{gb}, D_s, and D_g. Historically, however, the major emphasis has been placed on the ability of the additive to alter D_l through its effect on the defect chemistry of the host.

To determine how an additive will influence D_l, the defect chemistry of the host must be known. Specifically, we must know the nature of the rate-controlling species (anion or cation), the type of defect (vacancy or interstitial), and the state of charge of the defect. However, this information is known in only a very few cases. To illustrate the approach, let us consider Al_2O_3, a system that has been widely studied. The intrinsic defect structure is believed to consist of cationic Frenkel defects [39]:

$$O \rightarrow Al_i^{\cdot\cdot\cdot} + V_{Al}''' \tag{11.60}$$

It has also been suggested that the diffusion of aluminum ions controls the rate of densification, with the faster diffusion of oxygen ions occurring along the grain boundary [36]. Based on Eq. (11.60), we may assume that the rate-controlling species is the diffusion of triply charged aluminum interstitial ions. A possible defect reaction for the incorporation of MgO into the Al_2O_3 lattice is

$$3MgO \xrightarrow{Al_2O_3} 3Mg'_{Al} + Al_i^{\cdot\cdot\cdot} + 3O_O^x \qquad (11.61)$$

According to this equation, the effect of MgO would be to increase the densification rate through an increase in the concentration of aluminum interstitial ions, $[Al_i^{\cdot\cdot\cdot}]$.

Let us now consider an alternative situation, by supposing that the densification rate is controlled by the diffusion of triply charged aluminum vacancies, V'''_{Al}. In this case, the creation of aluminum interstitials by Eq. (11.61) leads to a reduction in the concentration of aluminum vacancies, $[V'''_{Al}]$, by Eq. (11.60) and a corresponding reduction in the densification rate. However, if instead of MgO we use TiO_2 as a dopant, a possible defect reaction for the incorporation into Al_2O_3 is

$$3TiO_2 \xrightarrow{Al_2O_3} 3Ti^{\cdot}_{Al} + V'''_{Al} + 6O_O^x \qquad (11.62)$$

According to this equation, if the rate-controlling species is V'''_{Al}, the addition of TiO_2 would increase the densification rate through an increase in $[V'''_{Al}]$.

As this example of Al_2O_3 shows, if we know the intrinsic defect structure of the host, it is possible to select additives to increase D_l. In fact, a few rules have been put forward to assist in the selection of additives using this approach [40]:

1. Pick an additive with an ionic radius similar to that of the host (to assist formation of a solid solution).
2. Add this in the concentration close to the solid solution limit (to maximize the effect).
3. Pick as additive a relatively volatile species (to ensure that distribution is effected during sintering).
4. Pick an additive with a valence one unit different from that of the host (to affect the defect concentration and yet give reasonable solubility). As observed for Al_2O_3, if the intrinsic defect structure of the host is known, we can determine whether the valence should be greater or smaller than that of the host.

Although these rules may appear to be consistent with the use of MgO in Al_2O_3, they have proved generally ineffective in the selection of additives. This single factor of an increase in D_l is insufficient to guarantee an effective additive. There are cases where an additive increases D_l, e.g., TiO_2-doped Al_2O_3, but the attainment of theoretical density is not realized. Other possible functions of the additive must be examined.

Problems of Sintering

For the attainment of high density, an additive can also act favorably to decrease the surface diffusion coefficient D_s. However, rules for the selection of additives to accomplish this decrease in D_s are not known. The reason we require a decrease (rather than an increase) in D_s can be argued as follows. A decrease in D_s reduces the rate of coarsening, thereby increasing the densification rate. Therefore, the pore size r will be smaller at any given stage of microstructural development (e.g., grain size). Because of its strong dependence on r [see Eq. (9.84)], the pore mobility M_p increases considerably. The large M_p prevents separation of the pores from the boundary and hence eliminates abnormal grain growth. The diffusion distance for matter transport into the pores is kept small, thereby improving the probability of achieving high density.

An early explanation for the effectiveness of MgO as an additive in the sintering of Al_2O_3 was expressed in terms of its ability to increase M_p through an increase in D_s [see Eq. (9.84)]. However, this suggestion goes against our argument for a decrease in D_s. Furthermore, it neglects the multiplicity of roles that an additive can display. As we shall discuss later, while recent experiments do indicate a small increase in D_s, the major function of MgO in Al_2O_3 is not its ability to increase D_s. As a result, this early explanation for the role of MgO in Al_2O_3 is not generally believed.

To summarize at this stage, a favorable change in only one of the kinetic factors (e.g., an increase in D_l or a decrease in D_s) does not guarantee the attainment of theoretical density. Favorable changes in two or more kinetic factors may be sufficient. However, the information required to produce a favorable change in the kinetic factors is generally lacking.

C. Effect of Additives on the Thermodynamic Factors

The effects of solid solution additives on the surface and grain boundary energies, γ_{sv} and γ_{gb}, have been investigated in only a few instances. In one of the most detailed studies, Handwerker et al. [41] measured the effect of MgO doping on the distribution of dihedral angles in Al_2O_3. The angles were measured from grain boundary grooves on polished and thermally etched samples. The dihedral angle [Eq. (8.4)] provides a measure of the ratio γ_{gb}/γ_{sv}. It was found that MgO reduced the width of the distribution without significantly affecting the mean value (117°). This reduction in the spread of the dihedral angles can be interpreted qualitatively to mean that local variations in the driving force for sintering and coarsening are reduced so that a more homogeneous microstructure is favored. As outlined earlier in this book, an improvement in the microstructural homogeneity favors the attainment of a higher density at a given stage of microstructural development and reduces the possibility of initiating abnormal grain growth. The dihedral angle results therefore indicate that a function of MgO in Al_2O_3 is to alleviate the consequences of microstructural inhomogeneity.

D. Segregation of Additives and the Boundary Mobility

In Chapter 9, we considered the solute drag mechanism proposed by Cahn and found that it can be very effective for reducing the intrinsic grain boundary mobility M_b. However, the effectiveness of the method was seen to depend critically on the ability of the solute (i.e., the dopant) to segregate at the grain boundaries. We will recall that, in segregation, the concentration of the dopant is enhanced without the formation of a second phase. For the same concentration of dopant, a greater reduction in M_b occurs with increasing effectiveness of segregation. In order to determine the ability of a dopant to reduce M_b, we must first examine the factors that control segregation at a grain boundary.

Electrostatic Interaction: The Space Charge Concept

In ionic crystals, the energy of formation of cation defects is usually different from that of anion defects. On heating, the defect concentration at the grain boundaries (or, in general, surfaces and interfaces) will be determined by the energy of formation of the defects. However, the bulk of the crystal must be electrically neutral, so the defect concentration in the bulk will be determined by the principle of electroneutrality. Because of the different requirements, it would be expected that the defect concentration in the bulk will be different from that at the grain boundary. Since defects in ionic crystals usually carry an electric charge, it would be expected that the electrostatic potential will also vary with distance from the boundary. This potential difference can cause segregation of charged solute ions at the grain boundary.

As an example, let us consider sodium chloride, for which such effects have been examined in some detail [42]. In the simplest formulation, it is assumed that the grain boundaries act as perfect sources and sinks for vacancies, i.e., vacancies can be readily created or destroyed. If the intrinsic defects are of the Schottky type (see Chapter 7), then

$$O \rightarrow V'_{Na} + V^{\bullet}_{Cl} \qquad (11.63)$$

For NaCl, the energy required to form a cation vacancy has been estimated to be about half of that for an anion vacancy. The consequence of this difference in formation energy can be visualized as, initially, a tendency for an excess of cation vacancies with a negative charge to form at the grain boundaries and migrate into the interior of the crystal. This

Problems of Sintering

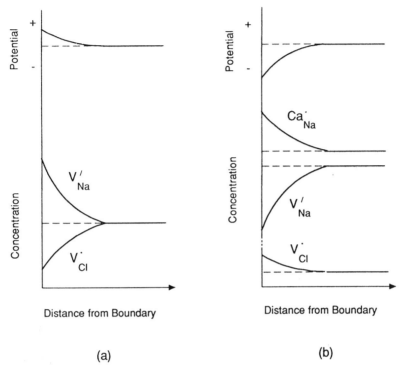

Figure 11.39 Schematic variations of the electrostatic potential and defect concentration as a function of distance from the grain boundary for (a) undoped NaCl and (b) NaCl doped with $CaCl_2$.

will lead to a negative space charge layer extending some distance into the crystal, leaving a positive charge at the boundary. The corresponding electrostatic potential will impede migration of the cation vacancies into the interior but will enhance migration of anion vacancies, leading to the establishment of a stationary state. The variation of the electrostatic potential and the defect concentration is summarized in Fig. 11.39a.

The formation energies $g_{V'_{Na}}$ and $g_{V^{\cdot}_{Cl}}$ for cation and anion vacancies in NaCl have been estimated at 0.65 and 1.21 eV, respectively. If e is the electronic charge, the electrostatic potential between the grain boundary and the bulk of the crystal is given by

$$V_\infty = \frac{1}{2e}[g_{V'_{Na}} - g_{V^{\cdot}_{Cl}}] \approx -0.28 \text{ V} \tag{11.64}$$

where e is the electronic charge. This estimate shows that the potential is not trivial in magnitude. The space charge has been estimated to extend a depth of ≈ 2–10 nm into the crystal.

Turning now to the case where an aliovalent dopant is present, such as $CaCl_2$ in NaCl, the defect reaction for the incorporation of the dopant can be written

$$CaCl_2 \xrightarrow{NaCl} Ca_{Na}^{\cdot} + V'_{Na} + 2Cl_{Cl}^{x} \tag{11.65}$$

Additional cation vacancies are generated. As discussed in Chapter 7, the defect concentrations produced by impurities and dopants are normally much larger than those for thermally generated defects. According to Eq. (11.63), increasing $[V'_{Na}]$ leads to a decrease in $[V_{Cl}^{\cdot}]$. This in turn leads to a decrease in the concentration of Na^+ ions and an increase in the concentration of Cl^- ions at the boundary. In this case, the boundary will be negatively charged. The negatively charged boundary will cause the Ca dopant with the positive effective charge to segregate at the grain boundary as a space charge to maintain the bulk of the crystal electrically neutral. The variation of the electrostatic potential and the defect concentration for Ca-doped NaCl are summarized in Fig. 11.39b.

To summarize, grain boundaries in an ionic solid can have an electric charge. The difference in electrostatic potential between the grain boundary and the interior of the grain can cause segregation of aliovalent dopants at the grain boundary as a space charge extending ≈ 2–10 nm from the boundary.

The space charge concept has not found definitive support for solute segregation in MgO or Al_2O_3, presumably because of the low solubility of many cations in these two materials. However, it has been reported as the controlling factor in the segregation of divalent and trivalent cations in tetragonal zirconia polycrystals, despite the presence of a thin glassy layer at the grain boundaries [43]. The space charge concept has also been invoked to explain the segregation of a variety of dopants at the grain boundaries of $BaTiO_3$ and $SrTiO_3$ [44].

Elastic Strain Energy
When the radius of the solute ion is different from that of the host, substitution leads to a misfit in the lattice. As a consequence, there is an increase in the elastic strain energy of the lattice. If r_i is the ionic radius of the solute ion and r_0 that of the host, the misfit Δr is

$$\Delta r = r_i - r_0 \tag{11.66}$$

Problems of Sintering

The magnitude of the elastic strain energy is related to the misfit and is given by [45]

$$U_0 = \frac{6\pi r_0^3 K_i (\Delta r/r)^2}{1 + 3K_i/4G_h} \tag{11.67}$$

where K_i is the bulk modulus of the solute and G_h is the shear modulus of the host crystal. According to this equation, the strain energy is proportional to the square of the misfit. Table 11.2 provides a list of ionic radii that can be used in the estimation of the misfit.

The grain boundary, as we outlined in Chapter 9, is widely regarded as a thin region of disorder between two crystalline regions (the grains). There will be some fraction of sites at the grain boundary where the energy associated with the addition of solute ions is small. Segregation of the solute ions to the distorted grain boundary region provides partial relaxation of the elastic strain energy. The segregation to the grain boundary depends on the difference in energy $\Delta G_a = E_l - E_b$, for the solute at the boundary, E_b, and in the lattice, E_l. The term ΔG_a is sometimes referred to as the free energy of adsorption at the grain boundary. It is evident that segregation is increasingly favored for larger values of ΔG_a because this represents a greater reduction in the free energy of the system. It is expected that ΔG_a will have some spatial dependence because the distortion of the grain boundary varies with distance from the center of the boundary. Details of the spatial dependence are not known, but it is believed to extend up to a few lattice spacings from the center of the boundary.

McLean [46] showed that the segregation of solute to the grain boundary is given by

$$C_{gb} = \frac{C_0 \exp(\Delta G_a/RT)}{1 + C_0 \exp(\Delta G_a/RT)} \approx C_0 \exp\left(\frac{\Delta G_a}{RT}\right) \tag{11.68}$$

where C_{gb} and C_0 are the solute concentrations at the grain boundary and in the lattice, respectively, R is the gas constant, and T is the temperature. In Eq. (11.68), the approximation applies when the denominator is close to 1. By annealing Al_2O_3 samples at several temperatures, quenching the samples, and using Auger electron spectroscopy to measure the Ca content at the grain boundaries, Johnson [47] found that Eq. (11.68) provided a good description of Ca segregation in Al_2O_3 for $\Delta G_a \approx 117$ kJ/mol.

If it is assumed that ΔG_a is proportional to U_0, i.e., proportional to $(\Delta r/r_0)^2$, then we can use the ΔG_a value for Ca adsorption to estimate

Table 11.2 Ionic Crystal Radii (in units of 10^{-10} m)

Coordination number = 6

Ion	r	Ion	r	Ion	r	Ion	r	Ion	r	Ion	r	Ion	r	Ion	r	Ion	r	Ion	r	Ion	r
Ag^+	1.15	Al^{3+}	0.53	As^{5+}	0.50	Au^+	1.37	B^{3+}	0.23	Ba^{2+}	1.36	Be^{2+}	0.35	Bi^{5+}	0.74	Br^-	1.96	C^{4+}	0.16	Ca^{2+}	1.00	Cd^{2+}	0.95
Ce^{4+}	0.80	Cl^-	1.81	Co^{2+}	0.74	Co^{3+}	0.61	Cr^{2+}	0.73	Cr^{3+}	0.62	Cr^{4+}	0.55	Cs^+	1.70	Cu^+	0.96	Cu^{2+}	0.73	Dy^{3+}	0.91	Er^{3+}	0.88
Eu^{3+}	0.95	F^-	1.33	Fe^{2+}	0.77	Fe^{3+}	0.65	Ga^{3+}	0.62	Gd^{3+}	0.94	Ge^{4+}	0.54	Hf^{4+}	0.71	Hg^{2+}	1.02	Ho^{3+}	0.89	I^-	2.20	In^{3+}	0.79
K^+	1.38	La^{3+}	1.06	Li^+	0.74	Mg^{2+}	0.72	Mn^{2+}	0.67	Mn^{4+}	0.54	Mo^{3+}	0.67	Mo^{4+}	0.65	Na^+	1.02	Nb^{5+}	0.64	Nd^{3+}	1.00	Ni^{2+}	0.69
O^{2-}	1.40	P^{5+}	0.35	Pb^{2+}	1.18	Pb^{4+}	0.78	Rb^+	1.49	S^{2-}	1.84	S^{6+}	0.30	Sb^{5+}	0.61	Sc^{3+}	0.73	Se^{2-}	1.98	Se^{6+}	0.42	Si^{4+}	0.40
Sm^{2+}	0.96	Sn^{2+}	1.18	Sn^{4+}	0.69	Sr^{2+}	1.16	Ta^{5+}	0.64	Te^{2-}	2.21	Te^{6+}	0.56	Th^{4+}	1.00	Ti^{2+}	0.86	Ti^{4+}	0.61	Tl^+	1.50	Tl^{3+}	0.88
U^{4+}	0.97	U^{5+}	0.76	V^{2+}	0.79	V^{5+}	0.54	W^{4+}	0.65	W^{6+}	0.58	Y^{3+}	0.89	Yb^{3+}	0.86	Zn^{2+}	0.75	Zr^{4+}	0.72				

Coordination number = 4

Ion	r	Ion	r	Ion	r	Ion	r	Ion	r	Ion	r	Ion	r	Ion	r	Ion	r	Ion	r	Ion	r	Ion	r
Ag^+	1.02	Al^{3+}	0.39	As^{5+}	0.34	B^{3+}	0.12	Be^{2+}	0.27	C^{4+}	0.15	Cd^{2+}	0.84	Cr^{4+}	0.44	Cu^{2+}	0.63	F^-	1.31	Fe^{2+}	0.63	Fe^{3+}	0.49
Ga^{3+}	0.47	Ge^{4+}	0.40	Hg^{2+}	0.96	Li^+	0.59	Mg^{2+}	0.49	N^{5+}	0.13	Na^+	0.99	Nb^{5+}	0.32	O^{2-}	1.38	P^{5+}	0.33	Pb^{2+}	0.94	S^{6+}	0.12
Se^{6+}	0.29	Si^{4+}	0.26	V^{5+}	0.36	W^{6+}	0.41	Zn^{2+}	0.60													

Source: R. D. Shannon and C. T. Prewitt, *Acta Cryst.*, B25: 925 (1969).

Problems of Sintering

those for other cations in Al_2O_3. For example, the ΔG_a for segregation of cation M can be approximated by

$$\frac{(\Delta G_a)_M}{(\Delta G_a)_{Ca}} \approx \frac{(\Delta r/r_0)_M^2}{(\Delta r/r_0)_{Ca}^2} \tag{11.69}$$

Using the ionic radii given in Table 11.2 and $(\Delta G_a)_{Ca} \approx 117$ kJ/mol, the adsorption energies calculated according to Eq. (11.69) and the enrichment factor C_{gb}/C_0, determined at 1600°C from Eq. (11.68) are listed in Table 11.3 for various cations in Al_2O_3. Observed enrichment factors from the literature [47] are given in the final column.

The large enrichment observed for Ca^{2+} and Y^{3+} and the apparent lack of segregation of Mg^{2+}, Ni^{2+}, and Si^{4+} have sometimes been cited as evidence to support the applicability of the lattice strain model to solute segregation in Al_2O_3. However, measurements conducted recently with sapphire (single-crystal Al_2O_3) reveal that Mg segregates as strongly as Ca to free surfaces [48,49]. The lack of Mg segregation found in the earlier studies was attributed to the poor sensitivity of the techniques to Mg. The observation of segregation to free surfaces suggests that Mg may also segregate to grain boundaries in polycrystalline Al_2O_3. However, the unambiguous detection of Mg segregation at the grain boundaries in polycrystalline Al_2O_3 remains a source of controversy.

Coupling of Electrostatic and Elastic Strain Effects

Aliovalent dopants in ionic solids, as outlined earlier, have an excess charge. When grain boundary segregation of these dopants is induced by the elastic strain energy, the charge density in the grain boundary will be affected. A new electrostatic potential will be set up by the modified

Table 11.3 Estimated Segregation Energies (ΔG_a), Estimated Grain Boundary Enrichment $(C_{gb}/C_0)_{est}$ at 1600°C, and Observed Enrichment $(C_{gb}/C_0)_{obs}$ for Solute Ions in Alumina

Ion	Radius (10^{-10} m)	ΔG_a (kJ/mol)	$(C_{gb}/C_0)_{est}$	$(C_{gb}/C_0)_{obs}$ (Ref. 46)
Al^{2+}	0.53	—	—	—
Ca^{2+}	1.00	117	1850	120
Y^{3+}	0.89	69	85	150–200
Mg^{2+}	0.72	19	3	0
Ni^{2+}	0.69	14	2	3
Si^{4+}	0.40	9	2	0

charge distribution. Although the elastic strain energy may not directly affect other defects, the new electrostatic potential would be expected to change the equilibrium spatial distribution of the defects in the grain boundary region. Thus the electrostatic potential and strain energy interactions are coupled. Calculations [50] show that when $|U_o/eV_\infty| \gg 1$ the elastic strain energy has a significant effect on the electrostatic potential but that the effect is negligible when $|U_o/eV_\infty| \ll 1$.

Electric Dipole Effects

Charged solute ions may combine with defects with an opposite charge to form solute–vacancy complexes. This association of defects leads to a decrease in the free energy. In the simplest case, the complex is electrically neutral but has a dipole moment. Because of the changing electrostatic potential (Fig. 11.39), an electric field exists in the region of space charge. (The electric field, say, in the x direction is defined as $E_x = -dV/dx$, where V is the potential). The electric field is nonuniform; that is, it varies with distance in the space charge region. Because of this nonuniformity, an attractive force is exerted on the dipole of the solute–vacancy complex. However, calculations show that in comparison with the electrostatic potential and elastic strain energy interactions, the dipole effects can be neglected.

To summarize, electrostatic interaction, dependent on the effective charge of the dopant, and elastic strain effects, dependent on the size (ionic radius) of the dopant, are the major factors that control segregation at the grain boundaries.

E. The Role of MgO in Al_2O_3

As we outlined earlier, since the work of Coble in 1961, MgO-doped Al_2O_3 has been the subject of numerous investigations. The dopant role in this system is now believed to be understood, at least at a phenomenological level. In a thorough review, Bennison and Harmer [51] described the main obstacles that hampered progress over the years and the critical experiments and theoretical contributions that helped to overcome these obstacles.

One of the main obstacles has been the use of powders containing impurities that have masked the true effect of the MgO. A major advance was made by Harmer and coworkers (see Ref. 51), who used very high purity Al_2O_3 in a detailed investigation of the sintering and grain growth kinetics. From the results, they were able to isolate the effects of the MgO on the kinetic factors of densification and on the grain boundary mobility. At a concentration giving an Mg/Al atomic ratio of 0.025%, i.e., less than

Problems of Sintering

the solid solubility limit of ≈0.03%, it was found that the MgO (i) increased the densification rate by a factor of ≈3 through an increase in D_l, (ii) increased D_s by a factor of ≈2.5, and (iii) reduced M_b by a factor of ≈25. By combining the density–grain size diagram (Fig. 9.38) with the grain size–pore size diagram (Fig. 9.20), Harmer and coworkers produced a new microstructural map that proved to be valuable in analyzing the separate and combined effects of MgO. An example of such a map is shown in Fig. 11.40 for undoped Al_2O_3 and MgO-doped Al_2O_3. The map shows two features: a grain size versus density trajectory and a separation region where abnormal grain growth occurs. According to this map, the increase in D_l and D_s produced no change in the grain size–density trajectory. The most significant effect was produced by the reduction in M_b, which significantly reduced the separation region, moving it toward larger grain size. As a consequence, sintering can proceed to full density without entering the separation region. The dominant role of MgO is therefore seen as a reduction in M_b.

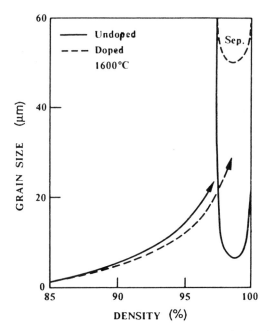

Figure 11.40 Predicted effect of simultaneously increasing lattice diffusion by a factor of 3, increasing surface diffusion by a factor of 2.5, and decreasing grain boundary mobility by a factor of 25, via doping with MgO solute, on microstructure development in Al_2O_3. (From Ref. 51, reprinted by permission of the American Ceramic Society.)

Mechanism for the Reduction in M_b

The mechanism by which MgO reduces the boundary mobility in Al_2O_3 is still the subject of some controversy. Two major mechanisms have been proposed: (i) solute drag due to Mg^{2+} segregation at the grain boundaries and (ii) pinning of the grain boundaries by fine second-phase particles of $MgAl_2O_4$. While the solute drag mechanism is believed to be operating below the solid solubility limit, a major problem is that segregation of Mg at the Al_2O_3 grain boundaries has not been detected unambiguously.

In a classic experiment, Johnson and Coble [52] provided a critical test of the two mechanisms. In the experiment, an undoped Al_2O_3 powder compact was sintered in close proximity to a preequilibrated two-phase mixture of $MgAl_2O_4$ spinel and Al_2O_3. The experiment was designed to allow transfer of MgO from the spinel-alumina compact to the undoped Al_2O_3 compact in concentrations below the solid solubility limit. It produced an Al_2O_3 pellet with a fully dense, fine-grained, MgO-doped outer surface and a core devoid of MgO and consisting of abnormal grains and entrapped porosity. Analysis of the Al_2O_3 pellet by scanning Auger microprobe failed to detect segregation of Mg at the grain boundaries of the dense outer zone: the Mg concentration at the grain boundaries was almost identical to that in the bulk of the grains. However, the analysis revealed segregation of Ca (present as an impurity in the powder) at the grain boundaries of both the dense outer zone and the porous core of the pellet, with the concentration of the segregated Ca being approximately the same in the two regions. From the experiment of Johnson and Coble, we can conclude that

1. A second phase is not necessary for MgO to function effectively as a sintering aid, and
2. The beneficial effect of MgO as a sintering aid is not due to Ca segregation.

As outlined earlier, the segregation of Mg to surfaces in sapphire (single-crystal Al_2O_3) has been revealed by more sensitive analysis using Auger spectroscopy [48]. The observation of segregation to free surfaces suggests that Mg may also segregate to grain boundaries in polycrystalline Al_2O_3 and that a solute drag mechanism may explain the reduction in M_b. However, the unambiguous detection of Mg segregation at the grain boundaries in polycrystalline Al_2O_3 is still a controverial issue.

The Ineffectiveness of Ca as a Dopant

Calcium has the same valence as magnesium and therefore should produce the same electrostatic effects. Because of its larger ionic radius, Ca should segregate more effectively at the grain boundaries in Al_2O_3. In fact, as we discussed earlier, considerable enrichment of Ca at the grain boundary is observed in most Al_2O_3 samples. On the basis of ionic charge and ionic size considerations alone, Ca should be more effective than Mg as a grain growth inhibitor. However, high-purity Al_2O_3 doped with Ca typically undergoes abnormal grain growth and cannot be sintered to theoretical density.

The distribution of Ca at the grain boundaries of polycrystalline Al_2O_3 was measured recently by Baik and Moon [53]. In their experiments, high-purity Al_2O_3 powder compacts were doped with controlled amounts of CaO (100 ppm) or MgO (300 ppm) or with a combination of CaO (100 ppm) and MgO (300 ppm). The samples were fired for 12 h at 1300, 1400, and 1500°C. Analysis of the fired samples by Auger electron spectroscopy revealed that with CaO alone some of the grain boundaries were enriched with an exceptionally high concentration of Ca. For the sample doped with both CaO and MgO, the distribution of the Ca in the grain boundaries became more homogeneous. These observations suggest that the ineffectiveness of CaO as a grain growth inhibitor may be due to its anisotropic segregation. However, further work is needed to provide a clearer understanding.

MgO as a Microstructural Homogenizer

Observations of partially sintered Al_2O_3 powder compacts typically show a fairly inhomogeneous microstructure. An example of such a microstructure is shown in Fig. 11.41, where we see dense regions in a highly porous matrix. Two categories of grain boundaries can be distinguished: those that intersect pores and are therefore active in contributing to densification (referred to as type A) and those that are entirely connected to other grain boundaries (type B). The existence of type B boundaries is most likely due to inhomogeneities in the powder compact. Densely packed regions of the powder compact undergo local densification, leading to the development of dense, pore-free regions in an otherwise porous microstructure. Normally, the dense regions will be better able to support grain growth because the drag exerted by porosity is absent.

In a careful study, Shaw and Brook [54] measured the pore surface area and the grain boundary area of undoped and MgO-doped Al_2O_3 powder compacts at various stages of sintering. The presence of the additive had no effect on the pore surface area. However, an interesting finding was that the additive increased the grain boundary area at any given density (Fig. 11.42). The raising of the grain boundary area can be interpreted as being due to the inhibition of grain growth in the densified regions; that is, the grains in these densified regions remain small. This interpretation is consistent with the ability of MgO to reduce the grain boundary mobility

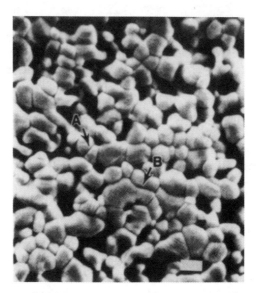

Figure 11.41 Microstructure of partially sintered Al_2O_3 (relative density 0.75) showing boundaries (A) linked to porosity and (B) linked to other grain boundaries. (Bar = 0.5 μm.) (From Ref. 54, with permission.)

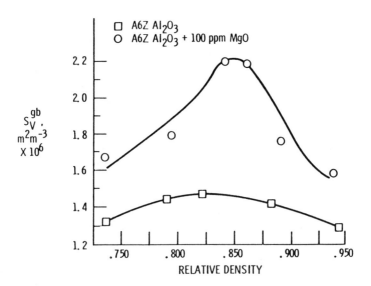

Figure 11.42 Grain boundary area per unit volume, S_v^{gb}, versus relative density during the sintering of undoped and MgO-doped Al_2O_3. The data were obtained by counting grain boundary intersections with random lines on polished sections. (From Ref. 54, with permission.)

Problems of Sintering

in Al_2O_3. In the absence of the additive, the grains in the densified regions can grow rapidly, so the grain boundary area is lowered and abnormal grain growth is promoted. In this interpretation, the function of the additive is to restrain grain growth in the densely packed regions of the powder compact until the less densely packed regions have had an opportunity to densify. The additive can be seen to act as a homogenizer of the microstructure in that it can smooth out the consequences of inhomogeneity.

In summary, MgO additions to Al_2O_3 influence all the parameters (D_l, D_{gb}, D_s, γ_b/γ_s, and M_b) to some extent. Generally the additive is effective because it influences these parameters in a favorable way. In this sense, the role of MgO in Al_2O_3 appears somewhat special. However, the most important function is the reduction in M_b. The mechanism by which M_b is reduced is not clearly understood but is most probably that of solute drag due to segregation of the dopant. The benefits of the additive for sintering are a reduction in the tendency for abnormal grain growth and a smoothing out of the consequences of inhomogeneities. Based on our understanding of the MgO-doped Al_2O_3 system, a primary consideration in the selection of additives in other systems is the effectiveness of the additive in reducing M_b.

11.6 SINTERING WITH CHEMICAL REACTION: REACTION SINTERING

Reaction sintering, sometimes referred to as *reactive sintering*, is a firing process in which the chemical reaction of the starting materials (the reactants) and the densification of the powder compact are both achieved in a single heat treatment step [55,56]. It should not be confused with *reaction bonding*, which is a fabrication process in which chemical reactions lead to the production of monolithic bodies (e.g., Si_3N_4) from powders but in which shrinkage of the sample is not a primary consideration (see Chapter 1). In reaction sintering, depending on the processing conditions such as particle size, temperature, heating rate, and applied pressure, reaction and densification can occur in sequence or concurrently or in some combination of the two. An understanding of the influence of the process variables on the rate of reaction and densification is essential for the control of the microstructure of the product.

A. Reaction Sintering Systems

Systems that undergo reaction sintering can be divided into two main classes, depending on whether single-phase solids or composites are produced.

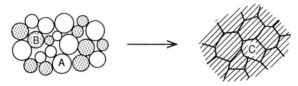

Figure 11.43 Schematic diagram illustrating the production of a single-phase solid (C) by reaction sintering of a compacted mixture of two powders (A and B).

Production of Single-Phase Solids

For a powder compact consisting of a mixture of reactant powders, the simplest example of reaction sintering is shown in Eq. (11.70) and Fig. 11.43. During firing, reaction between two starting materials (A and B) and densification occur to produce a polycrystalline, single-phase solid (C):

$$A \text{ (powder)} + B \text{ (powder)} \rightarrow C \text{ (solid polycrystal)} \qquad (11.70)$$

In most cases, A and B are themselves compounds. An example of the process described by Eq. (11.70) is the formation of zinc ferrite from a compacted powder mixture consisting of equimolar quantities of zinc oxide and ferric oxide:

$$ZnO + Fe_2O_3 \rightarrow ZnFe_2O_4 \qquad (11.71)$$

The reaction sintering process is not used industrially to produce single-phase ceramics because of difficulties in controlling the microstructure. The origins of these difficulties are examined later.

Production Multiphase Solids or Composites

A more complex example of reaction sintering in a compacted mixture of reactant powders is shown in Eq. (11.72) and Fig. 11.44. During firing,

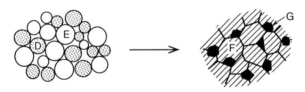

Figure 11.44 Schematic diagram illustrating the production of a composite solid, consisting of a matrix (F) and inclusions (G), by reaction sintering of a compacted mixture of two powders (D and E).

Problems of Sintering

reaction between two starting materials (D and E) and densification occur to produce a polycrystalline solid consisting of two phases (F and G):

$$D \text{ (powder)} + E \text{ (powder)} \rightarrow F \text{ (solid)} + G \text{ (solid)} \tag{11.72}$$

An example of this equation is the reaction between alumina and zircon to produce zirconia-toughened mullite:

$$3Al_2O_3 + 2(ZrO_2 \cdot SiO_2) \rightarrow 3Al_2O_3 \cdot 2SiO_2 + ZrO_2 \tag{11.73}$$

In the process, a dense product of mullite with fine zirconia inclusions is obtained. This is an attractive method for the production of zirconia-toughened mullite owing to the relatively low cost of alumina and zircon.

A variation of the process described by Eq. (11.72) is obtained when one of the phases of the product, G, say, is a liquid. It occurs when the firing temperature is above the eutectic temperature. This corresponds to the process of liquid-phase sintering that we discussed in detail in Chapter 10. Here, we will confine our discussion of reaction sintering to the case where the products are solids at the firing temperature.

B. General Features of Reaction Sintering

From the point of view of fabrication costs, reaction sintering has the benefit of eliminating the prereaction (or calcination) step in the formation of solids with complex composition. Taking $ZnFe_2O_4$ as an example, the conventional route for the production of a solid polycrystalline material involves the following steps. A loose mixture of ZnO and Fe_2O_3 powders is calcined at $\approx 800°C$ to produce $ZnFe_2O_4$ powder. The reaction normally leads to agglomeration of the product, so a milling step is also necessary prior to compaction. The compacted powder is finally sintered to produce a dense solid. In reaction sintering, the reaction and densification occur in the same heating cycle, so the calcination and associated milling steps in the conventional route are eliminated.

In practice, reaction sintering has several shortcomings. As a result, it finds little use in the production of single-phase solids. The shortcomings include (i) the risk of producing chemically inhomogeneous materials due to incomplete reaction, (ii) significantly reduced densification rates caused by microstructural changes accompanying the reaction, and (iii) difficulties in controlling the microstructure as a result of the added complexity introduced by the reaction.

As we discussed earlier in this book, the sintering of a pure single-phase powder can be a complex process, particularly when a carefully controlled microstructure is required. The addition of a powder reaction, which itself can be fairly complex, leads to further difficulties in understanding the reaction sintering process. Powder reactions, as we discussed in Chapter 2, are considerably more complicated than the idealized reactions between single crystals of the reactants. Another factor is that the driving force for a solid-state chemical reaction is normally much greater than the driving force for sintering (see Chapter 8). The occurrence of the reaction can therefore bring about considerable microstructural changes, which may be difficult to control. Furthermore, the mechanism of a solid-state chemical reaction depends on the chemistry of the reactants and will differ from one system to another.

In view of the complexity, it is to be expected that no single model of the reaction sintering process will have wide applicability. Nevertheless, as discussed by Yangyun and Brook [56], some qualitative predictions can be made about the significance of changes in the principal processing parameters such as particle size, temperature, and applied pressure. These predictions can be useful in the design of reaction sintering processes.

C. Effect of Processing Parameters

Particle Size of Reactants

Herring's scaling laws predict the effect of change of scale (particle size) on the rate of matter transport during sintering (Chapter 8). Depending on the mechanism, the densification rate varies with particle size R according to

    Densification rate $\sim 1/R^4$ grain boundary diffusion
    Densification rate $\sim 1/R^3$ lattice diffusion

For the reaction, if the product forms coherently on the surface of the particles [see Fig. 2.10 and Eq. (2.9)], the reaction rate varies as $1/R^2$. If the product does not form coherently, then the rate is expected to vary as $1/R$. The dependence of the densification rate and the reaction rate on the particle size is sketched in Fig. 11.45. Both the densification rate and the reaction rate increase with decreasing particle size. However, because of the stronger size dependence, the densification mechanism is more strongly influenced than the reaction mechanism. A reduction in the particle size increases the densification rate relative to the reaction rate.

Firing Temperature

Although densification and reaction will most likely occur by different paths (mechanisms), they both involve diffusion and are therefore expected to be thermally activated; that is, the rates have an Arrhenius dependence on temperature. The dependence of the densification rate and the reaction rate on temperature is sketched in Fig. 11.46 for the more typical situation in which the activation energy for densification, Q_d, is

Problems of Sintering

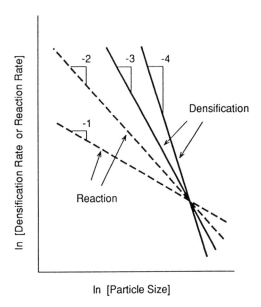

Figure 11.45 Predicted effect of particle size on the rates of densification and reaction during reaction sintering of a powder mixture. Smaller particle size enhances the rate of densification relative to the rate of reaction. (From Ref. 56.)

greater than that for the reaction, Q_r. In this case, increasing the temperature leads to an increase in the densification rate relative to the reaction rate.

Applied Pressure

The application of an external pressure p_a increases the driving force for densification. According to Eq. (8.119), for matter transport by diffusion the densification rate increases with p_a according to

Densification rate $\sim p_a + \Sigma$

where Σ is the sintering stress. For applied pressures of 20–40 MPa typically available in hot pressing, p_a is normally more than 10Σ for micrometer-size powders. The effect of applied pressure is therefore very strong. The effect of applied pressure on the reaction rate is difficult to predict. However, because of the strong effect on the densification process, the role of applied pressure is likely to lie in its ability to accelerate the densification process rather than in any effect on the reaction. In practice, hot pressing provides one of the surest ways to adjust the relative rates of densification and reaction in the desired direction.

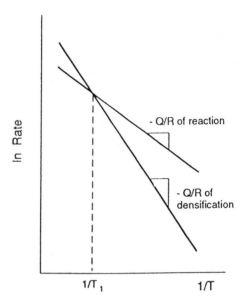

Figure 11.46 Predicted effect of sintering temperature on the rates of densification and reaction during reaction sintering of a powder mixture when the activation energy for densification is greater than that for the reaction. Higher sintering temperature enhances the densification rate relative to the reaction rate. (From Ref. 56.)

D. Processing Trajectories in Reaction Sintering

Three different trajectories in reaction sintering are sketched in Fig. 11.47. When the reaction rate is much faster than the densification rate (curve A), the reaction occurs predominantly before any significant densification. Densification must be achieved for a microstructure consisting of the fully reacted powder. Curve C shows the trajectory when the reaction rate is much slower than the densification rate. In this case, densification occurs without any significant reaction. The reaction process must be carried out in a fully dense microstructure. Curve B represents the trajectory for a system in which the reaction rate is comparable to the densification rate.

The principal processing parameters, as we discussed earlier, control the relative rates of densification and reaction, which in turn determine the processing trajectory. Faced with the problem of achieving a desired microstructure by reaction sintering, a useful approach is to determine the most appropriate processing trajectory and then choose the processing

Problems of Sintering

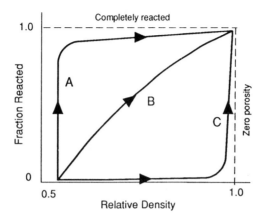

Figure 11.47 Sketch showing three different processing trajectories in reaction sintering. Curve A is the trajectory for a system in which the reaction rate is much higher than the densification rate; curve B is for a system where the reaction rate and densification rate are of roughly the same magnitude; for the system following trajectory C, the densification rate is much higher than the reaction rate. (From Ref. 56.)

conditions that favor this trajectory. However, it is likely that many systems will not be amenable to such process control. For such systems where favorable conditions are difficult to achieve, we would have to determine whether reaction sintering can still be exploited.

E. Experimental Observations of Reaction Sintering

The fabrication of single-phase and composite systems, e.g., those described by Eqs. (11.71) and (11.73), form the most significant application of reaction sintering. The microstructural requirement for each system is very distinct. In one system, a chemically homogeneous, single-phase product with high density is normally required. In the other system, in addition to high density, composites require inhomogeneity as a microstructural characteristic.

Fabrication of Single-Phase Ceramics

Systems in which two simple oxides react to form a complex oxide have been studied in detail [57]. Although the reactions in powder mixtures are not clearly understood, this system forms a good model for investigating the production of single-phase ceramics. Two examples are the formation of zinc ferrite by Eq. (11.71),

$$ZnO + Fe_2O_3 \rightarrow ZnFe_2O_4 \tag{11.71}$$

and the formation of zinc aluminate spinel by the reaction between equimolar quantities of ZnO and Al_2O_3 powders,

$$ZnO + Al_2O_3 \rightarrow ZnAl_2O_4 \tag{11.74}$$

The two systems have important similarities and differences. For both systems, the molar volume of the mixture of reactants is almost equal to that of the product. At temperatures used in sintering, the reaction between ZnO and Fe_2O_3 occurs by a Wagner counterdiffusion mechanism where the cations migrate in opposite directions and the oxygen ions remain essentially stationary. The reaction between ZnO and Al_2O_3 is not as clear. It is believed to occur by a solid-state mechanism in which the diffusion of Zn^{2+} ions through the $ZnAl_2O_4$ product layer is rate-controlling. However, the rate constant for the gas–solid reaction between ZnO vapor and Al_2O_3 is in excellent agreement with the measured reaction rate [57].

In both systems, the reaction is relatively rapid, so the mixture is almost fully reacted prior to densification. The processing trajectory therefore is close to curve A in Fig. 11.47. For the zinc ferrite system, the reaction is accompanied by very little change in volume: the powder compact expands typically by less than 3 vol %. For this system, the densification rate of the reaction-sintered powder mixture is roughly similar to that for the single-phase $ZnFe_2O_4$ powder, and final densities in excess of 95% of the theoretical can be obtained (Fig. 11.48). Furthermore, the microstructure of the reaction-sintered mixture evolves in roughly the same way as that of the single-phase $ZnFe_2O_4$ powder. Reaction sintering appears to be a viable process for the fabrication of zinc ferrite materials [58].

For the zinc aluminate system, a large expansion (typically 25–30 vol %) accompanies the reaction between ZnO and Al_2O_3. The densification rate of the reaction-sintered mixture is drastically lower than that of a single-phase $ZnAl_2O_4$ powder (Fig.11.49). The microstructure of the reaction sintered mixture also evolves very differently from that of the single-phase $ZnAl_2O_4$ powder. Reaction sintering does not appear to be viable for the production of zinc aluminate materials unless the microstructure can be improved.

A comparison of the reaction sintering of the zinc ferrite and zinc aluminate systems indicates that changes in the microstructure of the compact caused by the reaction play a critical role. If the microstructure is

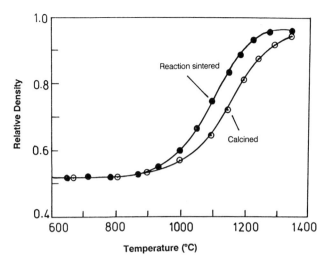

Figure 11.48 Data for the relative density versus temperature for the reaction sintering of a mixture of ZnO and Fe_2O_3 powders (reaction sintered) and for the sintering of a single-phase $ZnFe_2O_4$ powder (calcined). The samples were heated at a constant rate of 5°C/min. (From Ref. 58.)

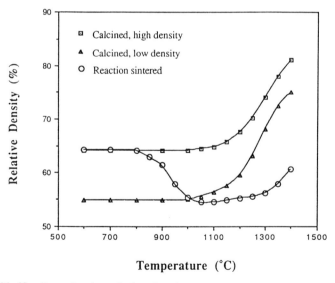

Figure 11.49 Data for the relative density versus temperature for the reaction sintering of a mixture of ZnO and Al_2O_3 powders (reaction sintered) and for the sintering of a single-phase $ZnAl_2O_4$ powder (calcined). The samples were heated at a constant rate of 5°C/min. For the reaction-sintered sample, note the expansion between 800 and 1000°C caused by the reaction and the subsequent low sinterability.

not disrupted significantly by the reaction, as in the zinc ferrite system, then the occurrence of the reaction prior to densification has little effect on the subsequent densification. However, if the reaction produces unfavorable microstructural changes, as in the zinc aluminate system, the reaction prior to densification can severely inhibit the subsequent densification. In the reaction between ZnO and Al_2O_3, the $ZnAl_2O_4$ product forms predominantly on a percolative network of the original Al_2O_3 particles. Following the reaction, the network is difficult to densify. The inhibition of sintering in the ZnO-Al_2O_3 mixture can, however, be relieved. One way is to use an applied pressure during firing (e.g., hot pressing). Another way is to modify the initial powder packing of the mixture so that the disruption of the microstructure caused by the reaction is prevented. Coating the alumina particles with zinc oxide (as opposed to the normal mechanical mixing of the powders) leads to an improved densification rate, which can approach that for the single-phase powder (Fig. 11.50).

To summarize, in the formation of single-phase ceramics by reaction sintering of compacted powder mixtures, the occurrence of a reaction prior to densification and any associated volume change are not the dominant factors. Instead, the microstructural change produced by the reaction is the dominant factor. The microstructural change is itself influenced by the reaction transport mechanism.

Fabrication of Composites

In light of the difficulties encountered in the production, by reaction sintering, of a chemically homogeneous single-phase product with high density, it appears that the process might be better applied to the fabrication of composites, i.e., materials that require inhomogeneity as a microstructural characteristic. Displacement reactions such as that described by Eq. (11.73) have been the subject of research for the fabrication of ZrO_2-toughened materials:

$$3Al_2O_3 + 2(ZrO_2 \cdot SiO_2) \rightarrow 3Al_2O_3 \cdot 2SiO_2 + ZrO_2 \qquad (11.73)$$

Provided that the reaction is not accompanied by a significant volume change, densification should occur prior to the reaction if composites with high density and a controlled distribution of second-phase particles are required. Therefore for the fabrication of composites, a processing trajectory close to curve C in Fig. 11.47 should be followed. After the attainment of high density, the precipitation of the reinforcing phase [i.e., ZrO_2 particles in Eq. (11.73)] can be carefully controlled to yield the desired size of precipitate particles. To achieve a trajectory close to curve C, processing should be carried out with fine powders and in a temperature regime chosen to maximize the ratio of the densification rate to the reaction rate.

Problems of Sintering

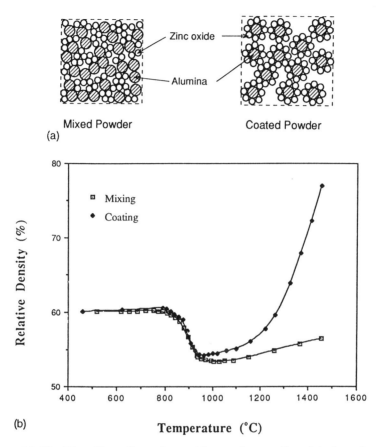

Figure 11.50 The effect of powder packing on the reaction sintering of a compacted mixture of ZnO and Al_2O_3 powders. (a) Sketch illustrating the distribution (cross-sectional view) of the ZnO and Al_2O_3 powders in compacts formed from mechanically mixed powders and from coated powders; (b) data for the sintering of the powder compacts formed from the mechanically mixed powder and from the coated powder. Note the significantly improved densification of the sample formed from the coated powder.

In the case where the reaction occurs prior to densification, the reaction may disrupt the microstructure, as outlined earlier, making it difficult to attain high density. However, even if adequate densification can be achieved, the precipitate particles may coarsen during the densification stage, making it difficult to control the microstructure.

For the reaction between zircon and alumina described by Eq. (11.73), different accounts of the reaction and densification sequence have been reported. The data of Di Rupo et al. [59] show that the densification and reaction processes occur simultaneously during hot pressing and sintering at 1450°C. However, Claussen and Jahn [60] found that densification and reaction occurred separately, with densification completed after 1 h at 1450°C prior to any reaction (Fig. 11.51). Subsequent heat treatment above 1500°C allowed the precipitation of ZrO_2 to occur rapidly, giving a distribution of fine precipitates as shown in Fig. 11.52. The mechanical strength of the composite (400 MPa) was superior to that of unreinforced mullite prepared by hot pressing (270 MPa). Finally, Yangyun and Brook [56] report a type of behavior that is intermediate between that of Di Rupo et al. and that of Claussen and Jahn. The differences in results obtained in the three studies are most probably due to differences in the chemical composition and particle size of the starting powders. The powders used were of low purity and had been milled to different extents. The powders that densified first, i.e., in the work of Claussen and Jahn, had a finer particle size and, because of severe attrition milling, a higher possible level of impurity.

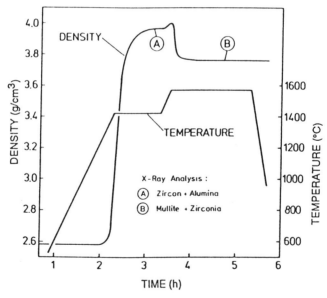

Figure 11.51 Typical sintering and reaction schedule for the reaction sintering of zircon and alumina powder mixtures. At A, only zircon and alumina are detected by X-ray analysis; at B, the reaction to produce mullite and zirconia is completed. (From Ref. 60, with permission.)

Problems of Sintering

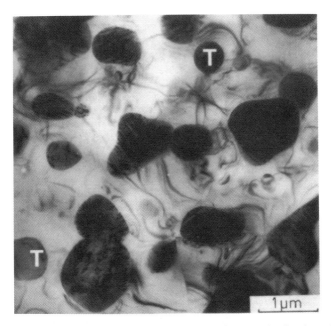

Figure 11.52 Bright field transmission electron micrograph of a zirconia-mullite material produced by reaction sintering of zircon-alumina mixtures. The material was annealed for 1 h. Larger twinned particles are monoclinic ZrO_2, and the smaller particles (T) are tetragonal zirconia. Note the considerable stress contours in the mullite matrix. (From Ref. 60, with permission.)

11.7 VISCOUS SINTERING WITH CRYSTALLIZATION

Sol-gel techniques, as discussed in Chapter 5, can be used to prepare porous dried gels (in the form of a film or monolithic solid) or fine powders, which are normally amorphous. The dried gel or compacted powder is fired to produce a dense material. Assuming that the composition is crystallizable, it is possible that crystallization and densification can occur in sequence, in combination, or in some combination of the two. This sounds similar to the situation in reaction sintering discussed in the previous section, with the process of crystallization now taking the place of the chemical reaction. By analogy with reaction sintering, we must now understand how the process variables influence viscous sintering and crystallization in order to produce a material with the desired density and structure.

Compared to reaction sintering, the objectives in the sintering of an amorphous powder appear to be clearer. For the same chemical compound, the viscosity of a crystalline material is considerably higher than that of an amorphous material. The densification of a polycrystalline material will therefore be considerably more difficult than that of an amorphous powder. This difference in sinterability suggests that the most favorable processing trajectory for amorphous materials is to achieve full density prior to crystallization. In Fig. 11.47, if we replace reaction with crystallization, the processing trajectory should be close to path C.

Crystals that form during sintering can also be considered to be inclusions in a viscous sintering matrix. This suggests that we may be able to adapt our earlier analysis of sintering with rigid inclusions to determine how the nucleation and growth of crystals influence the sintering of the material. Since the crystals grow, the increase in the volume fraction of the crystals must also be taken into account.

A. Effect of Process Variables

Viscosity

The viscosity can be altered by changes in the firing temperature or in the chemical composition of the glass. It is more meaningful to consider the influence of the viscosity instead of the separate variables of temperature or composition. In Chapter 8, we found that the rate of viscous sintering is inversely proportional to the viscosity of the glass, η. If η is constant (e.g., sintering at a fixed temperature), the extent of densification after a time t is proportional to t/η. When η changes with time (e.g., sintering at a constant heating rate), the extent of densification is proportional to $\int dt/\eta$.

The effect of viscosity on the kinetics of crystallization was considered by Zarzycki [61]. For a fixed number of nuclei formed and when the thermodynamic barrier to formation of a nucleus is constant, Zarzycki showed that the extent of crystallization also depends on the quantity t/η, or if η changes with time, on $\int dt/\eta$. Since the extents of densification and crystallization both depend on the same function of the viscosity, any factor that changes the viscosity influences both processes in such a way that the volume fraction of crystals is exactly the same by the time a given density is reached. Changes in the viscosity cannot therefore be used to vary the extent of densification relative to crystallization.

Pore Size

The crystallization process does not depend on the microstructure of the glass. However, according to the theories of viscous sintering (Chapter 8), the densification rate is inversely proportional to the pore size r. Small pores favor rapid sintering without affecting the driving force for crystallization. The reduction in the pore size provides one of the most effective methods for increasing the ratio of the densification rate to the crystallization rate; in practice, however, the pores cannot be made too small (e.g., less than ≈5 nm). This is because the sintering rate may become so rapid that full density is achieved prior to complete burnout of the organic constituents present in materials prepared, for example, by sol-gel processing. The impurities remaining in the dense solid can lead to a substantial deterioration of the properties.

Applied Pressure

Applied pressure can be very effective for increasing the densification rate if it is significantly higher than the sintering stress Σ. For pores with a radius r, the sintering stress can be taken to be $\approx \gamma_{sv}/r$, where γ_{sv} is the specific surface energy of the solid/vapor interface. Assuming $\gamma_{sv} \approx 0.25$ J/m^2 and $r \approx 10$ nm gives $\Sigma \approx 25$ MPa. This is comparable to the pressures available in hot pressing. We would therefore expect that hot pressing would have a significantly greater effect for larger pore sizes.

Compared to its effect on densification, applied pressure generally has little effect on crystallization. However, in some cases the effect of pressure on the crystallization process cannot be ignored. Crystallization usually leads to a reduction in specific volume, and the application of an applied pressure can accelerate the process.

Other Variables

As discussed later, increasing the heating rate at lower temperatures where the nucleation rate of the crystals is fast can delay the onset of crystallization. In this way, higher density can be achieved for a given amount of crystallization.

B. Analysis of Viscous Sintering with Crystallization

TTT Diagrams

One approach to the competition between viscous sintering and crystallization was put forward by Uhlmann et al. [62], who used a time–temperature–transformation (TTT) diagram to represent the results. TTT diagrams have been used by metallurgists for a long time to show the influence of thermal history (e.g., cooling rate) on phase transformations, particularly those in steels.

An example of a TTT diagram for the crystallization of a glass is sketched in Fig. 11.53. The curve is typically C-shaped and shows the times for the beginning (and sometimes the end) of crystallization at a given temperature. If sintering is to be completed prior to crystallization, the conditions must be chosen to remain to the left of the curve C_s repre-

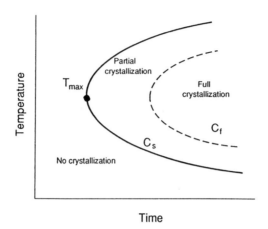

Figure 11.53 Sketch illustrating the regions in a time–temperature–transformation (TTT) diagram for the crystallization of an amorphous material.

senting the onset of crystallization. The nose of the curve, denoted by T_{max}, represents the maximum crystallization rate.

Following Brinker and Scherer [63], we outline the main steps in the construction of a TTT diagram. Assuming that the nucleation and growth rate are constant, the volume fraction of crystals produced by a given thermal history can be calculated by Avrami's equation [64],

$$v = 1 - \exp\left(-\frac{\pi I_v u^3 t^4}{3}\right) \approx \frac{\pi I_v u^3 t^4}{3} \quad (11.75)$$

where I_v is the rate of crystal nucleation, u is the rate of crystal growth, and t is time. In Eq. (11.75), the approximation applies when the volume crystallized is small. The determination of v relies on theories or measured values for I_v and u. According to the classical theory, the nucleation rate, defined as the number of stable nuclei produced per unit volume of untransformed material per unit time, varies according to [65]

$$I_v \sim \exp\left[-\frac{\Delta G^* + \Delta G_n}{kT}\right] \quad (11.76)$$

where ΔG^* is the free energy for the formation of a nucleus, ΔG_n is the free energy of activation for atomic migration across the interface between the nucleus and the untransformed material, k is the Boltzmann constant, and T is the temperature. As the temperature decreases below the liquidus, the nucleation rate increases rapidly because the thermodynamic driving

Problems of Sintering

force increases. The rate eventually decreases again because of the decreasing atomic mobility. This variation leads to the C-shaped TTT curve (Fig. 11.53) with a maximum crystallization rate at T_{max}. Below T_{max}, the crystal growth rate can be approximated by

$$u \sim T \exp(-\Delta G_c/kT) \tag{11.77}$$

where ΔG_c is the free energy of activation for atomic migration across the interface between the crystal and the uncrystallized material. It is generally assumed that $\Delta G_n \approx \Delta G_c$ and that both are equal to the activation energy Q for viscous flow [see Eq. (8.18)], i.e.,

$$\eta = \eta_0 \exp(Q/RT) \tag{11.78}$$

Substituting for Q in Eqs. (11.76) and (11.77) gives

$$I_v \sim \frac{1}{\eta} \exp\left(-\frac{\Delta G^*}{kT}\right) \tag{11.79}$$

and

$$u \sim T/\eta \tag{11.80}$$

If the viscosity of the glass is known at various temperatures, then I_v and u can be found from these two equations.

The time for completion of sintering by viscous flow is given approximately by [see Eq. (8.70)]:

$$t_f \approx \frac{\eta}{\gamma_{sv} N^{1/3}} \approx \left(\frac{4\pi}{3}\right)^{1/3}\left(\frac{\eta r}{\gamma_{sv}}\right) \tag{11.81}$$

According to this equation, t_f can be found if the pore radius r, the specific surface energy of the glass/vapor interface, γ_{sv}, and η are known.

Figure 11.54 shows an example of a TTT curve for silica gel. In the construction of the curve, I_v and u were calculated using the viscosity data obtained from sintering studies by Sacks and Tseng [66]. These values were then substituted into Eq. (11.75) to find the condition for a given volume fraction v of crystals. In Fig. 11.54, $v = 10^{-6}$ was taken to represent the onset of crystallization. The figure shows two TTT curves, one for "wet" silica with a high hydroxyl content and the other for dry silica. The viscosity of the "wet" silica is so much lower than that of the dry silica that the TTT curve for the wet silica is shifted significantly to shorter times at any temperature. Using Eq. (11.81) for t_f, the sintering curves were plotted for two values of the pore size, 5 nm and 500 nm. The curves show reasonable agreement with the crystallization data of Sacks and Tseng for "wet" silica.

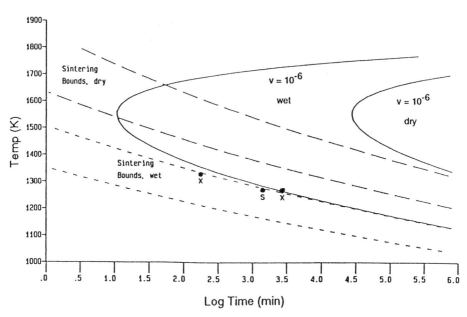

Figure 11.54 Sintering (dashed line) and crystallization (solid line) curves for dry silica and for "wet" silica with a higher hydroxyl content. The lower viscosity of the "wet" silica leads to a shift of the curves to shorter times at any temperature. X marks treatments that resulted in crystallization with little sintering, and S marks treatments that produced sintered glasses free of crystallinity. (From Ref. 62.)

Analysis in Terms of Sintering with Rigid Inclusions

The approach of Uhlmann et al. described above, where the Avrami equation is combined with the theory of viscous sintering to construct a TTT diagram is valid if the volume fraction of crystals is small. For higher volume fraction, the effect of the crystals on the densification of the material must be taken into account.

Scherer [67] analyzed the effect of crystallization on the sintering of an amorphous matrix as a case of sintering with rigid inclusions. Compared to our earlier analysis of the problem, the increase in the size and volume fraction of the inclusions (i.e., the crystals) must now be taken into account. In his analysis, Scherer combined Eq. (11.43) for the linear densification rate of a composite with the Avrami equation.

The effect of crystallization on sintering depends on where the crystals are formed and on the nature of the crystallization. Scherer considered in detail the case where the nucleation rate is relatively low and the growth rate is high. As sketched in Fig. 11.55, the crystals grow to a size that is

Problems of Sintering

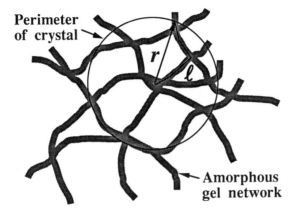

Figure 11.55 Schematic diagram illustrating the crystallization of an amorphous gel network when the nucleation rate is low and the growth is fast. Individual crystals are large compared to the scale of the gel structure. For a crystal to grow to a radius of r, it must trace a path of length l. (From Ref. 67.)

large compared to the scale of the microstructure of the gel, and the volume fraction of the crystals formed can be easily calculated. Even before the analysis, we can see an important effect from Fig. 11.55: If crystallization occurs prior to significant densification, considerable porosity will be trapped within the crystalline zone and the achievement of high density will be severely limited.

Because the effects become less difficult to predict than the more realistic case of constant heating rate sintering, we shall limit our discussion to isothermal sintering. The linear densification rate of the free matrix (i.e., the glass) can be found from the models for viscous sintering described in Chapter 8. The Mackenzie and Shuttleworth equation is used because of its simple form:

$$\dot{\epsilon}_{fm} = \frac{1}{2} \left(\frac{4\pi}{3} \right)^{1/3} \left(\frac{\gamma_{sv} N^{1/3}}{\eta} \right) \left(\frac{1}{\rho} - 1 \right)^{2/3} \quad (11.82)$$

where ρ is the relative density of the glass and the other terms have been defined earlier. The linear densification rate of the composite containing a volume fraction v of crystals is assumed to follow Eq. (11.43) for the self-consistent model, i.e.,

$$\dot{\epsilon}_c = \frac{(1 - v) \dot{\epsilon}_{fm}}{1 + v(4G_c/3K_m)} \quad (11.83)$$

The shrinkage of the body results entirely from the densification of the matrix, so that

$$\dot{\epsilon}_c = (1 - v) \frac{1}{3\rho} \left(\frac{d\rho}{dt}\right) \tag{11.84}$$

From Eqs. (11.83) and (11.84), $d\rho/dt = 3\rho\dot{\epsilon}_{fm}/[1 + v(4G_c/3K_m)]$. This and Eq. (11.82) give

$$\frac{d\rho}{d\theta} = \frac{3}{2}\left(\frac{4\pi}{3}\right)^{1/3} \left(\frac{\rho^{1/3}(1 - \rho)^{2/3}}{1 + v(4G_c/3K_m)}\right) \tag{11.85}$$

where

$$\theta = \left(\frac{\gamma_{sv}N^{1/3}}{\eta}\right) t \tag{11.86}$$

The evaluation of Eq. (11.85) requires making some assumptions about the rheology of the system. Scherer assumed a relationship that follows from the self-consistent theory, i.e.,

$$1 + v\left(\frac{4G_c}{3K_m}\right) = \frac{(1 + v_c)(1 - v)}{1 + v_c - 3(1 - v_c)v} \tag{11.87}$$

where v_c is the Poisson's ratio of the composite, given approximately by

$$v_c = \frac{v}{5} + \left(\frac{1 - v}{2}\right)\left(\frac{\rho}{3 - 2\rho}\right)^{1/2} \tag{11.88}$$

Assuming homogeneous nucleation where I_v and u are constant, Scherer derived the following equation for v in terms of a parameter C and a function of the density:

$$v = 4C\rho \int_0^\theta \exp\left[-4C \int_{\theta'}^\theta \rho(\theta'')\theta''^3 \, d\theta''\right] \theta'^3 \, d\theta' \tag{11.89}$$

where

$$C = \frac{(\pi/3)I_v u^3}{\gamma_{sv}N^{1/3}/\eta} \tag{11.90}$$

Using Eq. (11.81) for t_f, C is also given by

$$C = \frac{\pi}{3} I_v u t_f^4 \tag{11.91}$$

Problems of Sintering

Comparison of this equation with Eq. (11.75) shows that C is the volume fraction of crystals that appear during the time required for sintering of the free matrix.

Taking fixed values of C, the parameters v and ρ can be found by numerically evaluating Eqs. (11.85) and (11.75). The results show that the densification behavior is highly dependent on C. For low values of C, the porosity can be eliminated (Fig. 11.56a). However, for larger values of C, substantial porosity gets trapped within the crystals (Fig. 11.56b). When v reaches $\approx$50 vol %, densification of the matrix virtually stops because the glass is held in the interstices of the network of crystals. The matrix crystallizes completely, trapping residual porosity of $\approx$10 vol %. Densification is predicted to stop when $v \approx$ 50 vol % because Eq. (11.87) gives a sort of percolation threshold at $\approx$ 50 vol %. However, we now understand that the threshold occurs earlier, so densification is arrested by a smaller volume fraction of crystals.

To summarize, while the analysis of Scherer considers a simple model of sintering with crystallization, the results show that crystallization prior to full densification can severely limit the final density of the sample. To reduce the problem, slow heating through the low-temperature region where the nucleation rate is high should be avoided.

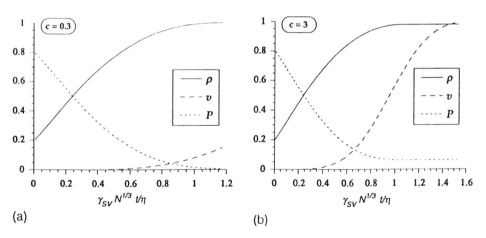

Figure 11.56 Predicted effect of crystallization on the sintering of an amorphous gel. The density of the matrix (ρ), volume fraction of crystals (v), and the porosity (P) are plotted versus reduced time for a gel with initial relative density of 0.2. Calculations are shown for two values of the parameter C: (a) $C = 0.3$ and (b) $C = 3.0$. (From Ref. 67.)

C. Experimental Observations of Sintering with Crystallization

The inhibiting effects of crystallization prior to full densification have been observed for many compositions [63]. In the case of alkali silicate gels, the crystallization rate is so fast that it is easier to prepare dense glasses by the more traditional melting route. Mullite aerogels with an initial density $\rho_0 \approx 0.05$ prepared by supercritical drying were found to be difficult to densify [68]. The density increased to $\rho \approx 0.50$ after sintering at 5°C/min to 1250°C. However, when the supercritically dried gel was compacted to a green density of $\rho_0 \approx 0.50$, a final density of $\rho \approx 0.97$ was obtained under the same sintering conditions. The difficulty of densifying the low-density aerogel can be attributed to the lower densification rate compared with the crystallization rate. Appreciable crystallization occurred and prevented the sample from reaching a high density.

Amorphous TiO_2 films prepared by sol-gel processing attained a density $\rho \approx 0.70$ after sintering at 1°C/min to 750°C [69]. The low fired density was attributed to the inhibiting effects of crystallization. When the film was heated rapidly to avoid excessive nucleation at lower temperatures, densification was completed prior to extensive crystallization. A dense crystalline material was obtained. Further characterization of the microstructure of the films revealed that densification stopped when the volume fraction of the crystals reached a certain value, most likely corresponding to the percolation threshold [70]. The shrinkage of a cordierite-type glass was observed to decrease markedly when crystallization occurred. Furthermore, by applying an axial load to the sample, it was shown that the viscosity increased significantly as the sintering rate dropped [71,72].

11.8 CONCLUDING REMARKS

In this chapter, we considered several important issues that are not taken into account in the simple sintering models. For a single phase powder, inhomogeneities in the powder compact are a common source of problems for sintering. They lead to differential densification which can severely retard the sintering rate and create microstructural defects. A normal feature of such inhomogeneities is that their effects become magnified during sintering. While a few techniques such as a limited amount of coarsening and the use of dopants may reverse the deterioration in the microstructure, the most practical approach is to minimize the extent of inhomogeneities in the powder compact. In composites, rigid inclusions such as reinforcing particles, platelets, or fibers can severely retard the sintering rate of a powder matrix, particularly in the case of a polycrystalline matrix. Realis-

Problems of Sintering

tic models predict that the stresses due to differential densification between the inclusions and the sintering matrix are relatively small. The major impediments to densification can be explained in terms of processing-related effects such as network formation of the inclusion phase and packing inhomogeneities. The sintering of thin films attached to a rigid substrate is impeded by transient stresses caused by the constrained sintering in the plane of the film. The constrained film is always predicted to sinter more slowly than the free film. Continuum models provide an adequate explanation of the sintering of amorphous films. The sintering of polycrystalline films is observed to be qualitatively different from that of amorphous films and is not adequately described by the models.

We discussed the present understanding of the role of solid solution additives (dopants) in the sintering of ceramics. A dopant can potentially influence all of the kinetic and thermodynamic factors in sintering. However, when effective, the major dopant role appears to be a reduction in the grain boundary mobility. The effectiveness of dopants is dependent on their ability to segregate at the grain boundaries. Segregation in turn is controlled by electrostatic effects due to a space charge and elastic strain energy effects due to size misfit between the dopant ion and the host.

The principles governing the use of reaction sintering for the production of ceramics and composites were considered. The principal processing parameters (e.g., particle size, temperature, and applied pressure) control the relative rates of densification and chemical reaction. Successful fabrication depends on the ability to achieve a processing path in which full densification occurs prior to the reaction.

Finally, we discussed the major issues in the densification of amorphous materials that can crystallize during firing. Crystallization prior to full densification can lead to the retention of considerable porosity in the fabricated article. The processes of crystallization and densification can normally be decoupled by changes in the microstructure (pore size) and the applied pressure. High density can be achieved by increasing the rate of densification relative to crystallization or by delaying the onset of crystallization through careful manipulation of the key processing parameters.

PROBLEMS

11.1 Is grain growth beneficial for sintering? Discuss.
11.2 Compare the sintering of a powder compact with a broad particle size distribution with that for a compact with a narrow size distribution if the average particle size of the powders is the same.
11.3 Derive Eqs. (11.2) through (11.4).

11.4 For a powder matrix containing rigid inclusions, if the volume fraction, v_f, of the inclusions in the fully dense composite is 0.2, plot the inclusion volume fraction, v_i, as a function of the relative density of the matrix, ρ_m, for values in the range $0.5 \leq \rho_m \leq 1.0$.

11.5 The sintering of an Al_2O_3 powder compact (Fig. 11.8) formed from a mixture of coarse and fine powders approximately follows the predictions of a rule of mixtures. However, the sintering of a compact consisting of a fine Al_2O_3 powder matrix and coarse, rigid inclusions of SiC is significantly lower than the predictions of a rule of mixtures. Discuss the difference in sintering behavior between the two systems.

11.6 Derive Eq. (11.6).

11.7 Schematically sketch the variation of the electrostatic potential and defect concentrations as functions of the distance from the grain boundary for ZrO_2, assuming that anion vacancies are the dominant point defects and that the formation energy of an anion vacancy is smaller than that of a cation vacancy. Repeat the procedure for ZrO_2 doped with CaO, assuming that the defect concentrations produced by the dopant are much larger than those generated thermally.

11.8 Consider the reaction sintering of a compact consisting of equimolar mixtures of Al_2O_3 and ZnO powders with the same particle size. If the $ZnAl_2O_4$ reaction product forms on the network of Al_2O_3 particles, estimate the change in volume accompanying the reaction process. In practice (see Fig. 11.49), a volume expansion of 20–30% is observed. Suggest an explanation for the difference between the estimated and observed values. The densities of Al_2O_3, ZnO and $ZnAl_2O_4$ are 3.96, 5.61 and 4.58 g/cm^3, respectively.

REFERENCES

1. A. G. Evans, *J. Am. Ceram. Soc.*, 65:497 (1982).
2. R. Raj and R. K. Bordia, *Acta Metall.*, 32:1003 (1984).
3. F. F. Lange, *J. Am. Ceram. Soc.*, 66:396 (1983).
4. H. E. Exner, *Rev. Powder Metall. Phys. Ceram.*, 1:1 (1979).
5. M. W. Weiser and L. C. De Jonghe, *J. Am. Ceram. Soc.*, 69:822 (1986).
6. T-S. Yeh and M. D. Sacks, *Ceram. Trans.*, 7:309 (1990).
7. F. F. Lange, *J. Am. Ceram. Soc.*, 67:83 (1984).
8. M. P. Harmer and J. Zhao, in *Ceramic Microstructures '86:Role of Interfaces* (Mater. Sci. Res. Vol. 21) (J. A. Pask and A. G. Evans, Eds.), Plenum, New York, 1987, pp. 455–464.
9. M. Lin, M. N. Rahaman, and L. C. De Jonghe, *J. Am. Ceram. Soc.*, 70: 360 (1987).

10. M.-Y. Chu, L. C. De Jonghe, M. K. F. Lin, and F. J. T. Lin, *J. Am. Ceram. Soc.*, 74:2902 (1991).
11. J. P. Smith and G. L. Messing, *J. Am. Ceram. Soc.*, 67:238 (1984).
12. L. C. De Jonghe, M. N. Rahaman, and C. H. Hsueh, *Acta Metall.*, 34:1467 (1986).
13. M. N. Rahaman and L. C. De Jonghe, *J. Am. Ceram. Soc.*, 70:C-348 (1987).
14. S. P. Timoshenko and J. N. Goodier, *Theory of Elasticity*, 3rd ed., McGraw-Hill, New York, 1970.
15. R. M. Christensen, *Theory of Viscoelasticity: An Introduction*, 2nd ed., Academic, New York, 1982.
16. G. W. Scherer, *Relaxation in Glass and Composites*, Wiley-Interscience, New York, 1986.
17. N. G. McCrum, C. P. Buckley and C. B. Bucknall, *Principles of Solid Polymer Engineering*, Oxford Univ. Press, Oxford, 1989, Chap. 4.
18. C. H. Hsueh, A. G. Evans, R. M. Cannon and R. M. Brook, *Acta Metall.*, 34:927 (1986).
19. R. K. Bordia and G. W. Scherer, *Acta Metall.*, 36:2393, 2399, 2411 (1988).
20. G. W. Scherer, *J. Am. Ceram. Soc.*, 70:719 (1987).
21. G. W. Scherer, *J. Am. Ceram. Soc.*, 67:709 (1984).
22. G. W. Scherer, *J. Non-Cryst. Solids*, 34:239 (1979).
23. R. Zallen, *The Physics of Amorphous Solids*, Wiley, New York, 1983, Chap. 4.
24. G. W. Scherer, *Ceram. Bull.*, 70:1059 (1991).
25. G. W. Scherer and A. Jagota, in *Ceramic Transactions*, Vol. 19 (M. D. Sacks, Ed.), American Ceramic Society, Westerville, OH 1991, pp. 99–109.
26. A. Jagota and G. W. Scherer, *J. Am. Ceram. Soc.*, 76:3123 (1993).
27. R. E. Dutton and M. N. Rahaman, *J. Am. Ceram. Soc.*, 75:2146 (1992).
28. F. F. Lange, *J. Mater. Res.*, 2:59 (1987).
29. O. Sudre and F. F. Lange, *J. Am. Ceram. Soc.*, 75:519 (1992).
30. O. Sudre, G. Bao, B. Fan, F. F. Lange, and A. G. Evans, *J. Am. Ceram. Soc.*, 75:525 (1992).
31. C. L. Hu and M. N. Rahaman, *J. Am. Ceram. Soc.*, 75:2066 (1992); 76:2549 (1993); 77:815 (1994).
32. G. W. Scherer and T. J. Garino, *J. Am. Ceram. Soc.*, 68:216 (1985).
33. R. K. Bordia and R. Raj, *J. Am. Ceram. Soc.*, 68:287 (1985).
34. T. J. Garino and H. K. Bowen, *J. Am. Ceram. Soc.*, 73:251 (1990).
35. K. T. Miller, F. F. Lange and D. B. Marshall, *J. Mater. Res.*, 5:151 (1990).
36. R. L. Coble, *J. Appl. Phys.*, 32:793 (1961).
37. R. L. Coble, U.S. Patent 3,026,210 (March 1962).
38. R. J. Brook, in *Ceramic Fabrication Processes* (Treatise Mater. Sci. Technol., Vol. 9) (F. F. W. Wang, Ed.), Academic, New York, 1976, pp. 331–364.
39. F. A. Kroger, *J. Am. Ceram. Soc.*, 66:730 (1983).
40. R. J. Brook, *Proc. Br. Ceram. Soc.*, 32:7 (1982).
41. C. A. Handwerker, J. M. Dynys, R. M. Cannon, and R. L. Coble, *J. Am. Ceram. Soc.*, 73:1365, 1371 (1990).
42. W. D. Kingery, *J. Am. Ceram. Soc.*, 57:1, 74 (1974).

43. S.-L. Hwang and I-W. Chen, *J. Am. Ceram. Soc.*, *73*:3269 (1990).
44. Y.-M. Chiang and T. Takagi, *J. Am. Ceram. Soc.*, *73*:3278 (1990).
45. J. D. Eshelby, in *Solid State Physics*, Vol. 3 (F. Seitz and D. Turnbull, Eds.), Academic, New York, 1956, p. 79.
46. D. McLean, *Grain Boundaries in Metals*, Clarendon Press, Oxford, 1957.
47. W. C. Johnson, *Metall. Trans.*, *8A*:1413 (1977).
48. S. Baik, *J. Am. Ceram. Soc.*, *69*:C101 (1986).
49. S. Baik and C. L. White, *J. Am. Ceram. Soc.*, *70*:682 (1987).
50. M. F. Yan, R. M. Cannon, and H. K. Bowen, *J. Appl. Phys.*, *54*:764 (1983).
51. S. J. Bennison and M. P. Harmer, *Ceram. Trans.*, *7*:13 (1990).
52. W. C. Johnson and R. L. Coble, *J. Am. Ceram. Soc.*, *61*:110 (1978).
53. S. Baik and J. H. Moon, *J. Am. Ceram. Soc.*, *74*:819 (1991).
54. N. J. Shaw and R. J. Brook, *J. Am. Ceram. Soc.*, *69*:107 (1986).
55. D. Kolar, *Sci. Ceram.*, *11*: 199 (1981).
56. S. Yangyun and R. J. Brook, *Sci. Sintering*, *17*:35 (1985).
57. H. Schmalzried, *Solid State Reactions*, Academic, New York, 1974, pp. 101–102.
58. M. N. Rahaman and L. C. De Jonghe, *J. Am. Ceram. Soc.*, *76*:1739 (1993).
59. E. Di Rupo, E. Gilbart, T. G. Carruthers, and R. J. Brook, *J. Mater. Sci.*, *14*:705 (1979).
60. N. Claussen and J. Jahn, *J. Am. Ceram. Soc.*, *63*:228 (1980).
61. J. Zarzycki, in *Advances in Ceramics*, Vol. 4, American Ceramic Society, Columbus, OH, 1982, pp. 204–216.
62. D. R. Uhlmann, B. J. Zelinski, L. Silverman, S. B. Warner, B. D. Fabes, and W. F. Doyle, in *Science of Ceramic Chemical Processing* (L. L. Hench and D. R. Ulrich, Eds.), Wiley, New York, 1986, pp. 173–183.
63. C. J. Brinker and G. W. Scherer, *Sol-Gel Science*, Academic, New York, 1990, Chap. 11.
64. M. Avrami, *J. Chem. Phys.*, *7*:1103 (1939); *8*:212 (1940); *9*:177 (1941).
65. J. W. Christian, *The Theory of Phase Transformations in Metal and Alloys*, 2nd ed., Pergamon, New York, 1975.
66. M. D. Sacks and T.-Y. Tseng, *J. Am. Ceram. Soc.*, *69*:532 (1984).
67. G. W. Scherer, *Mater. Res. Soc. Symp. Proc.*, *180*:503–514 (1990).
68. M. N. Rahaman, L. C. De Jonghe, S. L. Shinde, and P. H. Tewari, *J. Am. Ceram. Soc.*, *71*:C338 (1988).
69. J. L. Keddie and E. P. Giannelis, *J. Am. Ceram. Soc.*, *74*:2669 (1991).
70. J. L. Keddie, P. V. Braun, and E. P. Giannelis, *J. Am. Ceram. Soc.*, *77*:1592 (1994).
71. P. C. Panda and R. Raj, *J. Am. Ceram. Soc.*, *72*:1564 (1989).
72. P. C. Panda, W. M. Mobley, and R. Raj, *J. Am. Ceram. Soc.*, *72*:2361 (1989).
73. A. Jagota and C. Y. Hui, *Mechanics of Materials*, *9*:107 (1990); *11*:221 (1991).
74. R. K. Bordia and A. Jagota, *J. Am. Ceram. Soc.*, *76*:2475 (1993).

12
Densification Process Variables and Densification Practice

12.1 INTRODUCTION

During firing, the microstructure and hence the engineering properties of the final article are developed. Processing prior to firing, as discussed in detail earlier, has a significant effect on the microstructure. However, assuming that proper powder preparation and consolidation procedures are in effect, successful fabrication is still dependent on the ability to control the microstructure through manipulation of the process variables in the firing stage. In this chapter, we will describe the main process variables, methods, and technology associated with the densification of the powder compact during firing.

The methods used in firing can be divided into two broad classes: (i) firing without the application of an external pressure to the sample, referred to as *conventional*, *free*, or *pressureless sintering*, and (ii) firing with an externally applied pressure, referred to as *pressure sintering*. Conventional sintering is the preferred method because it is more economical. However, the additional driving force available in pressure sintering normally guarantees the attainment of high density and a fine-grained microstructure.

Heating of the powder compact is commonly achieved with electrical resistance furnaces that allow temperatures as high as $\approx 2500°C$ to be attained. The main process variables in conventional sintering are the firing schedule (temperature as a function of time) and the atmosphere surrounding the powder compact. The control of these two parameters can have a decisive effect on the ability to achieve the desired microstructure and properties. In the firing schedule, the heating rate, firing temperature, and, in some cases, the cooling rate are the variables that must be carefully controlled. For some ceramics, annealing at temperatures below the firing temperature is also necessary. The atmosphere is important for controlling processes such as decomposition, evaporation of volatile constituents, and vapor transport. It can also influence the oxidation number of atoms (particularly those of the transition elements) and the stoichiometry of the solid. Ferrites, for example, form an important class of ceramics for which atmospheric control is essential for developing optimum magnetic properties.

Recently, the use of microwave energy for heating and sintering ceramics has attracted considerable attention. This method of heating is fundamentally different from that in conventional furnaces in that heat is generated internally by interaction of microwaves with the material. Conventional sintering using microwave heating is commonly referred to as *microwave sintering*. Studies have shown considerable enhancement of sintering with microwave heating. However, the control of microwave sintering is more complicated than that of conventional sintering.

Pressure sintering is commonly used for the production of high-cost ceramics and prototype ceramics, for which high density must be guaranteed. It also provides an effective additional variable for the study of sintering mechanisms. In ceramics, the commonly used pressure sintering methods are (i) hot pressing, during which pressure is applied uniaxially to the powder in a die, and (ii) hot isostatic pressing, during which pressure is applied isostatically by means of a gas. Although hot pressing is simpler and more widely used, hot isostatic pressing has seen considerable increase in its use in recent years.

12.2 CONVENTIONAL SINTERING

Sintering without the use of an externally applied pressure is referred to as conventional, free, or pressureless sintering. No term has won general acceptance, and we shall make no distinction between them. Two methods can be used to heat the consolidated powder form to the sintering temperature: (i) heating in a furnace and (ii) microwave heating. In the commonly used method, the sample is placed in a furnace where heat (radiation) from a hot source (e.g., the electrically heated furnace element) is transferred to the sample (Fig. 12.1a). A different principle is used in microwave heating, where the absorption of microwaves causes the sample to generate heat internally (Fig. 12.1b). The use of microwave heating in the sintering of ceramics is a relatively recent development; work is still going on to achieve a better understanding of the interaction of microwaves with ceramics and to achieve better control of the process. We shall consider microwave sintering in a separate section later.

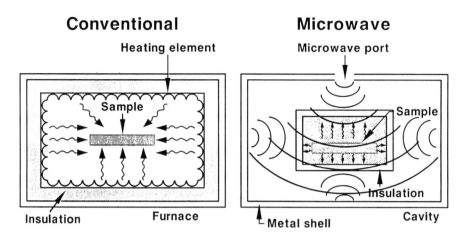

Figure 12.1 Heating patterns in conventional and microwave furnaces. (From Ref. 20, with permission.)

A. Furnaces

A variety of furnaces are available commercially to provide the temperatures used in sintering. Sizes range from those for small, research-type powder compacts to large furnaces that can accommodate parts several feet in diameter for industrial production. The most common types are electrical resistance furnaces in which a current-carrying resistor, commonly called the *furnace element* or *winding*, provides the source of heat. In addition to size and cost, important considerations in the selection of a furnace are (i) its maximum temperature capability and (ii) the atmosphere in which it can be operated for extended periods (Table 12.1). Several metal alloys (e.g., nichrome) can be used as furnace elements for temperatures up to $\approx 1200°C$ in both oxidizing and reducing atmospheres. For extended use in air, other furnace elements can deliver higher temperatures: e.g., tungsten, up to $\approx 1400°C$; silicon carbide (SiC), 1450°C; and molybdenum disilicide ($MoSi_2$), 1600°C. Lanthanum chromite ($LaCrO_3$) can be used in oxidizing atmospheres for temperatures up to $\approx 1750°C$.

For higher temperatures, furnace elements consisting of various refractory metals (e.g., Mo, W, and Ta) can be used in vacuum, inert, or reducing atmospheres (e.g., Ar, He) up to $\approx 2000°C$. At these high temperatures, the metallic elements must not be exposed to atmospheres containing carbon or nitrogen because they readily form carbides and nitrides,

Table 12.1 Heating Elements Commonly Used in Electrical Resistance Furnaces

Material	Maximum temperature (°C)	Furnace atmosphere
Nichrome	1200	Oxidizing; inert; reducing
W	1400	Oxidizing
SiC	1450	Oxidizing
$MoSi_2$	1600	Oxidizing
$LaCrO_3$	1750	Oxidizing
Mo, W, Ta	2000	Vacuum; inert; reducing[a]
C (graphite)	2800	Inert; reducing[b]

[a] Atmosphere containing nitrogen or carbon must be avoided above ≈1500°C.
[b] Vacuum may be used below ≈1500°C.

which reduce the life of the element considerably. Graphite furnace elements, heated electrically or by induction, can provide temperatures up to ≈2800°C in inert or reducing atmospheres. For extended use of graphite furnace elements above ≈2000°C, high-purity helium from which the trace impurities of oxygen have been removed (e.g., by passing the gas through an oxygen getter prior to entering the furnace) provides a useful atmosphere.

A temperature controller forms an integral part of most furnaces to take the system through the required heating cycle. In many cases it is necessary to have fairly precise control of the sintering atmosphere (i.e., around the powder compact) or to use a sintering atmosphere that is different from the atmosphere around the furnace element. This requirement is most easily achieved with the use of a "tube furnace" (Fig. 12.2). Silica

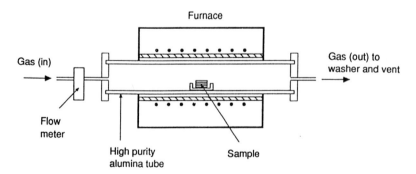

Figure 12.2 Schematic diagram of a tube furnace used for controlled atmosphere sintering.

Densification Process

tubes can be used for temperatures up to ≈1100°C, and high-purity alumina tubes can be used up to temperatures in the range of ≈1600–1800°C (depending on whether the axis of the tube is arranged horizontally or vertically). Because of alumina's low resistance to thermal shock, alumina tubes limit the rate at which the furnace can be heated or cooled (typically less than 10–20°C/min, depending of the diameter of the tube). A convenient way to heat the powder compact faster is to preheat the furnace to the required temperature and push the compact into the hot zone at a controlled rate.

B. Measurement of the Sintering Kinetics

As observed in the earlier chapters, kinetic data for the densification of the powder compact during sintering are important practically and theoretically. They can be obtained as functions of time or temperature by two methods: (i) intermittently, by measuring the density of a powder compact fired to a given temperature or for a given period of time, or (ii) continuously, by the technique of dilatometry. The intermittent method is fairly time-consuming. A different sample is used for each run, and the density of the fired sample is determined from its mass and dimensions (for a regular geometry), by Archimedes' method, or by both of these methods.

Dilatometry provides continuous monitoring of the shrinkage of the powder compact over the complete firing schedule. A variety of dilatometers are available commercially for providing the required accuracy in shrinkage measurement as well as control of the temperature and atmosphere. The equipment is identical to that commonly used to measure the coefficient of thermal expansion of solids (Fig. 12.3). The change in length

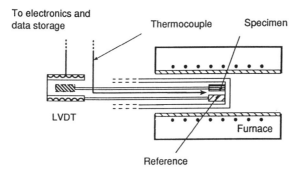

Figure 12.3 Schematic diagram of a dual pushrod dilatometer used for continuous monitoring of shrinkage kinetics.

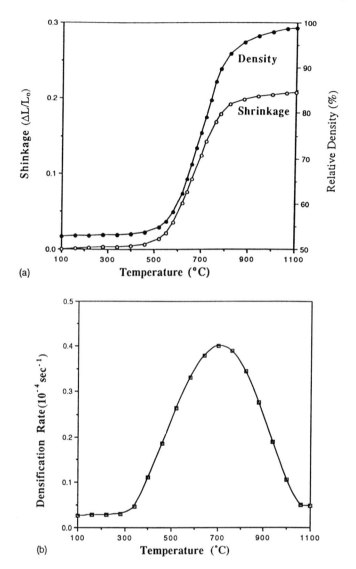

Figure 12.4 Sintering curves for ZnO in air during constant heating rate sintering at 5°C/min obtained from the dilatometer curve, showing (a) the shrinkage and density as a function of temperature and (b) the densification rate as a function of temperature.

Densification Process

of the powder compact is monitored by a linear voltage displacement transducer (LVDT), the output of which is plotted on a recorder or stored in a computer. Commonly, the shrinkage, $\Delta L/L_0$, is plotted as a function of time or temperature (where ΔL = change in length and L_0 = initial sample length). If the mass of the powder compact remains constant and the shrinkage is isotropic, the relative density can be determined from the shrinkage data using the equation

$$\rho = \frac{\rho_0}{(1 - \Delta L/L_0)^3} \qquad (12.1)$$

where ρ_0 is the initial relative density of the compact. It is sometimes useful to determine the densification rate $\dot\rho$, defined as $(1/\rho)d\rho/dt$. The procedure involves fitting a smooth curve through the data for density versus time and differentiating to find the slope at any given time. As an example, Fig. 12.4 shows the data obtained by dilatometry for a ZnO powder compact (ρ_0 = 0.52) that was sintered at a constant heating rate of 5°C/min to ≈1100°C. The data provide a valuable source of information for the densification behavior. In this case they show that shrinkage (or densification) starts at ≈500°C and is almost complete at ≈1000°C (Fig. 12.4a). The maximum densification rate occurs at ≈800°C, corresponding to a density of ≈0.80 (Fig. 12.4b).

C. Isothermal and Constant Heating Rate Sintering

In the simplest situations, there are two heating schedules commonly used in sintering experiments (Fig. 12.5). In one case the powder compact is

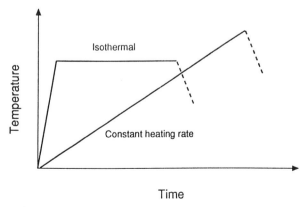

Figure 12.5 Sketch of the temperature–time schedule in (a) isothermal sintering and (b) constant heating rate sintering.

heated rapidly to a fixed sintering temperature, held at this temperature for the required time, and finally cooled to room temperature. This schedule is referred to as *isothermal sintering*. In the other case, referred to as *constant heating rate sintering*, the compact is heated at a fixed rate to the required temperature, after which it is cooled.

For powder systems in which the coarsening rate is higher than the densification rate at lower temperatures (see Fig. 9.40), isothermal sintering may have an advantage from the point of view of achieving high density; the powder compact spends less time at lower temperatures where coarsening dominates over densification. We outlined this argument earlier in Chapter 9; the practical realization of the advantage will be described later when we come to the topic of fast firing. Theoretically, isothermal sintering also has an advantage in that the data can be more easily compared with the predictions of the sintering models. However, the sintering equipment normally places a limit on the maximum heating rate. If the rate is relatively slow, considerable densification can occur during the heat-up stage and the data for this nonisothermal stage will be unsuitable for comparison with the predictions of the models.

Constant heating rate data are more difficult to analyze theoretically because (i) the mechanisms of sintering may change over the temperature range and (ii) the activation energies required for compensation of the temperature changes are normally unavailable. However, the constant heating rate schedule is more relevant to the situation in industrial fabrication in that the large furnaces or large parts cannot be subjected to fast heating rates (i.e., rates typical of the heat-up stage in isothermal sintering).

The two simple heating schedules of isothermal and constant heating rate sintering find more use in research experiments. As we shall see later, the schedule can be more complicated in practice.

D. Sintering Process Variables

Two of the most important variables in conventional sintering are the firing schedule and the atmosphere surrounding the sample. These two variables, in particular the firing schedule, have received considerably less attention than other processing variables (e.g., particle size). In many cases an arbitrary schedule is used in some convenient atmosphere. However, the effects of these two variables are fairly well understood, and their control can provide an effective and important approach for altering the microstructure of the fabricated article in the desired direction.

E. Control of the Heating Schedule

For single-phase powders that do not undergo a phase transformation and have been consolidated without a binder, the heating schedule can be relatively simple. The powder compact is heated rapidly or at a controlled rate to an isothermal sintering temperature, held at this temperature for the required time to accomplish the desired density, and finally cooled (Fig. 12.5). In general, the heating schedule for more complex systems can consist of several stages. An example of a multistage heating schedule is shown in Fig. 12.6. However, for many industrial systems, it is possible to find a more complex schedule.

Stage 1. Binder Burnout. As shown in Fig. 12.6, stage 1 may consist of a binder burnout stage during which the sample is heated slowly (typically <5°C/min) and held at a temperature in the range of 400–500°C for the required time. As we outlined in Chapter 6, the time required for binder burnout depends on the size of the article and the amount of binder. It can vary from ≈1 h for small pellets used in research to several days for large injection-molded parts. During this stage, an oxidizing atmosphere is beneficial for complete removal of the binder. Carbonaceous residues can react with the powder to form unwanted phases (e.g., oxycarbides in some oxide ceramics) and microstructural flaws in the fabricated article. If the powder compact contains phases that can undergo significant oxidation at the binder burnout temperature, removal of the binder must be accomplished in an inert atmosphere (e.g., Ar). In this case, a binder with good burnout characteristics in a nonoxidizing atmosphere should be selected at the outset of processing.

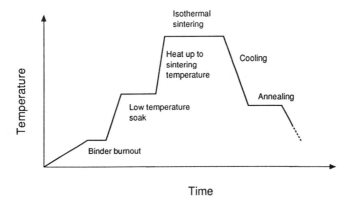

Figure 12.6 Sketch of a general heating schedule.

Stage 2. Low-Temperature Soak. The second stage of the heating schedule may involve a low-temperature soak. The powder compact is heated to a temperature below that of the onset of densification but sufficiently high to cause chemical homogenization or to modify the microstructure. Chemical homogenization may, for example, involve a solid-state reaction in which a small amount of dopant is incorporated into the powder or a reaction leading to the formation of a liquid phase. The precoarsening of a heterogeneous powder compact described in Chapter 11 provides an example of the type of microstructural modification that can be accomplished during this stage. However, this type of microstructural modification is rarely performed.

Stage 3. Heat-Up to the Sintering Temperature. The heat-up stage involves heating the powder compact to the isothermal sintering temperature. The rate of sintering should be carefully controlled to develop the required microstructure in the final article. It is often observed that a faster heating rate in this stage enhances the densification in the subsequent isothermal sintering stage. A possible explanation of this observation is that the coarsening of the powder during heat-up is reduced due to the shorter time taken to reach the isothermal sintering temperature. Thus, the powder compact has a finer microstructure at the start of isothermal sintering. Fast heating rates can be achieved by pushing the powder compact at a controlled rate into a furnace preheated to the desired temperature. However, for large articles it is important to limit rapid temperature changes so as to avoid large temperature gradients between the surface and the interior of the article. Large gradients can lead to cracking. They can also lead to the premature densification of a surface layer that hinders the densification of the interior of the compact.

Stage 4. Isothermal Sintering. The isothermal sintering temperature is chosen to be as low as possible yet compatible with the requirement that densification be achieved within a reasonable time (typically less than a ≈ 5 h). Longer sintering times may be needed if the sintering process involves a chemical homogenization between reactant powders. Higher sintering temperatures lead to faster densification, but the rate of coarsening also increases. The increased coarsening rate may lead to abnormal grain growth with pores trapped inside large grains. Thus, although densification proceeds faster, the final density may be limited. Figure 12.7 shows qualitatively some possible sintering profiles for different sintering temperatures.

Stage 5. Cooling. After densification, the article is cooled to room temperature or, in some cases, to an annealing temperature at which thermal stresses are reduced or some modification of the chemical composition

Densification Process

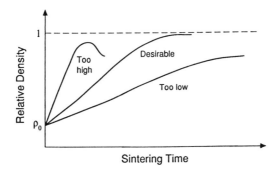

Figure 12.7 Sketch of sintering curves to illustrate the selection of a desirable isothermal sintering temperature.

or of the microstructure is carried out. The cooling rate can be fairly fast for relatively small articles. However, for large articles the cooling rate is much slower to prevent large temperature gradients that can lead to cracking. When compositional or microstructural modification must be achieved, the cooling rate needs to be carefully controlled. As discussed later, some functional ceramics (e.g., ferrites) also require that the cooling be carried out in a carefully controlled atmosphere to control the stoichiometry and microstructure.

The cooling rate can influence the precipitation of second phases and their distribution in the fabricated article. Strontium-doped TiO_2 provides a good example of this type of effect [1]. The solubility limit of Sr in TiO_2 is slightly less than 0.5 mol % at 1400°C. Even for concentrations below the solubility limit, cooling from 1400°C leads to the precipitation of $SrTiO_3$ at the grain boundaries. This precipitation is probably due to the segregation of Sr ions, which have a large size misfit in the TiO_2 lattice. However, the morphology of the precipitates depends on the cooling rate. Well below the solubility limit (0.2 mol % Sr), slow cooling from 1400°C leads to the precipitation of fine discrete particles. However, if the same composition is cooled rapidly to room temperature, no second phase is formed. Near the solubility limit (0.5 mol %), slow cooling leads to the precipitation of a continuous second phase between 1400 and 1250°C, but this breaks up on further cooling below 1200°C to produce discrete particles that have a length several times their width. Rapid cooling of this composition to room temperature leads to a continuous second phase. Well above the solubility limit (2 mol % Sr), the $SrTiO_3$ is precipitated as fine particles at the grain boundaries and within the grains regardless of the cooling rate. The effect of cooling rate on the microstructures of Sr-doped TiO_2 is shown in Figs. 12.8 and 12.9.

In certain devices (e.g., varistors and some capacitors), it is desirable to have a continuous layer of second phase along the grain boundaries. The results of the Sr-doped TiO_2 system indicate that the cooling rate can sometimes provide an important parameter for such control of the microstructure. For example, in this system we would use a strontium concentration slightly less than 0.5 mol % to prevent precipitation within the grains. After sintering at 1400°C, an optimum heat treatment will involve slow cooling between 1400 and 1300°C to allow sufficient precipitation of a continuous $SrTiO_3$ phase along the grain boundaries, followed by rapid cooling below 1300 °C.

Stage 6. Annealing. Many materials require an isothermal heat treatment or annealing at an elevated temperature (but below the sintering temperature) to (i) reduce thermal stresses or (ii) modify the chemical composition or the microstructure. Annealing to reduce thermal stresses is common in systems that contain a glassy matrix or that undergo a crystallographic transformation involving a fairly large volume change. Modification of the chemical composition and the microstructure is fairly common in many

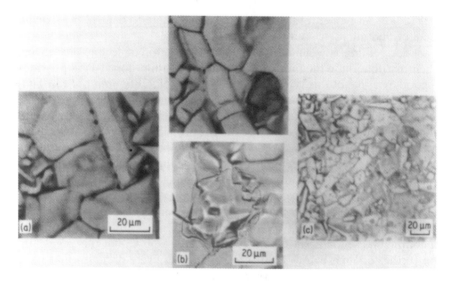

Figure 12.8 Optical micrographs of Sr-doped TiO_2 with (a) 0.2, (b) 0.5, and (c) 2.0 mol % Sr after sintering for 2 h at 1400°C and slow cooling to room temperature. (From Ref. 1.)

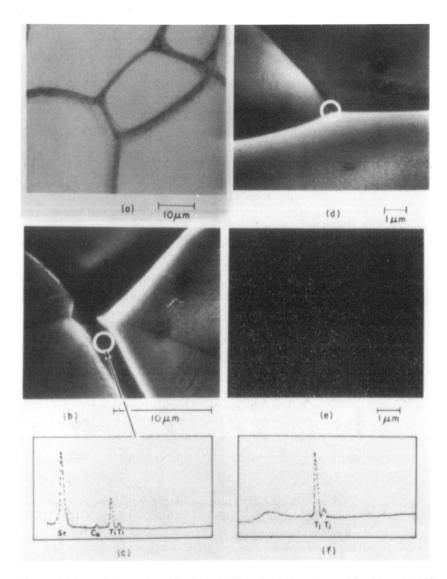

Figure 12.9 Micrographs of Sr-doped TiO_2 after being sintered for 2 h at 1400°C and quenched to room temperature. (a) 0.5 mol % Sr, optical micrograph; (b) 0.5 mol % Sr, scanning electron micrograph; (c) the associated X-ray analysis; (d) 0.2 mol % Sr, scanning electron micrograph; (e) strontium mapping of the same area; (f) X-ray analysis of the triple junction circled in (d). (From Ref. 1.)

functional ceramics. Annealing of ferrites (discussed later) and the Sr-doped TiO_2 system discussed above provide good examples of this type of modification. In structural ceramics, a good example of the use of an annealing step to modify the microstructure is the crystallization of the grain boundary phase in Si_3N_4 to improve the high-temperature creep resistance (see Chapter 10).

Effect of Heating Rate on Sintering
Most sintering is carried out at a constant heating rate (typically in the range of 1–10°C/min) because many furnaces (particularly large industrial furnaces) cannot be heated very fast. Even in the so-called isothermal sintering, considerable densification may occur during the rapid heat-up to the isothermal sintering temperature. Furthermore, as we outlined in Chapter 9, rapid heating (referred to as fast firing) has been shown to be very effective for the production of some ceramics with high density and fine grain size. Structural changes in amorphous materials (e.g., gels) can also be influenced by changes in temperature (see Chapter 5). The heating rate is therefore an important part of sintering, used either as a parameter to influence the course of microstructural evolution or as a contribution to the starting conditions for isothermal sintering. For convenience, we consider the effects of heating rate in amorphous and polycrystalline materials separately.

Amorphous Materials. For amorphous materials where coarsening is absent and viscous flow is the dominant densification mechanism, the behavior is very dependent on the structure of the material. In the case of glasses produced by melting and casting, the initial stage sintering kinetics of the compacted powder are in good agreement with the predictions of Frenkel's model (see Chapter 8) if the viscosity of the glass, η, is described by a thermally activated process with a constant activation energy Q [2]:

$$\eta = \eta_0 \exp(Q/RT) \tag{12.2}$$

where η_0 is a constant, R is the gas constant, and T is the absolute temperature. According to this equation, the viscosity of the glass is a single-valued function of temperature.

The densification of gels during constant heating rate sintering depends on the gel structure [3]. Particulate gels show behavior similar to that of powder compacts of glasses produced by melting and casting, and the data can be well represented by the theories. With increasing heating rate, the shrinkage shifts to higher temperatures. This is because at a higher heating rate the gel has less time for densification at a given temperature. Polymeric gels do not show good agreement with the theoretical predictions, regardless of the geometrical model used. As an example,

Densification Process

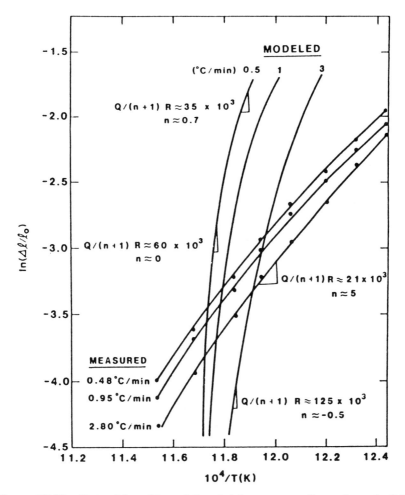

Figure 12.10 Natural logarithm of the shrinkage versus the reciprocal of the temperature for a multicomponent gel during constant heating rate sintering: calculated curves (Modeled) and measured data (Measured). (From Ref. 3.)

Fig. 12.10 shows a comparison of the shrinkage data with the predictions of Scherer's cylinder model (see Chapter 8) for a multicomponent gel sintered at heating rates of 0.5, 1, and 3°C/min [3]. The predicted values were calculated by assuming a constant activation energy for viscous flow ($\approx$500 kJ/mol) throughout the densification process. The approach fails because the viscosity of the gels is not a single-valued function of the temperature. The apparent activation energy depends on the hydroxyl content of the gel, which varies with temperature and time. As a result, the viscosity changes with temperature and time. The densification kinetics therefore depend on the thermal history and not just on the current temperature.

The effect of heating rate is of considerable interest and importance for the sintering of gels. Particulate gels, as outlined earlier, show normal behavior where the shrinkage curves shift to higher temperatures with increasing heating rate. For polymeric gels, the behavior depends on the interplay between the hydroxyl content of the gel (which influences the viscosity) and the time available for sintering. It is commonly observed that above a certain value of the heating rate the shrinkage curves shift to *lower* temperatures with increasing heating rate. This is because the more rapidly heated gel retains more hydroxyl groups, which lower the viscosity and offset the shorter time available for sintering. Isothermal sintering of polymeric gels leads to a rapid increase in the viscosity, which causes the densification rate to decrease drastically. It is therefore advantageous to increase the temperature continuously during sintering so that the rising temperature compensates for the loss of hydroxyl groups and the corresponding increase in the viscosity. The faster the heating rate, the lower the temperature at which densification is completed. However, this applies when crystallization does not interfere with the densification. Extremely fast heating rates can lead to undesirable effects such as incomplete burnout of the organics and trapped gases that expand and cause cracking or bloating of the article. The maximum heating rate must be determined empirically for each gel.

Polycrystalline Materials. The constant heating rate sintering of polycrystalline materials is complicated, even for the initial stage, by the occurrence of multiple mechanisms. Some of these mechanisms, we will recall, lead to coarsening of the microstructure. A realistic representation of the data would take into account the interplay between densification and coarsening. To examine the effects of heating rate, let us consider the sintering of ZnO powder compacts. This powder system was studied in detail by Chu et al. [4], and, in contrast to earlier studies in which only the initial stage of sintering was examined, the experiments covered almost the entire densification process. Figure 12.11 shows the data for the relative density ρ versus temperature T for compacts with approximately the same green density ($\rho_0 \approx 0.50$) sintered at constant heating rates ranging from 0.5 to 15°C/min. Generally, the curves are shifted to slightly higher temperatures with increasing heating rate. By fitting smooth curves to the data and differentiating, the temperature derivative of the densification strain, defined as $(1/\rho)(d\rho/dT)$, can be determined. Interestingly, the data fall approximately on a common curve for the range of heating rates

Densification Process

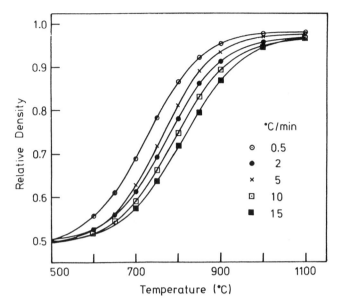

Figure 12.11 Relative density versus temperature for ZnO powder compacts with the same initial density (0.50 ± 0.01), sintered at constant rates of heating (0.5–15°C/min). (From Ref. 4.)

(Fig. 12.12). The sintering temperature T and the sintering time t are related through the heating rate by the equation

$$T = \alpha t + T_0 \tag{12.3}$$

where T_0 is an initial temperature. Differentiating Eq. (12.3) gives

$$dT = \alpha dt \tag{12.4}$$

Compared to the temperature derivative of the densification strain, the volumetric densification rate, i.e., $(1/\rho)(d\rho/dt)$, is a more meaningful parameter because it can be directly related to the predictions of the sintering theories. According to Eq. (12.4) the densification rate can be found from the data of Fig. 12.12 by multiplying by α. As Fig. 12.13 shows, the densification rate increases with increasing heating rate. Although the curves do not have identical shapes, some systematic trends can still be observed:

1. The maximum of each curve shifts to higher temperature with increasing heating rate.
2. Above about 700°C the densification rate at a given temperature is

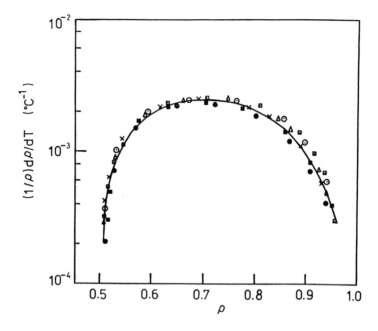

Figure 12.12 Change in the densification strain per unit change in temperature as a function of density, calculated from the data of Fig. 12.11. The data for the different heating rates lie on a common curve. (From Ref. 4.)

approximately proportional to the heating rate, but below this temperature there is a smaller variation with heating rate.
3. In the initial stage of sintering, the densification rate is not independent of heating rate.

The final grain size after sintering to 1100°C decreases with increasing heating rate. The results can be fitted approximately by a relation in which the cube of the grain size, G^3, increases linearly with the inverse of the heating rate, $1/\alpha$ (Fig. 12.14). As an indication of this variation, scanning electron micrographs of the compacts sintered at heating rates of 0.5 and 5°C/min are shown in Fig. 12.15. The grain size at 0.5°C/min is almost twice that at 5°C/min.

As outlined earlier, a representation of constant heating rate data should take into account the interplay between densification and coarsening. As a first approximation, a modification of the isothermal sintering equations to account for the changing temperature can be considered.

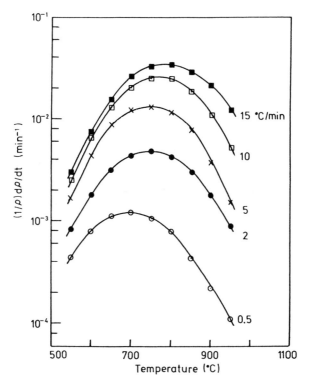

Figure 12.13 Densification rate versus temperature calculated from the data of Fig. 12.11. Above ≈700°C, the rate is approximately proportional to the heating rate. (From Ref. 4.)

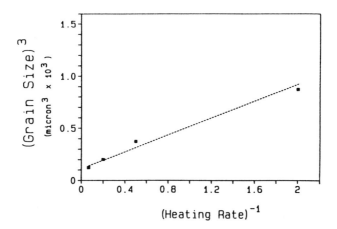

Figure 12.14 Cube of the final grain size versus the inverse of the heating rate. The data can be fitted by a straight line (correlation coefficient 0.995). (From Ref. 4.)

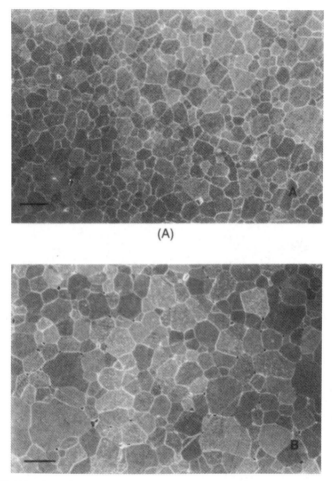

Figure 12.15 Scanning electron micrographs of polished and etched surfaces of ZnO powder compacts sintered to 1100°C at constant rates of heating: (A) 5°C/min and (B) 0.5°C/min. (From Ref. 4.)

Following Eq. (8.119), the linear densification rate (equal to one-third the volumetric densification rate) can be written

$$\dot{\epsilon}_\rho(T, t) = \frac{AD(T)\phi^{(m+1)/2}\Sigma}{G(T, t)^m kT} \tag{12.5}$$

where $\dot{\epsilon}_\rho$ is now a function of both temperature and time, A is a constant, the diffusion coefficient D is now a function of temperature, ϕ is the

Densification Process

stress intensification factor, Σ is the sintering stress, $G(T, t)$ represents a coarsening function that depends on temperature and time, k is the Boltzmann constant, and m is an integer that depends on the diffusion mechanism ($m = 2$ for lattice diffusion and $m = 3$ for grain boundary diffusion). The diffusion coefficient can be written

$$D(T) = D_0 \exp\left(-\frac{Q_d}{kT}\right) \qquad (12.6)$$

where D_0 is a constant and Q_d is the activation energy for the densification process. As discussed in Chapter 9, coarsening is normally described in terms of the evolution of the grain size. For dense solids this is undoubtedly the feature of importance, but for porous solids the mean distance between the pores would be expected to provide a more realistic representation. In this representation, coarsening of a porous compact can result from (i) densification (removal of the porosity) and (ii) nondensifying processes such as surface diffusion and evaporation/condensation (Fig. 12.16). The coarsening function must be expected to have the same form as the grain growth equation, Eq. (9.35). Thus, for isothermal sintering at a given temperature T, the coarsening function can be written

$$G^n(t) = G_0^n + \left[B \exp\left(-\frac{Q_d}{kT}\right) + C \exp\left(-\frac{Q_{nd}}{kT}\right)\right]t \qquad (12.7)$$

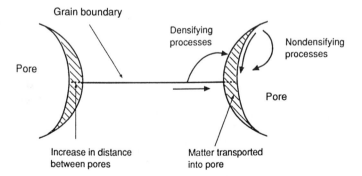

Figure 12.16 Microstructural coarsening is normally described in terms of an increase in the average grain size. For porous solids, the average distance between the pores may provide a more realistic representation. As shown schematically in this representation, coarsening of a porous solid can result from densifying processes and nondensifying processes.

where B and C are constants for a specific powder system, Q_d is the activation energy for the densification processes, and Q_{nd} is the activation energy for the nondensifying processes. According to Eq. (12.7), the first term in the square brackets represents the contribution to coarsening from densifying processes and the second term the contribution from nondensifying processes.

For constant heating rate sintering where the sintering temperature changes with time, Eq. (12.7) can be modified to give the grain size at any temperature and time history:

$$G^n(T, t) = G_0^n + B \int_0^t \exp\left(-\frac{Q_d}{kT}\right) dt + C \int_0^t \exp\left(-\frac{Q_{nd}}{kT}\right) dt \tag{12.8}$$

When B, C, Q_d, and Q_{nd} are constant, this equation reduces to Eq. (12.7). Substituting for dt from Eq. (12.4) gives

$$G^n(T, t) = G_0^n + \frac{B}{\alpha} \int_{T_0}^T \exp\left(-\frac{Q_d}{kT}\right) dt + \frac{C}{\alpha} \int_{T_0}^T \exp\left(-\frac{Q_{nd}}{kT}\right) dt \tag{12.9}$$

Since the integrands are functions of T only, when $G^n \gg G_0{}^n$, Eq. (12.9) can be written in the form

$$G^n \approx \frac{1}{\alpha} F(T) \tag{12.10}$$

where $F(T)$ is a function of temperature only. At any temperature, G^n is proportional to $1/\alpha$. When $n = 3$, the grain size data in Fig. 12.14 satisfy this relation.

The densification rate $\dot{\epsilon}_\rho$ is obtained by substituting for G from Eq. (12.9) into Eq. (12.5). A numerical evaluation of the densification rate based on Eqs. (12.5) and (12.9) is shown in Fig. 12.17 for the assumed values of $B = C$ and $Q_d = 2Q_{nd}$. The calculations show many of the features observed for the data of Fig. 12.13.

The relative density ρ for any time and temperature history follows from the integration of Eq. (12.5) for the known relationship for $G(T, t)$. (The procedure is described in Ref. 4.) When $G^n \gg G_0{}^n$, Eq. (12.10) can be used and the densification rate can be written

$$\dot{\epsilon}_\rho \approx \alpha H(T) \tag{12.11}$$

where $H(T)$ is a function of temperature only. The densification rate at any temperature is predicted to be proportional to the heating rate α. The

Densification Process

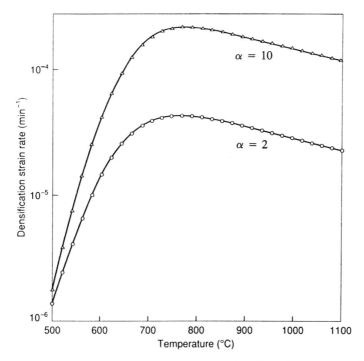

Figure 12.17 Calculated densification rate as a function of temperature for two different heating rates. α is the heating rate in °C/min. (Courtesy of L. C. De Jonghe.)

data for the densification rate shown in Fig. 12.13 are consistent with Eq. (12.11) for temperatures greater than $\approx 700°C$. To find the density, Eq. (12.11) must be integrated:

$$\frac{1}{3}\int_{\rho_0}^{\rho}\frac{d\rho}{\rho} = \alpha \int_{t_0}^{t} H(T)\,dt \qquad (12.12)$$

Substituting for dt from Eq. (12.4) gives

$$\frac{1}{3}\ln\left(\frac{\rho}{\rho_0}\right) = \int_{T_0}^{T} H(T)\,dT \qquad (12.13)$$

According to this equation, for compacts with the same green density ρ_0, the density at any temperature is independent of the heating rate and is a function of temperature only. This result shows good agreement with the density data in Fig. 12.11.

To summarize, densification data obtained in constant heating rate sintering of ZnO powder compacts show that the density is roughly independent of the heating rate and that the densification rate increases roughly with the heating rate. These data can be well described in terms of a model in which the coarsening process is represented in terms of two classes of microstructural coarsening processes: those associated with densifying processes and those associated with nondensifying processes.

Fast Firing

Fast firing is an example of heating cycle control where the powder compact is subjected to short (e.g., ≈ 10 min) firing at a high temperature (e.g., 1800–1900°C for Al_2O_3). The process has been shown to be effective for the attainment of high density and fine grain size in several ceramics. However, because of the rapid heating, the process is suitable only for the production of small or thin-walled samples. As outlined earlier, rapid heating of large or thick-walled samples leads to high temperature gradients that can limit densification and cause cracking.

In fast firing, the powder compact is pushed through a short (5–10 cm) hot zone in a tube furnace such as that shown in Fig. 12.2. The main features of the equipment consist of a tube furnace capable of achieving the desired high temperatures and a system for pushing samples through the furnace at constant speeds ranging from ≈ 0.25 to ≈ 20 cm/min. Temperature control in the furnace is achieved with a thermocouple or an optical pyrometer, and the sintering atmosphere can be controlled by flowing the required gases through the tube at a controlled rate. The temperature of the hot zone and the speed at which the sample is pushed through determine the rate of heating of the sample.

The concept underlying fast firing, put forward by Brook [5], was outlined in Chapter 9. If the activation energy for the densification mechanism, Q_d, is greater than that for the coarsening mechanism, Q_c, the use of high-temperature firing increases the ratio of the densification rate to the coarsening rate, $\dot{\rho}/G$ (see Fig. 9.40). Furthermore, the faster the sample is heated through the low-temperature region where the ratio $\dot{\rho}/G$ is unfavorable, the better the expected result. The argument therefore is for the use of rapid heating and short firing times at high temperature.

To predict which powder systems are expected to benefit from the use of fast firing, it is necessary to know the controlling mechanism for the processes of densification and coarsening and, in addition, to have reliable data for the activation enthalpies for the appropriate diffusion coefficients. In most cases, this information is unavailable or incomplete. The best way to determine the effectiveness of the fast firing technique is therefore to try it. However, some data can be found for a few systems, and it is worth considering the effectiveness of fast firing in these systems.

Densification Process

For Al_2O_3, the densification mechanism for fine powders at moderate to high temperatures is that of lattice diffusion of the cation with an activation energy of 580 kJ/mol. The surface diffusion energies derived from sintering studies of Al_2O_3 are relatively low (230–280 kJ/mol). On this basis, fast firing should work for Al_2O_3. In practice, fast firing has indeed been found to be effective for the production of Al_2O_3 with high density and fine-grained microstructure [6]. As shown earlier in Fig. 9.39, where the grain size versus density data for fast firing at 1850°C and for conventional sintering at 1560°C were compared, support for the expected benefit is obtained.

For MgO, activation energies for surface diffusion range from 360 to 450 kJ/mol, and those for cation lattice diffusion (believed to be the rate-controlling densification mechanism) range from 150 to 500 kJ/mol. The activation energies for lattice diffusion determined from sintering studies cover a narrower range (250–500 kJ/mol) but are still not precise enough to avoid an overlap with the surface diffusion values. As the grain size increases, densification by oxygen diffusion in the grain boundary is believed to become significant. The activation energy for the oxygen diffusion mechanism is in the range of 250–300 kJ/mol, i.e., significantly lower than that for surface diffusion. For densification controlled by the grain boundary diffusion of oxygen, firing to higher temperature will clearly lead to a reduction in the ratio $\dot{\rho}/\dot{G}$, i.e., an effect opposite to the one desired. Although the information for the rate-controlling mechanism and activation energies is fairly imprecise, experiments show that fast firing is ineffective for MgO. This finding is consistent with the general argument.

Fast firing has also been used successfully to produce dense $BaTiO_3$ ceramics with a fine-grained microstructure [7]. The fast-fired samples contained smaller grains at a given density than those prepared by conventional sintering. However, for this system very little information is available for the mechanisms of densification and coarsening and for the activation energies.

F. Control of the Sintering Atmosphere

The sintering atmosphere can have several important effects on densification and microstructure development during sintering. In many instances, the atmosphere can have a decisive effect on the ability to reach high density with controlled grain size. The important effects of the sintering atmosphere are associated with two phenomena: (i) gas solubility, particularly when the gas is trapped in isolated pores in the final stage of sintering, and (ii) chemical reactions with the powder system, particularly when the oxidation states of cations, the defect chemistry, or the volatility of the system can be modified.

Gases in Pores

Depending on its solubility in the solid, a gas trapped in the pores during the final stage of sintering can limit the density of the final article. As discussed earlier, initially the pores in a powder compact form continuous channels that are open to the sintering atmosphere. In this situation, as the porosity decreases, the gas easily flows out. However, in the final stage, when the pores become isolated the gas is trapped in the pores and can no longer escape into the atmosphere. At the point when the pores become isolated, the pressure of the gas trapped in the pores is equal to the pressure of the sintering atmosphere. The final density achieved in subsequent sintering is dependent on the solubility of the gas, i.e., its ability to diffuse into the solid and escape into the atmosphere.

At one extreme, we can consider a gas that is insoluble in the solid. As shrinkage of the isolated pores takes place, the gas is compressed and its pressure increases. When the gas pressure counterbalances the driving force for sintering, shrinkage stops. Assuming for simplicity that the pores are spherical, the limiting density is reached when

$$p_L = 2\gamma_{sv}/r_L \tag{12.14}$$

where p_L is the pressure in the pores when shrinkage stops, γ_{sv} is the specific energy of the solid/vapor interface, and r_L is the limiting pore size. Assuming a simple microstructural model in which the solid phase is a continuum and the pores are spherical and of the same size, the limiting porosity P_L, can be easily estimated. Applying the gas law, $p_1V_1 = p_2V_2$, to the initial situation (when the pores just become isolated) and to the limiting situation (when shrinkage stops) gives

$$p_0 N \left(\frac{4}{3}\pi r_0^3\right) = p_L N \left(\frac{4}{3}\pi r_L^3\right) \tag{12.15}$$

where N is the number of pores per unit volume, p_0 is the pressure of the sintering atmosphere, and r_0 is the radius of the pores when they become isolated. Substituting for p_L from Eq. (12.14) gives

$$p_0 r_0^3 = 2\gamma_{sv} r_L^2 \tag{12.16}$$

The porosity P_0 when the pores become isolated and the limiting porosity P_L are related by

$$\frac{P_0}{P_L} = \frac{r_0^3}{r_L^3} \tag{12.17}$$

Densification Process

Substituting for r_L from Eq. (12.17), after some rearrangement Eq. (12.16) yields

$$P_L = P_0(p_0 r_0/2\gamma_{sv})^{2/3} \qquad (12.18)$$

According to this equation, limiting porosity deceases with decreasing pressure of the sintering atmosphere and with decreasing pore radius. Since the pore radius is roughly proportional to the grain size, the limiting porosity decreases with decreasing grain size.

The other extreme is when the gas has a high solubility in the solid. The densification rate is unaffected by the gas in the pores because rapid diffusion through the lattice or along the grain boundaries can occur during shrinkage. At some intermediate solubility, the kinetics of the gas diffusion to the surface of the solid can, in principle, control the rate of final-stage densification. At some lower solubility, the diffusivity of the gas in the solid becomes low enough that the gas cannot escape to the surface in the time scale for sintering; it can only diffuse between neighboring pores.

In the case where only exchange of gas between neighboring pores can occur, the sintering in the final stage can be idealized as follows. After the pores become isolated, densification will become progressively slower because of the increasing gas pressure in the shrinking pores. Eventually a limiting density will be reached at which densification almost stops because the gas pressure in the pores will counterbalance the driving force for sintering. Any further changes during sintering will be controlled by the exchange of gases between neighboring pores. As discussed earlier in this book, there will be a distribution of pore sizes in the solid. Considering a small pore near a large pore (Fig. 12.18), initially, when the limiting

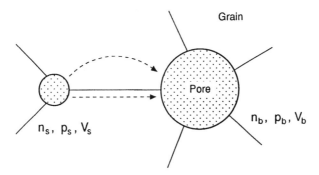

Figure 12.18 Sketch showing the parameters for two neighboring gas-filled pores; a net diffusion of gas occurs from the small pore to the large pore.

density is reached, the pressure p_b in the large pore is given by

$$p_b = \frac{n_b kT}{V_b} = \frac{2\gamma_{sv}}{r_b} \qquad (12.19)$$

where n_b is the number of molecules of gas in the large pore of radius r_b and volume V_b. Because of the higher pressure (or, more correctly, higher chemical potential), there will be a net diffusion of gas from the small pore to the large pore. If n_s molecules of gas are transferred to the large pore, the new gas pressure in the large pore will be

$$p'_b = \frac{(n_b + n_s)kT}{V_b} > \frac{2\gamma_{sv}}{r_b} \qquad (12.20)$$

Since the gas pressure becomes greater than the driving force for sintering, the large pore must expand. The expansion of the large pore is greater than the shrinkage of the small pore, so the overall result is an increase in porosity (or a decrease in density). In this case, in which exchange of gas between neighboring pores controls the final stage of sintering, the density goes through a maximum, as sketched in Fig. 12.19. This density decrease after the maximum is sometimes referred to as *desintering* or *swelling*.

The discussion so far has generally assumed that the solid phase of the powder compact is a continuum, so it is most appropriate to glasses or to solids with a significant amount of glassy grain boundary phase.

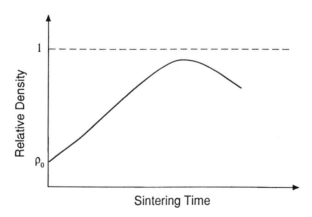

Figure 12.19 Schematic diagram showing the densification of a powder compact with a fairly insoluble gas in the pores. The density goes though a maximum with time as a result of sintering with the trapped gas in the pores.

Densification Process

For polycrystalline solids, additional effects must be considered. In these materials, the location of the pores will influence the gas diffusion path. When the pores are located within the grains, diffusion will depend on the gas solubility in the crystal lattice. For pores located at the grain boundaries, diffusion of the gas through the disordered grain boundary region provides an important additional path. The effect of grain growth must also be taken into account. If grain growth and pore coalescence occur, the average pore size will increase. The equality described by Eq. (12.19) cannot be maintained, and, as a result, the volume of the pore will increase. If the pore size increases faster than the grain size, this will lead to the sintering behavior sketched in Fig. 12.19 where the density goes through a maximum.

Information on the diffusivity of gases in ceramics is normally not available, so it is difficult to predict the effects of a given gas trapped in the pores. From the point of view of size, interstitial diffusion will be more favorable for smaller molecules such as H_2 and He than for larger ones such as Ar and N_2. Diffusion of larger molecules may occur by a vacancy mechanism, but this will depend on the stoichiometry of the solid. Diffusion along the disordered grain boundaries can provide an easier path for smaller as well as larger molecules. Whenever it can be used, vacuum sintering (at least to the onset of pore closure) eliminates the problem of trapped gases.

The significance of gases in pores was first highlighted by Coble [8] in his experiments on the sintering of MgO-doped Al_2O_3 to theoretical density. Coble observed that the complete elimination of porosity was possible if the sintering atmosphere was H_2, O_2, or vacuum. Porosity could not be completely removed when the atmosphere was He, Ar, or N_2 (or, therefore, air).

Chemical Effects of the Atmosphere

Chemically, the atmosphere can lead to a variety of important effects over the entire process of sintering. These effects include (i) reactions with additives and impurities, (ii) evaporation of volatile components from the solid, (iii) decomposition of the solid, and (iv) changes in the defect chemistry or stoichiometry of the solid.

Reactions with Additives and Impurities. Depending on the composition, the gas in the sintering atmosphere can react with dopants added to the powder for the enhancement of sintering, thereby reducing their effectiveness. The effect of atmosphere on sintering has been studied systematically for the silicon carbide powder system [9]. As discussed earlier, the bonding in SiC is highly covalent, so solid-state diffusion is slow. The powder must be sintered with dopants, typically a mixture of 0.5 wt %

carbon and 0.5 wt % boron, at high temperature ($\approx 2000°C$) to achieve high density. Not only can the gas in the atmosphere become trapped in the pores as described earlier, it can also react with the dopants, leading to their removal or the formation of undesirable compounds. For example, the use of an atmosphere of N_2 or CO leads to reaction and the removal of the boron; as a result, densification is drastically reduced. The reactions have been described as follows [9]:

$$B + N_2 \rightarrow 2BN \tag{12.21a}$$

$$B + CO \rightarrow BO + C \tag{12.21b}$$

Although vacuum eliminates the effect of gases trapped in the pores, it also leads to a small amount of decomposition of the SiC, with Si being removed from the surface region of the solid. Silicon carbide sintered in vacuum may have a fairly high density overall but have a fairly porous layer on the surface. Inert gases (such as helium and argon) provide the most effective sintering atmosphere but get trapped in the pores. However, as Eq. (12.18) indicates, the severity of the effect of trapped gases can be reduced by using a fine SiC powder and maintaining a fine pore (or grain) size during sintering.

The reaction of impurities in the powder with the gas from the sintering atmosphere can also lead to the presence of gases in the pores. For example, it is commonly observed that dense alumina fabricated by hot pressing will swell if annealed in an atmosphere containing a sufficient quantity of oxygen. Impurities in the solid become oxidized, generating gases such as CO, CO_2 and SO_2 that are insoluble in the solid. The pressure of the gases can be sufficiently high to produce voids in the solid.

Evaporation of Volatile Components. For some powders, evaporation of volatile components (e.g., Na and Pb) can occur during sintering, thereby making it difficult to control the chemical composition of the solid. The change in composition influences the evolution of the microstructure and can cause a deterioration of the properties of the solid.

If the amount of the component lost under a given set of conditions is known, the composition of the starting powder can be modified to compensate for the loss during sintering. In most cases, however, this would be difficult to do. A more common solution is to surround the powder compact with a coarse powder of the same composition as the compact. The coarse powder leads to the establishment of an equilibrium partial pressure of the volatile component, thereby reducing the tendency for evaporation from the powder compact. As an example of this approach, Fig. 12.20 shows a schematic illustration of the system used for the sintering of sodium β-alumina in a laboratory-scale experiment. The ceramic

Densification Process

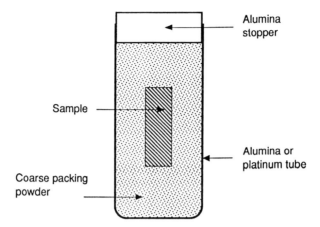

Figure 12.20 Schematic illustration of a laboratory scale apparatus for the sintering of sodium β-alumina.

is a good ionic conductor even at temperatures below 300°C and is used as a solid electrolyte in sodium-sulfur batteries that have electrodes of molten sulfur and sodium polysulfides. The sample surrounded by the coarse powder is encapsulated in an alumina or platinum tube. Alumina tubes are much cheaper than the platinum but, due to the reaction with the powder, they can be used only a few times before falling apart. For extended use, platinum may have an advantage in overall cost.

Figure 12.21 shows a more elaborate example of the method that is used for the sintering of lead lanthanum zirconium titanate (PLZT), a ceramic used in a variety of electronic applications [10]. Lead is volatile as well as poisonous, so its evaporation must be not only controlled but also contained. The surrounding powder, a mixture of lead oxide (PbO) and lead zirconate ($PbZrO_3$), maintains the appropriate partial pressure to minimize lead loss from the sample. With the controlled atmosphere apparatus, PLZT can be sintered to full density (Fig. 12.22), yielding materials with a high degree of transparency (Fig. 12.23).

Decomposition. For many systems, sintering must be carried out at temperatures at which the decomposition of the powder becomes important. If the rate of decomposition is high, the porosity generated by the weight loss limits the attainment of high density. Generally the weight loss during sintering should be kept below ≈2–4 % if high density is to be achieved.

Silicon nitride (Si_3N_4) is a good example of a system in which the effects of decomposition are important. The bonding in Si_3N_4 is highly covalent. As discussed in Chapter 10, additives that form a liquid phase

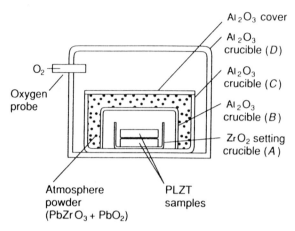

Figure 12.21 Apparatus for the sintering of lead lanthanum zirconium titanate (PLZT) in controlled atmosphere. (From Ref. 10, with permission.)

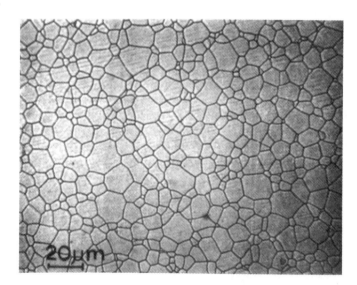

Figure 12.22 Typical microstructure of PLZT ceramic produced by sintering in controlled atmosphere (thermally etched at 1150°C for 1h). (From Ref. 10, with permission.)

Figure 12.23 Examples of sizes and optical quality of PLZT ceramics produced by sintering in controlled atmosphere. (From Ref. 10, with permission.)

must be used to aid the sintering process. However, even with the liquid phase, the temperatures used for sintering are still fairly high (typically in the range of 1700–1900°C). At such high temperatures, the decomposition of the powder is significant. The reaction can be written as

$$Si_3N_4 \rightleftharpoons 3Si + 2N_2 \tag{12.22}$$

The vapor pressure of the N_2 gas generated by the decomposition is ≈0.01 MPa (0.1 atm) at 1700°C and increases to ≈0.1 MPa (1 atm) at 1875°C. Because of the high vapor pressure, sintering in nitrogen gas at atmospheric pressure is accompanied by a significant weight loss. Apart from the difficulty in achieving high density, the furnace is contaminated by the silicon produced in the decomposition reaction: the silicon evaporates and solidifies on the cooler parts of the furnace. The problem of decomposition during the sintering of Si_3N_4 has been tackled in two ways:

1. By surrounding the powder compact with a powder of the same composition, as described earlier for suppressing the evaporation of volatile components
2. By raising the nitrogen gas pressure in the sintering atmosphere [so that the reaction described in Eq. (12.22) is driven to the left]

The approach of raising the nitrogen gas pressure is sometimes referred to as gas pressure sintering. The process employs nitrogen gas pressures of up to ≈7 MPa (70 atm) and must be carried out in a vessel capable of withstanding such pressures. Figure 12.24 shows the data for the density and weight loss [11] after sintering silicon nitride powder compacts for 30 min at various temperatures under ambient nitrogen pressures of 0.1 MPa (1 atm) and 2 MPa (20 atm). The drastic reduction in the weight loss and the increase in the density are evident at the higher pressure. Nitrogen gas has a fairly high solubility in silicon nitride, so the unfavorable sintering effects produced by gases trapped in pores are alleviated even for such high gas pressures. Refinements of the gas pressure sintering process and the use of appropriate additives have routinely led to the production of silicon nitride with densities greater than 99% of the theoretical.

Vapor Transport. Vapor transport, as discussed in Chapter 8, leads to coarsening and therefore has a significant effect on the sintering kinetics and microstructural evolution of the powder compact. For the attainment of high density, the rate of coarsening (more specifically, the ratio of the coarsening rate to the densification rate) and therefore the rate of vapor transport must be reduced. The rate of vapor transport can be varied by changing the composition or partial pressure of the gas in the sintering atmosphere.

The ability of vapor transport to produce significant changes in densification and microstructural evolution is clearly demonstrated by the work of Readey [12], who studied the sintering of several oxides in reducing atmospheres. The gas in the atmosphere reacts with the powder to produce gaseous species that enhance vapor transport. An example is the sintering of ferric oxide, Fe_2O_3, in an atmosphere of HCl gas. Firing was performed with the powder compacts sealed in quartz ampules to prevent weight loss. The reaction can be described as follows:

$$Fe_2O_3(s) + 6HCl(g) \rightarrow Fe_2Cl_6(g) + 3H_2O(g) \qquad (12.23)$$

The gaseous Fe_2Cl_6 produced in the reaction leads to enhanced vapor transport. Figure 12.25 shows the data for the shrinkage as a function of time at 1000°C for various partial pressures of HCl. Densification decreases as the partial pressure of the HCl and hence the rate of vapor

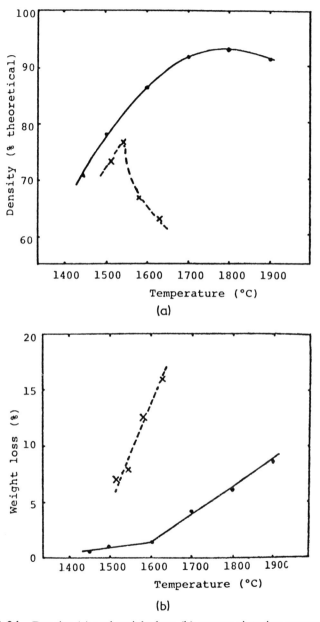

Figure 12.24 Density (a) and weight loss (b) versus sintering temperature for silicon nitride sintered for 30 min in nitrogen under atmospheric pressure (broken line) and under 20 atm pressure. (From Ref. 11.)

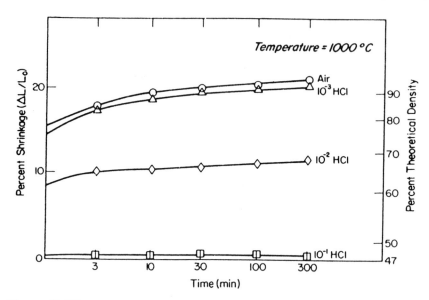

Figure 12.25 Shrinkage versus time for Fe_2O_3 powder compacts sintered at 1000°C in different partial pressures of HCl (in MPa). (From Ref. 12, reprinted by permission of the American Ceramic Society.)

transport increase. For HCl pressures in excess of 10^{-2} MPa (0.01 atm), shrinkage is almost completely inhibited. As Fig. 12.26 shows, the decrease in densification is due to the coarsening of the microstructure.

Water Vapor in the Atmosphere. Generally, sintering of advanced ceramics is carried out in a dry, controlled atmosphere. However, many studies have shown a significant effect of water vapor on the sintering of MgO powder compacts. Compared to sintering in a dry atmosphere (e.g., air, argon, or vacuum), the rates of densification and coarsening are both enhanced in an atmosphere containing water vapor.

The increased coarsening rate in MgO may be attributed to an enhancement of vapor transport or surface diffusion or both. Vapor transport appears to be ruled out by thermodynamic calculations. The calculations are partly confirmed by observations (based on an uncompacted powder) that show that the relative effect of water vapor on the coarsening is greater at 600 and 800°C than at 1050°C [13]. If the increase in coarsening were caused by an enhancement of vapor transport, the effect of water vapor would increase continuously with temperature. The increased rate of coarsening therefore appears to be related to an enhancement of surface diffusion.

Densification Process

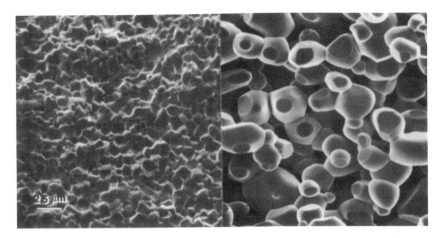

Figure 12.26 Scanning electron micrographs of fractured surfaces of Fe_2O_3 powder compacts sintered for 5 h at 1200°C in air (left) and in 10% HCl (right). (From Ref. 12, reprinted by permission of the American Ceramic Society.)

In contrast to the coarsening process, the increase in the densification produced by water vapor is more difficult to account for. A variety of explanations have been put forward, including rearrangement of the grains, changes in the equilibrium dihedral angle, and changes in the lattice defect structure. Figure 12.27 shows data for the porosity versus sintering time for MgO powder compacts at various temperatures in static dry air and in flowing water vapor [14]. A comparison of the data at any temperature (e.g., 1380°C), shows a faster densification in water vapor. The trajectories for grain size G versus relative density ρ obtained in the same experiments are shown in Fig. 12.28. Considering the data for the undoped MgO compacts in dry air and in water vapor, the most significant differences in the nature of the trajectories are

1. In the intermediate region (for ρ in the range of ≈ 0.72–0.85), the grain size of the sample sintered in water vapor increases faster with density than that of the sample sintered in dry air.
2. The rapid increase in the grain size in the late stage occurs at a higher density (≈ 0.87 compared to ≈ 0.82) for the sample sintered in water vapor.

Of the explanations outlined earlier for the enhanced densification in water vapor, rearrangement can be discounted because it is difficult to accomplish in solid-state sintering. Although invoked from time to time,

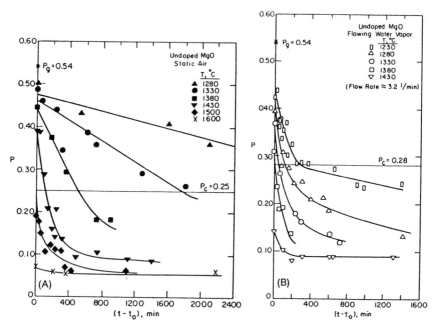

Figure 12.27 Porosity versus time for MgO powder compacts sintered at different temperatures in (A) static air and (B) flowing water vapor. (From Ref. 14, with permission.)

it has not found definitive support in solid-state sintering. In addition to the other two explanations (i.e., changes in the dihedral angle or the defect structure), it is likely that microstructural changes produced by the enhanced coarsening may also play a role. As discussed in Chapter 11, the ability of a limited amount of coarsening to homogenize a microstructure has been clearly demonstrated. Assuming that the enhanced coarsening of undoped MgO in water vapor leads to a more homogeneous microstructure, the onset of pore closure in the sample sintered in water vapor should be delayed to higher density than in the sample sintered in dry air. This is indeed observed from the data in Fig. 12.28 if the rapid rise in the grain size versus density trajectory is taken to represent the onset of pore closure. The trajectory for the undoped MgO sintered in water vapor crosses that for the sample sintered in dry air and starts to increase rapidly at a higher density.

To summarize, in the sintering of MgO powder compacts, water vapor in the sintering atmosphere is observed to lead to an increase in the rates

Densification Process

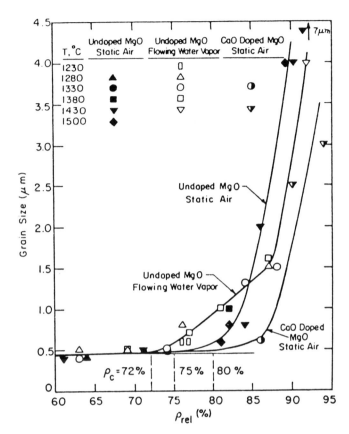

Figure 12.28 Grain size versus relative density for MgO powder compacts sintered in air and in flowing water vapor and for CaO-doped MgO powder compacts in static air. (From Ref. 14, with permission.)

of coarsening and densification. The increase in the coarsening rate appears to be due to an enhancement in surface diffusion. The mechanism for the enhanced densification is less clear. Explanations based on changes in the dihedral angle or the lattice defect structure have been put forward. However, microstructural changes produced by the enhanced coarsening may also play a role.

Oxidation Number and Defect Chemistry. The valence (or valency) of an atom is equal to the number of bonds it forms in the most satisfactory formulation of the substance. For example, the valence of the oxygen atom is 2. The oxidation number (or oxidation state) is based on an ionic view of the substance and is the charge on an atom in the most plausible ionic formulation of the substance. Often the valence and the magnitude of the oxidation number are the same; for example, in the substance MgO, the oxidation state of the oxygen atom is -2. However, many atoms form stable compounds in which the oxidation state is different from the valence. For example, titanium with a valence of 4 forms compounds in which its oxidation number is $+2$, $+3$, and $+4$.

The sintering atmosphere can influence the oxidation number of atoms, particularly those of the transition elements. Changes in the oxidation number produce changes in the properties of the substance and therefore in the sintering behavior. The oxides of chromium form a good system in which the control of the oxidation number is important for sintering [15]. The principal oxidation numbers of the chromium atom are $+6$ (as, for example, in CrO_3), $+4$ (CrO_2), $+3$ (Cr_2O_3), $+2$ (CrO), and 0 (Cr metal). The oxidation number of the Cr atom changes readily from $+6$ in an atmosphere with a high oxygen partial pressure to $+4$, $+3$, $+2$, and finally 0 as the oxygen partial pressure decreases. Each oxidation state, except for the $+3$ state, corresponds to an oxide with a fairly high vapor pressure.

As discussed earlier, a high rate of vapor transport leads to enhanced coarsening and a reduction in the driving force for densification. Of the chromium oxides outlined earlier, Cr_2O_3 has received most interest. The sintering of Cr_2O_3 powder compacts to high density must therefore be carried out in an atmosphere in which the chromium atoms are maintained in the oxidation state of $+3$.

The conditions for thermal stability of a substance can be calculated from standard thermochemical data. The results are most usefully plotted on an Ellingham diagram. Figure 12.29 shows an example of the calculations for the chromium oxide system [16]. The standard free energy of formation (per mole of oxygen) for three oxides is plotted versus temperature. Also shown (on the right) are (i) the oxygen partial pressure (in atmospheres) in equilibrium with the reactions and (ii) the partial pressures of the gas mixtures (H_2–water vapor and CO–CO_2) required to produce the desired oxygen partial pressure. For example, Fig. 12.29 indicates that at 1600°C (the vertical line), the oxygen partial pressure must not be below $10^{-11.7}$ (or 2×10^{-12}) atm if reduction of Cr_2O_3 to Cr is to be avoided (the lowest diagonal line).

For the sintering of Cr_2O_3, Fig. 12.30 shows the data for the relative density as a function of oxygen partial pressure after firing for 1 h at 1600°C. The powder cannot be densified to greater than $\approx 75\%$ of the theoretical density if the oxygen partial pressure exceeds 10^{-9} atm. The

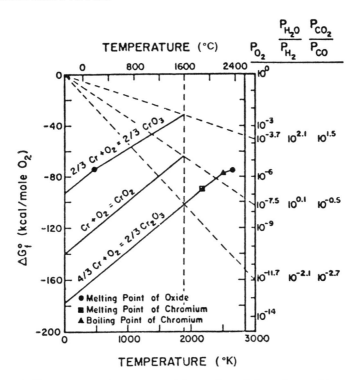

Figure 12.29 Standard free energy of formation of chromium oxides as a function of temperature. (From Ref. 16.)

maximum density (99.4%) is achieved when the oxygen partial pressure is $\approx 2 \times 10^{-12}$ atm, a value that corresponds to the equilibrium partial pressure of oxygen between Cr_2O_3 and Cr. The use of MgO (0.1 wt %) as an additive leads to an increase in the density, to near the theoretical value, by reducing the grain growth. The MgO reacts with the Cr_2O_3 to form fine $MgCr_2O_4$ particles at the grain boundaries, which results in the pinning of the grain boundaries by the Zener mechanism. However, even with the MgO addition, high density can be achieved only if the oxygen partial pressure is near the value (i.e., 2×10^{-12} atm) for the stabilization of the Cr_2O_3 phase. As shown in Fig. 12.31, the oxidation state control used for the sintering of Cr_2O_3 can also be applied to other chromites [17].

The ferrites are an important class of ceramics in magnetic applications, where control of the oxidation number and the stoichiometry of the compound are important [18]. Two compounds of technical and commercial interest are the nickel zinc and manganese zinc ferrites, both of which

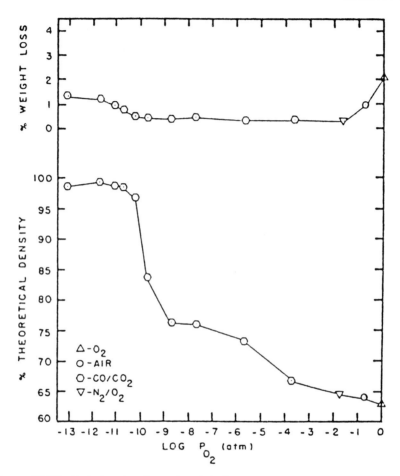

Figure 12.30 Relative density and weight loss versus oxygen partial pressure in the atmosphere for undoped Cr_2O_3 powder compacts after sintering at 1600°C for 1 h. (From Ref. 15, with permission.)

have the spinel crystal structure. The general formula of the stoichiometric compounds can be written $M_{1-x}Zn_xFe_2O_4$, where M represents Ni or Mn. However, commercial compositions are often nonstoichiometric and must be carefully formulated to produce the desired properties [18]. The nickel zinc ferrites, with the general formula $Ni_{1-x}Zn_xFe_{2-y}O_4$, are formulated with a deficiency of iron (with y in the range $0 < y < 0.025$) to keep the sensitivity high and the magnetic loss low. The oxidation number

Densification Process

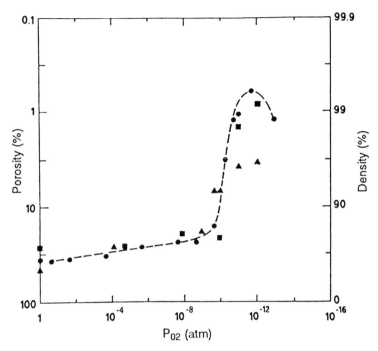

Figure 12.31 Final porosity versus oxygen partial pressure in the atmosphere for sintered chromites. ▲ $MgCr_2O_4$, 1700°C; ■ $La_{0.84}Sr_{0.16}CrO_{2.92}$, 1740°C; ● Cr_2O_3, 1600°C. (From. Ref. 17, with permission.)

of +2 is the most stable for both Ni and Zn. The +3 state for Fe is stable only if there is a slight deficiency of iron (or excess of oxygen). Nickel zinc ferrites can therefore be sintered under a wide variety of oxidizing atmospheres (e.g., air or oxygen).

The manganese-zinc system provides an additional degree of complexity in the sintering of ferrites with the spinel structure. The spinel phase is stable only over a certain range of atmosphere and temperature conditions. These ferrites, with the general formula $Mn_{1-x}Zn_xFe_{2+y}O_4$, are formulated with an excess of iron ($0.05 < y < 0.20$). The concentration of Fe in the +2 oxidation state (i.e., "ferrous" iron) is critical to the achievement of the desired properties of low magnetic loss and a maximum in the magnetic permeability. To produce the desired concentration of ferrous iron ($Fe^{2+}/Fe^{3+} \approx 0.05/2.00$), a two-stage firing schedule is employed (Fig. 12.32). The sample is first fired (at a temperature of 1250–1400°C) in an atmosphere of high oxygen partial pressure (0.3–1

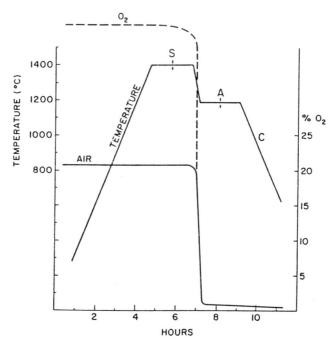

Figure 12.32 Schematic firing cycle for manganese zinc ferrites. (From Ref. 18.)

atm) to minimize the evaporation of zinc. In this stage the densification and grain growth are essentially completed. In the second stage, the sample is annealed at a lower temperature (1050–1200°C) in an atmosphere with fairly low oxygen partial pressure (50–100 ppm) to establish the desired concentration of ferrous iron. Figure 12.33 shows the compositional path followed in the firing schedule. During the first stage of the firing schedule, the ferrite has a composition corresponding to S. In the second stage, equilibrium is quickly reached because of the fairly high annealing temperature; the ferrite attains a different composition (corresponding to A) that depends on the temperature and atmosphere of annealing. Further cooling along C represents additional compositional changes if equilibrium is reached; in practice, however, with the cooling rates used and the decreasing temperature, the composition does not change significantly on cooling.

Densification Process

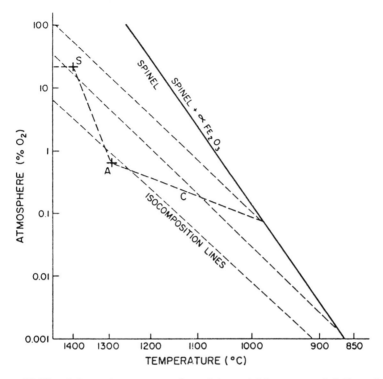

Figure 12.33 Schematic representation of the stoichiometry path followed during the firing of manganese zinc ferrite. (From Ref. 18.)

G. Production of Controlled Sintering Atmospheres

As observed earlier, atmospheres with a variety of compositions and oxygen partial pressures may be required for sintering. In research experiments, it may also be necessary to use flowmeters to control the rate of gas flow through the furnace. In most cases the required gas composition is available commercially (e.g., O_2, air, N_2, Ar, He, and H_2). Normally, Ar or He is used to provide an inert atmosphere. Depending on the purity, commercial sources of Ar or He contain a small amount of oxygen (ranging from ≈ 10 to 100 ppm) and may have to be purified by passing through an oxygen getter prior to entering the furnace. Nitrogen is a common atmosphere for the sintering of nitrides (e.g., Si_3N_4). However, some care must be exercised in its use at very high temperatures. It reacts with refractory metal windings and heat shields to form nitrides, thereby shortening the life of the furnace; it also reacts with graphite to produce poisonous gases. Hydrogen is fairly explosive and must be handled carefully. However, inert gases such as He and Ar containing less than 5% H_2 are not explosive and can be handled safely.

Atmospheres with high oxygen partial pressure (p_{O_2}) can be achieved with O_2 (1 atm) and air (0.2 atm). Mixtures of O_2 and an inert gas (He, Ar, or N_2) can provide atmospheres with a p_{O_2} down to $\approx 10^{-2}$ atm. Lower p_{O_2} (below $\approx 10^{-4}$ atm) is normally obtained by flowing mixtures of gases such as H_2–water vapor or CO–CO_2. Care must be exercised in the handling of both gas mixtures. As outlined earlier, H_2 is explosive; CO is poisonous and must be vented properly. CO–CO_2 gas mixtures are mixed prior to being admitted into the furnace (e.g., by passing the gases in the correct proportions through a column containing glass beads), and the total gas flow through the furnace must be adjusted to a fixed rate (e.g., 1 cm/s). For control of the oxygen partial pressure with H_2–H_2O mixtures, hydrogen is bubbled through a gas washer containing distilled water at a fixed temperature.

The equilibrium oxygen partial pressure of a mixture at any given temperature is calculated using standard thermodynamic equations and appropriate thermochemical data [19]. For example, if we require an atmosphere with an oxygen partial pressure of $10^{-11.7}$ atm at 1600°C (see Fig. 12.29), then this can be obtained relatively easily by establishment of the equilibrium

$$CO + \frac{1}{2} O_2 \rightarrow CO_2 \tag{12.24}$$

The reaction arises from the following two reactions:

$$CO(g) \rightarrow \frac{1}{2} O_2(g) + C(s), \qquad \Delta G^0 = 111{,}700 + 87.65T \quad J \tag{12.25}$$

$$C(s) + O_2(g) \rightarrow CO_2(g), \qquad \Delta G^0 = -394{,}100 - 0.84T \quad J \tag{12.26}$$

where ΔG^0 is the standard free energy change for the reaction. Hence, for the reaction described by Eq. (12.24), we obtain

$$\Delta G^0 = -282{,}400 + 86.81T \quad J \tag{12.27}$$

Using the relation $\Delta G^0 = -RT \ln K_p$, where K_p is the equilibrium constant for the reaction, gives, for $T = 1873$ K,

$$\ln K_p = 7.70 \tag{12.28}$$

Densification Process

The equilibrium constant is also given by

$$K_p = \frac{p_{CO_2}}{p_{CO}(p_{O_2})^{1/2}} = \frac{p_{CO_2}}{p_{CO} \times 10^{-5.85}} \qquad (12.29)$$

where the value of $p_{O_2} = 10^{-11.7}$ was inserted in Eq. (12.29). Using Eqs. (12.28) and (12.29) gives

$$\frac{p_{CO_2}}{p_{CO}} = 10^{-2.5} \qquad (12.30)$$

The value obtained in Eq. (12.30) is very close to that in Fig. 12.30. If we produce the same p_{O_2} at 1600°C using the reaction

$$H_2 + \frac{1}{2} O_2(g) = H_2O,$$
$$\Delta G^0 = -239{,}500 + 8.14T \ln T - 9.25T \quad J \qquad (12.31)$$

it is left as an exercise for the reader to show that

$$\frac{p_{H_2O}}{p_{H_2}} = 10^{-1.9} \qquad (12.32)$$

Thus, if in the H_2–H_2O mixture, $p_{H_2} = 1$ atm, then p_{H_2O} must equal 0.013 atm. Since 0.013 atm is the saturated vapor pressure of water at 11.1°C, the required gas mixture can be produced by bubbling H_2 gas through pure water at 11.1°C, thereby saturating the hydrogen with water vapor. When the reaction equilibrium is established at 1600°C, the oxygen partial pressure in the furnace will be $10^{-11.7}$ atm.

12.3 MICROWAVE SINTERING

Since about 1970, there has been growing interest in the use of microwaves for heating and sintering ceramics. As outlined earlier, microwave heating is fundamentally different from conventional heating in which electrical resistance furnaces are typically used (Fig. 12.1). In microwave heating, heat is generated internally by interaction of the microwaves with the atoms, ions, and molecules of the material.

Studies have shown significant enhancement of sintering by microwave heating, including very fast heating rates (rates greater than 1000°C/min are possible), much lower sintering temperatures, and enhanced diffusion rates. However, the control of microwave sintering is more complicated than that of conventional sintering. The heating depends not only on the operational parameters of the microwave source but also on the electrical and thermal properties of the material. An additional complication is that the electrical and thermal parameters of the material change with the temperature.

A. Interaction of Microwaves with Matter

Microwaves are electromagnetic waves with a frequency in the range of 0.3–300 GHz or, equivalently, with a wavelength in the range of 1 mm to 1 m. This wavelength range is of the same order as the linear dimensions of most practical ceramics. In common with other electromagnetic waves, microwaves have electric and magnetic field components, an amplitude and phase angle, and the ability to propagate (i.e., to transfer energy from one point to another). These properties govern the interaction of microwaves with materials and produce heating in some materials.

As sketched in Fig. 12.34, depending on the electrical and magnetic properties of the material, microwaves can be transmitted, absorbed, or reflected [20]. Microwaves penetrate metals only in a thin skin (on the order of 1 µm); metals can therefore be considered to be opaque to microwaves or to be good reflectors of microwaves. Most electrically insulating (or dielectric) ceramics such as Al_2O_3, MgO, SiO_2, and glasses are transparent to microwaves at room temperature. However, when heated above a certain critical temperature T_c, they begin to absorb and couple more effectively with microwave radiation. Other ceramics, such as Fe_2O_3, Cr_2O_3, and SiC, absorb microwave radiation more efficiently at room temperature. In some systems (e.g., composites) in which the ceramic is transparent to microwaves, an absorbing second phase can greatly enhance the interaction of the system with microwaves.

The effect produced by the microwaves depends on the degree of interaction between the material and the microwaves. The electric and magnetic field components of the microwave interact with the dielectric or magnetic material, and under these conditions energy is dissipated in the material by various mechanisms. The properties of the material that are most important for the interaction are the permittivity ϵ (for a dielectric material) and the permeability μ (for a magnetic material). Commonly the relative permittivity ϵ_r (also called the dielectric constant), equal to ϵ/ϵ_0, and the relative permeability μ_r, equal to μ/μ_0, are used, where ϵ_0 is the permittivity and μ_0 the permeability of vacuum. In alternating fields, the behavior is best described with the help of complex quantities ϵ_r^* and μ_r^* defined by the equations

$$\epsilon_r^* = \epsilon_r' - j\epsilon_r'' \tag{12.33a}$$

Densification Process

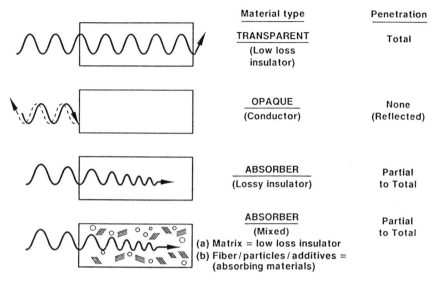

Figure 12.34 Schematic diagram illustrating the interaction of microwaves with materials. (From Ref. 20, with permission.)

and

$$\mu_r^* = \mu_r' - j\mu_r'' \tag{12.33b}$$

where ϵ_r^* is the complex relative permittivity, μ_r^* the complex relative permeability, the prime and double prime represent the real and imaginary parts of the complex quantity, and $j = \sqrt{-1}$. When microwaves penetrate the material (e.g., a dielectric material), the electromagnetic field induces motion in the free and bound charges (e.g., electrons and ions) and in dipoles. The induced motion is resisted because it causes a departure from the natural equilibrium of the system. This resistance, due to frictional, elastic, and inertial forces, leads to the dissipation of energy. As a result, the electric field associated with the microwave radiation is attenuated and heating of the material occurs. The loss tangent, tan δ, is used to represent the losses arising from all mechanisms; it is defined by

$$\tan \delta = \frac{\epsilon_r''}{\epsilon_r'} = \frac{\sigma_{ac}}{\omega \epsilon_0 \epsilon_r'} \tag{12.34}$$

where σ_{ac} is the dielectric or alternating current (ac) conductivity and ω is the angular frequency of the microwave radiation. The average power dissipated per unit volume of the material is given by

$$W = \frac{1}{2} E_0^2 \sigma_{ac} = \frac{1}{2} E_0^2 \omega \epsilon_0 \epsilon_r'' = \frac{1}{2} E_0^2 \omega \epsilon_0 \epsilon_r' \tan \delta \qquad (12.35)$$

where E_0 is the amplitude of the electric field, given by $E = E_0 \exp(j\omega t)$. According to Eq. (12.35), the power absorbed by the material depends on (i) the frequency and the square of the amplitude of the electric field and (ii) the dielectric constant and the loss tangent of the material. In practice, these quantities are interdependent and it is difficult to alter one without affecting the others. Nevertheless, Eq. (12.35) shows, in a general way, the important parameters that control the power absorbed. By analogy to the dielectric case outlined here, corresponding equations can be derived for the interaction of the microwave field with a magnetic material.

Since the electric field is attenuated as the microwaves propagate through the material, the depth of penetration of the microwaves into the material is an important parameter. If any linear dimension of the material is greater than the depth of penetration, uniform heating cannot be expected to occur. A few different parameters are used in the literature as a measure of the depth of penetration. The skin depth D_s commonly used for metals gives the distance into the material at which the electric field falls to $1/e$ of its original value, where e is the base of the natural logarithm (equal to 2.718). It is given by the equation

$$D_s = \frac{1}{[(1/2)\sigma \omega \mu]^{1/2}} \qquad (12.36)$$

where μ is the permeability and σ is the electrical conductivity. For metals, D_s is on the order of 1 μm at microwave frequencies. Insulating materials such as Al_2O_3 would have a "skin depth" on the order of 1 m, but this could hardly be described as a "skin." Nevertheless, Eq. (12.36) shows that for a given material the skin depth decreases with increasing frequency of the microwave radiation. A useful parameter is the depth D_p at which the power is reduced to half of its value at the surface of the material. It is given by the equation

$$D_p \approx \frac{\lambda_0}{10 \tan \delta (\epsilon_r')^{1/2}} \qquad (12.37)$$

where λ_0 is the wavelength of the incident microwaves (the wavelength in free space). According to this equation, for a given material D_p decreases with decreasing wavelength (or increasing frequency) of the microwave radiation.

Densification Process

Equations (12.35) and (12.37) show that for an insulating (or dielectric) material the most important parameters of the material that govern its interaction with microwave radiation are the relative permittivity or dielectric constant ϵ_r' and the loss tangent, $\tan \delta$. During microwave heating, both ϵ_r' and $\tan \delta$ change with temperature, and a knowledge of these changes is important for controlling microwave sintering. Figure 12.35 shows the dielectric constant for several ceramics during microwave heating at 8–10 GHz; except for Al_2O_3, the values do not change significantly with temperature. In contrast, the $\tan \delta$ values shown in Fig. 12.36 are far more affected by temperature. For materials such as Al_2O_3, BN, SiO_2, and glass ceramics, $\tan \delta$ initially increases slowly with temperature until some critical temperature T_c is reached. Above T_c, the materials couple more effectively with the microwave radiation and $\tan \delta$ increases fairly rapidly with temperature. The reason for the rapid rise in $\tan \delta$ above T_c is not clear. To overcome the problem of inefficient coupling with microwave radiation at lower temperatures, these materials can be heated conventionally to a temperature slightly above T_c, after which microwave heating can be used effectively to produce higher temperatures.

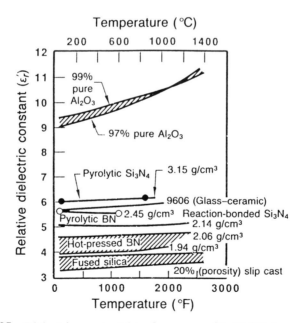

Figure 12.35 Dielectric constant (at a frequency of 8–10 GHz) versus temperature. (From Ref. 20, with permission.)

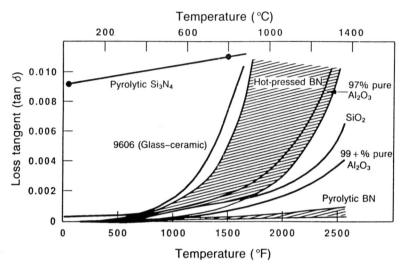

Figure 12.36 Dielectric loss tangent (at a frequency of 8–10 GHz) versus temperature. (From Ref. 20, with permission.)

B. Control of Microwave Sintering

The control of microwave sintering, as mentioned earlier, is more complicated than that of conventional sintering. A common problem is "thermal runaway," when the temperature of the sample increases rapidly with time. Thermal runaway can lead to melting of the powder compact or to the formation of local hot spots, with some regions having a much higher temperature than the rest of the sample. While it is clear that melting must be avoided, the formation of local hot spots also has severe consequences: It leads to differential densification and even to cracking of the sample.

The problem of thermal runaway is a direct result of nonuniformity of the microwave field and of the properties of the sample. Nonuniformities in the field lead to differences in power absorption within the sample [Eq. (12.35)] and therefore to different regions of the sample being heated at different rates. The problem becomes drastically more severe if the dielectric properties of the material change rapidly with temperature. For example, if ϵ_r'' increases rapidly with temperature, more power is absorbed, and this causes an increase in temperature and a corresponding increase in ϵ_r''. The flow of heat from the hotter regions would normally be slow because most ceramics are thermal insulators. The overall result of this repeating cycle of events is that the temperature increases rapidly. Successful microwave sintering depends on the ability to supply energy uniformly to the sample so that it heats uniformly. An important element of microwave sintering is the design of equipment that has the capability of producing uniform electric fields within a reasonable volume.

Accurate measurement of the temperature in microwave sintering also presents more difficulties than in conventional sintering. Because of interference from the microwave field, thermocouples cannot be relied upon to function properly. Unprotected thermocouple wires undergo electrical arcing and rapidly reach their melting point. A solution is to sheath the wires with a metallic protective tube (e.g., molybdenum). Because the sample is heated internally, temperature measurement with thermocouples must be made directly within the sample (e.g., by inserting the thermocouple into the sample) and not in the vicinity of the sample. Optical pyrometers provide a more practical approach to temperature measurement in microwave sintering. The pyrometer must be calibrated using a heated blackbody source and must be focused directly on the sample.

C. Microwave Sintering Equipment

Although the kitchen microwave oven can be used for preliminary screening and evaluation, it has serious shortcomings for use in controlled sintering experiments. Figure 12.37 shows a schematic diagram of the main components of laboratory-scale equipment for microwave sintering of ceramics [21]. The microwave generator is a key element in the system. Most generators are operated at a frequency of 2.45 GHz; however, a few studies have shown that higher frequencies (e.g., 28 GHz) lead to higher power absorption and, in some ceramics, to more uniform sintering. Important considerations in the selection of a microwave generator are (i) the power output, (ii) the capability for varying the output power and tuning the microwave field to the absorption maximum of the sample, and (iii) devices for protecting the magnetron tube (the source of the microwave field) from reflected microwave energy. The waveguide components are used to transmit the microwave energy to the applicator in which the sample is heated. A common applicator is a cavity consisting of an aluminum box that has been seam welded to prevent electrical arcing. The cavity can be fitted with a screened window to permit viewing and, to improve the uniformity of the microwave field experienced by the sample, a device (e.g., a turntable) for rotating the sample. The equipment should also include instruments to monitor and control the temperature (e.g., an optical pyrometer) and to control the sintering atmosphere.

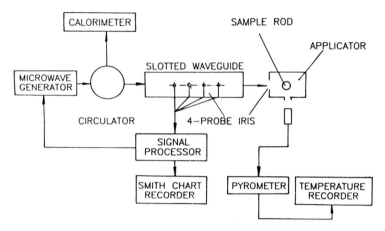

Figure 12.37 Schematic diagram of a microwave sintering system. (From Ref. 21, reprinted by permission of the American Ceramic Society.)

D. Microwave Sintering of Alumina

Microwave heating has been used in the sintering of a variety of ceramics. However, the most detailed studies so far have been carried out with Al_2O_3, both undoped and MgO-doped powders. The detailed nature of the results allows a meaningful comparison to be made between the sintering characteristics in microwave heating and those in conventional heating.

Figure 12.38 shows that data of Janney and Kimrey [22] for the density as a function of sintering temperature for Al_2O_3 doped with 0.1 wt % MgO obtained in microwave heating (at 28 GHz) and in conventional heating (tungsten furnace element). Apart from the method of heating, the processing and sintering conditions were almost identical in the two sets of experiments. The initial density of the powder compacts was $\approx 55\%$ of the theoretical, and heating was performed in a vacuum at a rate of 50°C/min to the required temperature followed by isothermal sintering for 1 h. The conventional sintering data are in good agreement with those obtained by others in similar experiments with the same powder. The powder starts to show measurable shrinkage above $\approx 1000°C$, and the densification rates are appreciable above $\approx 1250°C$. In contrast, the microwave-sintered compacts show significantly enhanced sintering rates; shrinkage starts at $\approx 900°C$, and fairly high relative densities (greater than 90% of the theoretical) are obtained at temperatures as low as 950°C. As mentioned earlier, accurate temperature measurement and control are normally more difficult in microwave sintering. However, it seems inconceivable that such a large reduction in the temperature (300–400°C) re-

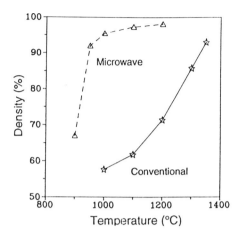

Figure 12.38 Relative density versus temperature for MgO-doped Al_2O_3 powder compacts during sintering by microwave heating and by conventional heating (in a resistance furnace). (From Ref. 22, reprinted by permission of the American Ceramic Society.)

quired to achieve a given density could be attributable to errors in temperature measurement. The data of Fig. 12.38 therefore provide strong evidence for a significant enhancement of densification by microwave heating.

By measuring the density as a function of time in isothermal sintering at several temperatures T, the activation energy Q_d for the densification process can be determined. If it is assumed that the densification process obeys an Arrhenius relationship, a plot of the natural logarithm of the densification rate versus $1/T$ has a slope equal to $-Q_d/R$, where R is the gas constant. In the standard procedure, the densification rate must be taken at fixed values of the density, and, for a given density, the rate must be compensated for differences in grain size. As the plots in Fig. 12.39 show, the activation energy in microwave sintering (170 kJ/mol) is drastically lower than that in conventional sintering (575 kJ/mol).

In a subsequent set of experiments, the grain growth kinetics of fully dense MgO-doped Al_2O_3 samples (fabricated by hot pressing) were measured during microwave heating (28 GHz) and conventional heating [23]. The general evolution of the microstructure was roughly similar in both types of heating; normal grain growth was observed, and the kinetics obeyed a cubic law. However, as Fig. 12.40 shows, the grain growth kinetics are considerably enhanced in microwave heating. The activation energy for grain growth in microwave heating (480 kJ/mol) is ≈20% lower than that in conventional heating (590 kJ/mol). [It should be realized that

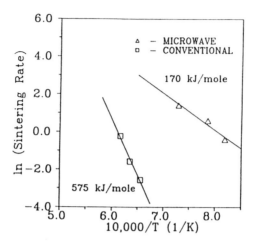

Figure 12.39 Arrhenius plot of the densification rate data (at a relative density of 0.80) for MgO-doped Al_2O_3 for sintering by microwave heating and by conventional heating. The activation energy, determined from the slopes of the lines, is considerably lower in microwave heating. (From Ref. 22, reprinted by permission of the American Ceramic Society.)

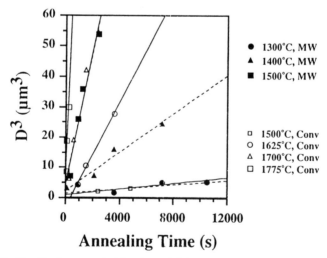

Figure 12.40 Grain growth kinetics of fully dense, hot-pressed MgO-doped Al_2O_3 during annealing by microwave heating and by conventional heating. The data follow a cubic growth law in both cases, but grain growth is considerably faster in microwave heating. (From Ref. 23, with permission.)

Densification Process

these activation energies are more appropriate to grain growth controlled by fine second-phase $MgAl_2O_4$ particles in the grain boundary because the MgO concentration (0.1 wt %) is significantly higher than the solid solubility limit.]

Measurement of the rate of oxygen diffusion in sapphire (single-crystal Al_2O_3) with an ^{18}O tracer revealed that the diffusion coefficient for microwave heating was ≈ 2 times that for conventional heating [22]. Furthermore, the activation energy for diffusion in microwave heating (410 kJ/mol) was considerably lower than that for conventional heating (710 kJ/mol).

To summarize, the results for the MgO-doped Al_2O_3 materials show that the diffusional processes leading to densification and grain growth are considerably enhanced by microwave heating. The enhanced oxygen diffusion in sapphire indicates that neither surfaces nor grain boundaries are essential for the enhancement of diffusion. The mechanisms for the enhanced diffusion are not clear. Microwave sintering currently forms an interesting area of sintering research.

12.4 PRESSURE SINTERING

Sintering with an applied pressure is commonly referred to as pressure sintering or pressure-assisted sintering. Here, the term *pressure sintering* will be used. It will be recalled that pressure sintering forms a very effective approach for the fabrication of ceramics with high density (usually near theoretical density) and fine grain size. The effectiveness of the approach, as outlined earlier, arises from the ability to significantly enhance the densification rate relative to the coarsening rate. Pressure sintering is particularly important for the fabrication of ceramic matrix composites where the reinforcing phase (particles or fibers) can severely limit the free sintering of the powder matrix.

Various ways can be used to transmit an external pressure to the powder compact during sintering. However, for the production of ceramics, there are two commonly used pressure sintering methods:

1. Hot pressing, where the pressure is applied uniaxially to the powder in a die
2. Hot isostatic pressing, where the pressure is applied isostatically by means of a gas

Of the two methods, hot pressing is simpler and more widely used. However, the use of hot isostatic pressing has been increasing steadily in recent years.

Hot pressing and hot isostatic pressing have similar capabilities in terms of the temperatures that can be used. Depending on the type of furnace, temperatures as high as 2500°C can be used with many types of commercial equipment. However, because of the different ways in which the pressure is applied to the sample, clear differences exist between the methods in terms of the pressure capability and ease of fabrication. In hot pressing, the maximum applied pressure is essentially limited by the strength of the die. Pressures up to ≈40 MPa are typical for the commonly used graphite dies. The use of specialty graphite and more expensive refractory metal (e.g., nimonic alloys) or ceramic (e.g., Al_2O_3 and SiC) dies increases the maximum pressure to ≈75 MPa. In hot isostatic pressing, the pressurizing gas is contained within a cooled pressure vessel, and pressures up to ≈200 MPa are routinely available in most equipment. Only disks and simple shapes can be produced by hot pressing. Hot isostatic pressing allows fairly complex shapes to be fabricated. In hot isostatic pressing, samples with a continuous network of open porosity must be encapsulated within a sealed metal or glass can that provides a medium for transmitting the pressure to the sample. Without the can, the pressurized gas enters the pores and resists the compressive sintering stress so that only highly porous materials are produced. The encapsulation of the powder compact in the can introduces an additional step that can be fairly tedious and expensive. As in die pressing at room temperature, die wall friction can lead to inhomogeneous densification in hot pressing. Although there is no die wall friction in hot isostatic pressing, distortion of the sample shape can be a problem.

A. Advantages and Disadvantages of Pressure Sintering

Advantages

The major advantage of pressure sintering is the ability to significantly increase the rate of densification $\dot{\rho}$ relative to the rate of coarsening $\dot{G}$, thereby guaranteeing, in most cases, the attainment of high density and fine grain size. Since most properties normally improve with high density and fine grain size, superior properties are achieved. The increase in the ratio $\dot{\rho}/\dot{G}$ leads to a reduction in the firing time or the firing temperature (on the order of a few hundred degrees). The significantly lower firing temperature is important for systems that contain volatile components or suffer from decomposition at higher temperatures.

In research studies, pressure sintering is well suited to the production of prototype materials for the investigation of microstructure–property relations. Pressure sintering also provides an additional variable that is very effective for the study of the mechanisms of sintering. As outlined

Densification Process

in Chapter 8, the kinetic data in conventional sintering are difficult to interpret because of the occurrence of multiple mechanisms. It will be recalled that some mechanisms lead to densification whereas others lead to coarsening and a reduction in the driving force for densification. The use of an applied pressure significantly enhances the densification mechanisms so that the interference from the nondensifying mechanisms becomes much less. Furthermore, as discussed later for the case of hot pressing, with suitable choice of temperature and particle size it is possible to isolate the mechanism of interest.

Disadvantages

The major disadvantage of pressure sintering is the high cost of production. Industrially, it is used only for the production of specialized, high-cost ceramics where high density must be guaranteed. Pressure sintering cannot be easily automated. In the case of hot pressing, the method is limited to the fabrication of relatively simple shapes. Although complex shapes can be produced by hot isostatic pressing, shape distortion can be a problem. Pressure sintering also has a size limitation. Large articles (e.g., greater than ≈ 1 m) have been produced industrially, but the equipment becomes highly specialized and expensive.

B. Hot Pressing

Equipment and Die Materials

Figure 12.41 shows a schematic diagram of the main features of a laboratory-scale hot press [24]. Typically, the powder or a consolidated powder form is placed in a die and heated while pressure is applied. Depending on the furnace, operating temperatures of up to $\approx 2500°C$ can be used. Typical operating pressures range from ≈ 10 to ≈ 75 MPa. Table 12.2 provides some information on the die materials that have been used in hot pressing [25]. Graphite is the most common die material because it is relatively cheap, is easily machined, and has excellent creep resistance at high temperatures. For pressures below ≈ 40 MPa, standard graphite can be used. For higher pressures, specialty graphite and more expensive refractory metal and ceramic dies can be used. Graphite oxidizes slowly below $\approx 1200°C$, so it can be exposed to an oxidizing atmosphere for short periods. Above $\approx 1200°C$, it must be used in an inert or reducing atmosphere. A common problem is the reactivity of graphite toward other ceramics, which leads to a deterioration of the contact surfaces of the die or to sticking of the sample to the die wall. This problem can be alleviated by coating the contact surfaces of the die with boron nitride.

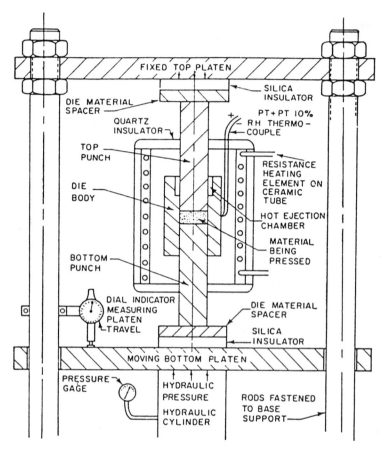

Figure 12.41 Schematic design of unit for pressure sintering (hot pressing) of ceramic materials. (From Ref. 24.)

Process Variables

The simplest procedure, commonly used in research studies when kinetic data are required, is to heat the die system rapidly to the isothermal firing temperature and quickly apply the pressure. However, when fabrication to high density is the primary requirement, the pressure is normally applied prior to or during heat-up. As discussed in Chapter 8, for ceramics where diffusional mechanisms are commonly observed, the densification rate (at a given density) increases linearly with the applied pressure. The rate also increases with temperature as described by the Arrhenius rela-

Densification Process

Table 12.2 Die Materials for Hot Pressing

Material	Maximum temperature (°C)	Maximum pressure (MPa)	Comments
Nimonic alloys	1100	High	Creep and erosion problems
Mo (or Mo alloys)	1100	20	Oxidation unless protected
W	1500	25	Oxidation and galling
Al_2O_3	1400	200	Expensive, brittle
BeO	1000	100	Expensive, brittle
SiC	1500	275	Expensive, brittle
TaC	1700	55	Expensive, brittle
WC, TiC	1400	70	Expensive, brittle
TiB_2	1200	100	Expensive, brittle
Graphite (standard)	2500	40	Severe oxidation above 1200°C
Graphite (special)	2500	100	Severe oxidation above 1200°C

Source: Ref. 25.

tion. In practice, the applied pressure and isothermal firing temperature are chosen to accomplish densification in less than ≈1 h.

As in conventional sintering, improvements in powder quality lead to clear benefits for densification. Powders for hot pressing should be fine (<1 μm), have a narrow particle size distribution, and contain no hard agglomerates. Die wall friction can lead to a reduction in the densification rate and to inhomogeneous densification. This problem can be alleviated (i) by reducing the reaction between the sample and the die walls (e.g., by coating the contact surfaces of the die with a layer of boron nitride) and (ii) by pressing fairly flat articles (e.g., disks or plates). In practice, hot pressing is best suited to the fabrication of flat articles.

Although the applied pressure increases the driving force for densification, it may still be difficult to achieve high density with some powders, in particular those of highly covalent materials where solid-state diffusion is slow. As in conventional sintering, it is necessary to use additives that aid the densification process by providing a high diffusivity path at the firing temperature (e.g., a liquid phase at the grain boundaries). However, because of the increased driving force due to the applied pressure, the amount of additive required for a given powder can be significantly lower than that in conventional sintering.

Because it is applied uniaxially, the pressure may cause the microstructure to evolve in an anisotropic manner. The most common effects are preferential orientation of the grains or enhanced grain growth in a specific direction. These effects are referred to as *texturing*. Normally, the direction of preferred orientation or enhanced grain growth is perpendicular to the direction of the applied stress. The origins of texturing can be seen as follows. Since the die walls are fixed, the compaction of the powder during hot pressing occurs only in the direction of the applied pressure. As sketched in Fig. 12.42, a representative element of the powder (e.g., three grains) will undergo the same relative shape changes as the overall compact. Thus, the grains become flatter in the direction of the applied pressure.

Insoluble gases trapped in the pores during hot pressing (or hot isostatic pressing) will be under very high pressures. When densification stops, the gas pressure p_i will be equal to the sum of the sintering stress Σ and the applied pressure p_a, i.e.,

$$p_i = \Sigma + p_a \tag{12.38}$$

If the hot-pressed article is subsequently heated to a high enough temperature, it will swell because p_i is much greater than Σ. Hot pressing in a soluble gas or in vacuum will alleviate the problem.

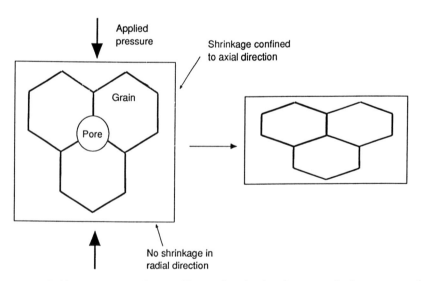

Figure 12.42 Schematic diagram illustrating the development of microstructural texturing during hot pressing.

Densification Process

Kinetics and Mechanisms

Kinetic data for the densification of the powder during hot pressing are valuable for studying the mechanisms of sintering. Such data can be obtained by measuring the change in thickness of the compacted powder as a function of time under known conditions of pressure, isothermal firing temperature, and particle size of the powder. The data for the change in thickness are best collected continuously by, for example, measuring the distance traveled by the movable punch of the hot press with a dial gauge or a linear voltage displacement transducer (see Fig. 12.41).

For a cylindrical die cavity of area A_0 and assuming that the mass of the powder, m_0, is constant, the density of the compacted powder is given by

$$d = \frac{m_0}{A_0 L}; \quad d_f = \frac{m_0}{A_0 L_f} \tag{12.39}$$

where d is the density and L is the length of the sample at any time t, and d_f and L_f are the final density and length, respectively. Combining the equations gives

$$d = \frac{d_f L_f}{L}; \quad \rho = \frac{d_f L_f}{d_\infty L} \tag{12.40}$$

where ρ is the relative density (at any time t) and d_∞ is the theoretical density of the sample. Since $L = L_f + \Delta L$, where ΔL is the change in length of the sample, the density as a function of time can be determined from the data for ΔL as a function of time and by measuring L_f and d_f. As discussed earlier for the sintering data, the densification rate, $\dot{\rho}$, equal to $(1/\rho)d\rho/dt$, can be determined by fitting a curve to the data for relative density versus time and differentiating to find the slope. Assuming that the experiments were performed at a fixed isothermal temperature and for different known pressures, the most useful representation of the data is in terms of (i) ρ versus log t (or t) and log $\dot{\rho}$ versus ρ for the pressures used and (ii) log $\dot{\rho}$ versus applied pressure for fixed values of the sample density. A detailed analysis of hot pressing data along these lines is given in Ref. 26.

As discussed in Chapter 8, the hot pressing models predict that the densification rate can be written in the general form

$$\dot{\rho} = \frac{1}{\rho}\frac{d\rho}{dt} = \frac{AD\phi^{(m+n)/2}}{G^m kT}(p_a^n + \Sigma) \tag{12.41}$$

where A is a constant, D is the diffusion coefficient for the rate-controlling species, G is the grain size, k is the Boltzmann constant, T is the temperature, ϕ is the stress intensification factor, p_a is the applied stress, and Σ is the sintering stress. The exponents m and n depend on the mechanism of sintering (Table 8.7). For a given powder, ϕ depends only on the density. Assuming that Σ is much smaller than $p_a{}^n$ and that G does not change significantly, a plot of the data for log ρ (at a fixed density) versus p_a allows the exponent n and thus the mechanism of sintering to be determined. For the commonly used pressure range in hot pressing ($\approx$10–50 MPa), data for most ceramics show that $n \approx 1$, i.e., the densification process is diffusion-controlled. This finding is not surprising in view of (i) the impediments to dislocation motion caused by the strong bonding and (ii) the fine grain size, which favors diffusional mechanisms.

For diffusional mechanisms, because of the larger grain size exponent ($m = 3$) and the generally lower activation energy, finer powders and lower hot-pressing temperatures favor grain boundary diffusion over lattice diffusion ($m = 2$). By careful selection of the process variables, the diffusion mechanism of interest can be isolated. As an example of such work, Fig. 12.43 shows the relationships between the various possible mechanisms for alumina [27]. The results are plotted on a map of temperature versus grain size. Maps such as Fig. 12.43 showing the relationships between the mechanisms and parameters of hot pressing are analogous to the sintering maps discussed in Chapter 8; they are referred to as hot pressing maps or deformation mechanism maps.

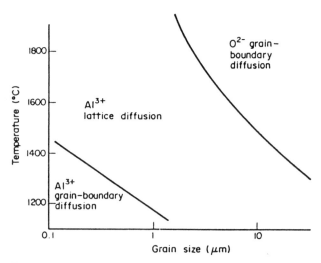

Figure 12.43 Hot pressing (deformation) map for pure α-alumina. (From Ref. 27.)

C. Hot Isostatic Pressing

Equipment

In hot isostatic pressing, commonly abbreviated HIP, the sample is placed in a pressure vessel, the gas pressure (a few thousand pounds per square inch) is applied, and the temperature is raised. Achieving the required pressure at the firing temperature relies on heat-up of the gas by the HIP furnace. As discussed below, the sample must be encapsulated in a sealed container if it contains continuous open porosity. A HIP system (Fig. 12.44) can be considered to consist of four basic units: (i) a pressure vessel, (ii) an internal furnace for heating the sample, (iii) a gas pressurization unit, and (iv) a control unit for controlling and monitoring the power and other variables (e.g., temperature) in the system [28]. The pressure vessel is typically of the cold-wall design, in which the furnace is thermally insulated from the wall of the vessel and flowing water provides external cooling (Fig. 12.45). It is penetrated, normally through the end closures, to supply power and control instrumentation to the furnace. Electrical resistance furnaces (e.g., graphite or molybdenum) and inert pressurizing gases (argon or helium) are commonly used. Depending on the furnace, HIP systems designed for use with inert gases can routinely be operated at temperatures up to $\approx 2500°C$ and pressures up to ≈ 200 MPa. Systems for which reactive gases (e.g., O_2 or N_2) can be used are also available, but the temperature and pressure capability are significantly lower.

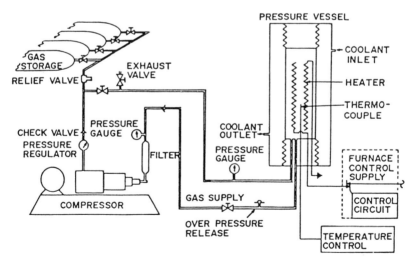

Figure 12.44 Simplified schematic diagram of a system used for a hot isostatic pressing. (From Ref. 28.)

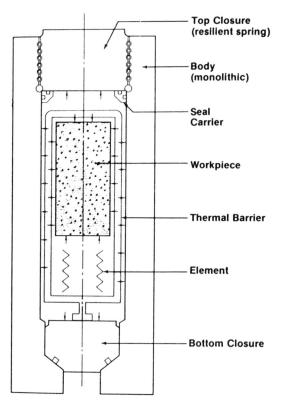

Figure 12.45 Schematic diagram of a pressure vessel with a sample (workpiece) for hot isostatic pressing. (Courtesy ASEA Autoclave Systems Inc., Columbus, OH.)

Process Variables

The method commonly used for densifying metal powders by HIP involves filling the loose powder into a deformable metal container (referred to as a *can*), evacuating and sealing the container, and then subjecting the system to the desired temperature and pressure. However, this method is not suitable for most ceramic powders. Because of their fine size and low poured (or tap) density, considerable deformation and distortion of the material occurs during HIP. To alleviate this problem, most ceramic powders are first consolidated (by one of the methods described in Chapter 6) to form a compact with the desired shape. If required, the powder compact can be lightly presintered to produce adequate strength for handling. Subsequent densification can be performed by two routes (Fig.

Densification Process

12.46); either (i) the powder compact is encapsulated in a can and then densified by HIP, or (ii) the powder compact is sintered conventionally to closed porosity (density greater than ≈90% of the theoretical) and then densified further by HIP.

In the first route, the can is required as a medium for transmitting the gas pressure to the porous compact. Compacts with open porosity, as outlined earlier, cannot be densified without the encapsulation because the pressurizing gas in the pores resists the sintering stress. The materials commonly used for encapsulation consist of a thin-walled metal (e.g., molybdenum or tantalum) can or a glass can. Although both types of cans are effective for transmitting the pressure to the powder compact, the trend in ceramics has been to use glass cans. Glass encapsulation is more economical, particularly when the compact has a complex shape. Gases trapped in the sealed can may limit the final densification; the system

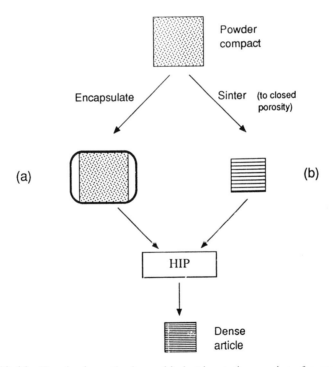

Figure 12.46 Two basic methods used in hot isostatic pressing of ceramic powder compacts: (a) encapsulation, for compacts with a continuous network of open porosity and (b) sinter/HIP, where the sample is sintered to closed porosity prior to hot isostatic pressing.

should be degassed and sealed in a vacuum. Following densification, the can is removed mechanically (or, less often, chemically).

The second route in which the powder compact is sintered to closed porosity and then subjected to HIP is referred to as *sinter plus HIP* or *sinter/HIP*. The simpler HIP equipment cannot be used for conventional sintering at high temperatures, and the sintering step must be carried out in a separate furnace. However, with improvements in design, commercial equipment is now available to allow the sintering and HIP stages to be carried out in the same pressure vessel. In sinter/HIP, the main objective is to remove the residual porosity. In conventional sintering, coarsening of the microstructure is normally very rapid in the final stage of sintering, thereby making the attainment of full density difficult. The significantly greater driving force for densification provided by HIP alleviates the coarsening problem.

Each route has its advantages and disadvantages. The selection of a route is usually governed by the cost or quality of the fabricated article. The sinter/HIP route eliminates the cost of encapsulation, but the time required for the overall sintering plus HIP stages is longer than that for HIP of an encapsulated powder compact. If significant coarsening occurs during the sintering stage of the sinter/HIP route, the microstructure can be considerably coarser than that of the article produced by the encapsulation route. Because the shrinkage during the HIP stage is relatively small in the sinter/HIP route, the dimensional control of the fabricated article is often better than that in the encapsulation route. Furthermore, reaction with the encapsulation (particularly in the case of ceramic oxides) can severely degrade the surface of the sample, and some surface machining may be necessary.

Because the powder compact can be shaped during forming (e.g., by slip casting or injection molding) or machined to the desired shape after consolidation, fairly complex shapes can be produced by HIP (Fig. 12.47). This ability to produce complex shapes provides one of the most important advantages over hot pressing. However, inhomogeneous densification can lead to undesirable shape changes (e.g., distortion) in the fabricated article. Several factors can lead to inhomogeneous densification, and they can be divided into two types: (i) factors due to inadequate processing prior to HIP and (ii) factors due to inadequate HIP practice. Factors due to inadequate processing include inhomogeneities (e.g., agglomerates) in the powder, inhomogeneous consolidation, and incomplete outgassing of the sample prior to encapsulation. As discussed earlier in the book, the problems associated with these factors are best alleviated through proper processing procedures.

Figure 12.47 Turbocharger impeller, formed by injection molding (right) and densifed to full density by hot isostatic pressing (left). (From Ref. 29.)

In the HIP stage, inhomogeneous densification can occur when the temperature and pressure are not uniform over the dimensions of the sample. Temperature gradients are a common source of problems, particularly for large articles. As in conventional sintering, when the temperature of the powder compact is raised, heat diffuses inward from the surface. If heat flow through the compact is slow (as in most ceramic powders), a temperature gradient is set up. As outlined earlier, in most HIP operations the compact is under pressure. This pressure causes the hotter surface layer to densify faster than the interior, giving rise to a dense skin. Heat conduction through the dense skin is faster than through the less dense interior, further adding to the temperature difference between the surface and the interior. Under certain circumstances, a densification front develops and propagates inward. This leads to large changes in the shape of the compact, which is no longer identical to that of the original powder compact.

An analysis of a one-dimensional model for coupled heat flow and densification in a powder compact [30] shows that the tendency to shape distortion can be characterized by a dimensionless quantity C^*, which measures the ratio of the densification rate to the heat transfer rate. When $C^* < 1$, densification is nearly uniform; however, when $C^* > 1$, there is a tendency for a densification front to form and propagate inwards from the surface. Extension of the ideas from one to three dimensions produces new complications. As sketched in Fig. 12.48 for a cylindrical shape, when densification is uniform the cylinder (as expected) shrinks uniformly.

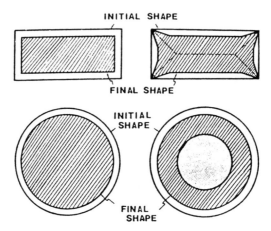

Figure 12.48 Consequences of inhomogeneous densification during hot isostatic pressing of an encapsulated cylindrical powder compact. (Top) Shape change and (bottom) density gradients with residual porosity and internal stress. (From Ref. 30, reprinted with permission from Elsevier Science Ltd.)

However, when nonuniform densification takes place, a dense shell forms around the cylinder. Pressure can be transmitted to the inner core only if the cylindrical shell creeps inwards, a process that can be slow. A larger part of the pressure is carried by the shell, and the densification of the core all but ceases.

The problem of shape distortion caused by temperature gradients can be alleviated by allowing the sample to equilibrate at the firing temperature before applying the full pressure. However, as outlined earlier, this may be impractical in the simpler HIP equipment. The problem can also be reduced by using the sinter/HIP route in which the sample is sintered conventionally to high density prior to HIP.

Kinetics and Mechanisms

Because of the nature of the HIP process, kinetic data are not easy to obtain. Data for most ceramics are limited to values for the final density after a specified isothermal firing time under a given pressure. As in the case of hot pressing, the nondensifying mechanisms are not enhanced by the applied pressure and can be neglected. Furthermore, particle rearrangement, which contributes to densification during the initial stage, is usually ignored because of the transient nature of the contribution and the difficulty of analyzing the process. The much higher applied pressure

Densification Process

means that, in metals, plastic deformation plays a more important role in HIP than in hot pressing. Instantaneous plastic yielding of the metallic powder particles at their contact points will be more significant in the early stages of HIP. However, plastic yielding is still expected to be unlikely for most ceramic powders. For polycrystalline ceramics, the possible processes that need to be considered are lattice diffusion, grain boundary diffusion, and plastic deformation by dislocation motion (power law creep).

Constitutive equations have been developed for the various mechanisms of densification during HIP [31]. With the aid of available data for the material parameters, these equations have been used to predict the relative contribution of each mechanism to densification. The procedure follows along the same lines as discussed in Chapter 8 for the construction of sintering diagrams. The results are plotted on HIP diagrams that show the conditions of dominance for each mechanism. Figure 12.49 shows the HIP diagram for Al_2O_3 powder with a particle size of 2.5 μm. As might be expected for this material, densification is predicted to occur predominantly by diffusion.

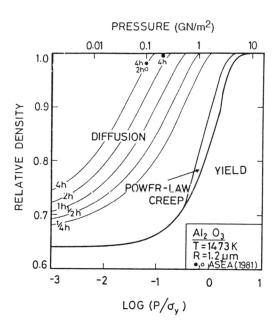

Figure 12.49 A hot isostatic pressing (HIP) map for α-alumina powder with a particle size of 2.5 μm at 1473 K. (From Ref. 31.)

12.5 CONCLUDING REMARKS

In this chapter, we considered the main methods for the densification of ceramic powder compacts. The methods can be divided into two broad classes: (i) conventional sintering, where the firing is performed without the application of an external pressure, and (ii) pressure sintering, where pressure is applied to the sample during firing. The firing temperature is normally achieved with electrical resistance furnaces, with heat transferred from the furnace to the sample. In recent years, microwave sintering, in which the sample is heated internally by microwave radiation, has attracted some interest. Although the control of microwave sintering is more complicated than that of conventional sintering, experiments have shown considerable enhancement of sintering when microwave heating is used. Compared to pressure sintering, conventional sintering is the preferred method for densification because it is economical. However, when it is difficult or impossible to produce high density by conventional sintering, pressure sintering is used. Even with proper powder preparation and consolidation procedures, successful fabrication is still dependent on the ability to control the microstructure through manipulation of the process variables in the firing stage. In conventional sintering, the main process variables are the heating cycle (temperature–time schedule) and the sintering atmosphere. Proper manipulation of these two variables can provide considerable benefits for densification. Applied pressure provides an additional variable that is effective not only in fabrication (e.g., the achievement of high density) but also in the study of the mechanisms of sintering.

PROBLEMS

12.1 Derive the following elementary relations for the sintering of a powder compact:
   (a) $\rho = \rho_0/[1 - (\Delta L/L_0)]^3$
   (b) $\Delta \rho/\rho = \Delta V/V_0$
   where ρ is the relative density, L is the length, and V is the volume of the compact at any time, ρ_0, L_0 and V_0 are the initial values, and $\Delta\rho$, ΔL, and ΔV represent the changes in the parameters.

12.2 In the constant heating rate sintering of a powder compact, discuss how an increase in the heating rate from 1°C/min to 20°C/min would be expected to influence the sintering behavior for the following two cases: (a) coarsening is insignificant, (b) coarsening is significant.

12.3 Using Eq. (12.18) and assuming that the grain size is approximately one-third of the grain size, plot the limiting porosity in a powder compact as a function of the grain size (in the range 0.1 to 100 μm) when sintering is carried out in an insoluble gas at atmospheric pressure. The specific energy of the solid-vapor interface is 1 J/m².

12.4 In the constant heating rate sintering of a compact of pure, fine CeO_2 powder, a student finds that a limiting density of 95% of the theoretical is achieved at ≈1300°C, after which the density decreases at higher temperatures. Discuss the factors that may be responsible for hindering the achievement of a higher final density. What steps can be taken to increase the final density?

12.5 In the sintering of Cr_2O_3 powder, the procedure adopted by a student calls for the powder to remain as Cr_2O_3 at all times above 800°C, never changing to Cr or CrO_2. Plot the limits of the H_2/H_2O ratio in the atmosphere as a function of temperature for the experiment.

12.6 A student is given the task of sintering a $BaTiO_3$ powder to a density greater than 98% of the theoretical and with a grain size not greater than 1 μm. The powder has an average particle size of 0.1 μm and a purity of greater than 99.9%. Discuss how the student should attempt to accomplish the task.

REFERENCES

1. H. C. Ling and M. F. Yan, *J. Mater. Sci.*, 18:2688 (1983).
2. I. B. Cutler, *J. Am. Ceram. Soc.*, 52:(1) 14 (1969).
3. C. J. Brinker, G. W. Scherer, and E. P. Roth, *J. Non-Cryst. Solids*, 72:345 (1985).
4. M. Y. Chu, M. N. Rahaman, L. C. De Jonghe, and R. J. Brook, *J. Am. Ceram. Soc.*, 74(6):1217 (1991).
5. R. J. Brook, *Proc. Br. Ceram. Soc.*, 32:7 (1982).
6. M. P. Harmer and R. J. Brook, *J. Br. Ceram. Soc.*, 80(5):147 (1981).
7. H. Mostaghaci and R. J. Brook, *Trans. Br. Ceram. Soc.*, 82(5):167 (1983).
8. R. L. Coble, *J. Am. Ceram. Soc.*, 45(3):123 (1962).
9. S. Prochazka, C. A. Johnson, and R. A. Giddings, in *Proceedings of International Symposium on Factors in Densification and Sintering of Oxide and Non-oxide Ceramics* (S. Somiya and S. Saito, Eds.), Tokyo Institute of Technology, Japan, 1979, pp. 366–381.
10. G. S. Snow, *J. Am. Ceram. Soc.*, 56(9):479 (1973).
11. P. Popper, Problems in Sintering Silicon Nitride, Res. Paper No. 699, British Ceramic Research Association, Stoke-on-Trent, United Kingdom, 1978.
12. D. W. Readey, *Ceram. Trans.*, 7:86–110 (1990).
13. P. J. Anderson and P. L. Morgan, *Trans. Faraday Soc.*, 60(5):930 (1964).
14. B. Wong and J. A. Pask, *J. Am. Ceram. Soc.*, 62(3/4):141 (1979).
15. P. D. Ownby and G. E. Jungquist, *J. Am. Ceram. Soc.*, 55:(9) 433 (1972).
16. P. D. Ownby, *Mater. Sci. Res.*, 6:431–437, (1973).
17. H. U. Anderson, *J. Am Ceram. Soc.*, 57(1):34 (1974).

18. T. Reynolds III, in *Treatise on Materials Science and Technology*, Vol. 9 (F. F. Y. Wang, Ed.), Academic, New York, 1976, pp. 199–215.
19. D. R. Gaskell, *Introduction to Metallurgical Thermodynamics*, 2nd ed., McGraw-Hill, New York, 1981.
20. W. H. Sutton, *Am. Ceram. Soc. Bull.*, *68*(2):376 (1989).
21. Y. L. Tian and D. L. Johnson, *Ceram. Trans.*, *1*:925–932 (1988).
22. M. A. Janney and H. D. Kimrey, *Ceram. Trans.*, *7*:382–390 (1990).
23. M. A. Janney, H. D. Kimrey, M. A. Schmidt, and J. O. Kiggans, *J. Am. Ceram. Soc.*, *74*(7):1675 (1991).
24. T. Vasilos and R. M. Spriggs, *Prog. Ceram. Sci.*, *4*:95–132 (1966).
25. J. Briggs, in *Concise Encyclopedia of Advanced Ceramic Materials* (R. J. Brook, Ed.), Pergamon, Oxford, 1991, pp. 219–222.
26. J. M. Vieira and R. J. Brook, *J. Am. Ceram. Soc.*, *67*(4):245 (1984).
27. M. P. Harmer and R. J. Brook, *J. Mater. Sci.*, *15*:3017 (1980).
28. R. R. Wills, M. C. Brockway, and L. G. McCoy, *Mater. Sci. Res.*, *17*:559–570 (1984).
29. H. T. Larker, *Mater. Sci. Res.*, *17*:571–582 (1984).
30. W.-B. Li, M. F. Ashby, and K. E. Easterling, *Acta Metall.*, *35*(12):2831 (1987).
31. E. Arzt, M. F. Ashby, and K. E. Easterling, *Metal. Trans.*, *14A*:211 (1983).

FURTHER READING

C. A. Handwerker, J. E. Blendell, and W. A. Kaysser (Eds.), *Sintering of Advanced Ceramics*, Ceramic Transactions, Vol. 7, The American Ceramic Society, Westerville, OH (1990).

D. Kolar, S. Pejovnik, and M. M. Ristic (Eds.), *Sintering Theory and Practice*, Elsevier, New York (1982).

W. H. Sutton, M. H. Brooks, and I. J. Chabinsky (Eds.), *Microwave Processing of Materials*, Mater. Res. Soc. Symp. Proc., Vol. 124, Materials Research Society, Pittsburg, PA (1988).

W. B. Snyder, Jr., W. H. Sutton, M. F. Iskander, and D. L. Johnson (Eds.), *Microwave Processing of Materials II*, Mater. Res. Soc. Symp. Proc., Vol. 189, Materials Research Society, Pittsburg, PA (1991).

Appendix

Physical Constants (SI Units)

Velocity of light, c	2.998×10^8 m s^{-1}
Permittivity of vacuum, ϵ_0	8.854×10^{-12} F m^{-1}
Permeability of vacuum, $\mu_0 = 1/\epsilon_0 c^2$	1.257×10^{-6} H m^{-1}
Elementary charge, e	1.602×10^{-19} C
Planck constant, h	6.626×10^{-34} J s
Avogadro number, N_A	6.022×10^{23} mol^{-1}
Atomic mass unit, $m_u = 10^{-3}/N_A$	1.661×10^{-27} kg
Mass of electron, m_e	9.110×10^{-31} kg
Mass of proton, m_p	1.673×10^{-27} kg
Mass of neutron, m_n	1.675×10^{-27} kg
Faraday constant, $F = N_A e$	9.649×10^4 C mol^{-1}
Rydberg constant, R_∞	1.097×10^7 m^{-1}
Bohr magneton, μ_B	9.274×10^{-24} J T^{-1}
Gas constant, R	8.314 J K^{-1} mol^{-1}
Boltzmann constant, $k = R/N_A$	1.381×10^{-23} J K^{-1}
Gravitational constant, G	6.67×10^{-11} N m^2 kg^{-2}
Stefan-Boltzmann constant, σ	5.670×10^{-8} W m^{-2} K^{-4}
Standard volume of ideal gas	22.414×10^{-3} m^3 mol^{-1}
Acceleration due to gravity (at sea level and zero degree latitude)	9.78 m s^{-2}

SI Units - Names and Symbols

Base Units

Quantity	Unit	Symbol
Length	meter	m
Mass	kilogram	kg
Time	second	s
Electric current	ampere	A
Temperature	kelvin	K
Amount of substance	mole	mol
Luminous intensity	candela	cd

SI Units - Names and Symbols

Derived Units with Special Names

Quantity	Unit	Symbol	Relation to Other Units
Frequency	hertz	Hz	s^{-1}
Force	newton	N	$kg\ m\ s^{-2}$
Pressure and stress	pascal	Pa	$N\ m^{-2}$
Energy	joule	J	$N\ m$
Power	watt	W	$J\ s^{-1}$
Electric charge	coulomb	C	$A\ s$
potential	volt	V	$J\ C^{-1}$
resistance	ohm	Ω	$V\ A^{-1}$
capacitance	farad	F	$C\ V^{-1}$
Magnetic flux	weber	Wb	$V\ s$
flux density	tesla	T	$Wb\ m^{-2}$
Inductance	henry	H	$V\ s\ A^{-1}$
Temperature	degree celsius	°C	$t\ (°C) = T(K) - 273.2$

Appendix

Conversion of Units

Length	1 inch = 2.54×10^{-2} m
	1 angström = 10^{-10} m
Mass	1 pound = 0.454 kg
Volume	1 liter = 10^{-3} m^3
Density	1 g cm^{-3} = 10^3 kg m^{-3}
Angle	1 radian = 57.3°
Force	1 dyne = 10^{-5} N
Pressure and stress	1 pound per square inch (psi) = 6.89×10^3 Pa
	1 bar = 10^5 Pa
	1 atmphere = 10^5 Pa
	1 torr = 1 mm Hg = 133.32 Pa
Energy	1 erg = 10^{-7} J
	1 calorie = 4.2 J
	1 electron volt = 1.6×10^{-19} J
Viscosity: dynamic	1 poise = 0.1 Pa s
kinematic	1 stokes = 10^{-4} m^2 s^{-1}

Index

Acheson process, 59
Activation energy, 343, 703
Activity, 360
Additives:
  for powder consolidation, 279–290
  for sintering, 637–653
Adsorption of ions, 154–155
Advanced ceramics, 1
Aerogel, 219
Agglomerates, 96
  effect on packing, 21, 40, 275, 302–303, 305
  effect on sintering, 585–593
Agglomeration, in powder synthesis, 52, 56, 71
Aggregates, 97
Aging of polymeric gels, 217
Agitated ball mill, 50
Alcogel, 217
Alkoxides (see Metal alkoxides)
Alumina:
  activation energy:
    for lattice diffusion, 703
    for surface diffusion, 703
  ambipolar diffusion, 370
  Bayer process, 67–68

[Alumina]
  composites, by directed metal oxidation, 10
  defect reactions, 349, 639–641
  dielectric properties, 729–730
  dihedral angle, 641
  fast firing of, 702
  grain growth:
    in liquid-phase sintering, 521
    in porous compacts, 494–496
  grain size versus density, 506–509
  hot isostatic pressing map, 749
  hot pressing map, 742
  isoelectric point, 156
  microstructure, 25, 31, 32, 391, 450, 451, 638
  microwave sintering, 732–735
  powders, from solution, 64
  pressure casting of, 305
  sintering:
    application of scaling laws, 388

[Alumina, sintering]
  effect of gases in pores, 707
  effect of powder packing, 589–591
  role of MgO, 648–653
  of thin films, 629, 632
  slip casting of, 303
  solute segregation, 647
  suspension stabilization, 192–197
  tape casting, 307–308
Ambipolar diffusion, 371
Annealing, 690
Aquagel, 217
Arrhenius relation, 58, 388
Atmosphere:
  effect on sintering, 703–725
  with controlled O_2, 723–725
Atomic polarizability, 149
Atomic spectroscopy, 124–125
Attractive surface forces, 148–152
  effect of intervening medium, 152
Attrition mill (*see* Agitated ball mill)
Auger spectroscopy, 134, 136–137
Avrami equation, 668
Azeotrope, 209

Ball milling, 46–50
Barium titanate:
  fast firing, 703
  powders:
    by coprecipitation, 69
    by solid-state reaction, 54, 58
  tape casting, 309
Bayerite, 213, 247
Bayer process, 67

BET surface area measurements, 114
Binder removal, 320–327, 687
  (*see also* Debinding)
Binders:
  cellulose, 283–284
  in injection molding, 312–314
  in powder consolidation, 280–286
  removal of, 320–327, 687
  synthetic, 282
  viscosity grade of, 285–286
Binding energies of electrons, 139
Bingham-type behavior, 184
Boehmite, 213, 247
Boltzmann distribution, 158
Boltzmann equation, 175
Bragg's law, 127
Bridging flocculation, 178, 189
Brouwer diagram, 356

Calcination, 54, 69
Capillary pressure, 224–225
Capillary rise, 223–224
Carbothermal reduction, 84
Carman-Kozeny equation, 227, 304
Casting of powder slurries, 298–309
Cation exchange capacity (CEC), 156
Cellulose binder, 283–284
Charge on colloidal particle, 153–156
Chemical composition, analysis of, 122–126
Chemical potential, 357–365
Chemical shift, 139
Chemical vapor deposition (CVD), 6–9, 83

Index

Chemical vapor infiltration (CVI), 8
Chromium oxides, sintering of, 717–721
Coagulation, 147, 168
Coalescence, 558
Coarsening, 375, 381, 455–458, 557, 562
Colloidal gel, 201
Colloidal solution, 146
Colloidal suspension, 146
Colloids, 97
  depletion flocculation, 180
  depletion stabilization, 172, 179–180
  electrostatic stabilization, 153–172
  electrosteric stabilization, 191–197
  lyophilic, 147
  lyophobic, 147, 152
  polymeric stabilization, 172–180
Comminution, 42–50
Composites, 8, 9, 27, 275
Composite sphere model, 602
Composition-depth profiling, 134
Constrained sintering:
  with rigid inclusions, 595–625
  of thin films, 626–632
Contact angle, 526
Contact flattening, 549
Coprecipitation, 69, 249
Coulter counter, 110
Counterions, 158
Creep, 424–426
Creep viscosity, 435
Critical moisture content, 317
Critical point:
  in drying, 222, 253
  of fluids, 237

Crystallization, effect on sintering, 665–674
Curvature:
  and chemical potential, 361–364
  and sintering stress, 433

Darcy's law, 227, 300
Debinding, 320–327
  mechanisms of, 321–322
  models for, 322–326
  oxidative degradation, 325
  stages, 321–322
  thermal degradation, 322–325
Debye forces, 150
Debye–Hückel approximation, 160, 162, 164
Debye length, 160
Decomposition, 54–56
Defects:
  chemistry, 347–357
  concentration, 349–357
  in crystals, 340
  extrinsic, 352–356
  Frenkel, 351–352
  intrinsic, 350–352
  Kroger–Vink notation, 348
  point, 340, 341
  Schottky, 350–351
Defects (flaws):
  in extrusion, 311
  in powder compaction, 296
Deflocculants (*see* Dispersants)
Densification (*see also* Sintering):
  and coarsening, 504–508
  of composites, 597–625
  differential, 585–587, 607
  of gels, 240–246
Densification viscosity, 433, 435

Densifying/non-densifying mechanisms, 376–379
Desintering, 706
Die compaction, 290–296
Die wall friction, 293–295
Differential densification, 585–587, 607
Diffusion:
  of charged species, 367–371
  coefficient, 227, 337, 345–347
  flux equations, 365–366
  mechanisms of, 340–345
  in solids, 336–347
  temperature dependence, 343
Diffusivity (*see* Diffusion, coefficient)
Dihedral angle, 381, 494, 500, 528
Dilatancy, of suspensions, 184
Dilatometry, 683
Dip coating, 256
Directed metal oxidation, 9–10
Directional grain growth, 559
Disjoining forces, 226
Dispersants, 280, 287–289
Dispersion forces (*see* London forces)
DLVO theory, 163
Doctor blading (*see* Tape casting)
Domains, in powder packing, 181, 266
Dopants (*see* Solid solution additives)
Double alkoxides, 210
Double layer (*see* Electrical double layer)
Double-layer thickness (*see* Debye length)
Dry-bag pressing, 297
Drying:
  critical point, 222, 233

[Drying]
  driving forces for, 223–227
  evaporation rate during, 236
  high humidity, 319
  stages of, 220–223
  stresses, 232–233
Drying control chemical additives (DCCA), 236
Drying of cast /extruded articles, 316–320
Drying of gels, 14, 202, 217–240
  cracking during, 233–236
  Scherer's theory of, 220, 233
  structural change during, 238–240
  supercritical, 236–238
Dupré equation, 526

Einstein photoelectric equation, 137
Elastic strain energy, in crystals, 645
Elastic–viscous analogy, 609
Electrical double layer, 157–166
Electrokinetic phenomena, 168–172
Electroneutrality principle, 349
Electrophoresis, 147, 168
Electrophoretic mobility, 168–172
Electrostatic stabilization, 153–172
Electrosteric stabilization, 191–197
Electroviscous effect, 187, 188
Encapsulation, 745
End capping, 296
Entropic effect (*see* Volume restriction effect)

Index

Equilibrium constant, 155, 336, 350
ESCA (*see* X-ray photoelectron spectroscopy)
Euler equation, 459
Evaporation/condensation (*see* Vapor transport)
Evaporation of solutions, 71–76
Evaporation rate, 220, 236
Evaporative decomposition (*see* Spray pyrolysis)
Extrinsic defects, 352–356
Extrusion, 309–311

Fabrication processes, 5
Fabrication routes for high density, 508–511
Faraday constant, 160
Fast firing, 510, 702–703
Ferrites, sintering of, 719–723
Fick's first law, 227, 337
Fick's second law, 337–340
Firing, 22 (*see also* Sintering)
Floc, 97
Flocculation, 147, 148, 167, 168, 176
Flux, 337
Flux equations, 365, 376
Freeze drying, 76–77
Frenkel:
  defect, 351–352
  energy balance, 378–379
  model for viscous sintering, 396–397
Functional ceramics, 1
Furnaces, 681–683

Gases in pores, effect on sintering, 704–707
Gates–Gaudin–Schuhman equation, 103
Gelation, 205, 215–217

Gels, 201
  aging of, 217
  colloidal, 210, 203–204
  crystallization of, 212, 245
  densification of, 203, 240–246
  drying of, 202, 217–240
  multicomponent, 202, 255
  polymeric, 201, 204–207
Gibbs free energy, 359
Glass ceramic process, 17
Glycine nitrate method, 78–80
Gouy–Chapman equation, 161
Grain boundary:
  diffusion, 345, 394–396, 406, 437–440
  effect on pore shape, 380–381
  energy, 381
  migration, 461
  mobility, 461, 462, 473
  segregation, 476–481, 639, 642–648
  sliding, 428
  structure, 447
  velocity, 461
Grain growth (*see also* Coarsening):
  Burke and Turnbull model, 460–462
  computer simulation of, 467–470
  definition, 447
  in dense solids, 460–473
  driving force for, 448
  effect of dopants, 476–481
  effect of inclusions, 473–476
  kinetics, 490–493
  mean field theories, 462–465
  normal/abnormal, 449
  and pore coalescence, 490
  in porous solids, 481–504

[Grain growth]
  topological analysis of, 465–467
  in very porous solids, 493–499
Grain shape accommodation, 521, 556–557
Granulation, 291
Granule, 97
Grinding (see also Comminution):
  media, 46–48
  rate, 46, 48

Hamaker constant, 151, 152
Hashin–Shtrikman equation, 613
Heating rate, effect on sintering, 688, 692–702
Heating schedule, control of, 687–692
Helmholtz–Smoluchowski equation, 169–170
Henry's equation, 170
Herring's scaling laws, 383–389
Heterogeneities (see Inhomogeneities)
High compression roller mill, 43
High humidity drying, 319
Hofmeister series, 288
Homogeneous nucleation, 60
Homogeneous precipitation, 64
Hooke's law, 601
Hot isostatic pressing (HIP):
  encapsulation, 745
  equipment, 743
  inhomogeneous densification, 747
  maps, 749
  mechanisms of, 748
  shape distortion, 747–748

Hot pressing:
  driving force for, 423
  equations, 426–428
  equipment and die materials, 737
  kinetics, 741
  with a liquid phase, 569–570
  maps, 742
  mechanisms of, 428–430
  models, 422–426
Hydrolysis reactions:
  for metal alkoxides, 13, 212–214
  for salt solutions, 63–71
  for TEOS, 61, 214–214
  for titanium isopropoxide, 63
  for yttrium chloride, 64–67
Hydrothermal synthesis, 69–71

Impurities, 40, 59, 74, 130
Impurity drag (see Solute drag)
Inclusions:
  effect on grain growth, 473–476
  effect on sintering, 595–624
Inhomogeneities, effects on sintering, 585–595
  control of, 587–589
  correction of, 589–595
Inhomogeneities, in particle packing, 24, 272–275, 585
Injection molding, 311–316
Interdomain porosity, 181
Interfacial energy:
  definition, 223, 381, 528
  value of, 230, 318, 518
Interstitialcy mechanism of diffusion, 341–342
Interstitial mechanism of diffusion, 341–342
Interstitials, 340, 341

Index

Intrinsic defects, 350–352
Intrinsic drag, on grain boundaries, 479
Intrinsic grain boundary mobility, 473
Ionic crystal radii, 646
Isoelectric point (IEP), definition, 155
  values, 156
Isomorphic substitution, 156
Isostatic compaction (pressing), 297–298

Jet mill, 44

Keesom forces, 148, 150
Kelvin equation, 116, 367
Kelvin–Voigt element, 606
Kroger–Vink notation, 348

LaMer diagram, 60
Langmuir adsorption isotherm, 114
Langmuir force method, 163
Laser heating of gases, 86–88
Lattice diffusion, 340–342, 403–405, 410–411, 441
Leatherhard moisture content, 317
Light scattering, 108–110
Liquid phase sintering:
  advantages and disadvantages, 520
  with applied pressure, 569–570
  capillary forces, 533–537
  coalescence, 558
  coarsening, 557, 562–569
  definition, 26
  densification:
    accompanied by Ostwald ripening, 553–556

[Liquid phase sintering]
  by contact flattening, 549
  driving force for, 518
  grain shape accommodation, 521, 556–557
  kinetic and thermodynamic factors, 525–537
  mechanisms of, 540–569
  Ostwald ripening, 525, 553–556
  penetration of grain boundaries, 529
  persistent, 519
  phase diagrams in, 570–574
  pore filling, 560–562
  rearrangement, 524, 545–547
  redistribution, of liquid, 524, 540–545
  solubility effects, 522–523
  solution-precipitation, 524–525, 549–559
  stages in, 521–525
  thickness, of liquid layer, 537–540
  transient, 519
Liquid precursor methods, 11
Log-normal distribution, 103
London forces, 148
LSW theory, 455–458, 562
Lubricants, in powder consolidation, 280, 289–290
Lyophilic colloid, 147
Lyophobic colloid, 147, 152

Mackenzie and Shuttleworth model, 412
Magnesium oxide (magnesia) activation energy:
  for lattice diffusion, 703
  for surface diffusion, 703

[Magnesium oxide]
  Al_2O_3-doped:
    Brouwer diagram, 356
    defect reactions, 354–356
    fast firing, 703
    liquid-phase sintering, 564–568
    powders, by decomposition, 54–56
    sintering:
      effect of water vapor, 714–717
      of inhomogeneous compacts, 592–593
Maxwell element, 605
Mechanisms
  of diffusion, 340–345
  of hot isostatic pressing, 748
  of hot pressing, 428–430
  of liquid-phase sintering, 540–569
  of pore motion, 482–483
  of solid-state sintering, 376
Mechanochemical synthesis, 50–54
Melt casting of glasses, 17
Mercury porosimetry, 118–120
Metal alkoxides:
  definition, 13, 207
  double, 210
  hydrolysis/condensation of, 13, 61, 63, 214–216
  molecular complexity, 211
  polymerization, 13, 214–218
  preparation of, 208–211
  properties, 211–214
Metal-organic compounds, 13, 207
Microscopy, 105
Microstructural flaws, 30–32, 585–586

Microstructure:
  of composites, 29
  control, 445–511
  definition, 2
  diagrams (maps), 486–490, 649
  of glass ceramics, 18
  in liquid-phase sintering, 26, 520–523, 565, 573
  of silicon nitride, 26, 573
  of single-phase ceramics, 25, 27, 28
  of triaxial whiteware, 28, 576–579
Microwaves, interaction with matter, 726–730
Microwave sintering, 725–735
  of alumina, 732–735
  control of, 730–731
  equipment, 731–732
Milling (see Comminution and Grinding)
Mills, 43–50
Mixing effect, in steric stabilization, 174
Mobility:
  of grain boundaries, 461, 462, 473
  of pores, 483–485
Moisture content:
  critical, 317
  definition, 220
  leatherhard, 317
Multicomponent gels, 202, 255

Newton's law of viscosity, 601
Nitridation, 83
Normal distribution, 102, 103
Nucleation:
  and growth, 60
  homogeneous, 60

Optical atomic spectroscopy, 123–126
Organometallic compounds, 207
Osmotic pressure, 163, 174, 225
Ostwald ripening:
  in dense solids, 465, 471
  in liquid-phase sintering, 525, 553–556
  LSW theory, 455–458
Oxidation number, 717–719

Packing:
  apparent volume, 271
  of binary mixtures of spheres, 269–271
  coordination number, 265–268
  dense random, 268
  density, 265–278
  inhomogeneities, 24, 272–275, 585
  loose random, 268
  of particles, 29, 61, 104, 265–278
  of particles and short fibers, 275–278
  random, 268–275
  regular, 266–268
Parabolic rate law:
  grain growth, 457
  reaction product, 56
  slip casting, 301
Parallel-plate capacitor, 163
Particle rearrangement (*see* Rearrangement)
Particles:
  aspect ratio, 104
  definition, 95–97
  packing, 29, 61, 104, 265–278
  porosity, 116
  shape, 104
  shape factor, 104

[Particles]
  size analysis techniques, 105–113
  size and size distribution, 98–103
  types of, 95–97
Particulate gel (*see* Colloidal gel)
Pechini method, 78–80
Peptization, 167, 248
Percolation, 616–621
  effect on sintering, 619
  threshold, 616
Permeability, 227, 300, 304
Persistent liquid-phase sintering, 519
Phase composition, analysis of, 127–130
Phase diagrams:
  in liquid-phase sintering, 570–574
  in vitrification, 576–579
Planck's law, 123
Plastic behavior, of suspensions, 184
Plasticizers, 280, 286–287
PLZT, atmosphere sintering of, 709–711
Point defects, 340–341
Point of zero charge (PZC), 155, 194
Poisson's equation, 158
Polarizability of an atom, 149
Poly(acrylic acid), 192, 282
Polycarbosilane, 16
Polyelectrolytes:
  adsorption, 194
  charging of, 192
  colloid stability map, 195
  effect on suspension rheology, 196–197
  stabilization of suspensions, 191

Polymeric stabilization, 172–180
Polymer pyrolysis, 15–16
Poly(methacrylic acid), 192, 282
Polysilazane, 16
Poly(vinyl alcohol), 188, 282
Poly(vinyl butyral), 282, 308
  decomposition of, 326–327
Pore:
  coalescence, 490
  coordination number, 499–500
  evolution during sintering, 496–499
  migration, 482–485
  mobility, 483–485
  separation (breakaway), 449, 485–490
  stability, 499–504
  velocity, 483
  volume and size distribution, 117–119
Pore-boundary interactions
  kinetics of, 485–490
  thermodynamics of, 499–504
Pore filling, in liquid-phase sintering, 560–562
Pore motion, 482–490
  mechanisms of, 482–483
Porosity, open to closed transition, 497–498
Powder characteristics, 38–40, 94, 95
Powder consolidation (compaction):
  additives for, 279–290
  defects in, 296, 311
  equation for pressing, 293
  methods, 279–316

Powder synthesis, 38–90
  by chemical methods, 54–89
  by gas–solid reactions, 82–84
  by mechanical methods, 42–54
  by solid-state reaction, 54–59
  from solutions, 59–82
  by vapor phase reactions, 82–89
Pressure casting (filtration), 303–306
Pressure sintering, 735–749
Primary particle, 95
Primary recrystallization, 448, 449
Pseudoplastic behavior, of suspensions, 184
Pycnometry, 120–122

Quartz, 156

Random packing of particles, 268–272
Reaction bonded silicon nitride (RBSN), 7
Reaction bonding, 6
Reaction kinetics, 55
Reaction rate for powders, 56–58
Reaction sintering, 653–665
  for composites, 654, 662–665
  processing parameters, 656–658
  processing trajectories, 658–659
  for single-phase solids, 654, 659–662
  of zircon-Al_2O_3, 662–665
  of ZnO-Al_2O_3, 660–662
  of ZnO-Fe_2O_3, 660–662

Index

Rearrangement:
  during hot isostatic pressing, 748
  during hot pressing, 428
  during liquid-phase sintering, 524, 545–547
  in powder compacts, 587
  in two-dimensional arrays of spheres, 587
Repulsion, between electrical double layers, 163
Retarded forces, 150
Reynolds number, 107
Rheology of suspensions, 183–191, 196–197
Rigidity threshold, 619
Ring mechanism of diffusion, 342
RMS end-to-end distance, 173
Rosin–Rammler equation, 103
Rule of mixtures, 598

Scaling laws (*see* Herring's scaling laws)
Scanning electron microscope (SEM), 131, 134
Scherrer equation, 111
Schottky defect, 350–351
Schulze–Hardy rule, 288
Secondary ion mass spectroscopy (SIMS), 127, 134, 139–142
Secondary recrystallization, 448, 449
Sedimentation, 106, 146, 275, 276
Segregation, at grain boundary, 476–481, 639, 642–648
Self-consistent model, 610
Self diffusion, 345
Semilogarithmic law, 417
Shear thickening/thinning, 184
Sieving, 106
Silanol, 214
Silicon carbide:
  CVD reactions, 7
  fibers, by polymer pyrolysis, 16
  powder synthesis, 59, 86–89
  sintering, atmosphere effects, 707–708
Silicon dioxide (silica):
  colloidal gels, 203
  CVD reactions, 7
  densification of gels, 242
  fibers, by sol-gel, 258–260
  isoelectric point, 156
  polymeric gels, 214–217
  powders, by Stober process, 61–62
  reduction, 59, 84
  soot, viscous sintering of, 409
  surface charging, 154
  suspensions:
    consolidation of, 182
    rheology of, 188–191
  TTT diagram, 669–670
Silicon nitride:
  by CVD reactions, 7
  decomposition, 709–712
  gas pressure sintering, 712
  hot pressing, 569–570
  liquid-phase sintering, 571–574
  microstructure, 26, 573
  phase composition, 128–130
  phase diagrams, 572
  by polymer pyrolysis, 16
  powder synthesis, 81–88
  reaction bonding, 10–11
  surface chemistry, 138–141
  surface structure, by TEM, 134–135

[Silicon nitride]
  vapor pressure, 711
Sintering (*see also* Firing):
  with applied pressure, 422–430, 569–570, 735–749
  atmosphere, effect of, 703–725
  constant heating rate, 686
  constrained, 595–632
  conventional, 680–725
  with crystallization, 665–674
  definition, 22
  diagrams (maps), 419–421
  effect of heating rate, 688, 692–702
  effect of inhomogeneities, 585–595
  effect of solid solution additives, 637–653
  isothermal, 686
  kinetics, 683–685
  liquid-phase, 515–582
  oxidation number control, 717–721
  reaction, 653–665
  with rigid inclusions, 595–624
  solid-state, 374–444
  of thin films, 624–632
  viscous, 396–397, 406–410, 412–413
Sintering stress, 433–436
Slip casting, 299–303
Sol, 201
Sol-gel processing, 12–15, 201–261
  for colloidal gels, 246–250
  for fibers, 258–260
  for films and coatings, 256–258

[Sol-gel processing]
  with metal alkoxides, 214–217, 250–256
  for monolithic ceramics and glasses, 260–261
  for polymeric gels, 250–256
Solid solution additives, 637–653
Solid-state reaction, 56, 69
Solid-state sintering:
  analytical models, 389–414
  diagrams (maps), 419–421
  driving forces, 334–336
  by grain boundary diffusion, 394–396, 406, 437–440
  by lattice diffusion, 403–405, 410–411, 441
  mechanisms of, 376
  numerical simulations of, 414–417
  parameters, 333
  phenomenological equations, 417–419
  stages of, 389–392
  theoretical analysis of, 382–419
Solubility, 532
Solute drag, on grain boundary, 476–481
Space charge, at grain boundary, 642–644
Spin coating, 256, 258
Spray drying, 72–76
Spray pyrolysis, 73, 74
Steric stabilization, 172–178, 189
Stern layer, 161
Stern potential, 170
Stober process, 61
Stokes' equation, 107
Stokes' law, 107, 169

Index

Stress intensification factor, 430–433
Structural ceramics, 1
Structural relaxation, in gels, 244
Supercritical drying, 236–238
Supersaturated solution, 60
Surface area (BET), 113–116
Surface charge, 161–163
Surface charge density, 161, 171
Surface chemistry, 130, 134–142
Surface curvature, 334
Surface diffusion, 344
Surface energy, 224, 381
Surface potential, 160
Surface structure, 130, 133–134
Surfactant, 290
Swelling, 548, 706 (*see also* Desintering)
Syneresis, 217

Tape casting, 306–309
Terminal velocity, 107
Tetraethylorthosilicate (TEOS), 61–63, 214–217
Tetrakaidechedron, 402
Texturing, 740
Thermal debinding, 320–327
Thermal runaway, 730
Theta temperature, 176
Thin films:
  breakup phenomena, 632–637
  constrained sintering, 626–632
  by sol-gel processing, 256–258
Thixotropy, 184–185
Titanium dioxide (titania):
  case study in processing, 33–35

[Titanium dioxide]
  compositions, by sol-gel, 252, 255
  CVD reactions, 7
  density and grain size, 505
  electrophoretic mobility, 171
  isoelectric point, 156
  powders:
    from solution, 20, 34, 63
    uniform packing of, 23, 35
  sintering and crystallization of films, 674
  Sr-doped, effect of cooling rate, 689–692
  surface charge density, 171
Titanium isopropoxide, 63
Tracer diffusion, 339
Traditional ceramics, 1
Transient liquid-phase sintering, 519
Transient stresses, during sintering, 601–609
Transmission electron microscope (TEM), 131, 134
Triaxial whiteware, 576–579
TTT diagrams, 667–670
Turbidimetry, 108
Two-sphere model (for sintering), 393
Types of particles, 95–97

Undercutting, 415
Uranium dioxide, 490–491, 497–499

Vacancies, 341
  chemical potential of, 360–365
  flux of, 365–366,
Vacancy mechanism of diffusion, 341–342

Van der Waals attraction, 148–152
  between atoms and molecules, 148–150
  effect of intervening medium, 152
  between macroscopic bodies, 150–152
Vapor pressure, over a curved surface, 366–367
Vapor transport, 377, 379, 712–714
Vibratory ball mill (vibro-mill), 49
Viscoelastic analogy, 605
Viscoelastic response, 601
Viscosity (*see also* Rheology of suspensions)
  Einstein equation, 185
  of glass, 387
  intrinsic, 186
  Newton's law of, 601
  of suspensions, 185–191, 196–197
  temperature dependence, 388
Viscous flow, 396
Viscous sintering:
  with crystallization, 665–674
  energy balance, 378–379
  Frenkel model, 396–397
  Mackenzie and Shuttleworth model, 412–413
  with rigid inclusions, 614–616
  Scherer model, 406–410
Vitrification, 27, 574–579

Vogel–Fulcher equation, 387–388
Volume faction, of inclusions, 596
Volume restriction effect, 175

Water vapor, effect on sintering, 714–717
Wet-bag pressing, 297
Wet-bulb temperature, 229
Wet milling, 47
Wetting, 526–527
Whiskers, 275

Xerogel, 217
X-ray fluorescence spectroscopy, 126, 136
X-ray line broadening, 111
X-ray photoelectron spectroscopy (XPS), 134, 137–139

Yield stress, 310
Young and Laplace equation, 119, 323
Young equation, 526
Yttrium chloride, hydrolysis of, 64–67

Zener relationship, 473–475
Zeta potential, 155, 168–172
Zircon-Al_2O_3, reaction sintering, 662–664
ZnO-Al_2O_3, reaction sintering, 659–662
ZnO-Fe_2O_3, reaction sintering, 659–662